T0324162

Sensitivity Analysis in Earth Observation Modelling

Sensitivity Analysis in Earth Observation Modelling

Edited by

George P. Petropoulos

*Geography & Earth Sciences, University of Aberystwyth,
Wales, United Kingdom*

Prashant K. Srivastava

*NASA Goddard Space Flight Center, Greenbelt, MD, United States;
Institute of Environment and Sustainable Development,
Banaras Hindu University, Varanasi, Uttar Pradesh, India*

ELSEVIER

AMSTERDAM • BOSTON • HEIDELBERG • LONDON • NEW YORK • OXFORD
PARIS • SAN DIEGO • SAN FRANCISCO • SINGAPORE • SYDNEY • TOKYO

Elsevier
Radarweg 29, PO Box 211, 1000 AE Amsterdam, Netherlands
The Boulevard, Langford Lane, Kidlington, Oxford OX5 1GB, United Kingdom
50 Hampshire Street, 5th Floor, Cambridge, MA 02139, United States

Notices
Knowledge and best practice in this field are constantly changing. As new research and experience broaden our
understanding, changes in research methods, professional practices, or medical treatment may become
necessary.

Practitioners and researchers must always rely on their own experience and knowledge in evaluating and using
any information, methods, compounds, or experiments described herein. In using such information or methods
they should be mindful of their own safety and the safety of others, including parties for whom they have a
professional responsibility.

To the fullest extent of the law, neither the Publisher nor the authors, contributors, or editors, assume any
liability for any injury and/or damage to persons or property as a matter of products liability, negligence or
otherwise, or from any use or operation of any methods, products, instructions, or ideas contained in the
material herein.

Library of Congress Cataloging-in-Publication Data
A catalog record for this book is available from the Library of Congress

British Library Cataloguing in Publication Data
A catalogue record for this book is available from the British Library

ISBN: 978-0-12-803011-0

For information on all Elsevier publications
visit our website at https://www.elsevier.com/

Working together
to grow libraries in
developing countries

www.elsevier.com • www.bookaid.org

Publisher: Candice Janco
Acquisition Editor: Marisa LaFleur
Editorial Project Manager: Marisa LaFleur
Production Project Manager: Paul Prasad Chandramohan
Designer: Greg Harris

Typeset by TNQ Books and Journals

I would like to dedicate this book to my partner, Nune Igityan, for her love, constant support, and all the experiences in life we have lived together so far, Thanks, Nunjan.
—George P. Petropoulos

I would like to dedicate this book to my parents, Bishwambhar N. Srivastava and Nirmala Devi, as well as to my beloved wife Manika for their continuous support.
—Prashant K. Srivastava

Contents

SECTION 1 INTRODUCTION TO SA IN EARTH OBSERVATION (EO)

SECTION 2 LOCAL SA METHODS: CASE STUDIES

SECTION 3 GLOBAL (OR VARIANCE)-BASED SA METHODS: CASE STUDIES

SECTION 4 OTHER SA METHODS: CASE STUDIES

SECTION 5 SOFTWARE TOOLS IN SA FOR EO

SECTION 6 CHALLENGES AND FUTURE OUTLOOK

List of Contributors

P.M. Atkinson
University of Southampton, Southampton, United Kingdom; Lancaster University, Lancaster, United Kingdom; Queens University Belfast, Belfast, United Kingdom

B.L. Barker
MGS Engineering Consultants, Olympia, WA, United States

B.A. Bryan
CSIRO Land and Water, Glen Osmond, SA, Australia

B. Cheviron
Irstea, UMR G-EAU, Montpellier, France

R. Ciampalini
INRA, UMR — LISAH, Laboratoire d'étude des Interactions Sol - Agrosystème — Hydrosystème, Montpellier, France; Cardiff University, Cardiff, Wales, United Kingdom

A. Couturier
INRA, Centre de recherche Val de Loire, Orléans, France

Q. Dai
Nanjing Normal University, Nanjing, China; University of Bristol, Bristol, United Kingdom

J. Dash
University of Southampton, Southampton, United Kingdom

P. Dietrich
Helmholtz Centre for Environmental Research (UFZ), Leipzig, Germany; Eberhard Karls University Tübingen, Tübingen, Germany

M. Dong
Beijing Institute of Surveying and Mapping, Haidian, Beijing, China

S. Follain
Montpellier SupAgro, UMR — LISAH, Laboratoire d'étude des Interactions Sol - Agrosystème — Hydrosystème, Montpellier, France

L. Gao
CSIRO Land and Water, Glen Osmond, SA, Australia

S. Golian
Shahrood University of Technology, Shahrood, Iran

H. Gupta
The University of Arizona, Tucson, AZ, United States

M. Gupta
USRA/NASA Goddard Space Flight Center, Greenbelt, MD, United States

D. Han
University of Bristol, Bristol, United Kingdom

E.A. Hernandez
Texas Tech University, Lubbock, TX, United States

M.C. Hill
University of Kansas, Lawrence, KS, United States

S. Hubbard
Intelligence at Airbus Defence and Space, Farnborough, United Kingdom

J. Indu
Indian Institute of Technology Bombay, Mumbai, India

J. Iwema
University of Bristol, Bristol, United Kingdom

M. Jabloun
Aarhus University, Tjele, Denmark

C. Kalaitzidis
MAICH/CIHEAM, Chania, Greece

A. Karim
Texas Tech University, Lubbock, TX, United States

M.C. Kennedy
Fera Science Ltd., York, United Kingdom

T. Lankester
Intelligence at Airbus Defence and Space, Farnborough, United Kingdom

Y. Le Bissonnais
INRA, UMR — LISAH, Laboratoire d'étude des Interactions Sol - Agrosystème — Hydrosystème, Montpellier, France

L. Lee
University of Leeds, Leeds, United Kingdom

A. Loschetter
BRGM, Orléans, France

K. Manevski
Aarhus University, Tjele, Denmark

T. Mannschatz
United Nations University, Dresden, Germany

Z. Micovic
BC Hydro, Burnaby, BC, Canada

S. Moazami
Islamshahr Branch, Islamic Azad University, Islamshahr, Tehran, Iran

G.L. Mountford
University of Southampton, Southampton, United Kingdom

R. Moussa
Montpellier SupAgro, UMR — LISAH, Laboratoire d'étude des Interactions Sol - Agrosystème — Hydrosystème, Montpellier, France

D. Nagesh Kumar
Indian Institute of Science, Bangalore, India

G.P. Petropoulos
University of Aberystwyth, Wales, United Kingdom

F. Pianosi
University of Bristol, Bristol, United Kingdom

D. Raucoules
BRGM, Orléans, France

S. Razavi
University of Saskatchewan, Saskatoon, SK, Canada

J.P. Rivera
Centro de Investigación Científica y de Educación Superior de Ensenada, Ensenada, Mexico

J. Rohmer
BRGM, Orléans, France

R. Rosolem
University of Bristol, Bristol, United Kingdom

F. Sarrazin
University of Bristol, Bristol, United Kingdom

M.G. Schaefer
MGS Engineering Consultants, Olympia, WA, United States

P. Sellitto
Laboratoire de Météorologie Dynamique/École Normale Supérieure, Paris, France

X. Song
Zhejiang University, Hangzhou, China

P.K. Srivastava
NASA Goddard Space Flight Center, Greenbelt, MD, United States; Banaras Hindu University, Varanasi, Uttar Pradesh, India

A.S. Tomlin
University of Leeds, Leeds, United Kingdom

V. Uddameri
Texas Tech University, Lubbock, TX, United States

J. Verrelst
Universitat de València, València, Spain

T. Wagener
University of Bristol, Bristol, United Kingdom

C. Walter
AGROCAMPUS OUEST, INRA, Rennes, France

M. Ye
Florida State University, Tallahassee, FL, United States

G. Zhao
University of Bonn, Bonn, Germany

T. Ziehn
CSIRO, Aspendale, VIC, Australia

Preface

Advances in computer science over the past decades have led to the increased use of sophisticated deterministic models in the simulation and prediction of processes, feedbacks, and mechanisms related to a number of science and engineering fields. These computer-based system models have become indispensable in many disciplines, ranging from finance to life sciences and from quantum physics to Earth and environmental sciences, including Earth observation (EO) technology. More recently, new methods utilizing EO data synergistically with models simulating Earth's physical processes have started to emerge in the scientific literature. These techniques endeavor to provide improved predictions of land surface processes by combining the horizontal coverage and spectrally rich content of EO data with vertical coverage. However, before applying a computer model in performing any kind of analysis or operation, a variety of validation tests are needed to evaluate the adequacy of the developed computer model in terms of its ability to reproduce the desired mechanisms with the necessary reality. A common strategy is to examine the model's simulated outputs versus actual observations using common statistical methods proposed in the classic literature. In addition to this approach, a sensitivity analysis (SA) has been identified as a necessary part of any such model building and validation.

SA can help to understand the behavior of a model and the coherence between the model and the real world as well as help to verify whether the model concept corresponds to the natural system's behavior in an appropriate manner. In addition, SA can assist in establishing the dependency of the model outputs on its input parameters in how different parts of the model interplay, as well as in identifying possible region(s) in the space of model input parameters where the model variation is maximum or divergent. As a result, SA provides a valuable method to identify critical input parameters and rank them in order of importance, offering guidance to the design of experimental programs as well as to more efficient model coding or calibration. This is because by means of an SA, irrelevant parts of the model may be dropped or a simpler model can be built or extracted from a more complex one (so-called model lumping), reducing, in some cases significantly, the required computing power. Today, there are a wide range of practices implemented for performing SA analyses extensively applied in a variety of applications in the field of EO modeling, including the development of operational products aiming to offer products at different observation scales. The inclusion of SA reveals the importance of more rigorous model calibration, thus facilitating a good modeling practice for environmental predictions, and it is now required for all modeling works.

After the launch of many sophisticated satellites in space, in recent decades the development of EO-based modeling system has gained considerable momentum among the Earth and environmental science communities for solving and simplifying the various complex problems. However, when developing a modeling technique, an understanding of various model parameters and their sensitivity is required for a rigorous model calibration. In essence, a comprehensive book is needed to put together a collection of the recent developments and rigorous applications of the SA techniques using the EO data sets. Therefore, this book is motivated by the desire to provide a better understanding of SA in a cost-effective and timely way for EO modeling.

In this context, this book highlights the state-of-the-art ongoing research investigations and new applications of SA particularly in the field of EO modeling. In this framework, original works concerned with the development or exploitation of diverse SA methods applied to different types of EO data have been included. An overview of SA methods and their principles is provided first, followed by case studies from the application of SA methods selected. The case studies included cover a wide spectrum of SA techniques including those on operational products. The book also discusses the key challenges in this field and the future prospects with regard to SA implementation in EO modeling.

This book becomes possible due to the extensive and valuable contributions from interdisciplinary experts/communities from all over the world in the field of SA and EO. Based on contributions, this book has been divided into six sections: Section I contains an Introduction to SA in EO. Section II details the local SA methods. Section III provides the global (or variance)-based SA methods. Sections IV and V deal with the miscellaneous nonconventional methods and the related software tools in SA, respectively, while Section VI provides the challenges and future outlook in SA use in EO.

Chapter 1 in the introductory section written by Lindsay and team provides an overview of the most likely appropriate SA methods for EO science with examples of their use in EO to date. Chapter 2 by Mannschatz et al. provides a discussion on the use of remote sensing data as input for environmental modeling and associated uncertainties. The second section of the book contains chapters related to local SA with several case studies. Chapter 3 of this section by Ciamapalini et al. provides a case study of implementing a LSA to the LandSoil model. Chapter 4 by Mountford et al. provides the sensitivity of vegetation phenological parameters from satellite sensors to spatial resolution and temporal compositing period. Chapter 5, Dai et al. describes the radar-rainfall SA. Chapter 6 by Manevski and team provides field-scale sensitivity of vegetation discrimination to hyperspectral reflectance and coupled statistics.

Section III of the book refers to the different methodologies for global sensitivity analysis (GSA). Chapter 7 of this section by Pianosi et al. provides a variance- and density-based GSA to support the calibration and evaluation of land surface models. Chapter 8 authored by Rohmer et al. briefly describes the GSA for supporting history matching of geomechanical reservoir models using satellite InSAR data: a case study at the CO_2 storage site of In Salah, Algeria. In Chapter 9, Sellito provides artificial neural network as a tool for spectral SA to optimize inversion algorithms for satellite-based EO, while Ye and Hill in Chapter 10 provides the GSA for uncertain parameters, models, and scenarios.

Section IV of the book deals with some other techniques as well as case studies. Chapter 11 by Micovic et al. summarizes the SA and uncertainty analysis for stochastic flood hazard simulation. Chapter 12 of this section by Uddameri et al. provides an SA of wells in a Large Groundwater Monitoring Network and its evaluation using GRACE Satellite—Derived Information. In Chapter 13, Nagesh and Indu present several sampling techniques that can be used in SA implementation. Chapter 14 by Moazami et al. demonstrates the ensemble-based multivariate SA of satellite rainfall estimates using the copula model.

Section V includes a series of chapters focusing on open source software tools in SA for EO. Chapter 15 in this section by Ziehn and Tomlin furnishes a detailed overview of efficient tools for GSA based on high-dimensional model representation. Verrelst and Rivera in Chapter 16 presents the ARTMO's GSA toolbox. Chapter 17 authored by Kennedy and Petropoulos presents the usefulness of Gaussian Emulation for Sensitivity Analysis software tool demonstration. Chapter 18 written by Fanny et al. provides the use of SAFE Matlab toolbox with practical examples and guidelines.

Finally, in Section VI chapters are presented on challenges and future outlook in SA use in EO. Chapter 19 written by Song et al. describes the sensitivity in ecological modeling from local to regional scales, while Chapter 20 authored by Gupta and Razavi points out the current limitations, challenges, and future outlook of SA use in EO.

In summary, the book is designed to advance the scientific understanding, development, and application of numerous SA techniques and its applications for various environmental problems. This book aims at promoting the synergistic and multidisciplinary activities among scientists. Therefore, it may be considered as the first book to this effect among scientists and users working in the field of SA and EO. We believe that the book would be of interest to the readers with an interest in EO, simulation modeling, geospatial technology, sustainable technology development, applications, and other diverse backgrounds within Earth and environmental sciences. We hope the book to also be beneficial for academicians, scientists, environmentalists, meteorologists, environmental consultants, and computing experts.

Last but not the least, we, the editors, are grateful to all the contributing authors and anonymous reviewers for their time, talents, and energies to support this endeavour for adherence to a strict timeline and to the staff at Academic Press, Elsevier, particularly Ms Marisa LaFleur and Paul Prasad Chandramohan, for their patience and support throughout the publication process.

ABOUT THE COVER

Sophisticated Soil Moisture and Ocean Salinity (or SMOS) satellite shown on the cover is provided by the European Space Agency. © ESA-Pierre Carril.

Wales, UK   George P. Petropoulos
Maryland, USA   Prashant K. Srivastava

INTRODUCTION TO SA IN EARTH OBSERVATION (EO)

OVERVIEW OF SENSITIVITY ANALYSIS METHODS IN EARTH OBSERVATION MODELING

1

L. Lee[1], P.K. Srivastava[2,3], G.P. Petropoulos[4]

University of Leeds, Leeds, United Kingdom[1]; NASA Goddard Space Flight Center, Greenbelt, MD, United States[2]; Banaras Hindu University, Varanasi, Uttar Pradesh, India[3]; University of Aberystwyth, Wales, United Kingdom[4]

CHAPTER OUTLINE

Sensitivity Analysis in Earth Observation Modelling. http://dx.doi.org/10.1016/B978-0-12-803011-0.00001-X

1. INTRODUCTION

There are many methods of sensitivity analysis, and their appropriateness depends on (1) the computer model (or simulator) being used and its complexity, (2) the resources available to carry out sensitivity analysis, and most importantly (3) the scientific question that is to be addressed. The methods range from simply changing one model factor to simultaneously changing many model factors and assessing the change in model outputs. The aim of all methods though is simply to assess the relative changes in model response from perturbation of different model factors, setting sensitivity analysis apart from the uncertainty analysis that simply aims to quantify the resulting uncertainty range in the model output. Sensitivity analysis helps the modeler to understand model behavior and provides information on how to reduce model uncertainty. Further reasons for conducting sensitivity analysis can be found in French (2003), Ireland et al. (2015), and Petropoulos et al. (2015). The extent to which the sensitivity analysis can be successful depends on whether the appropriate method was chosen and whether the model simulations were designed appropriately for the method and the question of interest. It is therefore important that sensitivity analysis is part of a carefully designed model study rather than a posthoc analysis of available model simulations. A more comprehensive discussion of the various methods of sensitivity analysis than are introduced here can be found in Saltelli et al. (2000).

1.1 DEFINING THE MODEL OUTPUTS AND INPUTS FOR SENSITIVITY ANALYSIS

The simple model can be defined as:

$$Y = f(X), \tag{1.1}$$

where capital letters denote random variables (unknown quantities) in the multiple factors X and therefore necessarily in the scalar model output Y. Any single model run that has been carried out is thus defined as:

$$y = f(x_1, x_2, \ldots, x_n) \tag{1.2}$$

for a single output value y with n factor values $x = x_1, x_2, \ldots, x_n$. It should be noted that X only refers to the unknown factors to be considered in the sensitivity analysis rather than an exhaustive list of all model uncertainties, and so its elements must be well defined.

1.1.1 Defining Factor (or Parametric) Uncertainty

In sensitivity analysis, X is used to represent the uncertain model factors that are being perturbed. The elements of X can be perturbations to unknown model parameters (unknown due to incomplete information or model scale), uncertain processes or model structures, and boundary and initial conditions. Typically, the individual components of X are scalar values and represent the perturbation or scaling of a single value. It is common to refer to X as parameters and its uncertainty as parametric uncertainty, although this extends beyond the traditional use of parameters to describe process settings in a computer model. Here we will use the term uncertain factors to refer to uncertain model elements X that form our sensitivity analysis.

To study the uncertainty in Y it is first necessary to define the uncertainty in X, which we denote as G, and for each unknown factor X_i is denoted by G_i. There are numerous ways through which G_i could be derived depending on the definition of the uncertain factor X_i and its uncertainty. When X_i is a model factor whose inclusion in the model is because the model resolution is too coarse, the uncertainty may be defined as the spatial variability in X_i that can be estimated from observations. For example, X_i could be the average cloud cover in any single Earth observation (EO) model grid box and its uncertainty could be the spatial variation in cloud cover in the grid box as measured by a satellite. When X_i is a factor that describes a process in which we have little scientific knowledge, its uncertainty may be used to describe the range of possible values that have been shown during model development to provide reasonable values of model output. In reality, despite the precise definition of a factor X_i its uncertainty will usually arise from a number of sources including scientific knowledge uncertainty, spatial and temporal variability, and factor representation uncertainty.

A common method for assigning uncertainty to the model factors is via expert elicitation in which experts are asked to define the model factor and use all their knowledge and available evidence, including satellite data, to produce an estimate of quantile values from which the uncertainty distribution G_i for X_i can be produced. Fig. 1.1 shows some possible probability distributions G_i to represent uncertainty in X_i. The Gaussian distribution is used when most of the uncertainty is thought to be symmetric around a central value, whereas the uniform distribution is used to describe equal uncertainty across the range of X_i. The beta distribution can be used to represent a skewed distribution when uncertainty is concentrated at one end of the range of uncertainty, and the trapezoid distribution is used to represent

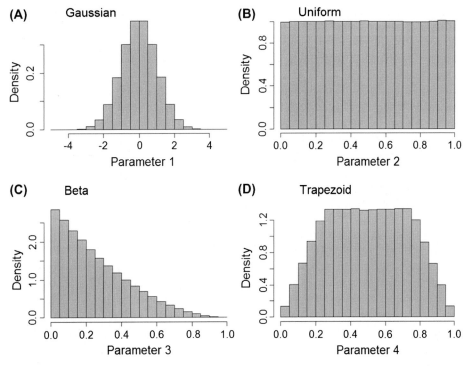

FIGURE 1.1

Commonly used uncertainty distributions. (A) The Gaussian distribution; (B) the uniform distribution; (C) the beta distribution; and (D) the trapezoid distribution.

equal uncertainty across some central range tapering toward the edge of the range. In expert elicitation, the experts must be satisfied that the distribution is a realistic representation of the factor uncertainty.

In a sensitivity analysis setting it is preferable that the uncertain model factors are statistically independent, that is, the value of any single factor does not determine the value or possible range of another model factor. When the value of any model factor X_i leads to a change in the defined probability distribution of X_j, independence cannot be assumed compromising the results of some sensitivity analysis methods and it may be preferable to redefine model factors so that they are statistically independent. The joint distribution of the uncertainty in the inputs is not usually considered beyond the question of independence due to the difficulties in its interpretation and precise definition.

2. LOCAL SENSITIVITY ANALYSIS

Local sensitivity analysis looks at the sensitivity of the model output to small changes in the model inputs. The slope, or derivative, of the model response in a very small neighborhood is used to estimate the local sensitivity of a function using

$$\frac{\partial y}{\partial \boldsymbol{x}}\bigg|\boldsymbol{x_0} \tag{1.3}$$

where $\boldsymbol{x_0}$ specifies the neighborhood in which the local sensitivity is to be estimated and ∂ represents the small change or partial derivative of the function. For example, $\boldsymbol{x_0}$ may represent the median value of all uncertain factors, and the derivative will estimate the sensitivity by looking at the change in y given a very small change away from $\boldsymbol{x_0}$. In this case, the small change can be applied to all elements of $\boldsymbol{x}$ rather than a single element. Local sensitivity analysis can be a very quick and informative method of understanding how model output responds to its uncertain factors but only for small factor changes and so does not directly provide the sensitivity of the model response over the whole of the range of uncertainty. This can be particularly useful as a first estimate of sensitivity for factor screening and to build the response function, when the model is computationally expensive. Care must be taken when local sensitivity shows values close to 0, as this also indicates saddle points in more complicated functions. To avoid ruling out factors that have saddle points in their response, surface local sensitivity analysis should be carried out in a number of different neighborhoods.

Analyzing the local sensitivity across the model space will give a comprehensive view of the model behavior but the interpretation of the many local sensitivity values can be difficult. Local sensitivity analysis across the uncertainty space can, however, represent very complex model behavior that a function may have with respect to some of its factors, and so if the exploratory data analysis suggests any complex model response, local sensitivity alongside a global sensitivity method could produce the most informative results. Further methods of local sensitivity analysis are explored in Saltelli et al. (2000) and Petropoulos et al. (2015).

2.1 CORRELATION ANALYSIS

The simplest quantitative assessment of a model output's sensitivity to its uncertain factors is the correlation coefficient. The correlation coefficient measures the strength of the statistical dependence between the model output and each of the uncertain factors. The appropriate correlation coefficient to use depends on the form of the relationship between the output and its uncertain factors. Once a sample has been collected, scatter plots should be produced. This will show any obvious sensitivities, but importantly, it will indicate the form of relationship between the factor and model output. When the relationship between the factors and model output appears linear the Pearson correlation coefficient can be used to measure sensitivity. When the relationship appears monotonic but not necessarily linear, the Spearman rank correlation coefficient should be used instead to measure sensitivity (Srivastava et al., 2014a). When nonlinearity and nonmonotonicity are present, then correlation coefficients are not a suitable measure of sensitivity (Srivastava et al., 2013a). Sensitivity of input variables can be represented by using correlation matrix plots as shown in Fig. 1.2. The figure indicates the correlation between Soil Moisture and Ocean Salinity (SMOS) satellite—derived soil moisture relation and reference evapotranspiration (ETo), European Centre for Medium Range Weather Forecast (ECMWF)

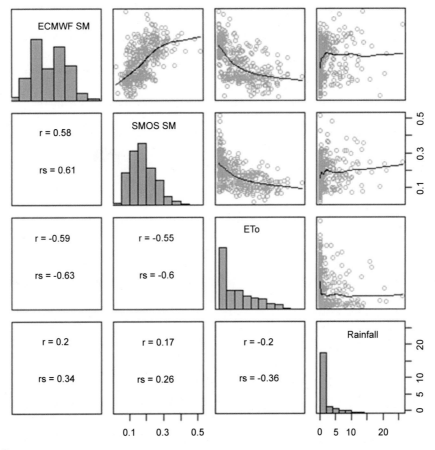

FIGURE 1.2

Example correlation matrix plot between satellite soil moisture and hydrometeorological variables.

soil moisture, and rainfall. The analysis indicates that for ETo estimation, rainfall is the least important when compared with SMOS and ECMWF soil moisture (Srivastava et al., 2013c).

2.2 REGRESSION ANALYSIS

Scatter plots suggest that linear relationships between the model output and the uncertain quantitative measures of sensitivity can be found using regression analysis (Srivastava et al., 2012). A simple linear regression model is found of the form

$$\mathbf{y} = \beta_0 + \beta_1 x_1 + \beta_2 x_2 + \ldots + \beta_{1,\ldots,n} x_{1,\ldots,n} + \varepsilon \tag{1.4}$$

which can include as many or as few terms as necessary. The model may contain all the terms associated with individual factors x_i and their interactions $x_{ij}, x_{ijk}, x_{i\ldots n}$ to produce the full model. In EO modeling, it might be expected that interactions of >2 factors will be redundant and that not all factors or their interactions contribute significantly to the model output, in which case the full model will be overfit. A technique called stepwise regression available in most statistical packages will test various models and include only those terms that contribute to a better model fit removing all inactive factors and their interactions. The terms in the model give the first indication of the sensitivity of the model outputs to its factors, but it is the coefficients β_i that are the more useful measures of sensitivity. In sensitivity analysis, the model should not contain terms that are functions of the model factors as interpretation becomes increasingly difficult; therefore regression analysis should only be used when the model is linear with respect to its factors and all their interactions. When a well-fitted linear model is found it will have a high R^2 value (the percentage of variance explained by the model fit) and the standardized residuals from the fitted model will show random scatter with ∼95% of the values lying in $[-2,2]$. Any departure from this suggests that the linear model is not suitable for sensitivity analysis. Regression-based sensitivity analysis is discussed further in Helton and Davis (2000).

3. GLOBAL SENSITIVITY ANALYSIS

Global sensitivity analysis aims to explore the response of a model output to the model factors quantitatively throughout the uncertain factor space (Fig. 1.3). In an uncertainty setting, this means we have a function, as in Eq. (1.1), $Y = f(X)$ with n uncertain factors $X = \{X_1, X_2, \ldots, X_n\}$ and we wish to understand how the uncertainty in X leads to uncertainty in $\mathbf{Y}$ and, in particular, how the individual elements of $X, \{X_1, X_2, \ldots, X_n\}$ lead to uncertainty in $\mathbf{Y}$ across the range of G_i. Global sensitivity analysis for nonlinear models has long been a topic of research as shown in Cukier et al. (1978), Iman and Helton (1988, 1991), Homma and Saltelli (1996), and Petropoulos et al. (2009c).

3.1 ONE-AT-A-TIME SENSITIVITY ANALYSIS METHODS

As the name suggests, one-at-a-time (OAT) tests are a way of testing the effect of perturbation of uncertain model factors on the model one at a time (Daniel, 1973). A single model factor X_i is varied within its defined uncertainty distribution G_i. OAT tests are usually carried out during the model development stage to test whether any development affects various model outputs $\mathbf{Y}$ as expected (Srivastava et al., 2012).

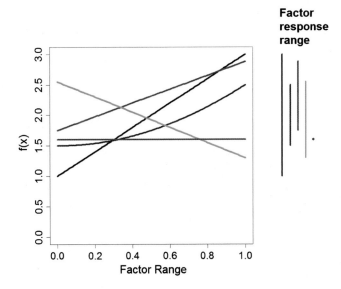

FIGURE 1.3

The model response to five factors showing the range of the response. The range of the factor response represents the relative sensitivity of each factor. The modeled output is most sensitive to the factor shown in black and not at all sensitive to the factor shown in purple. It is difficult to see whether the modeled output is most sensitive to the factor in green or red, but the global sensitivity analysis will quantify this, even for nonlinear responses.

OAT tests can be used for exploratory data analysis and as a screening method to more comprehensive methods of sensitivity analysis. OAT tests themselves do not exhaust the uncertainty space G but are a very useful first step in sensitivity analysis and should be designed as part of a sensitivity study. OAT tests carried out during model development should not be used as part of the sensitivity study as further model developments could invalidate previous results.

OAT tests should ideally include more than a single perturbation to the model factor X_i within G_i to assess the form of the model response. When only two perturbations are used at the range of the uncertainty distribution G_i, any response that is quadratic will suggest no model sensitivity to X_i. Fig. 1.4 shows that at least three perturbations (usually the end points and midpoints) to the model factor are required to show model sensitivity but that the shape of the response surface will be more closely defined with more than three perturbations to each model factor. It is also clear in Fig. 1.4 that the range of the perturbations is important to the result of the sensitivity test; for example, in Fig. 1.4B and C the sensitivity would scale directly with the range of the factor that is perturbed. When there is a priori information on the form of the factor—model response, it can be used to choose the appropriate perturbations to any factor X_i as part of the OAT testing.

When a factor X_i is shown to produce no response to the model outputs Y of interest, it can be removed from any further analysis as long as the assumption that it does not act together with any factor X_j to cause a model response (an interaction) is considered plausible. The greatest advantage of

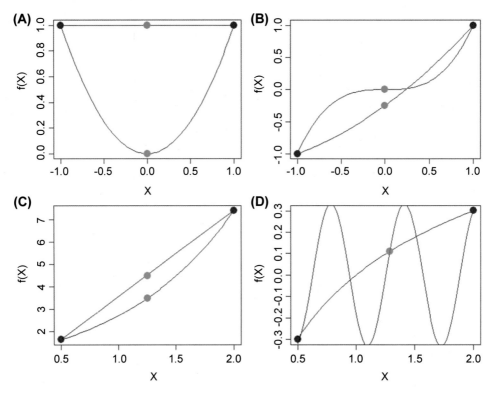

FIGURE 1.4

Some model responses with the results of three factor perturbations, at the edges and in the middle of the uncertainty distribution G_i. (A) It is clear that two perturbations can be misleading in sensitivity analysis; (B and C) three perturbations fail to capture some underlying features in the response, and in the presence of noise would not reveal a difference in response; (D) it is clear here that more than three perturbations would be required to reveal the response surface.

OAT tests from a single baseline is that the results are very easy to understand and can provide valuable information to the modelers straight away.

3.2 THE MORRIS METHOD FOR FACTOR SCREENING

When OAT tests are to be used for factor screening, it is preferable that the Morris method is used. With EO models the initial list of uncertain factors can be very long. Most methods of sensitivity analysis in EO models are limited by the number of factors that can be included due to computational burden. In such cases, a screening experiment may be preferred. The idea behind a screening experiment is to ensure that any factors that show no effect on the model outputs of interest are not included in further analysis, thus reducing the computational burden.

OAT tests will reveal any strong relationship between the model outputs and its uncertain factors, but it can be difficult to rule out factors from further analysis using simple OAT tests because there may be some joint relationship between factors causing model response (an interaction) that cannot be found by varying them from a single baseline. Consequently, the Morris method (Morris, 1991) was developed and is recommended for factor screening in a model such as that used in EO. The Morris method is still an OAT test, so the computational burden of screening is not increased; however, it does not perturb factors from a single baseline, so existing factor interactions may be seen if they exist, making it a more reliable tool for factor screening.

In the Morris method, a starting point is chosen and a single factor is perturbed. For the second factor perturbation, the first factor perturbation is also applied, and so on, so that by the time the final factor perturbation is applied, all factors are being perturbed away from the starting point. If this was carried out just once it would still be difficult to identify any possible interactions, so it is recommended that the Morris method be carried out using a number of different starting points, r, so that the total number of runs needed to do the Morris method of screening is $r(n + 1)$, where n is the number of factors to be perturbed and r is recommended to be between 5 and 15. Fig. 1.5 shows how the samples

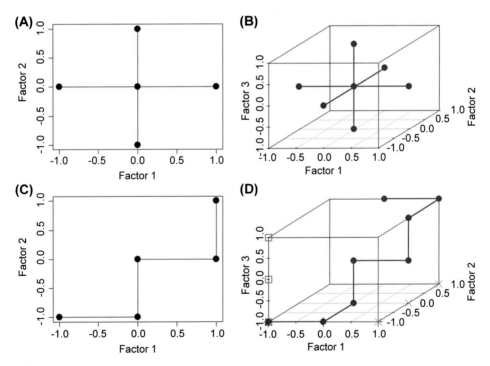

FIGURE 1.5

The baseline OAT design compared with the Morris method for screening in two- and three dimensions. In all cases, the factor has been chosen to have three levels of perturbation. In the simple OAT tests shown in (A and B), the perturbations are all applied from a single baseline points. For the Morris methods in (C and D) the perturbations are applied in turn without returning to the baseline. In (C) only a single starting point is shown. In (D) two starting points are shown.

can differ between an OAT test with a known baseline and the Morris method for OAT sampling where the factors are perturbed over three levels. Only two starting points of the Morris design are shown in Fig. 1.3 for clarity. Further methods for factor screening can be found in Saltelli et al. (2000).

3.3 VARIANCE-BASED SENSITIVITY ANALYSIS

More robust measures of sensitivity in the presence of nonlinearity and nonmonotonicity are the variance-based sensitivity indices. Based on the total law of variance and in the presence of statistically independent factors, the variance of the output can be decomposed into the variance associated with each individual factor and all the associated interactions. This decomposition is based on Cox (1982) and Sobol' (1993).

$$Var(Y) = \sum_{i=1}^{n} Var[E(Y|X_i)] + \sum_{i<j} W_{i,j} + \dots W_{i,j,\dots,n} \tag{1.5}$$

where $W_{i,j} = Var[E(Y|X_{i,j})] - Var[E(Y|X_i)] - Var[E(Y|X_j)]$. Dividing through by $Var(Y)$ yields the sensitivity measures. The main effect sensitivity is given by

$$S_i = \frac{Var[E(Y|X_i)]}{Var(Y)} \tag{1.6}$$

and can be interpreted as the fractional reduction in variance if the factor X_i were learned precisely. The first-order interaction sensitivity is given by

$$S_{i,j} = \frac{Var[E(Y|X_{i,j})]}{Var(Y)} \tag{1.7}$$

for each factor pair X_i, X_j. Higher order sensitivities follow similarly but are often not informative in complex models such as EO models. Another important sensitivity measure is the total sensitivity measure

$$S_{-i} = 1 - \frac{Var[E(Y|X_{-i})]}{Var(Y)} \tag{1.8}$$

where $-i$ indicates that all factors except for X_i are included. This is interpreted as the uncertainty remaining if all factors except X_i could be learnt. The total effect sensitivity for each factor measures the variance associated with each factor and its interactions. The difference between the total and main effect sensitivity indices across all model factors will also suggest between which factors interactions exist and which interaction terms should be investigated. The summation of the main effect indices will also suggest whether interaction terms are important—when the total main effect indices are close to 1, the variance associated with all other terms in Eq. (1.5) are necessarily small. Often the sensitivity indices are reported as percentages rather than fractions. Since variance-based measures often require large samples for their computation, it is recommended that, even with surrogate models, only the main effect indices and the total effect indices are calculated in the first instance with interaction terms only calculated when shown to be important. More description of this approach is provided in Chapter 20.

3.4 SAMPLING METHODS FOR GLOBAL SENSITIVITY ANALYSIS

Once the uncertainty distributions have been specified, it is important to think about how they can be sampled to learn about the model output. Design of experiments was introduced by Fisher (1935) and remains an important area of statistical analysis. For global sensitivity analysis it is required that the whole multidimensional factor uncertainty space is represented by the sample. This can be achieved in multiple ways depending on the number of model runs that can be completed using a number of experimental designs that are available. The details of the following methods, with example, are presented in the context of Earth observation in Chapter 13.

3.4.1 Random Sampling

The simplest method of sampling from the uncertain factor space is to use random sampling. Random sampling simply draws random values of the factors from the uncertainty distributions and investigates the resulting model output. Random sampling is simple, but to ensure that the entire joint distribution *G* of the model factors is represented, a very large sample may be required. A large random sample covering the distribution *G* is a Monte Carlo sample. Comparisons of random sampling to other sampling techniques for sensitivity analysis can be found in McKay et al. (1979) and Sobol' (1993).

3.4.2 Stratified Sampling and the Latin Hypercube

Stratified sampling involves splitting the distribution *G* into a number of nonoverlapping sections and randomly sampling within each of the sections independently, perhaps with different weights between the sections. This is preferred if the uncertainty distribution is not considered uniform. In order to well represent the distribution *G*, a large sample is still required. A particular type of stratified sampling that is useful in EO modeling is Latin hypercube sampling (LHS; Fig. 1.6). LHS divides the range of all factor distributions G_i into equal intervals (perhaps on a transformed scale)—when *n* samples are required there will be *n* intervals. The sample then consists of *n* sets of factor values with precisely one

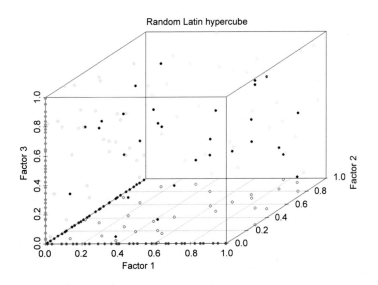

FIGURE 1.6

A maximin Latin hypercube in three dimensions showing the projection onto one and two dimensions. In all projections, there is good coverage of the uncertain factor space.

sample in every interval across all dimensions. As an example, a Sudoku puzzle consists of nine nonoverlapping Latin hypercube (LH) samples in two dimensions with nine samples—i.e., each sample contains the same digit falling precisely once in each of the nine rows and columns. The LH sample can be advantageous over other stratified samples, particularly in EO modeling where it is expected that many of the factors do not lead to significant model response—they are inactive. In the case of inactive factors, the samples will still provide coverage in all other dimensions and the marginal distributions of each individual factor remain well sampled. In particular, maximin LHS may be used where multiple LH samples are produced, and the one that has the largest distance between the closest pair of points is chosen to improve the space-filling properties of the design. When a LH sample is not comprehensive enough to directly derive accurate sensitivity indices, it is often used as the training sample for a surrogate model.

3.4.3 Sampling for Sensitivity Indices

Three sampling methods to calculate the main effect and total effect sensitivity indices, Eqs. (1.6) and (1.8), specifically have been developed: the Sobol' method, the Fourier Amplitude Sensitivity Test (FAST) method, and the extended-FAST method. The Sobol' method (Sobol', 1993 and Homma and Saltelli, 1996) requires a large sample to calculate each sensitivity measure and can be computationally burdensome. Another method to calculate the main effect sensitivity indices using a Fourier decomposition is the FAST (Cukier et al., 1973). To compute both the main effect and total sensitivity indices more efficiently the extended-FAST method was introduced (Saltelli et al., 1999). In any case, the computational burden of calculating the sensitivity measures is likely too high in EO models, and so surrogate models may be used to allow the samples to be collected.

3.5 SURROGATE MODELS FOR GLOBAL SENSITIVITY ANALYSIS

EO models are often too complex in terms of their CPU demands to produce a large enough sample to get reliable sensitivity indices so a surrogate model may be used. A surrogate model is based on machine learning and uses supervised learning to make predictions from a selection of EO model simulations. There are numerous methods to produce a surrogate model, and here we introduce some of the most commonly used in environmental science. The aim of the surrogate model is to adequately approximate the EO model across the whole uncertain factor space so that the sample required to calculate the sensitivity indices can be performed. In all cases, it is important that the surrogate model is validated to ensure that it can adequately estimate the EO model at points not used to build the surrogate. Surrogate models are often known as emulators. The surrogate models may include generalized linear models (GLMs), artificial neural network (ANN), and support vector machines (SVMs) (Srivastava et al., 2013b,c, 2015).

3.5.1 Generalized Linear Modeling

Generalized linear modeling (Nelder and Wedderburn, 1972) takes a number of model simulations and statistically models the response function in terms of the uncertain factors. Generalized linear modeling can handle factors that are categorical switches and can represent outputs that are not linear functions of the inputs—for example, binary outputs and count data. As with other linear models, the generalized linear model is evaluated by interrogation of the errors after fitting the statistical model.

The generalized linear model does not provide uncertainty on the point estimates, so its robustness for sensitivity analysis can be difficult to determine (Srivastava et al., 2016).

GLMs are a large class of statistical models for relating responses to linear combinations of predictor variables (Lindsey, 1997). One variable is considered to be an explanatory variable (x_i), and the other is considered to be a dependent variable (y_i) (Johnson and Wichern, 2002). They are represented by a family of regression model as:

$$y_i = x_i b^* + e_i, \tag{1.9}$$

where $i = 1...n$; y_i is a dependent variable, x_i is a vector of k independent predictors, b^* is a vector of unknown parameters, and e_i is stochastic disturbances. GLM models are characterized by stochastic component, systematic component, and link between the random and systematic components (McCullagh and Nelder, 1989). For a normal linear model $x_i b^*$ is an identity function of the mean parameter, whereas GLM is governed by some *link* function. The variance and link function used in this study all belong to the Gaussian family with link = "identity."

3.5.2 Neural Networks

The ANN (Haykin, 1994) was inspired by biological neural networks and uses a group of artificial neurons structured to represent the EO model. After development of sophisticated artificial intelligence techniques such as ANNs (Srivastava et al., 2013b,c), it is now possible to predict plant biochemical properties, using the hyperspectral remote sensing data sets (Gupta et al., 2014; Mulla, 2013). It has shown great promise for tackling complex pattern recognition problems (Petropoulos et al., 2012a,b, 2013) as well as prediction such as the modeling of biological variables as a function of multiple descriptors of the environment (Nagy et al., 2002; Schaap et al., 1998; Specht, 1991). In ANNs, the nodes are tuned using a cost function that requires some prior knowledge of the system (which can in fact represent little knowledge) and model simulations to minimize the cost function using some training algorithm, including expectation maximization, simulated annealing, or particle swarms. The neural network can be very effective in terms of representing the EO model, but producing the neural network can be difficult as its success depends on the cost function, the training algorithm, and the EO model simulations and can take some time to perfect. The major drawback with the ANNs are that they are "black box" because of little explanatory insight of the independent variables used in the prediction process (Olden and Jackson, 2002; Tzeng and Ma, 2005). Gevrey et al. (2003) resolve this issue by providing an all-inclusive comparison of various methodologies for estimating the variable importance in neural networks.

In ANNs, each input-to-node and node-to-node *connection* needs to be modified by a *weight*. The assembly contains an extra input in each node, which is assumed to have a constant value of one. The weight that modifies this extra input is termed as *bias*. Since a multilayered neural network containing one or more hidden layers, in theory, requires training to perform any regression or discrimination task, and therefore in this study, before the training process, the weights and biases were initialized to appropriately scaled values. For the output layer, a linear activation function is used with the relevant calculations as proposed by Aleksander and Morton (Aleksander and Morton, 1990):

$$O_a = h_{hidden} \left(\sum_{p=1}^{P} i_{a,p} w_{a,p} + b_a \right) \tag{1.10}$$

where $h_{hidden}(x) = \dfrac{1}{1 + e^{-x}}$

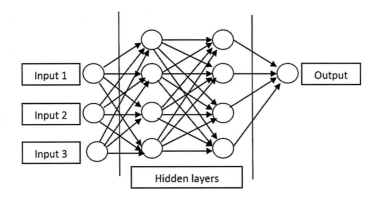

FIGURE 1.7

General architecture for neural network.

In this equation, O_a is the output of the current hidden layer node a, P is the number of nodes in the previous hidden layer, $i_{a,p}$ is an input to node a from either the previous hidden layer p or network input p, $w_{a,p}$ is the weight modifying the connection from either node p to node a or from input p to node a, and b_a is the bias. The general architecture of ANNs is provided in Fig. 1.7.

On the other hand, SVMs are supervised nonparametric statistical learning technique, which is implemented with the assumption of linear distribution of multispectral feature data in input space (Vapnik and Chervonenkis 1974; Fig. 1.8). With a set of predefined training pixels, SVMs algorithm aims to find a hyperplane that separates data sets into discrete predefined numbers of cover classes in a similar fashion.

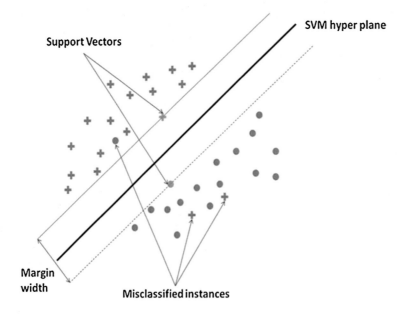

FIGURE 1.8

General architecture of support vector machines.

The term optimal hyperplane refers to iterative process of finding classifiers with optimal decision boundary to separate training patterns in high dimensional space. However, in general practice data sets of different class overlap to one another due to their closely resembling spectral features leading to misclassification and low accuracies. Such inseparability can be resolved by introducing soft margin or kernel tricks. There are a number of kernel functions available such as polynomial, radial basis function, linear, and sigmoid. Furthermore, the uncertain effect of choice of kernel function on the accuracy of classification studies is often considered a major setback concerning SVMs applicability.

3.5.3 Direct Sensitivity Analysis of Surrogate Models

Direct sensitivity analysis of GLM, ANNs, and SVMs can be performed to evaluate a relative importance of explanatory variables with a few other important distinctions. The relationship between explanatory and response variable can be explained by the model. For sensitivity analysis of GLM, ANNs, and SVMs, the popular Lek profile method can be used. Lek et al. (1996) formulated the "Lek profile method," which was later explained in detail by Gevrey et al. (2003). The principle of this algorithm is to construct a fictitious matrix pertaining to the range of all input variables. The generated matrix consists of values for explanatory variables where the number of rows are the number of observations, whereas the number of columns represents the explanatory variables. This matrix is then used to predict the values of each response variable from a fitted model object and then the process is repeated for different variables. Afterward, each variable is divided into a certain number of equal intervals between its minimum and maximum value (also called as scale). All variables except one are set initially, at their minimum values, then successively at their first quartile, median, third quartile, and maximum. In this way, five values for each of the scale's points are obtained for each variable. These five values are reduced to the median values to obtain the profile of the output variable (Lek et al., 1996). The same calculations can then be repeated for each of the other variables and to plot a curve. This gives a set of profiles of the variation of the dependent variable according to the increase of the input variables (see Fig. 1.9 with a scale of variation of 13). The final product is a set of response curves for one response variable across the range of values for one explanatory variable, while holding all other explanatory variables constant (Gevrey et al., 2003).

3.6 POLYNOMIAL CHAOS

Polynomial chaos (Wiener, 1938) aims to approximate the distribution of the modeled output using a polynomial basis with a germ (a new random variable, usually Gaussian, used to build the polynomial). The polynomial chaos approximation can be used to produce the sample required to calculate the sensitivity indices. The germ and the basis is nonunique and one depends completely on the other; as a result, the best polynomial chaos approximation can be difficult to find. Still it is possible to find a sufficient polynomial chaos expansion in many applications. The polynomial chaos approach to approximation is most common in applied mathematics rather than statistical applications.

3.7 GAUSSIAN PROCESS AND BAYES LINEAR EMULATION

Gaussian process (GP) emulation is another term for kriging, where the term kriging is normally reserved for spatial data, and is an interpolation method. Here the surface over which the kriging is applied is the uncertain factor space, and so the term GP emulation is used. Despite its name, GP does not

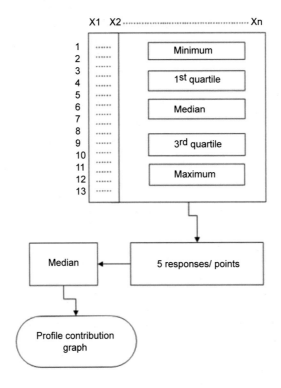

FIGURE 1.9

Explanatory schema of the profile method.

require that the model output be a Gaussian distribution. A priori information may be built into these functions if known, but a number of common functions are often chosen depending on the application. The details of calculating the GP can be found in O'Hagan (1994) and Kennedy and O'Hagan (2001). The GP in its simplest setting (and for most software packages) requires that the model output be a scalar and that there are no discontinuities in the function. Noise can be included via a nugget term in the emulator covariance function to represent uncertainty or variability in the modeled output. A particular advantage of the GP fitting process is that each estimated point in the uncertain factor space has a measure of uncertainty that can be used to validate the GP for use in sensitivity analysis. Work by Loeppky et al. (2012) suggests that a reasonable GP emulator can be built on a LH sample of model runs with 10 times the number of factors. GP emulation as a surrogate for complex computer codes has been the topic of much previous literature including Sacks et al. (1989), Currin et al. (1991), Welch et al. (1992), and O'Hagan (2006) with Oakley and O'Hagan (2004) concentrating particularly on sensitivity analysis using GP emulation.

Bayes linear methods (Goldstein and Wooff, 2007) have been developed to overcome the need to specify precise probability functions on the data and points in space. Instead, only the expectations, or means and variances, are required to estimate the function. In practical terms when used as a surrogate model, the Bayes linear method is very similar to the GP.

GP and Bayes linear emulation is conducted well in particular with SimSphere model outputs via a method based on Bayesian Analysis of Computer Code Outputs (BACCO; Kennedy and O'Hagan, 2001; Ireland et al., 2015; Petropoulos et al., 2014a,b) and is based on the use of the Gaussian Emulation Machine for Sensitivity Analysis tool for the purpose of model sensitivity and uncertainty analysis. The BACCO process is based on the use of a model "emulator"—a computationally cheap way of conducting the large set of model runs that are required to calculate the sensitivity indices, sometimes including the analysis of paired parameters. More details of this approach is provided in Chapter 17.

4. GRAPHICAL METHODS FOR GLOBAL SENSITIVITY ANALYSIS
4.1 SCATTER PLOTS

The most simple way to gauge sensitivity of a model output given some model input is to perform exploratory data analysis and inspect data graphics. Scatter plots of the model output versus each of the model inputs will show any clear relationship and whether any extreme value can be associated with any of the inputs. The linearity of the output response to the model inputs can also be assessed using scatter plots of available data. Exploring the data in this way can help to assess the most relevant method of sensitivity analysis when the sensitivity is to be quantified. Kleijnen and Helton (1999) discuss using scatterplots of Monte Carlo samples for sensitivity analysis.

Scatter dotty plots of the information measured can be used as shown in Fig. 1.10. Parameter sensitivity can be evaluated by estimating the spread of the cumulative parameter distributions and clear peak (see blue dots). In the example figure, in the case of K_f (*fast runoff factor*) a small variation is linked with the highest likelihood as a sharp and clear peak is observed with this parameter (Srivastava et al., 2014b). Similarly, the insensitive parameters are obtained by diffused peak represented by cumulative distributions. The dotty plot demonstrates that for each parameter, solutions with similarly good values of the objective function can be found within the complete prior range.

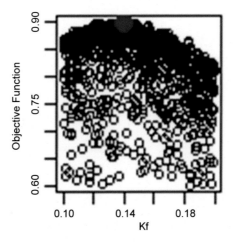

FIGURE 1.10

Example of dotty plot (Srivastava et al., 2014b).

4.2 PLOTTING THE RESPONSE SURFACE

Another common way to show the modeled output's dependence on each factor is to plot the response surface of the modeled output for each factor. This can be done in a few ways, and the most appropriate will depend on what is to be highlighted by the figure. When an emulator is used any sample of factor combinations can be achieved allowing different plots to be produced. Of course, it is important to be clear about what is shown in the figure, but common choices are to hold all factors at their central value and sample across the range of a single factor, sample all factors simultaneously, or sample across two factors holding all others at a central value.

4.3 PLOTTING THE SENSITIVITY INDICES

The simplest method of comparing the main and total effect indices is a bar chart in which both are plotted. Fig. 1.11 shows a bar chart of the sensitivity of a model output to five uncertain factors. It is clear that factor 4 dominates the uncertainty in the model output and that factors 1 and 2 are also important factors to learn to reduce the uncertainty in the model output. The main effect suggests that factor 5 can be ignored in terms of reducing the model output uncertainty but by comparing it to the total effect index, it can be seen that factor 5 interacts with another factor to affect the model output uncertainty and so should not be removed from the analysis. In fact, since only two factors show significant difference between the main and total effects, the interaction must occur between factors 1 and 5.

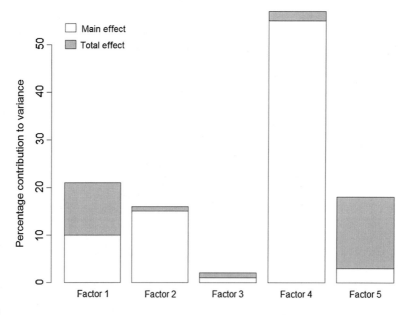

FIGURE 1.11

The typical representation of the main and total sensitivity indices for multiple factors. The white bar shows the main effect, and the gray bar shows the total effect additional to the main effect. A large white bar shows that the factor leads to a relatively large percentage of uncertainty in the modeled output. The appearance of the gray bar shows the amount each factor interacts with others to lead to uncertainty in the modeled output. Large gray bars may lead the modeler to investigate further interactions. Here the interaction between factors 1 and 5 would be worth further investigation.

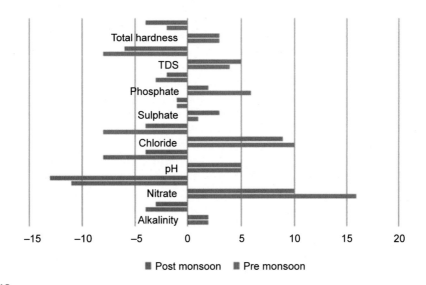

FIGURE 1.12

Tornado chart for sensitivity of physicochemical parameters (Srivastava et al., 2012).

Tornado diagrams or *tornado charts* (Fig. 1.12) are modified versions of bar *charts* and are also one of the classic tool of sensitivity analysis used by decision makers to have a quick overview of the risks involved. They reflect how much impact varying an input has on a particular output, providing both a ranking and a measure of the magnitude of the impact, sometimes given in absolute terms and sometimes in percentage terms (as in our example below). It can be possibly used for satellite input sensitivity analysis.

For EO models there is often interest in global model behavior. This could be achieved by calculating the sensitivity to global model outputs, but a far more informative method is to carry out the sensitivity on smaller model resolutions such as regions or grid boxes independently and then mapping the sensitivity indices to produce a picture of their effect across the globe.

Stacked bar charts of the main effect indices also provide useful information to the modelers regarding model behavior. The first piece of information comes from the height of the bars, because the height of the stacked bar is near 100% there are few model interactions and all the sensitivity information is present in the main effect indices. Second, by showing bars through time, seasonal behavior of the model can be seen. The graphics provide information on the physical behavior of the system, showing where particular uncertainties lead to uncertainty in the model prediction. There can be surprising results when all important uncertainties are taken into account and compared, as shown in Lee (2013).

5. CONCLUSIONS

Computer models are used in EO data processing as well as modeling with numerous uncertainties that must be comprehended to understand the robustness of any prediction. In the literature, many methods of sensitivity analysis are available with their uses depending on the scientific question of interest as well as the resources available. OAT testing is most likely to be used for factor screening and is easily understood due to the perturbations all being from a single baseline. Local sensitivity analysis can also be used to screen uncertain factors for removal before a global sensitivity analysis where computation demands are limiting. Although OAT testing and local sensitivity analysis are important for understanding the model, they are limited in providing information on the uncertainties in prediction and thus the robustness of EO models; in this case global sensitivity analysis is most useful.

Global sensitivity analysis can be used to understand how the uncertainties in the computer models feed through to its predictions revealing interesting physical behavior and providing information on the best data to be collected to improve the model predictions. It is important that the resources are not focused on obtaining data to reduce the largest factor uncertainties but to obtain data to reduce the prediction uncertainty. Due to computational demands of global sensitivity analysis, it is often necessary to more carefully consider sampling methods and surrogate modeling. Special sampling methods have been developed to obtain the results of a sensitivity analysis, but there is no agreed method for producing surrogate models. The surrogate model used also depends on the computational resources and the output to be estimated but is also chosen as a result of particular expertise or background; for example, a statistician might use a GP or Bayes Linear surrogate, a physicist might use a neural network surrogate, and an applied mathematician may use a polynomial chaos surrogate. It is most important that the surrogate is validated and is fit for purpose, whether in sensitivity analysis or in prediction.

In any case, the perturbations to the uncertain factors will be designed to explore the uncertainties based on expert's knowledge and what can be derived from EO data. The effect of the perturbations on the model predictions will be assessed, and those that lead to the largest prediction uncertainty will motivate further studies and further data collection. In the following chapters, examples of these techniques applied to EO modeling and data sets are presented.

REFERENCES

Aleksander, I., Morton, H., 1990. An Introduction to Neural Computing, vol. 240. Chapman and Hall, London.

Cox, D.C., 1982. An analytical method for uncertainty analysis of nonlinear output functions, with applications to fault-tree analysis. IEEE Transactions on Reliability 31, 265–268.

Cukier, R.I., Fortuin, C.M., Schuler, K.E., Petschek, A.G., Schaibly, J.H., 1973. Study of the sensitivity of coupled systems to uncertainties in rate coefficients. Journal of Chemical Physics 59, 3873–3878.

Cukier, R.I., Levine, H.B., Shuler, K.E., 1978. Nonlinear sensitivity analysis of multiparameter model systems. Journal of Computational Physics 26 (1), 1–42. Statist. Sci., 4, 409–435.

Currin, C., Mitchell, T., Morris, M., Ylvisaker, D., 1991. Bayesian prediction of deterministic functions, with applications to the design and analysis of computer experiments. Journal of the American Statistical Association 86 (416), 953–963.

Daniel, C., 1973. One-at-a-time plans. Journal of the American Statistical Association 68 (342), 353–360.

Fisher, R.A., 1935. The Design of Experiments. Oliver & Boyd, Edinburgh.

French, S., 2003. Modelling, making inferences and making decisions: the roles of sensitivity analysis. Top 11, 229–252.

Gevrey, M., Dimopoulos, I., Lek, S., 2003. Review and comparison of methods to study the contribution of variables in artificial neural network models. Ecological Modelling 160, 249–264.

Goldstein, M., Wooff, D., 2007. Bayes Linear Statistics, Theory and Methods, vol. 716. John Wiley & Sons.

Gupta, M., Srivastava, P.K., Mukherjee, S., Kiran, G.S., 2014. Chlorophyll retrieval using ground based hyperspectral data from a tropical area of India using regression algorithms. In: Remote Sensing Applications in Environmental Research. Springer International Publishing, pp. 177–194.

Haykin, S., 1994. Neural Networks: A Comprehensive Foundation.

Helton, J.C., Davis, F.J., 2000. Sampling-based methods. In: Saltelli, A., Chanand, K.E., Scott, M. (Eds.), Sensitivity Analysis. Wiley, New York, pp. 101–153.

Helton, J.C., Garner, J.W., McCurley, R.D., Rudeen, D.K., 1991. Sensitivity Analysis Techniques and Results for Performance Assessment at the Waste Isolation Pilot Plant. United States.

Homma, T., Saltelli, A., 1996. Importance measures in global sensitivity analysis of nonlinear models. Reliability Engineering & System Safety 52 (1), 1–17.

Iman, R.L., Helton, J.C., 1988. An investigation of uncertainty and sensitivity analysis techniques for computer models. Risk Analysis 8 (1), 71–90.

Ireland, G., Petropoulos, G.P., Carlson, T.N., Purdy, S., 2015. Addressing the ability of a land biosphere model to predict key biophysical vegetation characterisation parameters with Global Sensitivity Analysis Environmental Modelling & Software, 65, pp. 94–107. http://dx.doi.org/10.1016/j.envsoft.2014.11.010.

Johnson, R.A., Wichern, D.W., 2002. Applied Multivariate Statistical Analysis, vol. 4. Prentice Hall Upper Saddle River, NJ.

Kennedy, M.C., O'Hagan, A., 2001. Bayesian calibration of computer models (with discussion). Journal of the Royal Statistical Society. Series B 63, 425–464.

Kleijnen, J.P.C., Helton, J.C., 1999. Statistical analyses of scatterplots to identify important factors in large-scale simulations, 1: review and comparison of techniques. Reliability Engineering and System Safety 65, 147–185.

Lee, L.A., Pringle, K.J., Reddington, C.L., Mann, G.W., Stier, P., Spracklen, D.V., Pierce, J.R., Carslaw, K.S., 2013. The magnitude and causes of uncertainty in global model simulations of cloud condensation nuclei. Atmospheric Chemistry and Physics 13, 8879–8914. http://dx.doi.org/10.5194/acp-13-8879-2013.

Lek, S., Delacoste, M., Baran, P., Dimopoulos, I., Lauga, J., Aulagnier, S., 1996. Application of neural networks to modelling nonlinear relationships in ecology. Ecological Modelling 90, 39–52.

Lindsey, J.K., 1997. Applying Generalized Linear Models. Springer Verlag.

Loeppky, J.L., Sacks, J., Welch, W.J., 2012. Choosing the sample size of a computer experiment: A practical guide. Technometrics.

McCullagh, P., Nelder, J.A., 1989. Generalized Linear Models. Chapman & Hall/CRC.

McKay, M.D., Beckman, R.J., Conover, W.J., 1979. Comparison of three methods for selecting values of input variables in the analysis of output from a computer code. Technometrics 21 (2), 239–245.

Morris, M.D., 1991. Factorial sampling plans for preliminary computational experiments. Technometrics 33 (2), 161–174.

Mulla, D.J., 2013. Twenty five years of remote sensing in precision agriculture: key advances and remaining knowledge gaps. Biosystems Engineering 114, 358–371.

Nagy, H., Watanabe, K., Hirano, M., 2002. Prediction of sediment load concentration in rivers using artificial neural network model. Journal of Hydraulic Engineering 128, 588–595.

Nelder, J.A., Wedderburn, R.W.M., 1972. Generalized linear models. Journal of the Royal Statistical Society. Series A (General) 135 (3), 370–384. http://dx.doi.org/10.2307/2344614.

O'Hagan, A., 1994. Kendall's advanced theory of statistics. In: Bayesian Inference, vol. 2B. Arnold, London.

O'Hagan, A., 2006. Bayesian analysis of computer code outputs: a tutorial. Reliability Engineering & System Safety 91 (10), 1290–1300.

Oakley, J.E., O'Hagan, A., 2004. Probabilistic sensitivity analysis of complex models: a Bayesian approach. Journal of the Royal Statistical Society: Series B (Statistical Methodology) 66 (3), 751–769.

Olden, J.D., Jackson, D.A., 2002. Illuminating the "black box": a randomization approach for understanding variable contributions in artificial neural networks. Ecological Modelling 154, 135–150.

Petropoulos, G.P., Arvanitis, K., Sigrimis, N., 2012a. Hyperion hyperspectral imagery analysis combined with machine learning classifiers for land use/cover mapping. Expert Systems With Applications 39, 3800–3809.

Petropoulos, G.P., Kalaitzidis, C., Vadrevu, K.P., 2012b. Support vector machines and object-based classification for obtaining land-use/cover cartography from Hyperion hyperspectral imagery. Computers & Geosciences 41, 99–107.

Petropoulos, G.P., Vadrevu, K.P., Kalaitzidis, C., 2013. Spectral angle mapper and object-based classification combined with hyperspectral remote sensing imagery for obtaining land use/cover mapping in a Mediterranean region. Geocarto International 28, 114–129.

Petropoulos, G.P., Griffiths, H.M., Carlson, T.N., Ioannou-Katidis, P., Holt, T., 2014a. SimSphere Model Sensitivity Analysis Towards Establishing its Use for Deriving Key Parameters Characterising Land Surface Interactions. Geoscientific Model Development 7, 1873–1887. http://dx.doi.org/10.5194/gmd-7-1873-2014.

Petropoulos, G.P., Griffiths, H.M., Ioannou-Katidis, P., Srivastava, P.K., 2014b. Sensitivity Exploration of Sim-Sphere Land Surface Model Towards its Use for Operational Products Development from Earth Observation Data, Chapter 4, pages 31−56. In: Mukherjee, S., Gupta, M., Srivastava, P.K., Islam, T. (Eds.), Advancements in Remote Sensing for Environmental Applications. Society of Earth Scientists Series, Springer International Publishing, Switzerland. http://dx.doi.org/10.1007/978-3-319-05906-83.

Petropoulos, G.P., Ireland, G., Griffiths, H.M., Kennedy, M.C., Ioannou-Katidis, P., Kalivas, D.K.P., 2015. Extending the Global Sensitivity Analysis of the SimSphere model in the Context of its Future Exploitation by the Scientific Community. Water MDPI 7, 2101−2141. http://dx.doi.org/10.3390/w705210.

Petropoulos, G.P., Wooster, M.J., Kennedy, M., Carlson, T.N., Scholze, M., 2009c. A global sensitivity analysis study of the 1d SimSphere SVAT model using the GEM SA software. Ecological Modelling 220 (19), 2427−2440. http://dx.doi.org/10.1016/j.ecolmodel.2009.06.006.

Sacks, J., Welch, W.J., Mitchell, T.J., Wynn, H.P., 1989. Design and Analysis of Computer Experiments.

Saltelli, A., Chan, K., Scott, M.E., 2000. Sensitivity Analysis. Wiley, New York.

Saltelli, A., Tarantola, S., Chan, K., 1999. A quantitative model independent method for global sensitivity analysis of model output. Technometrics 41, 39−56.

Schaap, M.G., Leij, F.J., van Genuchten, M.T., 1998. Neural network analysis for hierarchical prediction of soil hydraulic properties. Soil Science Society of America Journal 62, 847−855.

Sobol', I.M., 1993. Sensitivity analysis for nonlinear mathematical models. Mathematical Modeling and Computational Experiment 1, 407−414.

Specht, D.F., 1991. A general regression neural network neural networks. IEEE Transactions on Neural Network 2, 568−576.

Srivastava, P.K., Han, D., Gupta, M., Mukherjee, S., 2012. Integrated framework for monitoring groundwater pollution using a geographical information system and multivariate analysis. Hydrological Sciences Journal 57, 1453−1472.

Srivastava, P.K., Han, D., Ramirez, M.A., O'Neill, P., Islam, T., Gupta, M., 2014a. Assessment of SMOS soil moisture retrieval parameters using tau-omega algorithms for soil moisture deficit estimation. Journal of Hydrology.

Srivastava, P.K., Han, D., Ramirez, M.A., Islam, T., 2013a. Appraisal of SMOS soil moisture at a catchment scale in a temperate maritime climate. Journal of Hydrology 498, 292−304.

Srivastava, P.K., Han, D., Ramirez, M.A., Islam, T., 2013b. Machine learning techniques for downscaling SMOS satellite soil moisture using MODIS land surface temperature for hydrological application. Water Resources Management 27, 3127−3144.

Srivastava, P.K., Han, D., Rico-Ramirez, M.A., Al-Shrafany, D., Islam, T., 2013c. Data fusion techniques for improving soil moisture deficit using SMOS satellite and WRF-NOAH land surface model. Water Resources Management 27, 5069−5087.

Srivastava, P.K., Han, D., Rico-Ramirez, M.A., Islam, T., 2014b. Sensitivity and uncertainty analysis of mesoscale model downscaled hydro-meteorological variables for discharge prediction. Hydrological Processes 28, 4419−4432.

Srivastava, P.K., Islam, T., Gupta, M., Petropoulos, G., Dai, Q., 2015. WRF dynamical downscaling and bias correction schemes for NCEP estimated hydro-meteorological variables. Water Resources Management 29, 2267−2284.

Srivastava, P.K., Yaduvanshi, A., Singh, S.K., Islam, T., Gupta, M., 2016. Support vector machines and generalized linear models for quantifying soil dehydrogenase activity in agro-forestry system of mid altitude central Himalaya. Environmental Earth Sciences 75, 1−15.

Tzeng, F.-Y., Ma, K.-L., 2005. Opening the black box-data driven visualization of neural networks. In: Visualization, 2005. VIS 05. IEEE. IEEE, pp. 383−390.

Vapnik, V., Chervonenkis, A., 1974. Theory of Pattern Recognition [in Russian]. Nauka, Moscow (German Translation: Wapnik, W., Tscherwonenkis, A., 1979. Theorie der Zeichenerkennung. Akademie-Verlag, Berlin).

Welch, W.J., Buck, R.J., Sacks, J., Wynn, H.P., Mitchell, T.J., Morris, M.D., 1992. Screening, predicting, and computer experiments. Technometrics 34 (1), 15−25.

Wiener, N., 1938. The homogeneous chaos. American Journal of Mathematics 60 (4), 897−936.

MODEL INPUT DATA UNCERTAINTY AND ITS POTENTIAL IMPACT ON SOIL PROPERTIES

2

T. Mannschatz[1],*, P. Dietrich[2,3]

United Nations University, Dresden, Germany[1]; Helmholtz Centre for Environmental Research (UFZ), Leipzig, Germany[2]; Eberhard Karls University Tübingen, Tübingen, Germany[3]
*Current address: ResearchGate, Berlin, Germany

CHAPTER OUTLINE

1. INTRODUCTION

Environmental processes are highly interconnected, complex, and additionally driven by natural as well as human-made factors (Liu et al., 2015). For instance, the resources water, soil, and waste are connected by, among other things, the hydrological, energy, and biogeochemical cycle, which can be described as a "nexus" (Lal, 2015). In order to comprehensively understand and predict the complex

Sensitivity Analysis in Earth Observation Modelling. http://dx.doi.org/10.1016/B978-0-12-803011-0.00002-1

interactions of nexus-related processes, it is required to implement an integrated environmental modeling approach using diverse software tools (Kelly Letcher et al., 2013; Mannschatz et al., 2016). To provide a guideline and to assure the highest achievable quality standards in modeling, several "good practices" guidelines for modeling have been published, such as Bennett et al. (2013), Black et al. (2014), Engel et al. (2007), Jakeman et al. (2006), Moriasi et al. (2007), and Refsgaard et al. (2005, 2007).

The general "good practice" modeling steps (Fig. 2.1) reveal that detailed information (especially spatially continuous data) about the environment where they are applied is needed even before the actual software tool is chosen. The data requirements include, for instance, information about the climate, soil, geology, and vegetation. For instance, evapotranspiration, important for modeling of the water balance, is a highly dynamic process since it depends on spatial and temporal variation of the vegetation. An indicator that can be used for evapotranspiration prediction in models is the leaf area index (LAI), which is represented by the ratio of the total one-sided area of photosynthetic tissue and unit ground surface area (Zheng and Moskal, 2009).

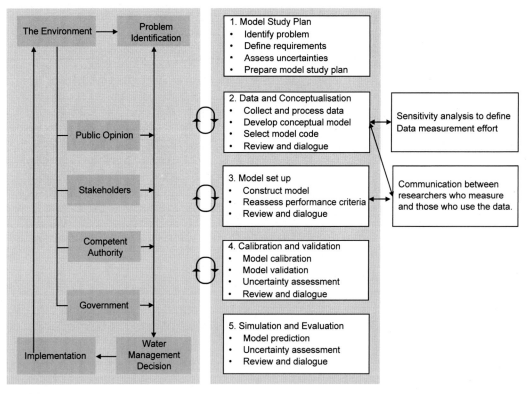

FIGURE 2.1

Workflow for environmental modeling.

Modified from Refsgaard, J.C., van der Sluijs, J.P., Højberg, A.L., Vanrolleghem, P.A., 2007. Uncertainty in the environmental modelling process — a framework and guidance. Environmental Modelling & Software 22, 1543–1556.

Conventional parameter assessment methods are often limited to information at a certain location (point information), time-consuming, and expensive. Therefore, alternative methods are increasingly often applied that are faster, more cost-efficient, and produce spatially distributed continuous data. Such methods typically draw on near-surface geophysics or remote sensing. For instance, multispectral satellite-based remote sensing has been proved to successfully derive information about soil, vegetation, and climate as input data for environmental modeling (Arora, 2002; Houser et al., 2012; Melesse et al., 2007). Near-surface geophysical methods such as visible near-infrared (vis-NIR) spectroscopy have been successfully applied to predict soil properties (e.g., Viscarra Rossel et al., 2011). However, the interpretation of geophysical measurement data is not a trivial task and is associated with uncertainties, which in turn might affect environmental modeling. For this reason, we can ask the more general question, if and when is it important to be aware of input data uncertainties and how might it affect modeling?

This chapter will address this question by drawing attention to the importance of appropriate consideration and communication of the uncertainty associated with input data used for modeling. In the field of environmental modeling, and modeling in general, this poses a considerable deviation from the usual approach. When using measurement data in a model, scientists start from the assumption that this data represents as an unquestionable "truth." However, as this chapter will illustrate, the methods used for collecting data are not without uncertainty, and it is crucial to take these uncertainties into consideration and communicate them appropriately, because this uncertainty can translate/propagate into considerable uncertainties in the model output.

To this end, the chapter starts by looking at modeling, uncertainties, and how these uncertainties can be assessed during the modeling process. A case study will then be used to illustrate the importance of taking into consideration the uncertainty that arises when we measure or predict model input data. The impact of input data uncertainty (i.e., texture estimated using a geophysical method) on environmental modeling will be shown by looking at the water balance.

2. A WORLD OF MODELS — HOW CAN THEY BE CLASSIFIED?

Reviewing the current literature, a standardized classification system for models could not be identified (Jakeman et al., 2006). A classification that is frequently applied is the division of environmental models into empirical, conceptual, and physical categories (Jakeman et al., 2006; Wainwright and Mulligan, 2004). The classification of models in this chapter will adopt this system.

Process description in *empirical models* is generally based on measured data for a particular study area or region. A mathematical empirical function is determined that best fits the measured data, where no physical processes need to be assumed. The model is then used to make assumptions about the observed effects and conditions of the specific environment. These established mathematical models are subsequently applied to new data to simulate and potentially predict system responses and behavior. This method generally proves successful for a specific study site, but cannot simply be generalized or even transferred to other environments.

Conceptual models offer a more generalized approach. A specific system is separated into the relevant processes and system components, and their conceptual relationship is described. The concept of interconnectedness of the different system components is then represented through data and empirical functions, or even models. For instance, a conceptual model of the hydrological cycle may describe the simplified water balance concept, distinguishing between precipitation, runoff, evaporation, transpiration, and storage change. Such conceptual models are partly transferable to other (similar) regions, although not all system components or processes that were relevant for the original region will be relevant in another region.

In contrast, *physically based models* attempt to fully explain the physical reality and comprehensively describe the processes of a system. These models, theoretically, are transferable to any region; because the physical assumptions and principles are equal everywhere and should be based on sound science. Generally, a large amount of constraints are required to parameterize the physical mathematical equations. In the case that the physical description of the system is correct and covers all relevant processes, the modeled values should match and reproduce the observations. In many cases this is not totally true. This may result from overlooking relevant processes; many physically based models also include some simplifications such as empirical relationships, for example, when the physical process is unknown. Physical models focus on the process understanding and can be used to get insights into the principles of the studied system. These models need to be calibrated against measured data.

3. CAN WE TRUST MODELS? — MODEL ACCURACY AND THEIR SENSITIVITY TO INPUT DATA UNCERTAINTY

The usefulness and reliability of models depend on the accuracy of the output of the modeling. The model output is subject to imprecisions, since all models are imperfect abstractions of the reality and 100% precise input data does, de facto, not exist (UNESCO, 2005).

Each model type requires different amount of data of various levels of quality. The gathering of these model parameters and data is often not easy, especially not in remote areas (Wainwright and Mulligan, 2013). A more realistic and holistic representation of an environmental system requires distributed data, optimally with a high spatial (e.g., soil information) and/or temporal resolution (e.g., climate data, streamflow data). Spatially distributed data has the advantage of being able to account for spatial and temporal heterogeneity having significant influence on environmental modeling (e.g., Loosvelt et al., 2013). This becomes, for instance, visible when looking at soil property variations that may change on small scale (see Section 7). Traditional measurement methods are often limited to point information and are therefore often very pricy. For this reason, the usage of geophysical methods such as near-surface geophysics and satellite-based remote sensing have the potential to overcome those problems (Mannschatz et al., 2015). An overview of the general workflow for environmental modeling, the problem of data scarcity, and how to make use of the available technologies to obtain spatially distributed data is given in Mannschatz et al. (2015).

Nevertheless, dealing with environmental systems we are often confronted with a weak link between the determination of input data and their use in models for prediction. Measurement data are not only important to represent and to describe the system in question (i.e., model parameterization) but also to calibrate and to validate the model output. Uncertainties related to model input data, therefore, are assumed to impact the accuracy and meaning of the model simulation. If we have an uncertainty in the model simulation then we will also have an impact on our understanding of the model-based simulation of the current studied ecosystem, as well as on the prediction of the future system condition (e.g., climate conditions) and its behavior (e.g., Lenhart et al., 2002; Song et al., 2015) (Fig. 2.2).

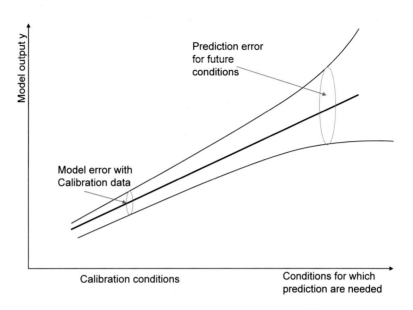

FIGURE 2.2

Dependency of the input data used for calibration and the model precision of predicted future (UNESCO, 2005).

This means that the model output values will depend and thus vary according to the uncertainty in the input data. This signifies that when we vary the input parameter within the uncertainty range of the data measurement of the particular input parameter, then the simulations will be influenced, resulting in an uncertainty range for the modeled output. The impact of the input data uncertainty on the modeled variables will depend on the sensitivity of the modeled processes to the particular input data, on the study site conditions where the model is applied, and on the model itself (i.e., model structure, model type) (e.g., Gould et al., 2014; Jakeman et al., 2006). Similarly, the calibration and validation of the model will only be possible if the measurement uncertainty of the data is taken in consideration.

Therefore, the accuracy of the model input data is of high importance. Measured data are often assumed to truly represent the studied system and as such a truth. However, it is important to be aware that those data are also estimation or model of the reality. This means that we actually try to parameterize and evaluate environmental models with modeled and estimated "hard (measured)" data. Both the modeled and measured data are thus associated with uncertainties. Gould et al. (2014) defines uncertainty as "a measure of unexplained variation that has three components in models of natural systems: (i) natural variability; (ii) measurement error; and (iii) incomplete knowledge about natural phenomena and complex processes." UNESCO (2005) summarized the sources of modeling uncertainty as related to informational uncertainty (i.e., boundary conditions, input data) and to modeling tool (i.e., model structure, equations used, parameter values, and numerical errors). Consequently, there are two main concepts that need to be understood, the analysis of uncertainty and sensitivity (Fig. 2.3) (UNESCO, 2005). The uncertainty analysis aims on assessing the total uncertainty related to the input data or the model output, whereas the sensitivity analysis aims at investigating where the

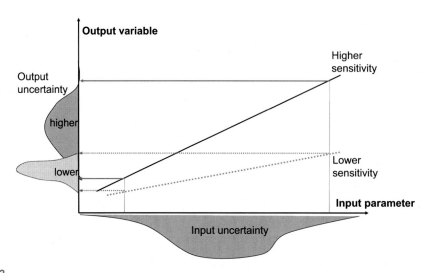

FIGURE 2.3

Relationship between the sensitivity of the model output to input data uncertainty under low and high model sensitivity to the input (UNESCO, 2005).

uncertainty comes from Song et al. (2015). In Fig. 2.3, we can explore the relationship between the sensitivity of the model output to input data uncertainty under low and high model sensitivity to the input (UNESCO, 2005). A brief overview of the main methods to accomplish a sensitivity analysis is shortly provided in Section 6.

This book chapter will focus on the uncertainty related to model input data and its impact on modeling. The main problem here is that the measurement uncertainties are often insufficiently communicated to modelers.

For instance, soil maps for different regions do not contain information about data distribution and quality, and in this manner it is not possible to evaluate the uncertainty of the soil maps. Consequently, they are often not taken into account during uncertainty evaluation of modeling prediction.

Taking the effect uncertainty of input data into consideration, steps must be taken to make the most out of the modeling results and to assure that results are useful and credible for stakeholders. The modeled results, such as current system state and predictions, should be communicated using appropriate visualization techniques and be accompanied by the uncertainty range instead of absolute values. This is important to allow an adequate "risk assessment," which quantifies and evaluates the impact of uncertainty on the modeling results (Black et al., 2014).

4. SELECTING THE MOST APPROPRIATE MODEL

The "best practices guidelines for modeling" argue (Section 1) that it is of high importance to select the most appropriate modeling tool for each project (e.g., Bennett et al., 2013; Black et al., 2011, 2014; Engel et al., 2007). However, a large amount of modeling tools are available. Additionally, the lack of a comprehensive inventory of available tools and their characteristics makes it a challenging task to identify the most appropriate tool. The appropriateness of a model depends on the project-specific requirements with regard to a modeling tool's capacity, the equations used, and the available input data (Mannschatz et al., 2016).

To address this challenge and advance the use of integrated environmental modeling, a comprehensive model overview and user-specific comparison tool is crucial. Knowledge about model capabilities, application range, and potential for coupling is a prerequisite for effective integrated modeling. Research review articles (e.g., Bagstad et al., 2013) and several online inventories (e.g., Hydrologic Modeling Inventory[1], IRRISOFT (Worldwide Web Database on Irrigation and Hydrology Software)[2], Register of Ecological Models[3], EBM Tools[4], Global Change Master Directory[5], EnviroLink[6], LIASE Kit[7]) take an important step in addressing this gap. Research papers reviewing modeling tools are mostly limited to a detailed review of three−four models and/or refereeing to others (e.g., Aksoy and Kavvas, 2005; Bagstad et al., 2013; Gentil et al., 2010; Köhne et al., 2009, 2009b; Ranatunga et al., 2008; and Srivastava and Migliaccio, 2007), which makes obtaining the required model information very time consuming and demanding—especially if one is starting the model selection process from scratch. However, to "the authors" knowledge, until today a model database that goes beyond a static model description list is not available. As such, it is nearly impossible for someone to be fully aware of all the available models on environmental resources and their capabilities; a lot of time is spent on the development of new models instead of recycling and further developing existing, well-established models. The "Nexus Tools Platform" (NTP),[8] a web-based model inventory platform launched in 2016, can be used to address the aforementioned information gap and to facilitate the identification of the most appropriate model (Mannschatz et al., 2016). The NTP allows the interactive statistical comparison of models. It compares the different modeling tools based on processes covered, application purpose, and any other full-text searchable term. The inventory includes modeling tools that are relevant for applying an integrated approach to the sustainable management of water, soil, and waste, considering global change and socioeconomic aspects. Neither the presented review of available models nor the database claims to be exhaustive. The NTP is, thus, a step forward to facilitate the selection of the most suitable modeling tool(s).

5. WHY AND HOW TO ACCOUNT FOR MODELING UNCERTAINTIES CAUSED BY DIFFERENT INPUT DATA SOURCES

Communication of the uncertainties related to modeling is often limited to the final modeling results, but it is usually impossible to know which uncertainties have been considered and included into the overall uncertainty. For instance, it remains unknown if uncertainty related to input data has been considered somewhere in the modeling process. Reflect the example of environmental modeling;

---

[1]http://hydrologicmodels.tamu.edu/.
[2]http://www.irrisoft.org/cms/.
[3]http://ecobas.org/www-server/index.html.
[4]http://www.ebmtools.org.
[5]http://gcmd.gsfc.nasa.gov/index.html.
[6]http://tools.envirolink.govt.nz/.
[7]http://www.liaise-kit.eu/exhibit/models.
[8]The current version can be found at data.flores.unu.edu/projects/ntp/.

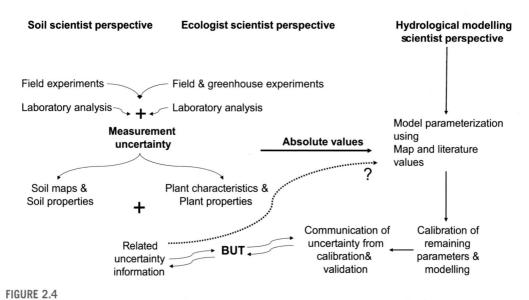

FIGURE 2.4

Different perspectives on data and modeling related uncertainties.

we can perceive that there are different perspectives about uncertainties. A soil scientist and ecologist are applying field and laboratory analysis, knowing that there are uncertainties related to the measurements (Fig. 2.4). Subsequently they produce maps representing their measurements. However, generally no information about the uncertainties of the maps are provided (e.g., on soil maps). Many environmental models are operating on larger scales such as watershed or basin scale requiring input data in the form of maps. The spatially distributed data are mandatory to obtain a good understanding of the spatially distributed environmental parameters through capturing its spatial heterogeneity and by decreasing uncertainties from interpolation between distant measurement points (Mannschatz, 2015; Mannschatz et al., 2015). The produced maps are used in the modeling process as input data, treated as absolute data—often unaware of the associated uncertainties and the influence it may have on the overall accuracy of the modeling output. This example shows once again that we need to communicate and consider data uncertainties during the whole modeling process: from data measurement, over data provision, modeling, and communication of model output. Harmel et al. (2014) recommends different steps for the communication of the uncertainty associated with modeling process. These steps include (1) the evaluation of initial model performance; (2) evaluation of outliers and extremes in observed values and bias in predicted values; (3) estimation of uncertainty in observed data and predicted values; (4) reevaluation of model performance considering accuracy, precision, and hypothesis testing; (5) interpretation of model results considering intended use; and finally, (6) the communication of the overall model performance (Harmel et al., 2014).

If the accuracy of the input data is of such importance, as seen previously, how can we improve it? An effective tool is making use of an effective sampling design for field measurements to collect samples that are representative for an investigated site or research problem. Details about the different methods for the definition of an appropriate sampling design are given for example in EPA (2002). In the context of the book chapter we would like to draw your attention on the usefulness of selecting a multistage sampling design where, for instance, a cost and time-efficient "rough and dirty" approach can be used. Such an approach allows for a good understanding of the heterogeneity of the parameter under investigation. Subsequently, a more expensive but more accurate sampling can be carried out using the a priori information (Mode et al., 1999). Where can we obtain these data? One way of realizing this approach is the application of remote sensing methods that allow a quasicontinuous spatially distributed parameter mapping that generally will measure the study parameters indirectly (see Mannschatz et al., 2015).

Satellite-based remote sensing can be used to derive spatially distributed model input parameters, such as LAI (e.g., multispectral remote sensing), precipitation (e.g., Tropical Rainfall Measuring Mission), evapotranspiration, soil moisture (e.g., Soil Moisture and Ocean Salinity (SMOS) Mission), or other parameters that may be requested by models (Lehmann et al., 2014). Near-surface geophysics might be applied to gather quasicontinuous information about horizontal or vertical soil layering and indirect properties such as electrical conductivity, which can be correlated to water content and clay content besides others. A comprehensive description of the working principle and application range of geophysical methods is given in Binley et al. (2010), Knödel et al. (2005), Rubin and Hubbard (2005), and Vereecken et al. (2006).

However, these methods measure indirect parameters, such as reflectance (e.g., remote sensing) or electrical conductivity (e.g., geoelectrics). Yet, reflectance or electrical conductivity does not provide us the information required for the modeling, such as LAI as a measure for plant productivity. For this reason, we are interested in finding a transfer function that describes best the relationship between the indirect (e.g., reflectance) and the actual target environmental property (e.g., LAI). Taking the LAI as an example, it might be derived from multispectral remote sensing images. The theory behind this practice is that green vegetation shows a specific response in the red and near-infrared spectra (band), which can be used to calculate the so-called vegetation indices (VIs). VI pixel values can be quantitatively correlated to the corresponding ground-based point measurements of LAI. If the mathematical relationship between VI and LAI is known for those pixel locations, then we can estimate additional LAI values for those pixels with calculated VIs. The measurement of indirect land surface characteristics and their mostly empirical relationship to the target parameter adds a larger level of uncertainty compared with measurements directly from a laboratory. Reviewing studies that investigated the retrieval of LAI from remote sensing found mean estimation uncertainties (root mean square error[9]) of 0.58 for corn and soybean (Vina et al., 2011), 0.63 for corn (Wu et al., 2007), 0.79 for potato (Wu et al., 2007), and 0.56 for deciduous forest (Wang et al., 2005).

Looking at geoelectrics for soil characterization as another example, the spatial distribution of electrical conductivity of the soil is used to derive indirectly a secondary parameter such as clay, salt, or water content. The electrical conductivity of the soil depends on different material-specific dielectric constants such as the water content, pore fluid chemistry, temperature, or clay content (Mannschatz, 2015). The geoelectrical field measurement produces "apparent" resistivity data, which indicate that it is not known from where the resistivity has actually been measured (Knödel et al., 2005). To obtain true depth information, the "apparent" data need to be converted using numeric inverse modeling. Once the true depth information is known, field measurements of the target parameter can be used to interpret the measured resistivity and to derive the spatial distribution of that target parameter. As we see the derivation of our target parameters involves inversion and modeling,

---

[9]Root mean squares error.

which is already an indication that uncertainties are related. These uncertainties can vary significantly depending on the measurement conditions and the instrument, which is true for all applied methods (see case study Chapter 7).

As we can see there are generally two types of data: direct data and indirect measurements (i.e., predictions). The indirect parameter measurements might be called "soft data," whereas direct measurements (e.g., soil laboratory measurements) might be called "hard data." For both "soft" and "hard" data, uncertainty can be minimized applying appropriate sampling designs to assure that property heterogeneity as well as representativeness are given to satisfy geostatistical requirements (EPA, 2002). The surface information can be completed by ground-based depth-stratified data, and the environmental processes can be simulated by the environmental models. However, every data assessment is associated with uncertainties. These uncertainties need to be communicated to the modeler, which, as can be see, for instance, with soil maps, is not typically the case.

6. ASSESSING SENSITIVITY OF ENVIRONMENTAL MODELS

If we look at the three different sources of uncertainty in applying models (see Section 5), we can respond and evaluate the uncertainty related to (1) input data and (2) model parameter, by using the so-called sensitivity analysis. Sensitivity analysis is able to turn apparent complex impacts or influences of parameters on the whole system, i.e., modeling process considering the algorithms (e.g., equations to calculate processes such as evapotranspiration) used for simulation. It is of high value to have a *feeling* about the data uncertainty and the sensitivity of the model to those uncertainties to achieve more confidence about the modeling results. The evaluation of the source of uncertainty associated with the (3) model structure can be achieved by a multimodel approach, which means that for the specific study area different models are applied. The comparison of the variation range of the simulation results produced by all different models when using the same parameterization (i.e., input data, model parameters) provides insights into differences due to model structure. A wide range of methods to evaluate uncertainties related to model structure are provided in, e.g., Refsgaard et al. (2006). A model is considered to be highly sensitive to a specific parameter when a variation of the parameter makes up a large portion of total variation (uncertainty) in model output. This means that small variations of the input parameter causes large changes in the model output (Song et al., 2015).

As stated in Section 5, this chapter focuses on the first model uncertainty source: input data. Sensitivity analysis is applied to systematically investigate the impact and influences of single or a combination of different input data (including model parameters) on the model output. The model sensitivity depends on model structure, model type, mathematical equations, and study site characteristics.

During sensitivity analysis, one or more input parameters are systematically varied around a specific value that is generally defined based on typical values for the application area or the input data. This analysis can be used to investigate the impact of a specific parameter or input data on the different single model outputs (Norton, 2015). The produced changes of modeled results are then reported and compared relatively to each other. We can basically distinguish between two types of sensitivity analysis: local and global sensitivity analyses (Norton, 2015; Song et al., 2015).

Briefly, in local sensitivity analysis the change of modeling output due to variations of one (one-at-a time perturbations, keeping other parameters fixed) or multiple (e.g., based on underlying probability function (Gwo et al., 1996) model parameters is compared with the change of those parameters around a specific parameter value (Lenhart et al., 2002; Norton, 2015). Even though being limited to the local effect of a single parameter, the advantages of this approach are its relative ease of use and interpretability, low computational effort (Song et al., 2015), as well as identification of nonlinear sensitivities (Loosvelt et al., 2013). The range of input parameter variation can be oriented by the estimated data input uncertainty range (e.g., measurement uncertainty). The analysis is often based on Morris method (Francos et al., 2003; Morris, 1991) or finite difference method (Lenhart et al., 2002; Loosvelt et al., 2013).

Model output changes might be expressed as absolute values or as percentage values, which is actually a normalized sensitivity [e.g., sensitivity index (SI), Lenhart et al. (2002)]. Global sensitivity analysis is preferable for various reasons; it is more scale independent. Since input data uncertainty may cross-influence, it is often advantageous to study model sensitivity to a large number of model parameter combinations simultaneously. This creates a high-dimensional problem and large computational requirements since all the interactions between the simultaneous change of various parameters and the model output need to be simulated and statistically evaluated. Therefore, it becomes necessary to apply an optimization algorithm that allows the automatic variation of input parameters simultaneously and to statistically investigate the model output changes due to input changes (Song et al., 2015). Common methods for global analysis include Sobol, Fourier amplitude sensitivity test, response surface method, or Monte Carlo—based methods (Loosvelt et al., 2013). Nevertheless, to prevent overoptimization, it is necessary to have a conceptual model of the studied system to consider only the most sensitive parameters and to exclude insensitive ones (Song et al., 2015). The sensitivity analysis allows ranking the most influential parameters, which in turn can be used to rank measurement effort of those parameters. Additionally, this might allow reducing model complexity if parameters can be discarded reducing the time for calibration (Song et al., 2015). The detailed description of the sensitivity analysis methods is beyond the scope of this chapter, but detailed descriptions can be found, e.g., in Kleijnen (2005), Norton (2015), Saltelli et al. (2000, 2004, 2008), and Song et al. (2015). Evaluation of model sensitivity can be based on statistical performance metrics and criteria, such as a SI (Lenhart et al., 2002) or the omnidirectional local SI for closed data (Loosvelt et al., 2013). The model uncertainties and sensitivity should be communicated and be visualized in an adequate accessible format. Closed data are an additional challenge for environmental modeling and are often overlooked when applying sensitivity analysis (Loosvelt et al., 2013). *Closed data* refer to the fact that environmental data are often correlated or otherwise connected to each other. A famous example is soil texture, where the texture fractions sand, clay, and silt always sum to 100%. Therefore, if one fraction is changed, then all remaining fractions need to be changed simultaneously (Loosvelt et al., 2013). An example of a local sensitivity analysis using closed data—soil texture—is provided in Section 7.

Sensitivity analysis is generally useful to make visible which data has the largest impact on modeling results. This information can be used to define the measurement effort and thus the data accuracy necessary for the model parameterization (Engel et al., 2007; Loosvelt et al., 2013). This means, if we look, for example, at a hydrological model used in an arid region, the accuracy of soil clay content data might be of less importance than it would be in humid regions. The reason is that infiltration processes become differently important depending on the soil and climate conditions

(Mannschatz, 2015; Mannschatz et al., 2014). To understand the uncertainties of modeling, it is also of importance to have a sound understanding of the behavior of the model (i.e., structure of the software tool), its sensitivity to data, and of the simulated system (Black et al., 2014). A second example, given in Mannschatz et al. (2014), estimated the LAI of bamboo in Northeast (NE) Brazil using multispectral remote sensing (RapidEye). The study showed that image processing such as atmospheric correction and selection of VIs had an impact on the estimation of LAI. This is also because the estimation of LAI is also a modeling (indirect measurement) and not a direct measurement as one would do directly on the plant. The mean LAI variation due to image processing was LAI ± 0.2. However, LAI variations especially in densely vegetated areas reached LAI ± 1.4. Subsequently, a local sensitivity analysis was carried out using a one-dimensional (1D) hydrological model (CoupModel, Jansson and Karlberg, 2010). The local sensitivity analysis investigated the change of CoupModel output (i.e., water balance components) relative to a reference parameterization using a typical LAI for bamboo of 3.2 (Mannschatz et al., 2014). The study revealed that this LAI uncertainty can translate into moderate CoupModel uncertainty that accounts for approximately ±1.7% of evapotranspiration, ±4.9% of interception, ±6.4% of evaporation, and ±1.4 of transpiration (Mannschatz et al., 2014). Nevertheless, if we investigate the results of the accomplished sensitivity analysis, we can see that LAI uncertainties (LAI ± 1.4) found in dense vegetation could strongly influence water balance modeling, especially for interception and evaporation (Fig. 2.5).

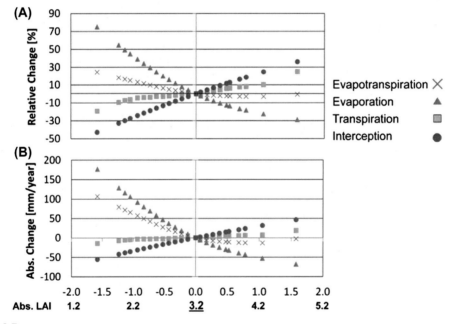

FIGURE 2.5

(A) Relative percentages and (B) absolute changes of interception, transpiration, evaporation, and evapotranspiration due to LAI variation are shown relative to the reference LAI value of 3.2.

Reprinted from Mannschatz, T., Pflug, B., Borg, E., Feger, K.-H., Dietrich, P., 2014. Uncertainties of LAI estimation from satellite imaging due to atmospheric correction. Remote Sensing of Environment 153, 24–39 with kind permission by Elsevier.

7. HOW SOIL TEXTURE MEASURED WITH VISIBLE-NEAR-INFRARED SPECTROSCOPY AFFECTS HYDROLOGICAL MODELING: A CASE STUDY

Most environmental models require detailed information about the soil. However, traditional soil mapping methods are cost- and time-consuming and mostly are not generally limited to point information. Therefore, alternative methods are nowadays applied to produce high capture soil spatial heterogeneity or to facilitate upscaling of point information. A method that has been proved to successfully predict soil physical and chemical properties is vis-NIR spectroscopy (e.g., Leone et al., 2012; Minasny and Hartemink, 2011; Stenberg et al., 2010; Viscarra Rossel et al., 2011). Vis-NIR spectroscopy is generally faster, cheaper, and has the potential to simultaneously predict various soil properties. Nevertheless, it still requires some, but a much smaller number, of soil laboratory analysis to establish a prediction model. Vis-NIR spectroscopy measures the spectral reflectance in dependency of the material characteristics. Soil samples are predicted based on the empirical relationship between soil samples and spectral response. The spectral response is mostly not totally specific for the soil property due to overlapping absorbance and scatter effects besides other sources, and thus the soil property—specific response is hidden and masked. Therefore, to find and extract spectral features relevant for predicting the target soil property, chemometric methods such as partial least squares regression (PLSR) are applied. Based on these assumptions, I can summarize that the use of vis-NIR spectroscopy general consists of (1) collecting soil samples across the study site for (2) establishment of a calibration and prediction model using chemometric methods and laboratory measurement results, and (3) using the prediction model to predict new soil samples based on their spectra only. Since the prediction of new samples is based on modeling, prediction uncertainty will be associated. Other sources of uncertainty are related to differences in the used measurement instruments, measurement setup, and study site, which includes differences in material composition.

However, coming back to environmental modeling, the question arises if this uncertainty in soil prediction is affecting modeling if these soil data are used as model input. Providing an idea about the importance of data input uncertainty during environmental modeling, I will first demonstrate the associated uncertainty related to prediction of soil properties using vis-NIR spectroscopy, and then illustrate that uncertainty of texture fractions prediction matters. To answer the latter, the impact of texture uncertainty on a Soil-Vegetation-Atmosphere-Transfer (SVAT) model is shown based on two different study sites with contrasting climates but similar soil type.

7.1 STUDY SITES AND INSTRUMENTS

The two study sites selected in this study are tropical equatorial NE Brazil and subtropical Mediterranean to semiarid Jordan (Kottek et al., 2006). The predominant soil type at both study sites is the vertisol. Vertisols are high in clay content, mainly consist of two horizons (A and C) and are strongly shrinking and swelling depending on the soil moisture (UEP et al., 2006). The Brazilian site was approximately 5 km^2 where the vegetation was composed of a bamboo plantation and a secondary forest. Yearly mean air temperature is about 14.4 °C in Jordan (Kraushaar et al., 2014) and about 24°C in Brazil (CEPLAC, 2012). The annual mean precipitation varies between 380 and 530 mm in Jordan (Kraushaar et al., 2014) and amounts to about 1600 mm in Brazil (CEPLAC, 2012). Major land use in Jordan is olive orchards, agricultural fields, and natural shrubs and bushes used for grazing (Kraushaar et al., 2014).

For the vis-NIR spectroscopic measurements two different spectrometers, an ASD FieldSpec3 device (PANalytical Company, Boulder, CO, USA) and a VERIS (shank-based, Veris Technologies, Salina, USA) have been used. Both instruments measure the spectral range between 400 and 2500 nm, but differ in spectral resolution and instrument architecture. The Fieldspec3 has a resolution of 3 nm in the visible range and 10 nm at 1400–2500 nm, and the VERIS has a resolution of 8 nm. The measurements have been carried out in the laboratory to provide more standardized conditions and to sources of perturbation (e.g., water content). Both instruments require calibration using a white or gray reference.

7.2 SOIL SAMPLES

For the establishment of the soil property prediction model, 94 soil samples have been collected on the study sites in Brazil, and 61 in Jordan (dots in Fig. 2.6). The collected soil samples have been dried and sieved to 2 mm. The samples were analyzed in the laboratory for texture fractions (clay, silt, sand), total organic carbon (TOC), total carbon (TC), total nitrogen (N), and total inorganic carbon (TIC) (Fig. 2.7).

The measurement with Fieldspec3 required soil samples of approximately 100–300 g, whereas approximately 5 g was needed for measurements with VERIS. Many measured single spectra have been averaged to obtain a representative spectrum for each soil sample.

7.3 CHEMOMETRICS

The measured data derived from both spectrometers was transformed using chemometric data transformation and spectroscopic modeling. Noise was reduced by omitting noisy bands (<400 nm, 1000–1100 nm, > 2220 nm). Investigating the raw spectral data measured with Fieldspec3 for both study sites reveals that the information encoded in the spectra depends on the study site (Fig. 2.8).

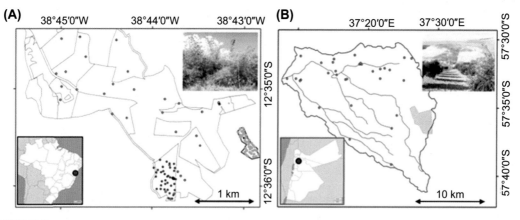

FIGURE 2.6

Study sites of the case study: (A) Brazil, bamboo plantation and (B) Jordan, olive orchards; points designate where soil samples have been collected (Mannschatz, 2015).

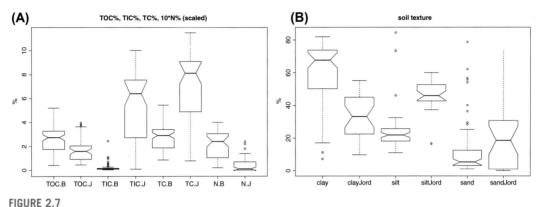

FIGURE 2.7

(A and B) Characteristics and results of laboratory analysis of soil samples from Brazil and Jordan.

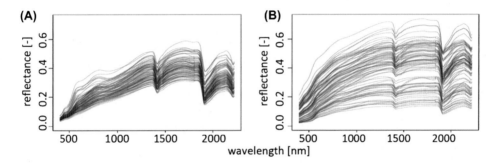

FIGURE 2.8

Measured reflectance of soil samples (A) from Brazil and (B) Jordan; measured with Fieldspec3.

There are visible differences in the spectral features between Brazilian and Jordan samples, which are a result of the different chemical and physical composition related to different soil-forming conditions. If we look on wavelength around 1400 nm (Fig. 2.8), we can see an example for the chemical differences, since this wavelength is sensitive to the clay minerals smectite and kaolinite (Stenberg et al., 2010). This is in line with the observations of Jarbas and Cunha (2000) and Ribeiro et al. (1990) who found smectite and kaolinite as the predominant clay minerals at the study site in Brazil. To show the potential influence of using different instruments and chemometric data preprocessing methods on soil property prediction, this case study will show the results of the investigation of both instruments for evaluating if they have similar soil prediction performance capabilities.

For reducing perturbations in the measured spectra and to improve spectral features, spectral transformation methods were applied to reduce disturbing influences arising from the measurement procedure (e.g., scattering). Diverse studies have shown the significant impact on prediction results depending on the applied spectral transformation method. To my knowledge, there are no clearly defined recommendations for specific soil properties (Stenberg et al., 2010). Therefore, this case study will show the influence of different data preprocessing methods. The methods applied include the conversion of the raw spectral data to either reflectance or absorbance (Osborne et al., 1993).

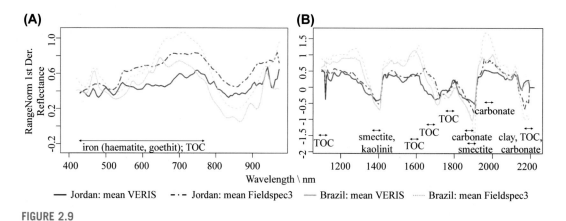

FIGURE 2.9

(A and B) Spectral signature after chemometric data transformation applying range normalization (Range-Norm) and first derivate of reflectance for Brazil and Jordan as well as for both instruments (VERIS, Fieldspec3).

Subsequently, on the raw spectral data were applied first or second derivates (both smoothed by Savitzky–Golay filter), third order (Savitzky and Golay, 1964), multiplicative scatter correction (Geladi et al. (1985)), or range normalization. The result of a data transformation from reflectance to first derivate shows that more specific features become visible (Fig. 2.9) that can be used in the PLSR algorithm to establish a calibration and prediction model needed for soil prediction. The quality assessment indicators were root mean square error of prediction (RMSEP) and an adjusted coefficient of determination of validation (Q^2) that takes into account the number of samples and principle components of the PLSR algorithm (see Mannschatz, 2015).

The soil prediction ability comparison was carried out using a bootstrapped robust linear regression with "leave-one-out" cross-validation. The term "bootstrapped" means here that 10,000 regression models were produced that were based on a randomly varying number of soil prediction pairs. Information obtained from analysis of these regression models provided the basis for the confidence intervals (95% quantile) of all possible regressions, in accordance with the assumption that the sample population is similar to the original population, which can be represented by multiple realizations from the sample population (Viscarra Rossel, 2007). Bootstrapping is based on resampling algorithms so that confidence intervals can be calculated, even if the sample population is not perfectly normally distributed or is small (Efron and Tibshirani, 1994). Prediction quality can be assessed using RMSEP.

7.4 IMPACT OF CHEMOMETRIC METHOD ON SOIL PREDICTION

In the following discussion, I aim to show that data processing can also strongly influence soil property prediction. Those data transformation methods are applied prior to the construction of a final calibration model. The "one-leave-out" cross-validation results as RMSEP are shown for clay and TOC predictions in Fig. 2.10. Comparing Figs. 2.10A and 2.10B, we see that the success of the data transformation method depended on soil property, study site, as well as instrument. For both instruments, the overall pattern of prediction capability of specific transformation methods was similar, but not identical at each of the study sites (Fig. 2.10). Considering the differences in detail,

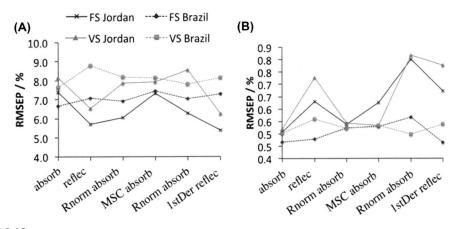

FIGURE 2.10

Comparison of data transformation methods based on RMSEP for (A) clay and (B) total organic carbon (TOC) for VERIS (VS) and Fieldspec3 (FS) at Jordan and Brazilian study sites.

it can be concluded that no particular data transformation method could be specified, which worked best for the prediction of a specific soil property. For these reasons, it can be recommended that several transformation methods should be tested at each individual study site. Similar conclusions are given in other studies (Peng et al., 2014; Ben-Dor and Banin, 1995). The reason for the varying success of transformation methods is related to the different effects of data transformation methods on collinearity between wavelengths. Differing collinearity influences the selection of informative wavelengths that are used for constructing a calibration model for soil prediction (Peng et al., 2014). In summary, we see that the preprocessing of data has already an important impact on the precision of soil property estimation, as we can see on clay prediction, where the amplitude of RMSEP variation for clay was between ±0.9 to ±1.2 for Brazil and between ±2.0 and ±2.1 for Jordan. Therefore, one should adequately consider those uncertainties in the error propagation investigation.

7.5 DIFFERENT INSTRUMENTS, DIFFERENT SOIL PREDICTIONS? WHAT WAS FINALLY THE BEST SOIL PREDICTION ACCURACY?

As it was seen earlier, the preprocessing method has an influence on soil property prediction accuracy. For this reason, I will show here the results of the "leave-one-out" cross-validation of both instruments of the best working preprocessing method. If we compare, at first, between both instruments, the statistical analyses revealed that the provided predictions were not significantly different (U-test, $p < .05$) from each other for TOC, TIC, TC, N, clay, silt, and sand at the specific study site. Even though there is no significant difference, the predictions of a single soil sample are not totally equal, which is indicated by the correlation coefficient of determination (R^2) (Fig. 2.11). One reason is the general uncertainty that is associated with the predicted soil property, which is also related to different instrument architecture and measurement setup.

Nevertheless, at the end one would be interested in the final results of soil property prediction accuracy. The accuracy of prediction depended on the study site, data processing, and thus the covered

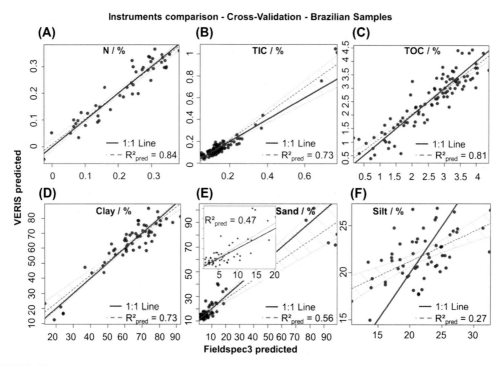

FIGURE 2.11

(A–F) Bootstrapped regression of "one-leave-out" cross-validation results for the soil samples from Brazil as predicted versus predicted for VERIS and Fieldspec3. R^2_{pred} (Spearman) calculated based on instrument predictions.

value range (Fig. 2.7) as mentioned before (see, e.g., Section 7.3). It would be beyond the scope of this chapter to discuss the findings in detail. The detailed results, however, including a discussion about the reasons for different prediction accuracy can be found in Mannschatz (2015).

To summarize, the precision (one-leave-out cross validation) of soil property prediction [SI unit (%)] averaged over both instruments. The precision evaluation is provided based on RMSEP compared with the mean value of prediction (mean$_P$ ± RMSEP, Fig. 2.7) and adjusted Q^2.

From the study site in Jordan, were obtained good validation results for TC ($Q^2 = 0.89$, 7.0 ± 1.0%) and TIC ($Q^2 = 0.87$, 5.3 ± 1.0%) and sufficient results for clay ($Q^2 = 0.76$, 33.4 ±5.6%) and TOC ($Q^2 = 0.72$, 1.7 ± 0.5%), whereas the results were low for N ($Q^2 = 0.42$, 0.05 ± 0.05%), silt ($Q^2 = 0.45$, 48.2 ± 3.7%), and sand ($Q^2 = 0.25$, 22.6 ± 16.9%). For the Brazilian site, good validation results were achieved for clay ($Q^2 = 0.81$, 62.3 ± 7.2%) and sand ($Q^2 = 0.81$, 13.2 ± 9.8%) and sufficient results for N ($Q^2 = 0.79$, 0.22 ± 0.04%), TOC ($Q^2 = 0.79$, 2.6 ± 0.5%) and TC ($Q^2 = 0.72$, 2.7 ± 0.5%). However, low-quality results were obtained for TIC ($Q^2 = 0.50$, 0.2 ± 0.2%) and silt ($Q^2 = 0.21$, 21.8 ± 4.3%). In Viscarra Rossel et al. (2006), average PLSR (cross-validation) prediction results are summarized for TIC ($R^2 = 0.96$), TC ($R^2 = 0.91$), TOC ($R^2 = 0.89$), N ($R^2 = 0.86$), clay ($R^2 = 0.86$), silt ($R^2 = 0.80$), and sand ($R^2 = 0.70$). In most cases, the case study results correspond

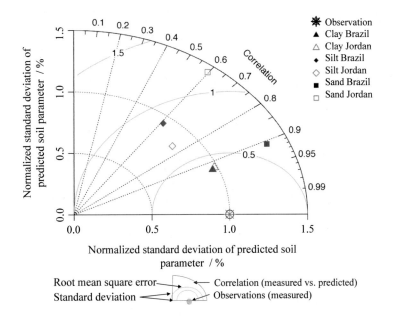

FIGURE 2.12

Comparison of (normalized) soil prediction ability of Fieldspec3 for clay, silt, and sand, and between both study sites (Brazil and Jordan).

well with the results communicated in the relevant literature (with the exception of Brazil: TIC, TC, and silt, and Jordan: N, sand) (Viscarra Rossel et al., 2006) Compared with traditional soil analysis methods, the precision achieved by spectroscopy is much lower [e.g., soil texture: pipette method precision = 1% (Minasny and McBratney, 2002)]. As can be seen on the Q^2 and RMSEP provided above, in some cases, the results are controversial. For instance, the prediction of sand in Brazil revealed a quite high Q^2 of 0.81; however, the RMSEP is large (9.8) compared with the mean value of prediction (13.2). The dependency of prediction quality on the study site conditions is shown in Fig. 2.12 based on the prediction results using Fieldspec3.

7.6 WHAT DOES THIS FINALLY MEAN FOR OUR ENVIRONMENTAL MODELING?

When we try to evaluate TOC spectral prediction uncertainty in the context of environmental modeling it would be useful to look at organic matter (OM) content instead. OM can be estimated by multiplying TOC by 1.724 (Chang et al., 2001). The prediction uncertainty (mean RMSEP of VERIS and Field-spec3) of 0.5% for Brazil and Jordan (see Section 7.5) corresponds to an OM uncertainty of 0.9%. Evaluating this result in the context of soil quality, the soil OM content can indicate healthy (OM = 4−6%), unhealthy (OM < 2% or > 8%), and neutral soil conditions (OM = 2−4% or 6−8%) (Schoenholtz et al., 2000). A prediction uncertainty of OM of −0.9% would move the mean Brazilian soil quality indicator from healthy (OM = 4.5% → OM = 3.6%) to neutral, whereas no change in soil quality class would occur at the Jordan site (OM = 2.9% → OM = 2.0%).

TIC is influencing soil pH through buffering of acidification processes (Shi et al., 2012). Since pH is an important soil quality indictor, TIC can be considered as an indirect soil quality indicator (Mukherjee and Lal, 2014; Sparling et al., 2008). However, to my knowledge, no soil quality-related thresholds for TIC have been defined.

Total N is a better soil nutrient with an optimal target range of total N of 0.10–0.70% for forestry production (Sparling et al., 2008). This total N range can be classified in the context of forestry as very depleted (<0.10%), depleted (0.10–0.20%), adequate (0.20–0.60%), ample (0.60–0.70%), and high (>0.7%) (Sparling et al., 2008). On the contrary, Mukherjee and Lal (2014) divided total N with respect to plant growth limitation into moderate limitation (0.20–0.30%) and slight to no limitation (>0.30%). Considering the total N content, we can conclude that the Jordanian soil is "very depleted" (mean $N = 0.05 \pm 0.05\%$). The Brazilian soil is considered "adequate," but the uncertainty from prediction (mean $N = 0.22 \pm 0.04\%$) might move it into the "depleted" class.

The soil texture is not a direct soil quality parameter, but it is crucial in the context of soil hydrology, such as it influences hydraulic conductivity (i.e., influence on soil infiltration, and surface runoff) and water storage. For these reasons, texture is an important input parameter into environmental models. Therefore, authors would like to show here the potential impact of the spectral prediction uncertainty on the water balance modeling. For this example, authors used the process-oriented 1D CoupModel, which explicitly focuses on all relevant processes related to water, heat, carbon, and nitrogen within the soil–plant–atmosphere continuum (Jansson and Karlberg, 2010). The model allows modeling at varying levels of complexity due to its modular structure and number of equations available for selection. The major model input requirements for the water balance module are climate data (e.g., precipitation), soil parameters (e.g., texture), and plant physiological parameters (Jansson and Karlberg, 2010).

What would be an effective way to demonstrate and evaluate the impact of model input on the modeling? The most common way is probably the accomplishment of a (local) sensitivity analysis that aims at investigating the changes in output due to systematic variation of the soil texture fractions around a reference value keeping all other parameters fixed. A second approach is to show the specific model output change due to the particular spectroscopic texture prediction uncertainty (e.g., based on RMSEP, Section 7.5). Authors would like to report the results of the latter approach, because it not only reveals the model's sensitivity to texture changes but also includes the ecohydrologic consequences. Therefore, I compared the change in model output of the separate components of the surface water balance to changes in texture fraction corresponding to the spectroscopic RMSEP for each texture fraction. As was mentioned earlier in this chapter, soil texture is a closed data and thus it is needed to change at least two soil fractions simultaneously. This means that if we change the clay content (i.e., to investigate model sensitivity to clay uncertainty), then we need to modify another fraction such as silt or sand or both. To keep the results easier to interpret, I used two scenarios, e.g., clay uncertainty impact on model; (1) clay is jointly changed with sand, where silt is fixed and (2) clay is jointly changed with silt, where sand is fixed. Subsequently, CoupModel was run twice, each one with the parameterization for each scenario. The results of the model output components for both scenarios are then averaged. This simplified approach is feasible, because we are not really interested in the "true" water balance for the study site, but in the relative changes of model output due to variations in texture. The evaluation was based on a simplified annual water balance equation as $WI = P - E - T - Q$, where WI is the accumulated (acc.) annual water input into the surface water balance, E is acc. annual evaporation, T is acc. annual transpiration, Q is acc. annual surface runoff, and P is the acc. annual precipitation. WI can be considered a measure of the acc. annual available water, where part of that might be lost through deep percolation. ΔWI is the percentage change in WI (WI_{var}) relative to the reference model parametrization (WI_{ref}): $\Delta WI = [(WI_{ref} - WI_{var})/WI_{ref}] \times 100\%$. I used ΔWI to define evaluation terms for the impact of input parameter uncertainty to model output. If ΔWI corresponds to $\leq 2\%$, then the impact of input parameter uncertainty to output is evaluated as negligible. Larger changes of modeled available water (ΔWI) are evaluated as small ($>2\%$, $<5\%$), medium ($>5\%$, $<15\%$), high ($>15\%$, $<25\%$), and very high ($>25\%$).

We first might have a look on the impact of the prediction uncertainty on single model output components. For the clayey loam soil from Jordan, the prediction uncertainty of cross-validation for clay (RMSEP = 4.9%) resulted in a CoupModel uncertainty for annual soil evaporation of 1.1% (2.1 mm/a), whereas the RMSEP for silt (4.1%) results in uncertainties of 1.2% (2.1 mm/a). Clay and silt uncertainty is noninfluential upon transpiration and annual surface runoff. Sand RMSEP (20.0%) causes variations in soil evaporation of 7.4% (13.4 mm/a) and transpiration of 0.3% (0.4 mm/a).

In contrast, for the clayey soil from Brazil, clay uncertainty (RMSEP = 6.7%) led to SVAT model uncertainty for annual soil evaporation of 4.7% (15.9 mm/a), transpiration of 0.2% (1.9 mm/a), and surface runoff of 42.4% (147.8 mm/a). Silt uncertainties (RMSEP = 4.1%) resulted in variations in soil evaporation of 1.7% (5.7 mm/a), in transpiration of 0.1% (0.6 mm/a), and in surface runoff of 18.2% (63.6 mm/a). Sand prediction uncertainty (RMSEP = 10.7%) led to model uncertainties of 6.3% (21.5 mm/a) for soil evaporation, 0.3% (2.8 mm/a) for transpiration, and 55.5% (193.7 mm/a) for surface runoff.

Considering the results, it seems like the impact of texture uncertainty, which I obtained using spectroscopy, had in general a relatively small impact on CoupModel water balance components (exception surface runoff in Brazil). However, the question came up what actually happens to the ecosystems water balance. To become a better sense for the system, we can fill out the annual water balance equation for (1) the reference parameterization and (2) for each of the texture fractions varied. But before that, I will shortly summarize the reference parameterization:

- Brazil
  - soil texture fractions: clay = 63%, silt = 24%, sand = 13%,
  - annual acc. precipitation = 1600 mm/a,
  - vegetation parameters for bamboo.
- Jordan
  - soil texture fractions: clay = 33%, silt = 46%, sand = 21%,
  - annual acc. precipitation = 455 mm/a,
  - vegetation parameters for grassland.

For the water balance equation, we see that the water input balance acc. for 1 year is on both study sites positive (i.e., water surplus) if we use the reference parameterization (WI_{ref}). WI changed (ΔWI) changed strongly on Brazilian site, whereas only small changes are observed on Jordan, if we consider an increase of texture fraction content equal to spectroscopic uncertainty (RMSEP) $WI_{var,texture}^{clay+RMSEP}$ (Table 2.1). The findings reveal that the sensitivity of water balance components to the texture variation depends on the initial soil texture type and the precipitation. The vegetation influences the transpiration, but it is not influenced by texture variations (Table 2.1). From the results, we also notice that the effect on the water balance is negligible in Jordan, whereas it is strong in Brazil, where it might even turn from a water surplus to a water deficit. In Brazil, the main reason for the change in WI

Table 2.1 Changes in water balance components due to increase of texture fraction corresponding to RMSEP ($WI_{var,texture}^{clay+RMSEP}$) relative to reference parameterization (WI_{ref})

	WI (Water Input) (ΔWI, %) =	P (Precipitation) mm/a	− E (Evaporation) mm/a	− T (Transpiration) mm/a	− Q (Surface Runoff) mm/a
Brazil	$WI_{ref} = +47.7\ (\pm 0)$	1600	−339.1	−864.1	−349.1
	$WI_{var,clay}^{clay+6.7\%} = -82.3\ (-272.5)$	1600	−323.2	−862.2	−496.9
	$WI_{var,silt}^{silt+4.2\%} = -9.6\ (-120.1)$	1600	−333.4	−863.5	−412.7
	$WI_{var,sand}^{sand+10.7\%} = +260.1\ (+445.3)$	1600	−317.6	−866.9	−155.4
Jordan	$WI_{ref} = +165.7\ (\pm 0)$	455	−180.9	−108.4	−0.0
	$WI_{var,clay}^{clay+4.9\%} = +163.6\ (-1.3)$	455	−183.0	−108.4	−0.0
	$WI_{var,silt}^{silt+4.1\%} = +163.6\ (-1.3)$	455	−183.0	−108.4	−0.0
	$WI_{var,sand}^{sand+20.0\%} = +178.7\ (+7.8)$	455	−167.5	−108.8	−0.0

is mainly related to changes in surface runoff. Under humid conditions with initial high clay content, a further increase in clay leads to clogging of the pore system, and thus to an increase in surface runoff. This effect is especially pronounced under humid tropical climate conditions where large amount of precipitation occurs on short periods of time. The increase of sand content in Brazil has the opposite effect on surface runoff. An increase in sand content strongly increases the hydraulic conductivity and ultimately causes a decrease in surface runoff.

Although the impact on a single component of the water balance might not be large, the overall water balance based on the surface water input could be strongly affected (Table 2.1). Depending on the study site conditions, the input data uncertainty propagation might turn a system with water surplus into a system with water deficit (e.g., under tropical humid conditions).

While certain aspects of the methodology applied in the case study can certainly be improved, the importance of communication between scientists who measure and scientists who model is undeniable.

8. WHAT DID WE LEARN?

The case studies teach us that "hard" data are also associated with uncertainties, and thus soil maps do not necessarily represent the total truth—they actually are not only an approximation. In the beginning of the chapter we stated the question: "if and when is it important to be aware of input data uncertainties" during the environmental modeling process? The first case study showed how one can make use of remote sensing techniques to obtain important plant parameters needed for hydrological modeling (Section 6). At the same time, it became visible that data coming from those measurement techniques are associated with uncertainties, which we need to be aware of.

In the second case study, in addition to the first case study, were seen the potential uncertainties associated with soil measurements using exemplary vis-NIR spectroscopy and their potential impact on a hydrological model (Section 7). These uncertainties, conclusively, propagate along the way of measurements to modeling. The input data uncertainties might cause negligible to large uncertainties in model output depending on study site conditions, applied environmental model, and initial model boundary conditions (e.g., general soil type).

Nevertheless, we can acknowledge the technical developments and advances in remote sensing and other geophysical methods since the beginning of the 21st century that allow us to obtain much better data in terms of spatial resolution, and also in temporal resolution, because those technologies offer a more time- and cost-effective measurement. We recommend using sensitivity analysis to explore the model response to the specific study site and to the input data to adjust the measurement effort and to define the required measurement precision. This data can help us to effectively parameterize, calibrate, and validate our environmental models and potentially due to a higher spatial resolution allow us to produce more accurate and detailed simulations of our environmental systems. Conclusively, the main objective of the chapter was to raise awareness that we need to be aware of the uncertainties in all steps of modeling earth surface and to communicate them in an adequate and effective manner that all related parties understand (including the scientists who use environmental models and those who produce the data they use, as well as other stakeholders). This will allow stakeholders to base their decisions on the modeling results taking into account the range of uncertainty.

REFERENCES

Aksoy, H., Kavvas, M.L., 2005. A review of hillslope and watershed scale erosion and sediment transport models. CATENA 64, 247–271. http://dx.doi.org/10.1016/j.catena.2005.08.008.

Arora, V.K., 2002. Modelling vegetation as a dynamic component in soil-vegetation-atmosphere-transfer schemes and hydrological models. Review of Geophysics 40 (2), 1006–1032. http://dx.doi.org/10.1029/2001RG000103.

Bagstad, K., Semmens, D., Waage, S., Winthrop, R., 2013. A comparative assessment of decision-support tools for ecosystem services quantification and valuation. Ecosystem Services 5, 27–39. http://dx.doi.org/10.1016/j.ecoser.2013.07.004.

Ben-Dor, E., Banin, A., 1995. Near infrared analysis as a rapid method to simultaneously evaluate, several soil properties. Soil Science Society of America Journal 59, 364–372.

Bennett, N.D., Croke, B.F.W., Guariso, G., Guillaume, J.H.A., Hamilton, S.H., Jakeman, A.J., Marsili-Libelli, S., Newham, L.T.H., Norton, J.P., Perrin, C., Pierce, S.A., Robson, B., Seppelt, R., Voinov, A.A., Fath, B.D., Andreassian, V., 2013. Characterising performance of environmental models. Environmental Modelling & Software 40, 1–20. http://dx.doi.org/10.1016/j.envsoft.2012.09.011.

Binley, A., Cassiani, G., Deiana, R., 2010. Hydrogeophysics: opportunities and challenges. Bollettino di Geofisica Teorica ed Applicata 51, 267–284.

Black, D., Wallbrink, P., Jordan, P., Waters, D., Carroll, C., Blackmore, J., 2011. Guidelines for Water Management Modelling. Towards Best-Practice Model Application. eWater Cooperative Research Centre 2011.

Black, D.C., Wallbrink, P.J., Jordan, P.W., 2014. Towards best practice implementation and application of models for analysis of water resources management scenarios. Environmental Modelling & Software 52, 136–148. http://dx.doi.org/10.1016/j.envsoft.2013.10.023.

CEPLAC, 2012. Climate Data Derived From CEPLAC Experimental Station Sóstenes de Miranda (ESOMI). Santo Amaro da Purificação, Brazil [WWW Document]. dataset. http://www.ceplac.gov.br.

Chang, C., Laird, D.A., Mausbach, M.J., Hurburgh, C.R., 2001. Near-infrared reflectance spectroscopy—principal components regression analyses of soil properties. Soil Science Society of America Journal 65 (2), 480–490. http://dx.doi.org/10.2136/sssaj2001.652480x.

Efron, B., Tibshirani, R.J., 1994. An Introduction to the Bootstrap — Monographs on Statistics and Applied Probability. CRC Press, 456 p.

Engel, B., Storm, D., White, M., Arnold, J., Arabi, M., 2007. A hydrologic/water quality model application. Journal of the American Water Resources Association 43, 1223–1236. http://dx.doi.org/10.1111/j.1752-1688.2007.00105.x.

EPA, 2002. Guidance on choosing a sampling design for environmental data collection. Environment Protection Agency. Washington. EPA/240/R-02/005. https://www.epa.gov/sites/production/files/2015-06/documents/g5s-final.pdf.

Francos, A., Elorza, F.J., Bouraoui, F., Bidoglio, G., Galbiati, L., 2003. Sensitivity analysis of distributed environmental simulation models: understanding the model behaviour in hydrological studies at the catchment scale. In: Reliability Engineering and System Safety, pp. 205–218. http://dx.doi.org/10.1016/S0951-8320(02)00231-4.

Geladi, P., MacDougall, D., Martens, H., 1985. Linearization and scatter-correction for near-infrared reflectance spectra of meat. Applied Spectroscopy 39, 491–500.

Gentil, E., Damgaard, A., Hauschild, M., 2010. Models for waste life cycle assessment: review of technical assumptions. Waste Management 30 (12), 2636–2648. http://dx.doi.org/10.1016/j.wasman.2010.06.004.

Gould, S.F., Beeton, N.J., Harris, R.M.B., Hutchinson, M.F., Lechner, A.M., Porfirio, L.L., Mackey, B.G., 2014. A Tool for Simulating and Communicating Uncertainty When Modelling Species Distributions Under Future Climates, pp. 1–14. http://dx.doi.org/10.1002/ece3.1319.

Gwo, J.P., Toran, L.E., Morris, M.D., Wilson, G.V., 1996. Subsurface stormflow modeling with sensitivity analysis using a latin-hypercube sampling technique. Ground Water 34 (5), 811–818. http://dx.doi.org/10.1111/j.1745-6584.1996.tb02075.x.

Harmel, R.D., Smith, P.K., Migliaccio, K.W., Chaubey, I., Douglas-Mankin, K.R., Benham, B., Shukla, S., Muñoz-Carpena, R., Robson, B.J., 2014. Evaluating, interpreting, and communicating performance of hydrologic/water quality models considering intended use: a review and recommendations. Environmental Modelling & Software 57, 40–51. http://dx.doi.org/10.1016/j.envsoft.2014.02.013.

Houser, P.R., Lannoy, G.J.M., De, Walker, J.P., 2012. Approaches to Managing Disaster – Assessing Hazards, Emergencies and Disaster Impacts. InTech, Berlin, Heidelberg. http://dx.doi.org/10.5772/1112.

Jakeman, A.J., Letcher, R.A., Norton, J.P., 2006. Ten iterative steps in development and evaluation of environmental models. Environmental Modelling & Software 21, 602–614. http://dx.doi.org/10.1016/j.envsoft.2006.01.004.

Jansson, P.-E., Karlberg, L. (Eds.), 2010. CoupModel Manual – Coupled Heat and Mass Transfer Model for Soil–Plant–Atmosphere Systems, Manual. KTH Stockholm, Stockholm.

Jarbas, T., Cunha, F., 2000. Boletim de Pesquisa – Estudo de correlação de solos para fins de classificação nas regiões do Reconcavo baiano e microregião de Irece Bahia. Available at: http://www.cnps.embrapa.br/publicacoes/pdfs/bp092000classirece.pdf.

Kelly Letcher, R.A., Jakeman, A.J., Barreteau, O., Borsuk, M.E., ElSawah, S., Hamilton, S.H., Henriksen, H.J., Kuikka, S., Maier, H.R., Rizzoli, A.E., van Delden, H., Voinov, A.A., 2013. Selecting among five common modelling approaches for integrated environmental assessment and management. Environmental Modelling & Software 47, 159–181. http://dx.doi.org/10.1016/j.envsoft.2013.05.005.

Kleijnen, J.P.C., 2005. An overview of the design and analysis of simulation experiments for sensitivity analysis. European Journal of Operational Research. http://dx.doi.org/10.1016/j.ejor.2004.02.005.

Knödel, K., Krummel, H., Lange, G. (Eds.), 2005. Handbuch zur Erkundung des Untergrundes von Deponien und Altlasten – Geophysik, 2. ed, Handbuch zur Erkundung des Untergrundes von Deponien und Altlasten. BGR – Springer, Berlin, Heidelberg.

Köhne, J., Köhne, S., Šimůnek, J., 2009. A review of model applications for structured soils: b) Pesticide transport. Journal of Contaminant Hydrology 104 (1–4), 36–60. http://dx.doi.org/10.1016/j.jconhyd.2008.10.003.

Köhne, J.M., Köhne, S., Šimůnek, J., 2009. A review of model applications for structured soils: a) Water flow and tracer transport. Journal of Contaminant Hydrology 104, 4–35. http://dx.doi.org/10.1016/j.jconhyd.2008.10.002.

Kottek, M., Grieser, J., Beck, C., Rudolf, B., Rubel, F., 2006. World map of Köppen-Geiger Climate Classification – (updated with CRU TS 2.1 temperature and VASClimO v1.1 precipitation data 1951 to 2000). Meteorologische Zeitschrift 15, 259–263.

Kraushaar, S., Herrmann, N., Ollesch, G., Vogel, H.J., Siebert, C., 2014. Mound measurements – quantifying medium-term soil erosion under olive trees in Northern Jordan. Geomorphology. http://dx.doi.org/10.1016/j.geomorph.2013.12.021.

Lal, R., 2015. The nexus approach to managing water, soil and waste und changing climate and growing demands on natural resources. In: Kurian, M., Ardakanian, R. (Eds.), Governing the Nexus: Water, Soil and Waste Resources Considering Global Change. Springer, Cham, Heidelberg, New York, Dordrecht, London, pp. 39–60. http://dx.doi.org/10.1007/978-3-319-05747-7.

Lehmann, A., Giuliani, G., Ray, N., Rahman, K., Abbaspour, K.C., Nativi, S., Craglia, M., Cripe, D., Quevauviller, P., Beniston, M., 2014. Reviewing innovative Earth observation solutions for filling science-policy gaps in hydrology. Journal of Hydrology. http://dx.doi.org/10.1016/j.jhydrol.2014.05.059.

Lenhart, T., Eckhardt, K., Fohrer, N., Frede, H.-G., 2002. Comparison of two different approaches of sensitivity analysis. Physics and Chemistry of the Earth 27 (9–10), 645–654. http://dx.doi.org/10.1016/S1474-7065(02)00049-9. Parts A/B/C.

Leone, A., Viscarra Rossel, R.A., Amenta, P., Buondonno, A., 2012. Prediction of soil properties with PLSR and vis-NIR spectroscopy: application to mediterranean soils from southern Italy. Current Analytical Chemistry 8 (2), 283–299. http://dx.doi.org/10.2174/157341112800392571.

Liu, J., Mooney, H., Hull, V., Davis, S.J., Gaskell, J., Hertel, T., Lubchenco, J., Seto, K.C., Gleick, P., Kremen, C., Li, S., 2015. Systems integration for global sustainability. Science 347 (6225), 1258832-1–1258832-9. http://dx.doi.org/10.1126/science.1258832.

Loosvelt, L., Vernieuwe, H., Pauwels, V.R.N., De Baets, B., Verhoest, N.E.C., 2013. Local sensitivity analysis for compositional data with application to soil texture in hydrologic modelling. Hydrology and Earth System Sciences 17, 461–478. http://dx.doi.org/10.5194/hess-17-461-2013.

Mannschatz, T., 2015. Site Evaluation Approach for Reforestations Based on SVAT Water Balance Modeling Considering Data Scarcity and Uncertainty Analysis of Model Input Parameters from Geophysical Data. Technische Universität Dresden (Ph.D. thesis).

Mannschatz, T., Buchroithner, M.F., Hülsmann, S., 2015. Visualization of water services in Africa: data applications for nexus governance. In: Kurian, M., Ardakanian, R. (Eds.), Governing the Nexus: Water, Soil and Waste Resources Considering Global Change. Springer, Cham, Heidelberg, New York, Dordrecht, London, pp. 189–217. http://dx.doi.org/10.1007/978-3-319-05747-7_9.

Mannschatz, T., Pflug, B., Borg, E., Feger, K.-H., Dietrich, P., 2014. Uncertainties of LAI estimation from satellite imaging due to atmospheric correction. Remote Sensing of Environment 153, 24–39. http://dx.doi.org/10.1016/j.rse.2014.07.020.

Mannschatz, T., Wolf, T., Hülsmann, S., 2016. Nexus Tools Platform: Web-based comparison of modelling tools for analysis of water-soil-waste nexus. Environmental Modelling & Software 76, 137–153. http://dx.doi.org/10.1016/j.envsoft.2015.10.031.

Melesse, A.M., Weng, Q., Thenkabail, P.S., Senay, G.B., 2007. Remote sensing sensors and applications in environmental resources mapping and modelling. Evolution 7 (12), 3209–3241. http://dx.doi.org/10.3390/s7123209.

Minasny, B., Hartemink, A.E., 2011. Predicting soil properties in the tropics. Earth-Science Reviews 106, 52–62. http://dx.doi.org/10.1016/j.earscirev.2011.01.005.

Minasny, B., McBratney, A.B., 2002. Uncertainty analysis for pedotransfer functions. European Journal of Soil Science 53, 417–430. http://dx.doi.org/10.1046/j.1365-2389.2002.00452.x.

Mode, N.A., Conquest, L.L., Marker, D.A., 1999. Ranked set sampling for ecological research: Accounting for the total costs of sampling. Environmetrics 10, 179–194. http://dx.doi.org/10.1002/(SICI)1099-095X(199903/04)10:2<179. ::AID-ENV346>3.0.CO;2-#.

Moriasi, D.N., Arnold, J.G., Liew, M.W., Van Bingner, R.L., Harmel, R.D., Veith, T.L., 2007. Model evaluation guidelines for systematic quantification of accuracy in watershed simulations. ASABE 50, 885–900. http://dx.doi.org/10.13031/2013.23153.

Morris, M.D., 1991. Factorial sampling plans for preliminary computational experiments. Technometrics 33 (2), 161–174. http://dx.doi.org/10.2307/1269043.

Mukherjee, A., Lal, R., 2014. Comparison of soil quality index using three methods. PLoS One 9, e105981. http://dx.doi.org/10.1371/journal.pone.0105981.

Norton, J., 2015. An introduction to sensitivity assessment of simulation models. Environmental Modelling & Software 69, 166–174. http://dx.doi.org/10.1016/j.envsoft.2015.03.020.

Osborne, B.G., Fearn, T., Hindle, P.H., 1993. Practical NIR Spectroscopy with Applications in Food and Beverage Analysis, second ed. Longman Scientific and Technical, Singapore, p. 227.

Peng, X., Shi, T., Song, A., Chen, Y., Gao, W., 2014. Estimating soil organic carbon using VIS/NIR spectroscopy with SVMR and SPA methods. Remote Sensing 6, 2699–2717.

Ranatunga, K., Nation, E., Barratt, D., 2008. Review of soil water models and their applications in Australia. Environmental Modelling & Software 23 (9), 1182−1206. http://dx.doi.org/10.1016/j.envsoft.2008.02.003.

Refsgaard, J.C., Henriksen, H.J., Harrar, W.G., Scholten, H., Kassahun, A., 2005. Quality assurance in model based water management − review of existing practice and outline of new approaches. Environmental Modelling & Software 20 (10), 1201−1215. http://dx.doi.org/10.1016/j.envsoft.2004.07.006.

Refsgaard, J.C., van der Sluijs, J.P., Brown, J., van der Keur, P., 2006. A framework for dealing with uncertainty due to model structure error. Advances in Water Resources 29, 1586−1597. http://dx.doi.org/10.1016/j.advwatres.2005.11.013.

Refsgaard, J.C., van der Sluijs, J.P., Højberg, A.L., Vanrolleghem, P.A., 2007. Uncertainty in the environmental modelling process − a framework and guidance. Environmental Modelling & Software 22, 1543−1556. http://dx.doi.org/10.1016/j.envsoft.2007.02.004.

Ribeiro, L.P., Volkoff, B., Melfi, A.J., 1990. Química e mineralogia do solo − evolucao mineralogica das argilas em solos verticos do reconcavo baiana. Revista Brasileria de Ciência do Solo 14, 263−268. http://dx.doi.org/10.1590/S0100-06832014000500004.

Rubin, Y., Hubbard, S.S. (Eds.), 2005. Hydrogeophysics, Water Science and Technology Library. Springer, Netherlands, ISBN 978-9401783309.

Saltelli, A., Chan, K., Scott, E.M. (Eds.), 2000. Sensitivity Analysis: Gauging the Worth of Scientific Models. Wiley, UK, ISBN 978-0-471-99892-1, p. 504.

Saltelli, A., Tarantola, S., Campolongo, F., Ratto, M. (Eds.), 2004. Sensitivity Analysis in Practice: A Guide to Assessing Scientific Models. Wiley, UK. http://dx.doi.org/10.1002/0470870958.

Saltelli, A., Ratto, M., Andres, T., Campolongo, F., Cariboni, J., Gatelli, D., Saisana, M., Tarantola, S., 2008. Global Sensitivity Analysis: The Primer. John Wiley & Sons. http://dx.doi.org/10.1002/9780470725184.

Savitzky, A., Golay, M.J.E., 1964. Smoothing and differentiation of data by simplified least squares procedures. Analytical Chemistry 36 (8), 1627−1639. http://dx.doi.org/10.1021/ac60214a047.

Schoenholtz, S., Miegroet, H.V., Burger, J., 2000. A review of chemical and physical properties as indicators of forest soil quality: challenges and opportunities. Forest Ecology and Management 138, 335−356. http://dx.doi.org/10.1016/S0378-1127(00)00423-0.

Shi, Y., Baumann, F., Ma, Y., Song, C., Kühn, P., Scholten, T., He, J.-S., 2012. Organic and inorganic carbon in the topsoil of the Mongolian and Tibetan grasslands: pattern, control and implications. Biogeosciences 9, 2287−2299. http://dx.doi.org/10.5194/bg-9-2287-2012.

Song, X., Zhang, J., Zhan, C., Xuan, Y., Ye, M., Xu, C., 2015. Global sensitivity analysis in hydrological modeling: review of concepts, methods, theoretical framework, and applications. Journal of Hydrology 523, 739−757. http://dx.doi.org/10.1016/j.jhydrol.2015.02.013.

Sparling, G., Lilburne, L., Vojvodic-Vukovic, M., 2008. Provisional Targets for Soil Quality Indicators in New Zealand, Science. Manaaki Whenua Press, Lincoln, Canterbury, New Zealand.

Srivastava, P., Migliaccio, K.W., 2007. Landscape models for simulating water quality at point, field and watershed scales. ASABE 50 (5), 1683−1693. http://dx.doi.org/10.13031/2013.23961.

Stenberg, B., Viscarra Rossel, R.A., Mouazen, A.M., Wetterlind, J., 2010. Visible and near infrared spectroscopy in soil science. In: Sparks, D.L. (Ed.), Advances in Agronomy. Academic Press, Burlington, pp. 163−215. http://dx.doi.org/10.1016/S0065-2113(10)07005-7.

UEP, Araújo Filho, J.C., de Batista, F.H.B., Barros, A.H.C., 2006. Embrapa Solos Recife − Galeria de Imagens.

UNESCO, 2005. Model sensitivity and uncertainty analysis − chapter 9. In: UNESCO (Ed.), Water Resources Systems Planning and Management. UNESCO, p. 255.

Vereecken, H., Binley, A., Cassiani, G., Revil, A., Titov, K. (Eds.), 2006. Applied Hydrogeophysics. NATO Science Series. Springer, Netherlands.

Viña, A., Gitelson, A., Nguy-Robertson, A.L., Peng, Y., 2011. Comparison of different vegetation indices for the remote assessment of green leaf area index of crops. Remote Sensing of Environment 115, 3468−3478. http://dx.doi.org/10.1016/j.rse.2011.08.010.

Viscarra Rossel, R., 2007. Robust modelling of soil diffuse reflectance spectra by bagging- partial least squares regression. Journal of Near Infrared Spectroscopy 15, 39–47. http://dx.doi.org/10.1255/jnirs.694.

Viscarra Rossel, R.A., Chappell, A., de Caritat, P., McKenzie, N.J., 2011. On the soil information content of visible–near infrared reflectance spectra. European Journal of Soil Science 62, 442–453. http://dx.doi.org/10.1111/j.1365-2389.2011.01372.x.

Viscarra Rossel, R.A., Walvoort, D.J.J., McBratney, A.B., Janik, L.J., Skjemstad, J.O., 2006. Visible, near infrared, mid infrared or combined diffuse reflectance spectroscopy for simultaneous assessment of various soil properties. Geoderma 131, 59–75. http://dx.doi.org/10.1016/j.geoderma.2005.03.007.

Wainwright, J., Mulligan, M. (Eds.), 2004. Environmental Modelling – Finding Simplicity in Complexity, first ed. John Wiley & Sons, Chichester, UK.

Wainwright, J., Mulligan, M. (Eds.), 2013. Environmental Modeling – Finding Simplicity in Complexity, second ed. Wiley-Blackwell, Chichester, UK.

Wang, Q., Adiku, S., Tenhunen, J., Granier, A., 2005. On the relationship of NDVI with leaf area index in a deciduous forest site. Remote Sensing of Environment 94, 244–255. http://dx.doi.org/10.1016/j.rse.2004.10.006.

Wu, J., Wang, D., Bauer, M., 2007. Assessing broadband vegetation indices and QuickBird data in estimating leaf area index of corn and potato canopies. Field Crops Research 102, 33–42. http://dx.doi.org/10.1016/j.fcr.2007.01.003.

Zheng, G., Moskal, L.M., 2009. Retrieving leaf area index (LAI) using remote sensing: theories, methods and sensors. Sensors 9 (4), 2719–2745. http://dx.doi.org/10.3390/s90402719.

LOCAL SA METHODS: CASE STUDIES

2

LOCAL SENSITIVITY ANALYSIS OF THE LANDSOIL EROSION MODEL APPLIED TO A VIRTUAL CATCHMENT

3

R. Ciampalini[1,2], S. Follain[3], B. Cheviron[4], Y. Le Bissonnais[1], A. Couturier[5], R. Moussa[3], C. Walter[6]

INRA, UMR — LISAH, Laboratoire d'étude des Interactions Sol - Agrosystème - Hydrosystème, Montpellier, France[1]; Cardiff University, Cardiff, Wales, United Kingdom[2]; Montpellier SupAgro, UMR — LISAH, Laboratoire d'étude des Interactions Sol - Agrosystème — Hydrosystème, Montpellier, France[3]; Irstea, UMR G-EAU, Montpellier, France[4]; INRA, Centre de recherche Val de Loire, Orléans, France[5]; AGROCAMPUS OUEST, INRA, Rennes, France[6]

CHAPTER OUTLINE

1. INTRODUCTION

Nowadays, models combine several elementary processes in space and time using larger numbers of variables and it becomes difficult to understand model behaviors and parameter hierarchies, owing to multiple interactions between elementary processes. Sensitivity analysis (SA) is widely used to clarify a model's behavior and variables' relevance by furnishing fundamental information to improve model performances. There are many different techniques to implement SA in environmental models (Pianosia et al., 2016); the most appropriate technique depends on the model complexity and on the variables interaction, which is expected and to be clarified. Among the methods, the best-known SA methods are local SA, which is also termed "one-factor-at-a-time" (O-A-T), and global SA (Saltelli, 1999).

Local SA, or deterministic methods, studies local model derivatives, i.e., the variation of the local response (O) to variations of a particular input factor P_i, at a specified point P_0, expressed as $S_i = (O_2 - O_1)/(P_2 - P_1)$. The validity of the method is ensured by an appropriate screening of the parameter space. Alternative methods can perturb the values with a fixed amount, or percent, or with a standard deviation fraction of the input values (Saltelli et al., 2000). This procedure is effective for localized analysis, but some difficulties arise when a general analysis of the model is needed. Limitations are evident when models involve nonlinear relationships and strong interactions between parameters and when the increment interval for parameters is too large (Breshears et al., 1992); in this case, the local SA becomes dependent on the magnitude of the considered interval.

Global SA analyzes the effect of the parameters across the whole parameter space relying on hypothesized statistical distributions of the parameter values over their predefined ranges. In the popular variance-based methods, global SA considers the contribution of each input factor to the total variance of the output. Some examples are the Fourier amplitude sensitivity test (Crosetto and Tarantola, 2001), regional SA (Spear and Hornberger, 1980), or multiobjective generalized SA (Liu et al., 2004), where a multicriteria sensitivity is implemented by examining whether a random distribution of the parameters can statistically divide under a specific classification of the objective function using a Kolmogorov–Smirnov test.

Local SA, in contrast to global SA, does not capture the whole effect of input variations, but its crucial merit is to allow analyzing parameter sensitivity for specific input scenarios, i.e., near selected points in parameter space, accounting for situations of known interest (Saltelli, 1999). For complex models, the effect of the parameters can be extremely localized, and an overall index could be neither applicable nor even relevant to qualify a minority of preidentified cases (e.g., localized regions of the input parameter space). Local SA is classically used in procedures such as the Morris method (Morris, 1991; Campolongo et al., 2007), the forward SA procedure, or the adjoint SA procedure (Ionescu-Bujor and Cacuci, 2004), relying on the so-called variational methods, often used for data assimilation and real-time control purposes. The Morris method, for instance, performs well to identify the key parameters (e.g., Saltelli et al., 2000; Francos et al., 2003). To do so, the procedure divides the range of the input parameter P_i into n levels using random sampling from the $n*I$ parameter space (I is the number of parameters) and calculates the elementary effect with ∂x_i as a predetermined multiple of $1/(n - 1)$.

Among SA on similar models (i.e., soil erosion models), few different approaches have been adopted. Cheviron et al. (2010, 2011) compared the sensitivity of four soil erosion models (MHYDAS, STREAM, PESERA, and MESALES) using a one-at-a-time methodology associated with a Latin hypercube sampling. Wei et al. (2007), using local SA, explored parameter behavior of the rangeland hydrology and erosion model. Nearing et al. (1990) used one-at-a-time local sensitivity to test behaviors of the WEPP model. Tiscareno-Lopez et al. (1994) improved WEPP SA applied to rangeland conditions using a Monte Carlo methodology.

Here, we tested the LandSoil model (LANDscape design for SOIL conservation under soil use and climate change; Ciampalini et al., 2012), a model designed for simulation of soil erosion and agricultural landscape evolution. Our purpose was to determine the most significant parameters and their influence within the variables space. Furthermore, adopting a virtual catchment we analyzed the influence of the hillslope morphology on soil erosion sensitivity to model parameters.

2. MATERIALS AND METHODS
2.1 MODEL DESCRIPTION

The LandSoil model (Ciampalini et al., 2012) is a spatially distributed expert system for the analysis of soil redistribution processes and landscape topography evolution in medium terms (10—100 years) at a fine spatial scale (1—10 m). Developed on the basis of the STREAM model (Cerdan et al., 2002) and the WaTEM/SEDEM tillage erosion model (Govers et al., 1994), LandSoil simulates both water erosion (rill and interrill) and tillage erosion processes. It operates at the field and small catchment scale, with an event scale for rainfall and tillage operations. After each erosive event (water or tillage), a new elevation grid is calculated to account for erosion or deposition processes.

The hypothesis that grounds the model is that the soil surface properties control water infiltration, runoff, and sediment concentration (Cerdan et al., 2001, 2002). Soil surface properties, namely, soil roughness, surface crusting, and vegetation cover (Singer and Le Bissonnais, 1998) have been embedded into the model after field-scale experimentations (Le Bissonnais et al., 1998, 2005) and calibration, on a specific monitored catchment (Ciampalini et al., 2012). Soil surface properties are assigned to each land use and soil tillage combination on a monthly scale, and then summarized as a calendar of monthly values. The model integrates raster maps (i.e., spatially distributed) and vector elements for the delimitation of the soil surface properties, as well as a representation of the drainage network in the catchment (parcels, channels). The model also allows modifying the spatial patterns of infiltration and calibrating the infiltrability in the channels network. Finally, it considers an infiltration/runoff balance based on single rainfall events characterized by the total rainfall amount, the maximum rainfall intensity, and the rainfall duration.

Runoff/infiltration balance is computed at each pixel by local infiltration decision rules as follows (Eqs. (3.1) and (3.2)):

$$B = T - W - ID \qquad (3.1)$$

$$\text{With } (I, W) = f(roughness, crusting, cover) \qquad (3.2)$$

where B is the infiltration/runoff balance (mm), T is the total rainfall event amount (mm), W is the residual water soil storage after the previous event (mm), I is the steady-state soil infiltration rate (mm/h), D is the runoff event duration (h), and *roughness, crusting, and cover* are, respectively, soil roughness, soil crusting (classes), and the vegetation cover (%). Runoff is then routed along the catchment network obtained by combining Digital Elevation Model and a tillage direction model (Jenson and Domingue, 1988; Souchère et al., 1998) accumulating and/or reinfiltrating the runoff. The diverse contributions to erosion are calculated by different internal modules, namely, rill, interrill, and tillage erosion modules.

Rill erosion module is based on rill erosion propensity integrating soil surface properties, slope, and accumulated runoff as developed by Souchère et al. (2003). The resulting soil loss (E, kg) is expressed by Eq. (3.3):

$$E = \rho \lambda K_r \qquad (3.3)$$

where ρ is the soil bulk density (kg/m^3), λ is the pixel dimension (m), and K_r (m^2) is an empirical calibration parameter for expected rill section (Souchère et al., 1998) as follows (Eqs. (3.4) and (3.5)):

$$K_r = f(E_s); \quad \text{where } E_s = f(V, \text{ friction} \times \text{cohesion} \times V \times S_l) \tag{3.4}$$

$$\text{And, } (\text{friction}, \text{cohesion}) = f(\text{roughness}, \text{crusting}, \text{cover}) \tag{3.5}$$

where E_s is the sensitivity to rill erosion, *friction* and *cohesion* represent the class factors related to the soil surface properties, V (m^3) is the local cumulated runoff, and S_l (m/m) is the local slope.

Interrill erosion module is based on expert rules evaluating splash erosion on soil surface properties and rainfall intensities (Cerdan et al., 2002) as follows (Eq. (3.6)):

$$\gamma = f(\text{roughness}, \text{crusting}, \text{cover}, Ip_{max}) \tag{3.6}$$

where γ (g/L) is the sediment concentration in runoff and Ip_{max} (mm/h) is the maximum rainfall intensity. Local sediment delivery (M_i, kg) is calculated considering a slope dependency as follows (Eq. (3.7)):

$$M_i = \gamma S_f V \tag{3.7}$$

where S_f represents the slope factor (Ciampalini et al., 2012) as evaluated in the WEPP model (Laflen et al., 1991). In the model, both rill and interrill modules are limited by sediment transport, controlled by several threshold functions with respect to the local topographical constraints: profile curvature concavity (i.e., ideally, the reciprocal of the radius of a circle locally fitting the topographic surface) > 0.055 m^{-1}, slope gradient <0.02 m/m, soil use type (woods, scrublands, etc.), and soil cover (>60%). Concentration limits (Cerdan et al., 2002) range from 2.5 to 10 g/L.

Tillage erosion module integrates soil translocation due to agricultural practices, and it is based on the equation developed by Govers et al. (1994). The effect of each tillage operation is evaluated as single event with the tillage parameters depending on the plowing tools used (Van Muysen et al., 2002, 2006). The unit soil transport rate is calculated as follows:

$$Q_s = -K_t S_l \tag{3.8}$$

where Q_s is the unit volumetric sediment flux (kg/m) and K_t is the tillage coefficient (kg/m).

2.2 SENSITIVITY ANALYSIS

Our purpose was to test the sensitivity of water erosion to model parameters in single rainfall events. For that, only the parameters governing runoff and water processes have been analyzed whereas the model calibration settings have been retained as defined in the original model calibration (i.e., Roujan catchment calibration, Languedoc-Roussillon, Ciampalini et al., 2012). The interactions between model parameters and local calibration settings are assumed to remain constant, affecting only the total amount of the model response. Hence, this study represents a general SA of the model. Tillage erosion is not included in the test because of the basic mechanistic rules adopted in their equations and because it is uninfluential in a single rainfall event.

2.2.1 Parameters

We adopted a multicategory SA procedure (e.g., Cheviron et al., 2010, 2011) testing rainfall, soil, and topography (i.e., slope). Among these categories, we distinguish: rainfall and transmitted fluxes influencing surface runoff (i.e., precipitations, P; antecedent soil moisture, R) and soil properties (i.e., soil surface properties, p_e), which determine both water fluxes and resulting soil loss, and topography, influencing all the routing fluxes and accounted as local slope (p_s) (Table 3.1). In hydrological processes, the catchment morphology is directly related to the hydraulic connectivity and, more than the simple slope, can interact with the inputs influencing erosion sensitivity to parameters determining significant differences in results. With the aim to implement a realistic approach, we included a parameter describing the catchment morphology. This hypothesis is verified on a few cases of concave,

Table 3.1 List of the model parameters and the discrete values adopted in the sensitivity analysis.

Rainfall (P)		Runoff (R)	Soil (p_e)				Topography (p_s)
Total Rainfall (P_t)	Rainfall Intensity (P_i)	Antecedent Rainfall (A_r)	Crusting (C_r)	Roughness (R_h)	Cover (C_v)	Slope (S_l)
20	10	0	0	0	1	1
35	30	10	1	1	2	3
50	60	30	12	2	3	5
65		50	2	3		7.5
80				4		15

Legend

Parameter	Value	Description
Total rainfall (P_t):	20 to 80	(mm)
Rainfall intensity (P_i):	10 to 60	(mm h^{-1})
Antecedent rainfall (A_r):	0 to 50	(mm)
Soil roughness (R_h):		Soil surface roughness (height difference between the deepest part of microdepressions and the lowest point of their divide)
	0	R0: 0−1 cm
	1	R1: 1−2 cm
	2	R2: 2−5 cm
	3	R3: 5−10 cm
	4	R4: >10 cm
Sol crusting (C_r):		Soil surface crusting stage
	0	F0: Initial fragmentary structure
	1	F11: Altered fragmentary state with structural crusts
	12	F12: Local appearance of depositional crusts
	2	F2: Continuous state with depositional crusts
Soil cover (C_v):		Soil surface percentage covered by canopy or litter
	1	0−20%
	2	21−60%
	3	61−100%
Slope (S_l):	1 to 15	(%)

convex, and mixed topographic profiles. For this, we adopted some modifications of the virtual catchment, as detailed in following section.

In LandSoil model, soil parameters have discrete values. To fit this peculiarity (i.e., performing a discrete SA), the few existing continuous parameters (i.e., rainfall) have been discretized covering a wide range of values.

The parameter discretization we adopted is efficient because: (1) it reduces the number of simulations of the parameters combinations for O-A-T SA and (2) all the parameters combinations can be analyzed.

Following these assumptions, precipitations (P_t) were chosen to cover rainfall amounts from 20 to 80 mm (i.e., 20, 35, 50, 65, 80 mm). Rainfall intensity (P_i), consistently with the threshold adopted by the model itself, was discretized in three classes (10, 30 and 60 mm/h); antecedent rainfall (A_r) (i.e., soil moisture) was classed to values ranging from 0 to 50 mm (i.e., 0, 10, 30, 50 mm).

Soil surface properties (p_e parameter) are represented by a matrix containing soil roughness R_h (4 classes), soil crusting C_r (5 classes), and soil cover C_v (3 classes), as described in Le Bissonnais et al. (2005). Local slope (S_l) was set to five classes representing common hillslope values and coherent with the thresholds used in the model (i.e., 1%, 3%, 5%, 7.5%, 15%) (Table 3.1).

2.2.2 Virtual Catchment

In this procedure, the virtual catchment (Fig. 3.1) is the topographical entity on which soil loss and sensitivity are calculated, and it is inspired by the prototype proposed by Cheviron et al. (2010, 2011) for SA of different soil erosion models.

The catchment topology consists of a 3-square pixel structure having total length of 150 m allowing to test the model in different morphological configurations; for this, we adopted two main patterns (Fig. 3.2B): (1) linear hillslope (A-type) having a simple flat geometry and (2) complex hillslope (B- type) suitable to analyze some simple and realistic morphologies. The topographic configurations respond to the following objectives: (1) to study model behaviors for some main heterogeneous settings referring to a small number of easily identifiable cases and related to similar field conditions and (2) to avoid local and unverifiable behaviors opting to a simple pseudosymmetrical multiple configuration of the catchment. All the configurations are symmetrical to the median line catchment ending at the outlet, and the pathway flows are defined by the slope pattern itself (Fig. 3.1).

Linear hillslopes have a constant slope of 1%, 7.5%, and 15%, respectively. Complex hillslopes are a combination of different slopes; they are artificially built merging 1%, 7.5%, and 15% slope patches as reported in Fig. 3.2B. Concavity—convexity is calculated and imposed to the grid. The subfamilies B1 and B2 simulate downstream gradients with different border effect; subfamily B3 simulates a north—south stripes configuration, and subfamily B4 adopts the same patterns using only extreme slope values (i.e., maximum and minimum).

From a geomorphological point of view, these configurations correspond (Fig. 3.2B): in A-cases to simple linear hillslopes; in CC1, CC2, and CC3 to full concave hillslopes; in VV1, VV2, and VV3 to full convex profiles; and in all the other CV-VC declinations to a composed convex—concave and concave—convex profile along the main hillslope direction. All these configurations can easily be related to the main classical classifications of the hillslope, as reported by several authors (Huggett, 1975; Pennock et al., 1987) and illustrated in Fig. 3.2A.

As described in the model section, the only factors influencing the sensitivity to soil erosion due to different morphological configurations are slope and concavity/convexity; slope values <2% associated with a concavity of more than 0.055 m^{-1} (i.e., red areas in Fig. 3.2C) are the required

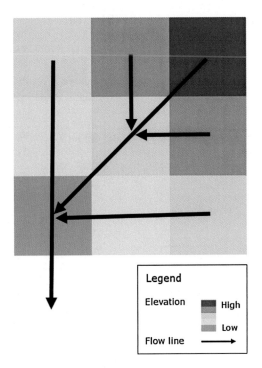

FIGURE 3.1

The virtual catchment adopted in the test. The grids are spaced of 50 m and the arrows represent the flow pathways.

conditions for topographically driven deposition, based on a maximum sediment concentration in runoff.

2.2.3 Sensitivity Calculation

In this SA, since we adopted a procedure with discretized data, we calculated the sensitivity to erosion in two ways.

First, to discern the direct influence of the parameters in homogeneous configurations, we analyzed the simple output variations (i.e., sediment delivery at the catchment outlet for both interrill and rill erosion processes) referred to a baseline value (i.e., the baseline reference is calculated on all the input parameters set to the medium value) as follows:

$$D_{i(x^0)} = \frac{Y\left(x_1^0, \dots, x_i^0 + \partial x_1^0, \dots, x_1^0\right) - Y\left(x^0\right)}{Y\left(x^0\right)} \tag{3.9}$$

where $D_{i(x^0)}$ is the variation of the output Y to the input factor x_i at point $x^0(x_1^0, \dots, x_i^0, \dots, x_I^0)$ and D_i is a nondimensional localized index that represents the normalized response of the output for each input value combination x_i. A positive (or negative) D_i indicates a positive (or negative) relationship between Y and x_i, i.e., an increase in x_i will cause an increase in Y.

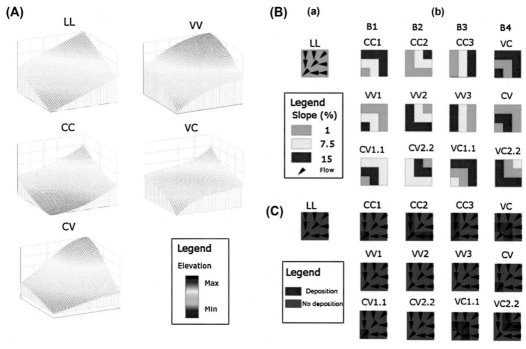

FIGURE 3.2

Representation of the catchment configurations adopted in the analysis. (A) The reference classical geomorphological forms (*LL*, linear; *CC*, full concave; *VV*, full convex; *CV*, convex/concave; *VC*, convex/concave). (B) The linear (a) and the complex patterns (B1, B2...). Each code (CC1, VV1, etc.) represents a different complex configuration. (C) Catchment patterns with a representation of the flow direction (*arrows*) and the cells where sediment deposition is allowed (in red).

Second, to check sensitivity among different geomorphological patterns, we calculated the local SA (S_i), which represents the normalized variation of an output with reference to the normalized variation of the input parameter (i.e., first derivative) with the following object equation:

$$S_{i(x^0)} = \frac{D_{i(x^0)}}{\frac{\partial x_i}{x_i^0}} \qquad (3.10)$$

where $S_{i(x^0)}$ indicates the degree of sensitivity of Y to x_i at point x^0. The value of $S_{i(x^0)}$ can be used to compare the sensitivity of the output to an input variable at its different magnitudes. It can also be used to compare the sensitivity of the output to different individual input factors, for example, the sensitivity of Y to x_i and x_j at point x^0.

Furthermore, we checked the erosion sensitivity to the global model behaviors aggregating as equivalent parameters rainfall and soil parameters. To test the maximal positive response of the parameter combinations, the normalized input parameter sequences have been kept monotonically growing as resulting from the single parameter analyzed (Figs. 3.4A, B and 3.5A, B); then their sensitivity outputs have been averaged (i.e., mean $S_{i(+PH, \ +PI, \ +A_r)}$; mean $S_{i(-R_h, \ -C_r, \ +Cv, \ +S_l)}$) for rill and interrill processes, respectively.

3. RESULTS AND DISCUSSION

3.1 LINEAR HILLSLOPE

Model parameters play different roles in SA outputs. Fig. 3.3 shows the normalized interrill variation over all the parameter space. Boxplots in Figs. 3.4 and 3.5 show the normalized variation of the outputs (interrill and rill erosion) for each of the seven parameters. Their relative influence is detailed on the left of the subplots 3.4A, B and 3.5A, B as input/output variations with reference to the baseline levels (D_i) for interrill and rill erosion and parameter type (rainfall parameters, Figs. 3.4A and 3.5A and slope with soil parameters Figs. 3.4B and 3.5B). Then, sensitivity to erosion (S_i) of all the parameters for both erosion processes is compared in Fig. 3.7A. Sensitivity to erosion for rainfall and soil aggregated parameters are shown in Fig. 3.6.

Concerning the rainfall parameters (Figs. 3.4A, 3.5A, and 3.7A), total rainfall plays a major role in both rill and interrill erosion processes, whereas rainfall intensity is relevant in interrill erosion. Antecedent rainfall is a second-order parameter in both cases, whereas rainfall intensity does not influence rill erosion processes since the parameter is not integrated in the equations describing the process. Sensitivity to slope and soil parameters (Figs. 3.4B, 3.5B, and 3.7A) results are more complex because of the interactions of all the involved processes (i.e., runoff/infiltration versus interrill and rill erosion).

In interrill erosion (Fig. 3.4B), after their respective sign, all the parameters appear to have the same order of influence, except slope, which is not included in the equations describing the process and used only for sediment redistribution at the parcel scale. Crusting always has a positive influence, whereas soil cover shows a negative trend. Soil roughness has a dual role because of its implication in all the equations governing interrill erosion. It has a clear positive influence in sediment load, leading to a general increase in sediment delivery, and, at the same time, it is a limiting factor for runoff reducing the amount of the overland flow that potentially exports sediments from the parcel. This low-level complexity is shown in Fig. 3.4B (left), where a general negative trend appears for low roughness values, whereas a positive trend is reported for high values of the parameter.

Rill erosion (Fig. 3.5B) is a complex process controlled by several subfactors that govern runoff and sediment load production. First, rill erosion depends on the sediment delivery of each cell itself. Second, it is influenced by the accumulated flow draining all the cells located uphill having different propensities to erosion. These factors associated with the local soil propensity to rill, produce differentiated output. The most influential parameters in this situation are soil roughness and soil cover. Both negatively influence runoff production and propensity to rill (Eqs. (3.3)−(3.5)). A direct relationship is shown instead with soil crusting; when it increases potential runoff, it locally reduces the cohesion, explicating a general global increasing erosion trend. On the contrary, high values of soil crusting produce a reduction in the sensitivity due to different process interactions.

3.1.1 Aggregated Parameters

A three-dimensional (3D) analysis of the aggregated parameters is reported in Fig. 3.6. The figure represents interrill (left) and rill erosion (right) sensitivity according to the aggregated rainfall and soil parameters. We notice an exponential trend of the sensitivity in both the cases. In rill erosion, the increment is more evident and we can hypothesize a sort of cutoff line after which the increment increases to exponential values; this is more evident for the rainfall component. This may be explained by the fact that the equations themselves (i.e., Eq. (3.4)) include a nonlinear relationship between rill erodibility and accumulated flow driving to higher increments for high cumulated flow values.

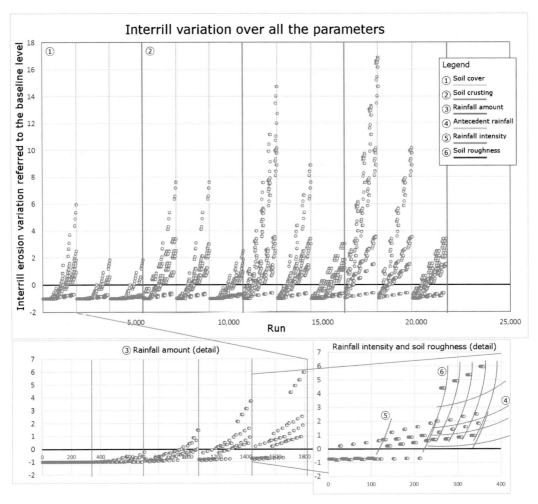

FIGURE 3.3

Interrill erosion expressed as variations referred to the baseline level over all the parameters values used in the model sensitivity analysis. The slope is not represented because it is uninfluential in interrill erosion.

3.2 COMPLEX HILLSLOPES

Due to hillslope complexity, soil erosion sensitivity in the geomorphological configurations varied, and therefore, the more interesting results were further examined (Table 3.2). The hillslope configurations relay to full concave, full convex, and several concave—convex mixed forms, as described in the methods section.

For the first major result, interrill erosion is not sensitive to the morphological variations; all the configurations show the same erosion sensitivity found for linear hillslopes (Fig. 7.1, left). This is explained by the fact that the slope and shape of the catchment are not involved in the equations describing the processes. On the contrary, as reported in Eqs. (3.3)—(3.5), the parameters governing rill erosion are very sensitive to changes in morphology, as it influences rill erodibility (Eqs. (3.3)—(3.5)),

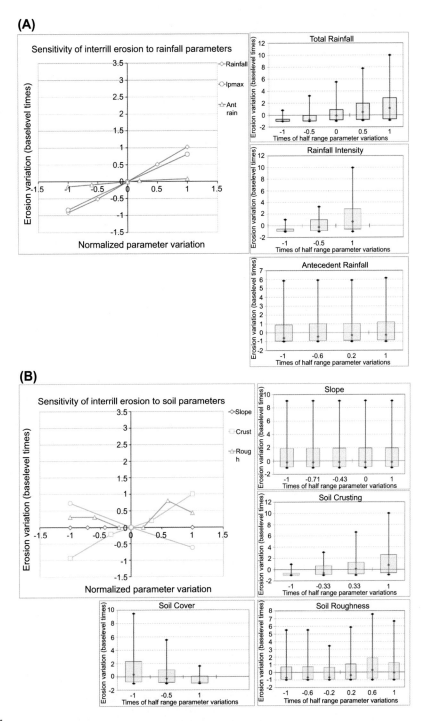

FIGURE 3.4

Sensitivity of interrill erosion to rainfall (A) and to soil parameters (B) expressed as variation referred to the baseline levels (on the left of each sub-plot) and boxplots representing the whole ranges variation (on the right).

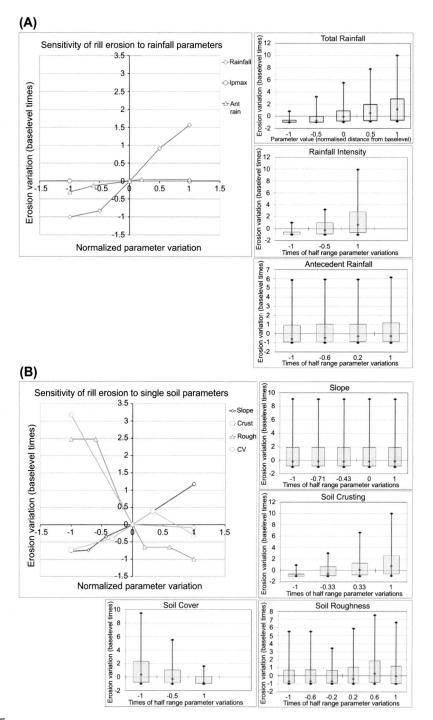

FIGURE 3.5

Sensitivity of rill erosion to rainfall (A) and to soil parameters (B) expressed as variation referred to the baseline levels (on the left of each sub-plot) and boxplots representing the whole ranges variation (on the right).

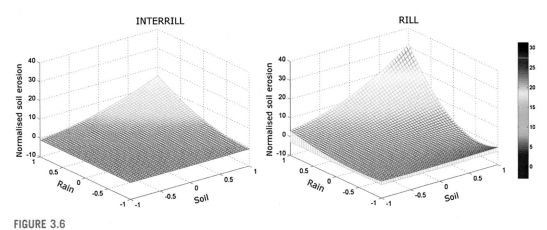

FIGURE 3.6

Sensitivity to erosion for rainfall ($P_t + P_i + P_a$) and soil ($-R_h - C_r + C_v + S_l$) aggregated parameters in interrill (left) and rill erosion (right).

which is a function of the local slope and sediment deposition when maximum transport capacity is reached (i.e., slope lower than 2% and concavity passing 0.055, red areas in Fig. 2.3.)

Because slope is not relevant in interrill erosion, in complex configurations (Fig. 3.7B–D) we compared only results for rill erosion. Furthermore, because total sediment production depends on the slope, the comparisons were done matching in rows configurations having the same average slope (i.e., CC1– VV2; CV11–CV22; VC11–VC22, etc.). In such cases, differences in sediment delivery are due only to the morphological patterns. For that, the slope parameter is not accounted as single parameter but in relation to the whole configuration. In full concave morphology (CC1, Fig. 3.7B), soil erosion is slightly sensitive to parameter variation and small differences are noticed among the three

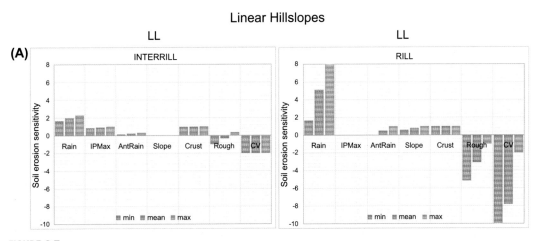

FIGURE 3.7

Sensitivity to erosion (S_i) of all the parameters for the more relevant morphological configurations: (A) linear hillslopes (LL), (B) full concave (CC) and full convex (VV) configurations, (C) concave-convex case (CV, downslope concavity), (D) convex-concave case (VC, upslope concavity).

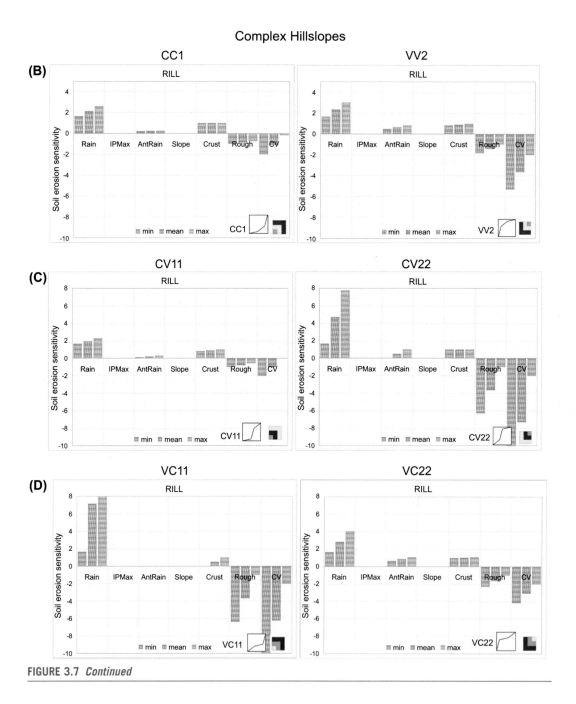

FIGURE 3.7 *Continued*

subconfigurations analyzed (CC1, CC2, and CC3; Fig. 3.2B). This is explained by the fact that the sediment deposition occurring at the bottom of a hillslope (Fig. 3.2C) determines a leveling of the sensitivity for all the parameters, and a deep smoothing is produced. Then, the variations become insignificant. In the full convex morphology (VV2, Fig 3.7B) soil erosion is instead more sensitive, both to rainfall (with the exception of rainfall intensity) and soil parameters. No deposition occurs in any configuration (VV1, VV2, and VV3; Fig. 3.2B). The more significant variations seem to depend on the total rainfall, roughness, and soil cover, enhancing and reducing the sensitivity power.

Different mixed morphologies (Figs. 3.7C and 3.7D), catchments presenting both concavities and convexities, were then compared. In CV patterns in Fig. 3.7C (CV11, left, and, CV22 right), the lower slopes are located at the bottom and at the top of the configurations, respectively (Fig. 3.2B). Consequently, deposition can occur only at the bottom of the hillslope (i.e., Fig. 3.2C; CC > 0.055, slope <2%), as the top is the location where the flow starts. We can argue that the first case is similar to a full concave configuration (CC) driven by the downslope depositional area, able to limit the sensitivity to all the parameters. Inversely, when no deposition occurs, such as in the second case, a strong sensitivity to rainfall, roughness, and vegetation cover is revealed. Furthermore, among all the CV1# and CV2# tested patterns, the erosion sensitivity to the parameters is confirmed to be proportional to the slope area located at the top contributing to the sediment amount routed in the catchment (i.e., high average slope → high routed sediment → high sensitivity) and inversely proportional to the surface of the depositional area located at the bottom.

Fig. 3.7D shows VC patterns where deposition occurs at the middle of the hillslope (Fig. 3.2C). On the left (VC11), the high contribution produced by the steep slope at the top is, in part, trapped by the depositional area and the remaining sediment is routed along the catchment preserving a large sensitivity in most of the parameters. On the contrary, on the right, at the top, the average slope gives a modest sediment contribution mostly retained by the depositional area. This implies a general low sensitivity for all the parameters because of the little amount of flux routed in the configuration. All these considerations are summarized in Table 3.2, where the main sensitivity parameters are reported for each morphological configuration.

This SA approach is comparable with several other cases of soil erosion models' SA, which can be found in the literature. O-A-T sensitivity is one of the most used methodologies due to its explicitness and capability to analyze locally the influence of all parameters. It was successfully adopted for erosion models such as SWAT, WEPP, and LISEM (e.g., van Griensven et al., 2006; Ascough et al., 2013; Sheikh et al., 2010), or as associated rather than as alternative to an LH sampling procedure or Monte Carlo sampling to reduce sample sizes and test wide parameter spaces (e.g., Feng and Sharratt, 2005; Veihe and Quinton, 2000a).

Concerning the geographical support on which the sensitivity is calculated, the use of a specific or reference catchments is common, and, in some cases, topography is included in the tested parameters [e.g., slope, dimension of the catchment (Ascough et al., 2013; Nearing et al., 1990; Veihe and Quinton, 2000b)]. Conversely, simulations are rarely done on an ideal catchment simulating different spatial configurations of the parameters or topographical configurations (Cheviron et al., 2010, 2011).

Similar to this study, Cheviron et al. (2011), comparing different erosion models—among them STREAM, the ascendant of the LandSoil model—found comparable responses between spatially homogeneous and distributed parameterizations. A difference between continuous (MHYDAS, PESERA) and parametric models (MESALES, STREAM) is noted showing more sharp variations in

Table 3.2 Summary of the results: degree of influence of the parameters in erosion sensitivity for all the morphological configurations we tested.

Interrill

Hillslope	Form	Rainfall Parameters	Soil Parameters
Linear hillslope	LL	Sensitive having ranking: R1 = 1, R2 = 1, R3 = 3	Sensitive having ranking: $C_v = 1$, $R_h = 2$, $C_r = 1$, $S_l = 0$
	CC	Same as linear	Same as linear
Complex hillslopes	VV	Same as linear	Same as linear
	CV	Same as linear	Same as linear
	VC	Same as linear	Same as linear

Rill

Hillslope	Form	Rainfall Parameters	Soil Parameters
Linear hillslope	LL	Sensitive with ranking: R1 = 1, R2 = 0, R3 = 3	Sensitive having ranking: $C_v = $ 1plus, $R_h = $ 1plus, $C_r = 2$, $S_l = 2$
	CC	All slightly sensitive having ranking: R1 = 2, R2 = 0, R3 = 3	All slightly sensitive having ranking: C_v, C_r, $R_h = 3$, slope: 0
Complex hillslopes	VV	Sensitive having ranking: R1 = 1, R2 = 0, R3 = 3	Sensitive having ranking: $C_v = 1$, $R_h = 2$, $C_r = 2$
	CV	All slightly sensitive having ranking: R1 = 2, R2 = 0, R3 = 3 (like CC)	Sensitive having ranking: $C_v = 3$, $R_h = 3$, $C_r = 3$ except with deposition at top
	VC	Sensitive having ranking: R1 = 1, R2 = 0, R3 = 3 (like VV)	Highly sensitive having ranking: $C_v = 1$, $R_h = 1$, $C_r = 3$ (like VV case)

Parameters: C_r, soil crusting; C_v, soil cover; R1, total rainfall; R2, rainfall intensity; R3, antecedent rainfall; R_h, soil roughness; S_l, slope. Influence ranking: 1, high; 2, medium; 3, low; 0, uninfluential.

parametric models, which is a characteristic that we observed in our SA for single-parameter analysis. In Cheviron et al. (2011) for STREAM model, soil parameters were more sensitive than slope; here, we reach a similar conclusion with the explication that slope is involved only in rill erosion, thus showing a nondominant relevance.

Similar to our study, Cheviron et al. (2011) found (Cheviron et al., 2010, p. 6), for total erosion sensitivity, a threshold in response to rainfall and soil wetness due to the different influences of rill and interrill erosion. Here, this peculiarity is revealed in Fig. 3.6 (right), where a marked nonlinear increasing trend is reported for rill erosion, which can lead to a comparable behavior if total erosion (rill + interrill) is considered.

Moving from homogeneous to nonhomogeneous patterns in the 3D maps of Cheviron et al. (2011, pp. 9 and 12), we note a progressive smoothing of the sensitivity response, probably because the outputs of the nonhomogeneous configurations have been aggregated. Using equivalent parameters (Fig. 3.6), we found similar smooth surfaces in our 3D analysis. Because of this, and to discern the single-parameter effects, we preferred to use single parameter outputs on different hillslope configurations.

4. CONCLUDING REMARKS

We developed an SA methodology for model parameters under different hillslope conditions (i.e., simulating various morphologies based on changes in slope patterns) using a virtual catchment.

This analytical approach is justified by the fact that the existence of rain and topography are the sine qua non conditions (hence, the forcing) for water and sediment fluxes. Topography controls all the water and sediment processes; this appears through the concept of sediment connectivity, which, in turn, dictates sediment dynamics. Sediments are routed along the catchment producing on- and off-site soil erosion.

Our results have shed some light on model parameters and on how morphological configurations influence soil erosion. In linear hillslopes, the most sensitive parameters are the total rainfall amount and soil cover for both rill and interrill erosion.

Slope is uninfluential in interrill erosion because its control on sediment redistribution acts only at parcel scale, as does rainfall intensity on rill erosion, because the parameter is not implemented in rill erosion equations.

The (3D) SA of the aggregated parameters (rainfall and soil, respectively) for rill and interrill erosion shows an exponential trend of the sensitivity in both the cases with increment and shape more marked in rill erosion.

In complex hillslopes (multiple slopes), we observe that interrill erosion parameters are not sensitive to changes in geomorphological patterns and present the same results reported for linear hillslopes. In concave cases, the sensitivity of all parameters is reduced by sediment deposition occurring at the bottom of the catchment. Conversely, sensitivity scores are enhanced in convex shapes outlining the influence of parameters such as total (event) rainfall, soil cover, and soil roughness. In mixed hillslopes, associating concave and convex shapes, the sensitivity of the parameters finds itself at intermediate levels between the observed concave and convex cases, mostly influenced by the slope's pattern itself, governing sediment erosion and deposition.

In our methodology, the O-A-T approach allowed us to easily inspect local parameter behaviors and clarify specific interferences among parameters and involved processes. The virtual catchment we adopted is a basic, efficient, approach to inspect soil erosion sensitivity on morphological patterns simulating natural conditions and achieving efficient results isolating single model behaviors. Anyway, some limitations arise because of its simple structure; a matter of future developments and researches, a few complex patterns, associated with a statistical methodology, could be suitable for the study of real cases. Finally, the local SA methodology we developed could be a useful approach for any environmental modeling SA where the modeled processes coexist with geographical constraints significantly interacting with the process.

ACKNOWLEDGMENTS

This research was financially supported by the French National Research Agency, "Vulnerability program: environments, climates and societies (VMCS)," Project "LandSoil" n. ANR-08-VULN-006.

REFERENCES

Ascough, J.C., Flanagan, D.C., Nearing, M.A., Engel, B.A., 2013. Sensitivity and first-order/Monte Carlo uncertainty Analysis of the WEPP hillslope erosion model. Transactions of the ASABE 56 (2), 437–452.

Breshears, D.D., Kirchner, T.B., Whicker, F.W., 1992. Contaminant transport through agroecosystems: assessing relative importance of environmental, physiological, and management factors. Ecological Applications 2 (3), 285–297.

Le Bissonnais, Y., Benkhadra, H., Chaplot, V., Fox, D., King, D., Daroussin, J., 1998. Crusting, runoff and sheet erosion on silty loamy soils at various scales and upscaling from m^2 to small catchments. Soil and Tillage Research 46, 69–80.

Le Bissonnais, Y., Cerdan, O., Lecomte, V., Benkhadra, H., Souchere, V., Martin, P., 2005. Variability of soil surface characteristics influencing runoff and interrill erosion. Catena 62 (2–3), 111–124.

Ciampalini, R., Follain, S., Le Bissonnais, Y., 2012. LANDSOIL: a model for the analysis of erosion impact on agricultural landscape evolution. Geomorphology 175, 176, 25–37.

Cheviron, B., Gumiere, S., Le Bissonnais, Y., Moussa, R., Raclot, D., 2010. Sensitivity analysis of distributed erosion models: framework. Water Resources Research 46. W08508.

Cheviron, B., Le Bissonnais, Y., Desprats, J.-F., Couturier, A., Gumiere, S., Cerdan, O., Darboux, F., Raclot, D., 2011. Comparative sensitivity analysis of four distributed erosion models. Water Resources Research 47. W01510.

Crosetto, M., Tarantola, S., 2001. Uncertainty and sensitivity analysis: tools for GIS-based model implantation. International Journal of Geographical Information Science 15 (5), 415–437.

Campolongo, F., Cariboni, J., Saltelli, A., 2007. An effective screening design for sensitivity analysis of large models. Environmental Modelling and Software 22, 1509–1518.

Cerdan, O., Le Bissonnais, Y., Couturier, A., Saby, N., 2002. Modelling interrill erosion in small cultivated catchments. Hydrological Processes 2002 (16), 3215–3226.

Cerdan, O., Souchere, V., Lecomte, V., Couturier, A., Le Bissonnais, Y., 2001. Incorporating soil surface crusting processes in an expert-based runoff model: sealing and transfer by runoff and erosion related to agricultural management. Catena 46, 189–205.

Francos, A., Elorza, F.J., Bouraoui, F., Bidoglio, G., Galbiati, L., 2003. Sensitivity analysis of distributed environmental simulation models: understanding the model behavior in hydrological studies at the catchment scale. Reliability Engineering and System Safety 79 (2), 205–218.

Feng, G., Sharratt, B., 2005. Sensitivity analysis of soil and PM10 loss in WEPS using the LHS-OAT method. Transactions of the ASAE 48 (4), 1409–1420.

Govers, G., Vandaele, K., Desmet, P., Poesen, J., Bunte, K., 1994. The role of tillage in soil redistribution on hillslopes. European Journal of Soil Science 45, 469–478.

van Griensven, A., Meixner, T., Grunwald, S., Bishop, T., Diluzio, M., Srinivasan, R., 2006. A global sensitivity analysis tool for the parameters of multi-variable catchment models. Journal of Hydrology 324, 10–23.

Huggett, R.J., 1975. Soil landscape systems: a model of soil genesis. Geoderma 13, 1–22.

Ionescu-Bujor, M., Cacuci, D.G., 2004. A comparative review of sensitivity and uncertainty analysis of large-scale systems: I. Deterministic methods. Nuclear Science and Engineering 147 (3), 189–203.

Jenson, S.K., Domingue, J.O., 1988. Extracting topographic structure from digital elevation data for geographic information system analysis. Photogrammetric Engineering and Remote Sensing 54, 1593–1600.

Liu, Y., Gupta, H.V., Soroochian, S., Bastidas, L.A., 2004. Exploring parameter sensitivities of the land surface using a locally coupled land-atmosphere model. Journal of Geophysical Research 109. D21101.

Laflen, J.M., Lane, L.J., Foster, G.R., 1991. WEPP: a new generation of erosion prediction technology. Journal of Soil and Water Conservation 46, 34–38.

Morris, M.D., 1991. Factorial sampling plans for preliminary computational experiments. Technometrics 33 (2), 161–174.

Van Muysen, W., Govers, G., Van Oost, K., 2002. Identification of important factors in the process of tillage erosion: the case of mouldboard tillage. Soil and Tillage Research 65, 77–93.

Van Muysen, W., Van Oost, K., Govers, G., 2006. Soil translocation resulting from multiple passes of tillage under normal field operating conditions. Soil and Tillage Research 87, 218−230.

Nearing, M.A., Deer-Ascough, L., Laflen, J.M., 1990. Sensitivity analysis of the WEPP hillslope profile erosion model. Transactions of the ASAE 33 (3), 839−849.

Pianosia, F., Bevenf, K., Freerc, J., Halld, J.W., Rougierb, J., Stephensone, D.B., Wagener, T., 2016. Sensitivity analysis of environmental models: a systematic review with practical workflow. Environmental Modelling & Software 79, 214−232.

Pennock, D.J., Zebarth, B.J., de Jong, E., 1987. Landform classification and soil distribution in hummocky terrain, Saskatchewan, Canada. Geoderma 40, 297−315.

Saltelli, A., 1999. Sensitivity analysis: could better methods be used? Journal of Geophysical Research 104 (D3), 3789−3793.

Saltelli, A., Tarantola, S., Campolongo, F., 2000. Sensitivity analysis as an ingredient of modeling. Statistical Science 15 (4), 377−395.

Spear, R.C., Hornberger, G.M., 1980. Eutrophication in peel inlet−II. Identification of critical uncertainties via generalized sensitivity analysis. Water Research 14, 43−49.

Singer, M.J., Le Bissonnais, Y., 1998. Importance of surface sealing in the erosion of some soils from a Mediterranean climate. Geomorphology 24, 79−85.

Souchère, V., King, D., Daroussin, J., Papy, F., Capillon, A., 1998. Effects of tillage on runoff directions: consequences on runoff contributing area within agricultural catchments. Journal of Hydrology 206, 256−267.

Souchère, V., Cerdan, O., Ludwig, B., Le Bissonnais, Y., Couturier, A., Papy, F., 2003. Modelling ephemeral gully erosion in small cultivated catchments. Catena 50, 489−505.

Sheikh, V., van Loon, E., Hessel, R., Jetten, V., 2010. Sensitivity of LISEM predicted catchment discharge to initial soil moisture content of soil profile. Journal of Hydrology 393, 174−185.

Tiscareno-Lopez, M., Lopes, V.L., Stone, J.J., Lane, L.J., 1994. Sensitivity analysis of the WEPP watershed model for rangeland: 1. Hillslope processes. Transactions of the ASAE 37 (1), 151−158.

Veihe, A., Quinton, J., 2000a. Sensitivity analysis of EUROSEM using Monte Carlo simulation. I: hydrological, soil and vegetation parameters. Hydrological Processes 14, 915−926.

Veihe, A., Quinton, J., 2000b. Sensitivity analysis of EUROSEM using Monte Carlo simulation II: the effect of rills and rock fragments. Hydrological Processes 14, 927−939.

Wei, H., Nearing, M.A., Stone, J.J., 2007. A comprehensive sensitivity analysis framework for model evaluation and improvement using a case study of the rangeland hydrology and erosion model. Transactions of the ASABE 50 (3), 945−953.

SENSITIVITY OF VEGETATION PHENOLOGICAL PARAMETERS: FROM SATELLITE SENSORS TO SPATIAL RESOLUTION AND TEMPORAL COMPOSITING PERIOD

4

G.L. Mountford[1], P.M. Atkinson[1,2,3], J. Dash[1], T. Lankester[4], S. Hubbard[4]

University of Southampton, Southampton, United Kingdom[1]; Lancaster University, Lancaster, United Kingdom[2]; Queens University Belfast, Belfast, United Kingdom[3]; Intelligence at Airbus Defence and Space, Farnborough, United Kingdom[4]

CHAPTER OUTLINE

Sensitivity Analysis in Earth Observation Modelling. http://dx.doi.org/10.1016/B978-0-12-803011-0.00004-5

1. INTRODUCTION

Vegetation phenology, the study of the timing of recurrent biological events of plants, is one of the most responsive and easily observable phenomena with which to assess the impact of climate change (White et al., 2003; Sparks et al., 2000). Moreover, vegetation phenology has the potential to influence both regional weather patterns and the global climate by influencing the seasonal patterns of surface atmosphere energy and water and trace gases exchanges (Richardson et al., 2013). Alterations in the growing season are also related to changes in gross primary productivity and net primary productivity (NPP), which are important for quantifying terrestrial carbon uptake and, therefore, heavily influence atmospheric carbon content (Piao et al., 2007; Jeong et al., 2011). Temperature affects NPP by changing the rates of photosynthesis, autotrophic respiration, nutrient mineralization, and the period of foliation and frost hardiness (Saxe et al., 2001).

Changes in the arrival of spring and leaf unfolding are a common focus of research into the impacts of climate change on vegetation. Recent warming trends have been linked with an earlier onset of vegetation greenness in spring and a longer growing season. However, there is less agreement on the implications of climate change on autumn phenology (Richardson et al., 2013). Previous studies have indicated an advancement of the start of season (SOS) by 0.2–8 days per decade, and a delayed end of season (EOS) by 0.5–6.1 days per decade, indicating a substantial increase in the length of the season (Zeng et al., 2011).

The timing of phenological events and their influence on ecosystem processes are assessed using five main methods: in situ observations, modeling, eddy covariance flux towers, global change experiments, and finally, remote sensing techniques. Remote sensing includes two broad approaches, one using near-surface remote sensing, e.g., digital camera photography, and the other using spaceborne remote sensing techniques. The term land surface phenology (LSP) is commonly used for remote sensing–based methods used to estimate dominant vegetation phenological parameters, and relates to the timing of changes at the scale of a satellite image pixel. Atmospheric contamination, snow and cloud cover, and bidirectional viewing effects may affect the estimation of LSP. Therefore, LSP is fundamentally different from vegetation phenology (Hamunyela et al., 2013). As vegetation phenology is an important controlling factor, assessment of the sensitivity of phenological parameters to variation in input parameters as part of the processing chain used to estimate phenology is vital to increase understanding of the effects of global climate changes (Schott, 2007).

Sensitivity analysis (SA) refers to the study of divergence in the output of a model related to sources of variance in the model input(s). Generally, the overall divergence in parameter estimates for each input requires detailed assessment. In any model representation, the model is a mathematical idealization used to predict the variable of interest. Therefore, all model outputs are approximations of the true values. Typically, more the complexity within a system, the less precise will the model output be, due to the sensitivity of all variables and parameters within the model. A parameter of a model is generally a fixed quantity that expresses a characteristic of the system. However, parameters can also be varied to analyze varying conditions. Observations are used to estimate model forms; this includes inductive and deductive reasoning, estimating parameters, and refining prior estimates through incorporating additional metrics (Cacuci, 2003).

In this chapter, the importance of assessing phenological changes using satellite sensor–derived data and the sensitivity of phenological parameters in remote sensing studies is introduced. A short case study highlights the sensitivity of phenological parameters, including SOS and EOS, to alterations to a selection of the input parameters for remotely sensed imagery.

2. MONITORING VEGETATION PHENOLOGY

Satellite-derived observations have been used in vegetation phenology analyses for over 30 years (Reed et al., 1994; Justice et al., 1985). The use of remote sensing in phenology studies allows seasonal changes in vegetation at regional, continental, and global scales to be assessed at regular, even daily, intervals (Schott, 2007; Justice et al., 1985; Xiao and Moody, 2005). Although remotely sensed imagery can expose large-scale phenological trends that would be difficult to detect from the ground, there are limitations; for example, limited ability to monitor individual plants or species, as well as capturing images generally from the canopy, which may obscure the understory vegetation.

Time series of remotely sensed data are an important source of data for characterizing vegetation dynamics over several time scales. This includes the seasonal development of vegetation including start, peak, duration, and end of growing season (Pouliot et al., 2011). Fig. 4.1 is a diagrammatic interpretation of a phenological time series indicating the relationship between vegetation index and phenological parameter. The most commonly used phenology metric within remote sensing is the Julian day of year (DOY) for the onset of greenness (green-up, SOS), maturity onset (onset of greenness maximum), senescence onset (onset of greenness decrease, onset of EOS), and end of senescence (the onset of greenness minimum or dormancy, EOS) (Zhang et al., 2009). Green-up or SOS is the date on which pixels start to green-up, or reach a defined percentage of the season maximum. The increase in greenness is quantified by an increase in the vegetation index as the photosynthetic tissue increases within a pixel (Delbart et al., 2015). In this chapter, SOS is used to reflect the DOY where there is a rapidly sustained increase in the vegetation index value after a period of dormancy, and EOS used to reflect the DOY of sustained decrease.

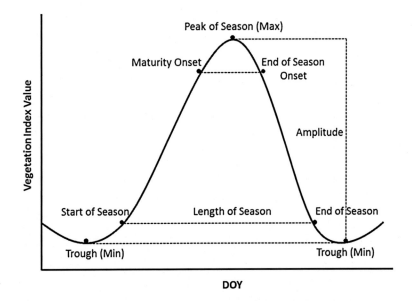

FIGURE 4.1

Diagrammatic interpretation of a vegetation index time series and related phenological parameters.

3. SENSITIVITY ANALYSIS

Uncertainty analysis (UA) and SA are necessary to explore the behavior of models and to assess the uncertainty and sensitivity of input parameters. Both UA and SA are integral assessments of the modeling process (Saltelli et al., 2000), and when conducted together, the model user is more informed about the confidence of results. There is a distinction between UA and SA. UA is performed to describe the range of possible outputs depending on the uncertainty (and variations) of the inputs and to investigate the effect of the lack of knowledge or errors of the model.

SA is conducted for several reasons: to determine which input parameters contribute the most variability in the output, which parameters are insignificant, if parameters interact which each other, and if the output(s) can be explained (Iooss and Lemaître, 2015). The variability associated with a sensitive input parameter is propagated to the output and can result in a large correlation between an input parameter and significant variation in the output (Hamby, 1994).

In models with multiple input parameters, SA assesses the main driver for variation in the output(s) and is an informative analysis on the influence of model inputs or parameters. SA defines the output parameter sensitivity to variation of individual input parameters, and is typically known (Hamby, 1994). The emphasis of an SA depends on the context and the questions that are being asked within the assessment, but centers on addressing changes to "optimal" metrics and the effects these have on parameter output values.

Initial screening tools, including descriptive statistics, scatterplots, and Pearson's correlation coefficient, identify parameters that have a significant influence on model outputs. Once complete, intensive methods can focus further on the sensitivity of the output(s). Local SA examines the local response of the output(s) by varying the input parameters one at a time while holding additional parameters constant. Global SA assesses the global response by quantifying the scale and shape of the input and output relationship and parameter interactions. There are several approaches to perform a SA, including one-at-a-time (OAT), deterministic, sensitivity coefficients, screening techniques, scatterplots, regression analysis, and variance-based methods (Satelli et al., 2000).

A deterministic SA varies each parameter of interest independently, holding other parameters constant, to assess how the output is sensitive to parameter values. The parameter values are changed individually, using upper and lower bounds, and the relative variance in the output is then assessed. The simplest and most commonly used form of an SA is the OAT approach, which assesses individual variances in output(s) to alterations in the value of one input metric (Crosetto et al., 2000). Importantly, OAT techniques do not explore input interactions and are limited to the nominal value sensitivities (Crosetto et al., 2000). However, this screening method allows fast exploration of the model and evaluation of influential parameters (Iooss and Lemaître, 2015). For the case study presented in this chapter, an OAT method was chosen to assess the variance of SOS and EOS.

4. SENSITIVITY OF REMOTELY SENSED PHENOLOGICAL PARAMETERS

While estimating LSP, the individual steps within the remote sensing process need to be considered, as they each impact the variation in the output. Changes in the input satellite sensor data, spatial resolution, composite period, correction, and smoothing techniques, all create a degree of divergence within the output data. The degree of divergence varies depending on the techniques used, as each processing technique has limitations and a predetermined bias and limits the entire process. The limitations of each input parameter and observed variation within phenological studies will now be discussed.

4.1 SATELLITE SENSOR

There are several satellite sensors used to estimate LSP; Landsat sensors, NOAA Advanced Very High Resolution Radiometer (AVHRR), Satellite Pour l'Observation de la Terre Vegetation (SPOT Vegetation), Terra and Aqua Moderate Resolution Imaging Spectroradiometer (MODIS), and Envisat Medium-Resolution Imaging Spectrometer (MERIS). With the launch of new sensors with a finer spatial and spectral resolution, such as the ESA Sentinel missions and the NOAA WorldView-3 satellite sensor, research has focused on achieving higher accuracy from satellite sensor imagery (Kross et al., 2011). The value of remote sensing for monitoring phenology is restricted by the overpass frequency and the amount of cloud cover, which often leads to insufficient data for phenology estimation (Justice et al., 1985). The sensor properties can also have implications, such as the number of spectral band detectors and the height of the satellite, as they define the signal-to-noise ratio.

4.2 VEGETATION INDEX

LSP is commonly estimated from remote sensing data through the use of satellite-derived vegetation indices (VIs). VIs were developed primarily for the remote sensing of vegetation and are generally composed of combinations of visible and near-infrared spectral measurements (Reed et al., 1994). The most common vegetation index used in remote sensing to study LSP is the Normalized Difference Vegetation Index (NDVI). NDVI Eq. (4.1) has successfully been used to assess stages in the growing season. However, NDVI has a negative bias due to atmospheric conditions and anisotropic bidirectional effects (Hird and McDermid, 2009). This is due to additive path radiance causing an increase in red reflectance, whereas lower atmospheric transmission reduces near-infrared reflectance (Beck et al., 2006). The NDVI value of vegetation is calculated by the following equation:

$$NDVI = \frac{NIR - RED}{NIR + RED} \tag{4.1}$$

Enhanced Vegetation Index (EVI) Eq. (4.2), improves the separation of bare soil from vegetation (Cleland et al., 2007) and was developed as a standard vegetation product for the Terra and Aqua MODIS satellites (Jiang et al., 2008). This is also a commonly used VI and is calculated as follows:

$$EVI = G * \frac{NIR - RED}{NIR + C1 * RED - C2 * BLUE + L} \tag{4.2}$$

where $G = 2.5$, $C1 = 6$, $C2 = 7.5$, and $L = 1$.

Another VI, the Soil-Adjusted Vegetation Index (SAVI) Eq. (4.3) (Huete, 1988), is a more stable VI, with a higher dynamic range at high end but less dynamic range at low end. This is calculated as:

$$SAVI = \frac{NIR - RED}{NIR + RED + L}(1 + L) \tag{4.3}$$

where $L = 0.5$.

Dash and Curran (2004) applied data from the MERIS sensor to create an alternative vegetation index, the MERIS Terrestrial Chlorophyll Index (MTCI) Eq. (4.4). MTCI is related directly to the canopy chlorophyll content and does not have some of the limitations related to NDVI, such as saturation at high biomass. In addition, the MTCI is less affected by noise from the atmosphere and soil background. NDVI has an atmospherically-induced negative bias, whereas for MTCI there is no overall bias due to the noise within the data and is assumed to be white with a zero mean value (Atkinson et al., 2012). This is formulated as follows:

$$\text{MTCI} = \frac{R_{Band10} - R_{Band9}}{R_{Band9} - R_{Band8}} \tag{4.4}$$

where R_{Band10}, R_{Band9}, and R_{Band8} are wavebands 753.75, 708.75, and 681.25 nm, respectively.

There are several other vegetation indices such as the Normalized Difference Water Index (NDWI), first proposed by Gao (1996) to designate a slightly different spectral index, which excludes the effects of snow. NDWI decreases with snowmelt and then increases during vegetation greening. For sparsely vegetated areas, the predominant VI used is the SAVI, as well as the Optimized Soil-Adjusted Vegetation Index and the Modified Soil-Adjusted Vegetation Index (Barati et al., 2011). Each VI has its own advantages and limitations, and since each VI is designed to measure a specific biophysical variable, the vegetation phenology metrics derived from these are not directly comparable. Thus, divergence in parameter estimates arises as a function of the choice of vegetation index.

4.3 SPATIAL RESOLUTION

Satellite sensors are limited when detecting phenological events due to spatial resolution and soil background characteristics (Reed et al., 1994). Pixels are not a true geographical object, and one pixel can cover a large geographical area (Duveiller and Defourny, 2010). Coarse spatial resolution sensors mean changes in a small area of the study may go unnoticed and not be fully represented in the data available (Fisher and Mustard, 2007), whereas with an increased spatial resolution it is possible to monitor vegetation dynamics in greater detail.

Aggregation of spatial coverage can cause a decrease in the degree of precision. With an increased spatial resolution (smaller pixel size), an increase in variance occurs: even within the same area a distinct increase is expected in the variance in phenological parameters such as SOS. Fig. 4.2 highlights the effect of spatial resolution on the variance of results. A relatively coarse spatial resolution can limit the spatial variability that is detectable within remotely sensed imagery. For example, the predictions from fine spatial resolution imagery will have a much greater variance in results than from coarse spatial resolution imagery, even when covering the same area. The effect of the variance of spatial resolution and the spatial aggregation implied should be quantified to highlight the sensitivity of phenological parameters to resolution within the same study area.

4.4 COMPOSITE PERIOD, SMOOTHING, AND FILTERING

Raw data are processed through filtering techniques to detect the dominant signal from a time-series of observations. The techniques fit curves to filter or smooth noisy time series to the estimate parameters, including compositing, smoothing, or screening, such that the data are integrated over several days, therefore, masking rapid changes in phenology (Dungan, 2002). The main issue with analysis of time series data is the period over which imagery is collected to build a composite (Foody and Dash, 2010).

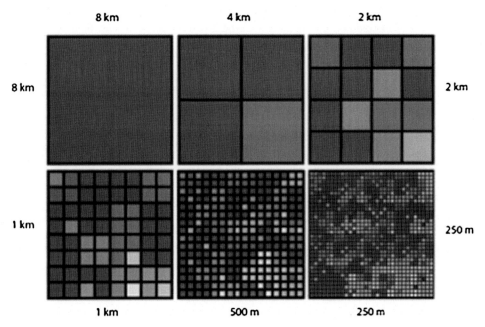

FIGURE 4.2

Diagrammatic interpretation of spatial aggregation of estimated phenological parameters from 250 m to 8 km. The dark grey represents an earlier estimate, whereas white represents an later estimate, thus highlighting the increase in the variance with an increase in spatial resolution.

Many studies require a short compositing period, which may be useful when assessing phenological events such as onset of greening. However, this may not be possible due to the requirement of at least one clear observation of the Earth's surface during the composite period (Foody and Dash, 2010).

4.4.1 Composite Period

Several earlier studies evaluated the sensitivity of phenological parameter estimation to composite period (Pouliot et al., 2011; Kross et al., 2011). Pouliot et al. (2011) suggested that a compositing period of between 7 and 11 days is appropriate to minimize random error and provide the best estimates of SOS. However, the most commonly used composite period for vegetation phenology studies is between 8 and 16 days. This is due to the effects of geometric registration, view angle, sun angle, atmosphere, and cloud cover (Zhang et al., 2009; Reed et al., 1994). Daily data have a greater temporal resolution than composite data. However, daily data are affected more by noise and data dropouts, which can affect the estimation of phenology parameters (Testa et al., 2015).

Zhang et al. (2009) assessed four common sampling methods to assess the effect of composite period on phenology estimation. The methods used were Maximum Value Compositing (MVC), the average value composite, the median value composite, and the random value composite. The composite periods assessed included 2, 4, 6, 8, 10, 12, 16, 18, and 20 days. To evaluate the effect of missing values on the detection of vegetation phenology, a set of Nadir Bidirectional Reflectance Distribution Function (BRDF) Adjusted Reflectance (NBAR) EVI data containing missing values was generated. Zhang et al. (2009) demonstrated that 71% of all error curves showed the lowest error at a 6- to 14-day composite period and the method with the greatest precision was the average value composite method, with least accurate methods being the random and median method.

4.4.2 Smoothing Techniques

Fitting functions simplify but may also introduce new errors and dampen important phenological features. Currently, there is no defined census between studies on which smoothing or filtering technique is best for calculating phenology transition times including; threshold-based, such as Best Index Slope Extraction (BISE); Fourier-based fitting methods; or asymmetric function fitting methods, including Gaussian and the weighted least squares linear regression. The most commonly used algorithms to estimate phenological parameters are based on least-square fitting of VI time series (Testa et al., 2015).

In all threshold-based methods the remaining noise in the remotely sensed data may cause false assessments of the SOS and EOS making temporal information uncertain (Jonsson and Eklundh, 2002). BISE has been used in numerous phenology studies. However, BISE requires a set sliding period and a threshold to assess the percentage increase in NDVI for regrowth during a sliding period based on an empirical strategy that is usually subjective and depends on the skills and experience of the analyst (Chen et al., 2004).

Fourier analysis is commonly used to smooth time series satellite-derived data. By approximating complicated curves as a sum of sinusoidal waves at multiple frequencies, Fourier analysis can be used to interpret vegetation growth cycles. Fourier uses one model parameter, the number of harmonics, to smooth noisy data and can be applied effectively with a few lines of software code (Atkinson et al., 2012). This method does have limitations as it requires a long time series or equally spaced observations. Fourier-based models may also generate false oscillations in the VI time series (Chen et al., 2004). Asymmetric Gaussian functions are arguably more flexible for high-quality VI data to represent vegetation phenological profiles. However, it may be difficult to identify a reasonable and reliable set of maxima and minima to fit the local functions. Also, this method being complex and time consuming, it is difficult to use for noisy data or areas with no clear seasonality (Chen et al., 2004).

White et al. (2009) compared 10 SOS methods, which averaged a difference of 60 days in the average SOS DOY estimate with a standard deviation of 20 days. SOS estimates were earlier in forested regions with annual snow cover, due to the distinct influx in VI value in the spring after the prolonged dormancy of winter. Harmonic Analyses of NVDI Time Series—Fast Fourier Transform Timesat (HANTS-FFT) is a conceptual mathematical technique that uses the maximum increase on Fourier approximation of NDVI, and the Midpoint$_{Pixel}$ is a local threshold technique that uses a per pixel threshold or an NVDI threshold. Compared with plant phenology estimates, both HANTS-FFT and the Midpoint$_{Pixel}$ method had greater statistical correlation, including the timing of first leaf. This equated to a65% acceptable SOS estimates, correlations >0.6, low offsets or bias, and regression slope near 1.

5. CASE STUDY

As highlighted previously, the most commonly used remote sensing techniques utilized within phenology studies have known limitations, which can impede the estimation of phenological parameters. The optimization of spatial resolution and compositing periods can increase the accuracy of LSP estimates. An accuracy assessment requires model-based characterization or external validation data, whereas an SA is performed for a different purpose, to assess the effect of a set of input parameters or variables on model estimates. Therefore, this case study aims to demonstrate the sensitivity of two estimated satellite-derived phenology parameters, SOS and EOS, to variation in the composite period and spatial resolution parameters.

An SA was applied utilizing several of the most common values of spatial resolution and composite period used within remote sensing phenological studies. The approach used here assesses the variance of the output when input metrics are altered on an OAT basis. The effect of the variation of a single metric output(s), while all other metrics are fixed at nominal values, enables the influence of the metric on the sensitivity of the phenological parameters to be assessed (Cariboni et al., 2007). This simplistic approach concentrates on two parameters that are known to influence the estimation of SOS and EOS (Zhang et al., 2009). By utilizing commonly used values for composite period and spatial resolution, this enables the most influential input parameter to be screened.

5.1 STUDY AREA

The chosen study area is the entire United Kingdom. The United Kingdom has a temperate maritime climate, due to the warming influence of the Gulf Stream on weather conditions, which equates to mild wet winters and relatively warm wet summers. For phenological studies, this means that there is a defined SOS and EOS due to seasonal climate changes. For this study, the vegetation class chosen is broad-leaved woodland. Broad-leaved woodland equated to 6% of the land over the United Kingdom and includes both native and nonnative broad-leaved and yew species.

5.2 DATA AND METHODOLOGY

The MERIS is a full-resolution geophysical product for ocean, land, and atmosphere studies. For this study, MERIS level 2 Full Resolution Full Swath (MER_FRS_2P) data with a spatial resolution of 300 m for the United Kingdom from January 1, 2005, to April 30, 2006, were used to estimate MTCI.

MERIS MTCI composites were created for periods of 4, 8, 10, and 16 days. The composites were created at a 250 m spatial resolution using a flux conversion algorithm. The compositing algorithm calculates the arithmetic mean of the input measurements within the composite period, with the additional option for weighting to be applied to the measurement. The 250 m composites were resampled using MATLAB, to represent average values at coarser spatial resolutions of 500 m, 1, 2, 4, and 8 km. First, dropouts were eliminated through an averaging process within a temporal neighbor prior to, and following, a dropout from 1 week to 2 months, thus, filling the gaps with the time series. Then the data set was smoothed using inverse Fourier transformation and LSP parameters (SOS and EOS) were estimated using the method from Dash et al. (2010). The Julian DOY was calculated as the median of the days within the assigned composite. The variance of estimated SOS and EOS due to the sensitivity to variation in spatial resolution and composite period was assessed using the mean DOY over the spatial data set and the overall variance within each output.

5.3 RESULTS AND DISCUSSION

Table 4.1 shows the results of the estimated mean SOS DOY across all chosen composite periods and spatial resolutions. The 16-day composite estimates a later SOS compared with the 4-day composite.

Table 4.1 Estimated SOS mean DOY for broad-leaved woodland

Spatial Resolution	Composite Period			
	16 d	10 d	8 d	4 d
250 m	99	82	73	71
500 m	101	85	73	73
1 km	99	78	69	66
2 km	97	78	61	63
4 km	94	66	55	51
8 km	95	71	47	38

DOY, *day of year;* SOS, *start of season.*

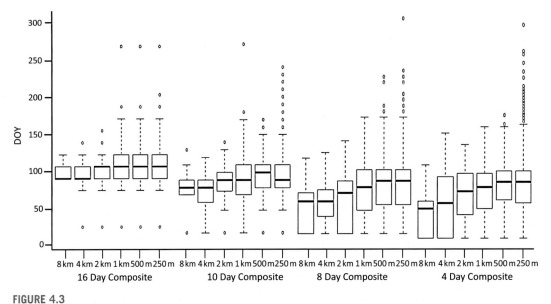

FIGURE 4.3

Boxplot showing the distribution of the estimated SOS DOY within the broadleaf and deciduous woodland vegetation class.

The coarse 8 km resolution estimates an earlier SOS when compared with the 250 m resolution, which is reflected across all composite periods. The range of estimated mean DOY is 63 days, indicating sensitivity to both spatial resolution and composite period.

The 4-day composite has the largest variance and increased outliers when compared with the 8-, 10-, and 16-day composites, of which the 16-day composite has the lowest variance (Fig. 4.3). As expected, there is an increase in outliers and increased variance from the mean of SOS estimates in the 250-m spatial resolution data, compared with the 8-km data. Within this vegetation class the earliest estimated SOS DOY is 38, estimated from the 4-day composite. This would place it as the February 7,

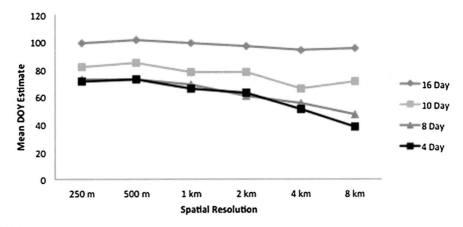

FIGURE 4.4

Mean SOS DOY estimates.

Table 4.2 Estimated EOS mean DOY for broad-leaved woodland					
		Composite Period			
Spatial Resolution		**16 d**	**10 d**	**8 d**	**4 d**
	250 m	386	392	393	397
	500 m	387	396	394	401
	1 km	382	382	390	402
	2 km	377	393	383	401
	4 km	371	382	371	383
	8 km	363	387	362	385
DOY, *day of year;* EOS, *end of season.*					

which would typically be too early for the seasonal change of spring to be occurring in the United Kingdom, whereas the later estimate from the 16-day composite, 101, would place it as April 11 (within the assumed period for spring).

Composite period has a greater influence on the sensitivity of estimated SOS date from remotely sensed observations, as there is a greater range in mean estimated DOY, 28–57 days, compared with that for spatial resolution of 7–35 days (Fig. 4.4). There is a distinct variation in the sensitivity of spatial resolution depending on the chosen composite period. The estimated DOY from the 16-day composite indicates a much lower variance with a variation in spatial resolution compared with the 4-day composite.

For EOS the 16-day composite estimated an earlier EOS (between 387 and 363), compared with the EOS estimations from the 4-day composite (between 402 and 383) (Table 4.2). In addition, the coarse spatial resolution of 8 km estimated an earlier EOS. The range of EOS DOY is from 362 to 402, indicating that the EOS occurred between late December 2005 to the beginning of February 2006,

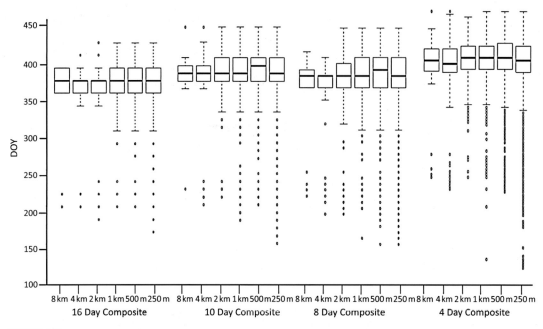

FIGURE 4.5

Boxplot highlighting the dispersion of estimated EOS DOY across all chosen values of spatial resolution and composite period within the broadleaf and deciduous woodland vegetation class. The bottom and top of the box shown by the 25th and 75th percentile, median is the centre line and outliers are represented by individual points.

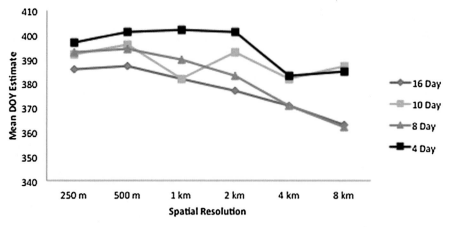

FIGURE 4.6

The estimated mean EOS DOY for the broadleaf and deciduous woodland vegetation class for each chosen metric (shown by the points). The line highlights the trend of the earlier estimation of EOS DOY with the decrease in spatial resolution.

which is very late in the winter season, and not in the assumed senescence of autumn in the United Kingdom (September to November). As with the SOS estimates, across all composite periods there is an increase in variance and outliers with an increase in spatial resolution (Fig. 4.5).

The variability of EOS estimation was 11- to 24-day due to the sensitivity of spatial resolution and between 14 and 32 days due to the sensitivity of composite period estimations. Although the degree of sensitivity is not as strong as with the SOS estimations, this does indicate that EOS phenological estimations are more sensitive to variations in composite period than to spatial resolution (Fig. 4.6).

One of the main observable limitations with the estimation of EOS is that it does not occur within the assumed senescence of autumn (September–November). The estimation of SOS within this study, particularly for the 16, 10, and 8-day composites, is within the window of the British spring time (March–May). For estimating SOS, the dramatic increase in MTCI values, indicating an increase in canopy chlorophyll, allows a more precise estimation of the arrival of spring within the time series. In contrast, EOS occurs gradually over an extended period within the time series, which causes difficulties in estimating remotely sensed LSP EOS. Nevertheless, the bias in the estimations is of less interest here than the sensitivity of the estimations to variation in the inputs.

The OAT approach used in this study to assess the variance in SOS and EOS DOY estimates is a simplistic method to assess the overall sensitivity of estimated phenological parameters. This enables the degree of divergence in the variance of SOS and EOS and, therefore, the impact of the sensitivity of variation in the metrics, to be assessed. With a 63 and 40-day difference in the mean SOS and EOS DOY, due to the sensitivity of phenological parameters to only two input parameters, this indicates the importance of accurate estimation for inclusion in other models (e.g., Global Climate Models (GCMs)). There are limitations with using this simplistic SA; it does not account for any relationship between the input variables, the variation in output variables is conditional on the linear behavior of the model, and it only assesses the selected input(s). However, it enables assessment of specific input parameter(s) and can identify noninfluential inputs before using more complex SA methods.

Global approaches are able to identify interactions with models, as they estimate the output(s) with all parameters varied (Cariboni et al., 2007). However, they are computationally extensive, whereas local methods require a lower number of derivatives. The utilization of a simplistic approach, by quantifying the importance and influence of input parameters, evaluates individual factors that need to be assessed further to establish a precise model output (Crosetto et al., 2000). Within this case study, the results highlight that composite period has an increased influence on the variation of both SOS and EOS estimations and that further assessment is required to assess the accuracy of results, and the relationship of all input parameters to variance in the results.

6. CONCLUSION

This chapter highlights the sensitivity of satellite-derived phenological parameter estimates to variation in input variables within the processing chain and, in particular, to variation in the choice of spatial resolution and composite period. Overall, remotely sensed phenological parameters were found to be more sensitive to variation in composite period, as less variation occurred between spatial resolution estimates than across all composite periods for both SOS and EOS.

For SOS, a coarser spatial resolution (8 km) led to an earlier SOS, whereas a coarser composite period (16 day) led to a later SOS. For EOS, the 16-day composite and the coarse spatial resolution led to an earlier EOS. However, there are known limitations in the detection of EOS from satellite-derived data. This is one area where LSP studies need to improve to provide precise estimates of the length of the growing season.

The simplistic nature of the SA used does not allow for the assessment of interactions between the factors within the model. However, this linear approach does highlight the variance of outputs as a function of variation in both spatial resolution and composite period. With a mean variance of up to 63 days for the estimated SOS and up to 40 days for EOS, due to the variance of the two input parameters, this highlights the possible inaccuracy of current phenological estimations. With the inclusion of variants across all steps within the remote sensing processing technique and all available methodologies, including the choice of satellite sensors, vegetation indices, and filtering techniques, the sensitivity of phenological parameters and the relationship of input parameters within the modeling process requires further and more in-depth assessment.

Future research should include variants of all input parameters used, and the comparison of satellite-based estimates to in-situ ground observations. By comparing satellite-derived SOS and EOS to ground data, the present SA can be extended to an assessment of the accuracy of the estimates conditional upon model choices. Specifically, bias can be assessed and the resolution of both parameters and other model choices can be optimized for phenological studies, as the precise and accurate assessment of phenological events is vital in assessing vegetation development, and the effects of climate and environmental change.

REFERENCES

Atkinson, P.M., Jeganathan, C., Dash, J., Atzberger, C., 2012. Inter-comparison of four models for smoothing satellite sensor time-series data to estimate vegetation phenology. Remote Sensing of Environment 123, 400–417.

Barati, S., Rayegani, B., Saati, M., Sharifi, A., Nasri, M., 2011. Comparison the accuracies of different spectral indices for estimation of vegetation cover fraction in sparse vegetated areas. The Egyptian Journal of Remote Sensing and Space Sciences 14, 49–56.

Beck, P.S.A., Atzberger, C., Hogda, K.A., Johansen, B., Skidmore, A.K., 2006. Improved monitoring of vegetation dynamics at very high latitudes: a new method using MODIS NDVI. Remote Sensing of Environment 100, 231–334.

Cacuci, D.G., 2003. Sensitivity and Uncertainty Analysis. CRC Press Company, US.

Cariboni, J., Gatelli, D., Liska, R., Saltelli, A., 2007. The role of sensitivity analysis in ecological modelling. Ecological Modelling 203 (1), 167–182.

Chen, J., Jonsson, P., Tamura, M., Gu, Z., Matsushita, B., Eklundh, L., 2004. A simple method for reconstructing a high-quality NDVI time-series data set based on the Savitsky–Golay filter. Remote Sensing of Environment 91, 331–344.

Cleland, E.E., Chuine, I., Menzel, A., Mooney, H.A., Schwartz, M.D., 2007. Shifting plant phenology in response to global change. Trends Ecology and Evolution 22, 357–365.

Crosetto, M., Tarantola, S., Saltelli, A., 2000. Sensitivity and uncertainty analysis in spatial modelling based on GIS. Agriculture, Ecosystems & Environment 81 (1), 71–79.

Dash, J., Jeganathan, C., Atkinson, P.M., 2010. Satellite sensor derived spatio-temporal variation in vegetation phenology over India. Remote Sensing of Environment 114, 1388–1402.

Dash, J., Curran, P.J., 2004. The MERIS terrestrial chlorophyll index. International Journal of Remote Sensing 25, 5403–5413.

Delbart, N., Beaubien, E., Kergoat, L., Toan, T.T., 2015. Comparing land surface phenology with leafing and flowering observations from the PlantWatch citizen network. Remote Sensing of Environment 160, 273−280.

Dungan, J.L., 2002. Toward a comprehensive view of uncertainty in remote sensing analysis. In: Foody, G., Atkinson, P. (Eds.), Uncertainty in Remote Sensing and GIS. Wiley, Chichester, pp. 25−37.

Duveiller, G., Defourny, P., 2010. A conceptual framework to define the spatial resolution requirements for agricultural monitoring using remote sensing. Remote Sensing of Environment 114 (11), 2637−2650.

Fisher, J.I., Mustard, J.F., 2007. Cross-scalar satellite phenology from ground, Landsat, and MODIS data. Remote Sensing of Environment 109, 261−273.

Foody, G.M., Dash, J., 2010. Estimating the relative abundance of C3 and C4 grasses in the great plains from multi-temporal MTCI data: issues of compositing period and spatial generalizability. International Journal of Remote Sensing 31, 351−362.

Gao, B.C., 1996. NDWI—a normalized difference water index for remote sensing of vegetation liquid water from space. Remote Sensing of Environment 58, 257−266.

Hamby, D.M., 1994. A review of techniques for parameter sensitivity analysis of environmental models. Environmental Monitoring and Assessment 32, 135−154.

Hamunyela, E., Verbesselt, J., Roerink, G., Herold, M., 2013. Trends in spring phenology of Western Europe deciduous forest. Remote Sensing 5, 6159−6179.

Hird, J., McDermid, G.J., 2009. Noise reduction of NDVI time series: an empirical comparison of selected techniques. Remote Sensing of Environment 113, 248−258.

Huete, A.R., 1988. A soil-adjusted vegetation index (SAVI). Remote Sensing of Environment 25, 295−309.

Iooss, B., Lemaître, P., 2015. A review on global sensitivity analysis methods. In: Meloni, C., Dellino, G. (Eds.), Uncertainty Management in Simulation-Optimization of Complex Systems: Algorithms and Applications. Springer, US.

Jeong, S.-J., Ho, C.-H., Gim, H.-J., Brown, M.E., 2011. Phenology shifts at start vs. end of growing season in temperate vegetation over the Northern Hemisphere for the period 1982−2008. Global Change Biology 17, 2385−2399.

Jiang, Z., Huete, A.R., Didan, K., Miura, T., 2008. Development of a two-band enhanced vegetation index without a blue band. Remote Sensing of Environment 112, 3833−3845.

Jonsson, P., Eklundh, L., 2002. Seasonality extraction by function fitting to time-series of satellite sensor data. IEEE Transactions on Geoscience and Remote Sensing 40, 1824−1832.

Justice, C.O., Townshend, J.R.G., Holben, B.N., Tucker, J.C., 1985. Analysis of the phenology of global vegetation using meteorological satellite data. International Journal of Remote Sensing 6, 1271−1318.

Kross, A., Fernandes, R., Seaquist, J., Beaubien, E., 2011. The effect of the temporal resolution of NDVI data on season onset dates and trends across Canadian broadleaf forests. Remote Sensing of Environment 115, 1564−1575.

Piao, S., Friedlingstein, P., Ciais, P., Viovy, N., Demarty, J., 2007. Growing season extension and its impact on terrestrial carbon cycle in the Northern Hemisphere over the past 2 decades. Global Biogeochemical Cycles 21, GB3018.

Pouliot, D., Latifovic, R., Fernandes, R., Olthof, I., 2011. Evaluation of compositing period and AVHRR and MERIS combination for improvement of spring phenology detection in deciduous forests. Remote Sensing of Environment 115, 158−166.

Reed, B.C., Brown, J.F., VanderZee, D., Loveland, T.R., Merchant, J.W., Ohlen, D.O., 1994. Measuring phenological variability from satellite sensor imagery. Journal of Vegetation Science 5, 703−714.

Richardson, A.D., Keenan, T.F., Migliavacca, M., Ryu, Y., Sonnentag, O., Toomey, M., 2013. Climate change, phenology, and phenological control of vegetation feedbacks to the climate system. Agricultural and Forest Meteorology 169, 156−173.

Saltelli, A., Chan, K., Scott, M., 2000. Sensitivity Analysis. John Wiley and Sons, New York.

Saxe, H., Cannell, M.G., Johnsen, Ø., Ryan, M.G., Vourlitis, G., 2001. Tree and forest functioning in response to global warming. New Phytologist 149, 369–399.

Schott, J.R., 2007. Remote Sensing: The Image Chain Approach. Oxford University Press, Oxford.

Sparks, T., Jeffree, E., Jeffree, C., 2000. An examination of the relationship between flowering times and temperature at the national scale using long-term phenological records from the UK. International Journal of Biometeorology 44, 82–87.

Testa, S., Boschetti, L., Borgogno Mondino, E., 2015. Modis EVI, NDVI, WDRVI, daily and composite: looking for the best choice to estimate phenological parameters from deciduous forests. 2015 IEEE International Geoscience and Remote Sensing Symposium (IGARSS) 4617–4620.

White, M.A., Brunsell, N., Schwartz, M.D., 2003. Vegetation phenology in global change studies. In: Schwartz, M.D. (Ed.), Phenology: An Integrative Environmental Science. Kluwer Academic Publishers, London.

White, M.A., de Beurs, K.M., Didan, K., Inouye, D.W., Richardson, A.D., Jensen, O.P., O'Keefe, J., Zhang, G., Nemani, R.R., van Leeuwen, W.J.D., Brown, J.F., de Wit, A., Schaepman, M., Lin, X., Dettinger, M., Bailey, A.S., Kimball, J., Schwartz, M.D., Baldocchi, D.D., Lee, J.T., Lauenroth, W.K., 2009. Intercomparison, interpretation, and assessment of spring phenology in North America estimated from remote sensing for 1982–2006. Global Change Biology 15, 2335–2359.

Xiao, J., Moody, A., 2005. Geographical distribution of global greening trends and their climatic correlates: 1982–98. International Journal of Remote Sensing 26, 2371–2390.

Zhang, X., Friedl, M.A., Schaaf, C.B., 2009. Sensitivity of vegetation phenology detection to the temporal resolution of satellite data. International Journal of Remote Sensing 30, 2061–2074.

Zeng, H., Jia, G., Epstein, H., 2011. Recent changes in phenology over the northern high latitudes detected from multi-satellite data. Environmental Research Letters 6, 1–11.

RADAR RAINFALL SENSITIVITY ANALYSIS USING MULTIVARIATE DISTRIBUTED ENSEMBLE GENERATOR*

5

Q. Dai[1,2], D. Han[2], P.K. Srivastava[3,4]

Nanjing Normal University, Nanjing, China[1]; University of Bristol, Bristol, United Kingdom[2]; NASA Goddard Space Flight Center, Greenbelt, MD, United States[3]; Banaras Hindu University, Varanasi, Uttar Pradesh, India[4]

CHAPTER OUTLINE

1. INTRODUCTION

Sensitivity analysis (SA) tries to explore and quantify the impact of errors associated with inputs on model outputs (Blasone et al., 2007). This term may be confused with uncertainty analysis (UA), although they can be distinguished from each other. SA tries to explore and quantify the impact of errors associated with inputs on model outputs. The latter is also an important issue in hydrological model, which is generally used alongside with SA. Model UA is the discipline that aims to investigate

---

*Portions of this chapter were previously published as a research article in the Elsevier journal *Atmospheric Research:* Dai, Q., Han, D., Zhuo, L., Huang J., Islam, T., Srivastava, P.K., 2015. Impact of complexity of radar rainfall uncertainty model on flow simulation. *Atmospheric Research*, 161–162: 93–101.

Sensitivity Analysis in Earth Observation Modelling. http://dx.doi.org/10.1016/B978-0-12-803011-0.00005-7

the degree of confidence in the mathematical model outputs and system performance indices given the model input conditions such as observations and parameters (Helton et al., 2006; Saltelli et al., 2000). It employs the probability distributions of model outputs derived by the probabilistic descriptions of model inputs (Blasone et al., 2007). The effect of input uncertainty and parameter sensitivity on output uncertainty is illustrated in Fig. 5.1. In the figure, one can observe that with the same input uncertainty, the output uncertainty with high sensitivity is greater than that with low sensitivity. The sensitivity is mainly reflected by the variation of parameters. Uncertainty involves the notion of randomness, indicating that one can only express a random variable in terms of the likelihood or probability within some specified range of values, instead of a certain value.

Currently, radar rainfall SA is mainly used to investigate the impact of radar rainfall uncertainty (RRU) on hydrological models. Since radar measures rainfall remotely and indirectly, it is challenging to quantify radar data uncertainty, and many methods have been proposed to describe the uncertainty of radar rainfall (Harrold et al., 1974; Kitchen et al., 1994; Villarini and Krajewski, 2010). The RRU is commonly modeled with the help of reference ground measurement such as rain gauge. The bias between radar and gauge rainfall is considered in almost all the published RRU models (Anagnostou et al., 1998; Borga and Tonelli, 2000; Ciach et al., 2000; Collier, 1986; Habib et al., 2008; Harrold et al., 1974; Seo et al., 1999; Smith and Krajewski, 1991). The error variance of radar rainfall was recognized and calculated in the early 1990s (Barnston, 1991; Ciach and Krajewski, 1999; Ciach, 2003; Habib et al., 2004). Considering the fact that the RRU increases with the growth of rainfall intensity, the rainfall intensity is regarded as an important term in some RRU models (Ciach et al., 2007; Habib and Qin, 2013; Villarini et al., 2009, 2010). Later, the spatial dependence (SD) of uncertainty between different radar grids was integrated to the models (AghaKouchak et al., 2010a,b;

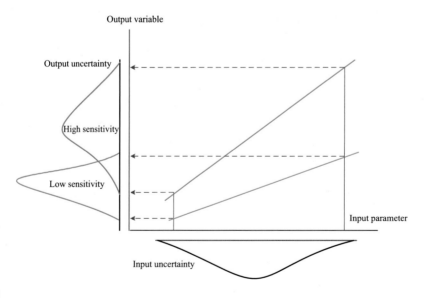

FIGURE 5.1

Schematic diagram showing relationship among model input parameter uncertainty and sensitivity to model output variable uncertainty.

Villarini et al., 2009), and temporal correlation has also been considered (Dai et al., 2014a,b; Germann et al., 2009; He et al., 2011). In the current study, the influence of synoptic regime such as seasons and wind on RRU is analyzed (Ciach et al., 2007; Dai et al., 2014c; Habib and Qin, 2013). The general assumption of the Gaussian distribution of rainfall residual errors is questioned, and the scheme to tackle non-Gaussian uncertainty is proposed (Dai et al., 2014a).

To integrate the RRU model with hydrological model, it is important that the RRU model should only reflect its most important features. Analysis of the uncertainty propagation of radar rainfall through hydrological models can demonstrate the RRU characteristics that flow simulation is sensitive to. In this chapter, we have designed a comprehensive experiment to analyze the sensitivity of flow simulation to the important features of RRU. We have implemented an RRU model named Multivariate Distributed Ensemble Generator (MDEG) with different model complexities and generated ensemble rainfall values under different designed situations. The generated rainfall ensemble members are then fed into a hydrological model to investigate the sensitivity of simulated flow to RRU.

2. DATA AND METHODS

2.1 STUDY AREA AND DATA SOURCES

A dense rain gauge network in the Brue catchment is used for the investigations carried out in this work. The Brue catchment is located in Somerset, South West of England. There are 49 Casella tipping bucket rain gauges deployed in a drainage area of 135 km^2 (see Fig. 5.2). The choice of catchment is based on the availability of quality data; furthermore, the characteristics of the Brue catchment are considered to be representative of rural UK catchments used for rainfall runoff modeling.

The rainfall is recorded up to a time resolution of 10 s with bucket size of 0.2 mm. The data sets are available from September 1993 to April 2000 (provided by the British Atmospheric Data Centre). It is not only because the rain gauge network is designed by considering the locations of radar pixels, thus ensuring the density and distribution of rain gauges are quite suitable for analyzing the radar-gauge rainfall relationship, but also the radar and gauge data sets are trustworthy with carefully quality control. A river gauging station located at Lovington provides the flow data. There is an automatic weather station within the catchment that records solar and net radiation, wet and dry bulb temperature, wind speed and direction, and atmospheric pressure every 15 min. The radar rainfall, gauge rainfall, flow, and other weather data from October 1993 to October 1998 are used as the calibration data, while the data sets covering the period from November 5, 1998, to February 7, 1999, are used to evaluate the proposed scheme. The data set used for the Xinanjiang (XAJ) model is shown in Table 5.1.

2.2 THE MULTIVARIATE DISTRIBUTED ENSEMBLE GENERATOR

The MDEG is an RRU model proposed by Dai et al., (2014b). It is designed to model the RRU and express it in the form of ensemble rainfall values. In general, the RRU model includes (1) calculating the probability distribution of "true" rainfall once given a radar estimate, which reveals the degree of confidence in the radar products; (2) investigating the relationship between radar-rainfall uncertainty and rainfall intensity; (3) analyzing the characteristics of the SD and temporal dependence (TD) of uncertainty model; and (4) investigating other possible factors that affect radar rainfall uncertainty such as

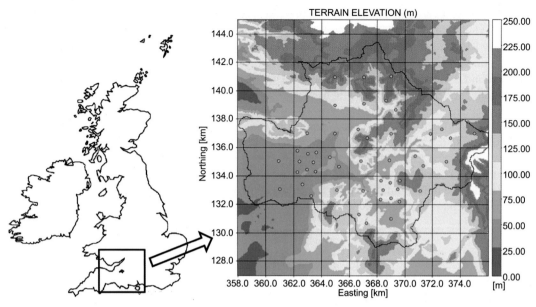

FIGURE 5.2

(Left) The map shows the locations of the study area. (Right) The map displays the rain gauge network and the study area with the terrain elevation in the background. The *dots* represent the rain gauge locations, and the *grids* represent the radar grids (2 km × 2 km) based on the British National Grid coordinate system.

Table 5.1 Data Set for the Flow Simulation Using the Xinanjiang Model

Datasets	Description
Rainfall	Radar: HYREX project, C-band radar
	Rain gauge: HYREX project, tipping bucket gauges
Stream flow	HYREX project, river gauging station
Digital elevation model	Ordnance Survey of Digimap Resource Centre
Ground weather data (e.g. solar radiation, temperature and wind)	HYREX project, automatic weather station
Land cover	Global Land Cover 2000 Project

wind, seasons, and storm types. In this chapter, we generate a large number of random fields to represent the radar-rainfall uncertainty based on the Monte Carlo method. In quantifying the uncertainties in a given radar-rainfall estimate, we assume that the true areal averaged pixel-scale rainfall is composed of two components, namely, the deterministic distortion component and the random component.

The MDEG model has three key advantages compared with other RRU models. First, the MDEG can be built under different situations. In addition, for most current RRU models, the assumption of the Gaussian distribution of residual errors is a fundamental premise, although this assumption cannot always be satisfied. MDEG can deal with both Gaussian and non-Gaussian situations. More importantly, MDEG is based on the Bayesian ensemble generator theory. In other words, the outcomes of MDEG are a number of time series ensemble rainfall values, which are convenient to be integrated into a hydrological model. For these reasons, the MDEG is a very useful tool for investigating the impact of RRU on flow simulation.

2.3 THE XINANJIANG MODEL

The XAJ model presented by Ren-Jun Zhao (Zhao et al., 1980) has been widely used in humid and semihumid regions (Zhao, 1992). The conventional XAJ model is a conceptual lumped rainfall-runoff model. There are three key components of the XAJ model. The first one is the evapotranspiration component. The model includes three soil layers (upper, lower, and deep). The second component is the runoff generation. Runoff at a point starts to appear only on repletion of the tension water storage at that point. The last one is the flow routing. The flow process is performed by the convolution of time through solving the Muskingum method. The spatially distributed XAJ model that integrates the features of a conventional lumped rainfall-runoff model and a physically based flow concentration scheme is used in this study. The model is calibrated with rain gauge measurements as input and spatially estimated using the Thiessen polygons method. More details regarding model calibration are provided in the article by Dai et al. (2014b). The hourly data from October 1993 to October 1998 are used to calibrate the XAJ model.

3. METHODOLOGY
3.1 EXPERIMENTAL DESIGN

There are three components of the MDEG model, namely, marginal distribution (MD), SD, and TD of RRU. The SD and TDs actually refer to the joint distribution among different lag distances or lag times. In this work, we have designed nine scenarios to analyze the following three issues (see Table 5.2).

The first task is to investigate the importance of the RRU model in flow simulation. To achieve this, we use the original radar data in Scenario 1. By comparing the results of Scenario 1 to those of Scenarios 2–9, we can find the difference of the simulated flows driven by the original radar data and postprocessed radar data, respectively.

The impact of the distribution of three components of MDEG on flow simulation is also investigated. In Scenarios 2–6, we use the Gaussian model in simulating the MD. In Scenarios 7–9, a non-Gaussian model based on the generalized extreme value (GEV) distribution is applied. For the distributions of SD and TD, the Gaussian copula and t-copula are used herein. Scenarios 3, 5, and 8 use the Gaussian copula, whereas Scenarios 4, 6, and 9 adopt the t-copula. It is worth remarking that MDEG uses the AR(2) model to generate the ensemble rainfall values with the desired temporal correlation. As AR(2) requires the residual error in the Gaussian distribution, there is no TD component when the MD is simulated using the GEV model.

Finally, it is useful to check if it is necessary to include the SD and TD components in the MDEG model. There are four subgroups designed to solve it, namely, Group A (Scenarios 2, 3, and 5), Group B (Scenarios 2, 4, and 6), Group C (Scenarios 7 and 8), and Group D (Scenarios 7 and 9). In Group A, the Gaussian MD and Gaussian copula are assumed in the MDEG model. Among these scenarios, we

Table 5.2 Configuration Scenarios of the Radar Rainfall Uncertainty Model With Different Components

Scenario	Marginal Distribution		Spatial Dependence		Temporal Dependence	
	Gaussian	GEV	Gaussian Copula	T-copula	Gaussian Copula	T-copula
Scenario 1						
Scenario 2	√					
Scenario 3	√		√			
Scenario 4	√			√		
Scenario 5	√		√		√	
Scenario 6	√			√		√
Scenario 7		√				
Scenario 8		√	√			
Scenario 9		√		√		

GEV, *generalized extreme value.*

only consider MD in Scenario 2, both MD and SD in Scenario 3, and TD in Scenario 5. In Group B, we use the same Gaussian MD but t-copula for the SD and TD components. The same as in Group A, we consider solely MD (Scenario 2), MD plus SD (Scenario 4) and all three components together (Scenario 6). Groups C and D are designed for the GEV MD. As mentioned already, the TD component cannot be used herein. To reduce the simulated times, SD and TD components are designed to use the same copula.

To implement these nine scenarios, we need to first build the MDEG model with the calibrated data as described earlier and then run it using the validated data. The MDEG model is built with four situations: (1) Gaussian MD with Gaussian copula for simulating SD and TD, (2) Gaussian MD with t-copula, (3) GEV MD with Gaussian copula, and (4) GEV MD with t-copula. After estimating the parameters of these situations, ensemble rainfall values of eight scenarios (except Scenario 1) are generated under the configurations of scenarios. To better describe the SD, the spatial correlation coefficients of rainfall uncertainty are parameterized with a three-parameter exponential function (Dai et al., 2014b,c):

$$\rho(\Delta s) = \rho_0 \exp\left(-\left(\frac{\Delta s}{R_0}\right)^F\right) \tag{5.1}$$

where Δs is the lag distance, ρ_0 refers to the immediate correlation jump, F is the shape coefficient of the parametric model, and R_0 is the long correlation radius.

3.2 VERIFICATION METHOD

The ensemble rainfall values generated under the designed eight scenarios (Scenario 1 uses the original radar rainfall) are input into the hydrological model to produce ensemble flows. Then the basic features of the uncertainty band of the ensemble flows are investigated. The uncertainty bands are demonstrated by two major indicators, spread of the band and ensemble bias. The former reveals the extension of the generated ensemble members, which is classified by the dispersion statistic δ_D and the normalized dispersion statistic δ_{ND}. Their definitions are given as (Dai et al., 2014c):

$$\delta_D = \frac{1}{nt} \sum_{i=1}^{nt} [P_{95}(t) - P_5(t)] \quad P_{50}(t) > 0 \tag{5.2}$$

$$\delta_{ND} = \frac{1}{nt} \sum_{i=1}^{nt} \left[\frac{P_{95}(t) - P_5(t)}{P_{50}(t)} \right] \quad P_{50}(t) > 0 \tag{5.3}$$

where nt represents the number of time steps when P_{50} greater than 0. $P_{95}(t)$, $P_5(t)$, and $P_{50}(t)$ correspond to the 95th, 5th, and 50th percentiles of the ensemble members at time t, respectively. The ensemble bias indicates the error of the medium ensemble members compared with the reference observed data P^{ref}. To evaluate the performance of the ensemble flows, we use three statistics to represent the ensemble bias, namely, mean error δ_{ME}, mean square error δ_{MSE}, and overall bias δ_{OS}, which are expressed as below:

$$\delta_{ME} = \frac{1}{n} \sum_{i=1}^{n} \left(P_{50}(t) - P^{ref}(t) \right) \tag{5.4}$$

$$\delta_{MSE} = \frac{1}{n} \sum_{i=1}^{n} \left(P_{50}(t) - P^{ref}(t) \right)^2 \tag{5.5}$$

$$\delta_{OB} = \frac{\sum_{i=1}^{n} P_{50}(t)}{\sum_{i=1}^{n} P^{ref}(t)} \tag{5.6}$$

where n is the number of time steps. Obviously, a better group of the ensemble flows should have a smaller mean error and mean square error. A value of 0 means that the reference flow is well encompassed in the ensemble members. For overall bias, a value of 1 corresponds to the best case.

4. RESULTS AND DISCUSSION
4.1 IMPLEMENTATION OF ENSEMBLE FLOW GENERATION

First, the ensemble rainfall members are generated using the MDEG model, and then they are input to the XAJ model to produce ensemble flow values. The parameters of MDEG are estimated for the MD, SD, and TD components and for different distributions. A large number of ensemble rainfall values for eight scenarios (except Scenario 1) are generated. A total of 500 realizations are used herein, which then produce 500 ensemble flows, respectively. The MDEG and XAJ models are driven to run from November 1998 to February 1999 with the first month as the warming up period. The data are not fully continuous because the radar data may be invalid in some time steps.

The uncertainty bands of the ensemble rainfall using the MDEG model with Gaussian MD and Gaussian copula for radar grid 1 are shown in Fig. 5.3. Radar grid 1 is the top left grid that contains a rain gauge in Fig. 5.2 (its center coordinate is Easting: 365 km, Northing 141 km). We can observe three uncertainty bands for Scenarios 2, 3, and 5. These scenarios include one, two and three components of the MDEG model respectively. The observed radar data are also drawn in black dots. The solid line refers to the deterministic component of radar rainfall for Scenario 5. In the MDEG

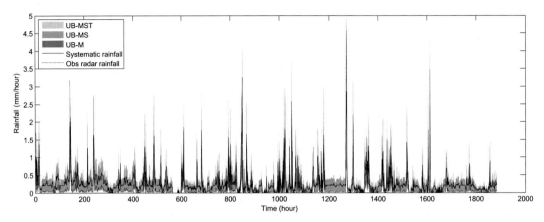

FIGURE 5.3

Uncertainty bands of ensemble rainfall using the MDEG model with Gaussian marginal distribution and Gaussian copula for radar grid 1. The observed radar data are also shown with *black dashes*. The *solid line* refers to the deterministic component of radar rainfall for Scenario 5. The *gray shaded* areas correspond to the uncertainty bands for Scenarios 2 (UB-M), 3 (UB-MS), and 5 (UB-MST).

model, the SD component mainly governs the relationship of the generated members between different radar grids. When considering the TD component, the generated value in the next time step should be constrained by the one from the previous time step. However, we can only find slight change of three uncertainty bands, indicating that the SD and TD components have little impact on the width of the RRU in a certain radar grid.

Figs. 5.4 and 5.5 display the uncertainty bands of simulated flows for the Gaussian and GEV MDs, respectively. A 95% confidence interval is chosen. In these figures, the black dots are the observed flows, whereas the solid line represents the simulated flow using the original radar data. In Fig. 5.4, the gray shaded areas correspond to the uncertainty bands for the MDEG model with only the MD component (UB-M), both MD and SD components (UB-MS), and all three components together (UB-MST). It is observed that the UB-MST has the largest width of uncertainty bands, whereas the UB-M corresponds to the narrowest one. The simulated flows by the original radar data significantly underestimate the realistic flow. The (see Fig. 5.5) characteristics of uncertainty bands of GEV are similar to those of the Gaussian distribution. As the TD component cannot be included in MDEG for GEV MD, we show the uncertainty bands for MDEG that have both MD and SD components using t-copula (UB-MS-T) and Gaussian copula (UB-MS-G), respectively.

4.2 IMPACT OF ERROR DISTRIBUTION ON MODEL OUTPUT

The impact of two kinds of uncertainty distributions on flow simulation is studied, namely, MD and joint distribution. The list of five indicators for features of uncertainty bands of ensemble flow are shown in Table 5.3. First, we compare the differences of uncertainty bands between Gaussian MDs (Scenarios 2–4) and GEV distribution (Scenarios 7–9). For the dispersion statistics, the differences between them are quite small, with an averaged difference of only 0.02. For example, the MDEG model with the Gaussian MD (Scenario 2) has a dispersion of 0.78, whereas that of the one with GEV MD (Scenario 7)

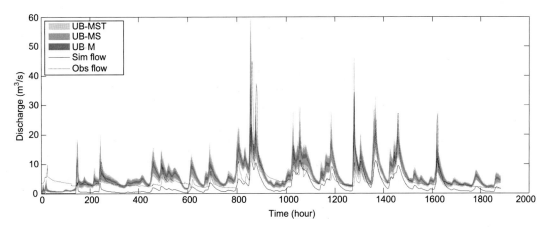

FIGURE 5.4

Uncertainty bands of the ensemble flow using the MDEG model with Gaussian marginal distribution and Gaussian copula. The observed flow data are also shown with *black dashes*. The *solid line* represents the simulated flow using the original radar data.

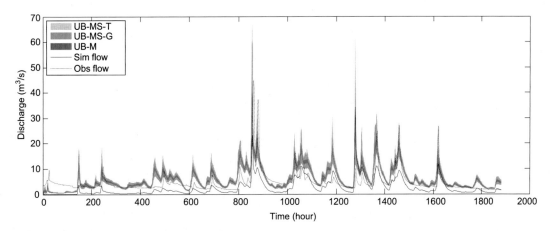

FIGURE 5.5

The same as Fig 5.4, but for GEV marginal distribution. The *gray shaded* areas correspond to the uncertainty bands for the Scenarios 7 (UB-M), 8 (UB-MS-G), and 9 (UB-MS-T).

is 0.80. The differences of the normalized dispersion are larger, but are also within a limited range (0.2). For the joint distribution, we can compare Scenario 3 with 4 (Gaussian MD with SD component), Scenario 5 with 6 (Gaussian MD with both SD and TD components), and Scenario 8 with 9 (GEV MD with SD component). There is no significant difference observed for these three situations.

In terms of ensemble bias statistics (mean error, mean square error, and overall bias), we find that the values in Scenario 1 are all remarkably worse than those in other scenarios. From this point of view, it is essential to apply the RRU model on flow simulation, at least to correct the deterministic distortion of

Table 5.3 List of Five Indicators That Describes the Features of Uncertainty Bands of Ensemble Flow for Different Scenarios

Scenario	Dispersion	Normalized Dispersion	Mean Error	Mean Square Error	Overall Bias
Scenario 1	–	–	−2.26	16.78	0.51
Scenario 2	0.78	0.33	0.88	14.52	1.19
Scenario 3	2.81	0.80	0.85	14.19	1.18
Scenario 4	2.83	0.71	0.87	14.58	1.19
Scenario 5	3.93	1.39	0.84	14.49	1.18
Scenario 6	4.07	1.29	0.82	14.14	1.18
Scenario 7	0.80	0.46	0.92	14.90	1.20
Scenario 8	2.80	0.76	0.81	14.22	1.18
Scenario 9	2.86	0.85	0.80	14.00	1.17

RRU. For other scenarios, the variation of ensemble bias statistics is not significant. Based on these evidences, it is concluded that the MD and joint distribution of RRU do not play an important role in flow simulation.

4.3 IMPACT OF SPATIOTEMPORAL DEPENDENCE ON MODEL OUTPUT

As mentioned previously, to analyze the impact of spatiotemporal dependence on flow simulation we reassign the scenarios to Group A (Scenarios 2, 3, and 5), Group B (Scenarios 2, 4, and 6), Group C (Scenarios 7 and 8), and Group D (Scenarios 7 and 9). For Group A, it is observed that the dispersion and normalized dispersion of Scenario 5 (three components are all considered in MDEG) are significantly larger than those of Scenario 3 (both MD and SD components are considered), whereas those of the latter are considerable larger than those of Scenario 2 (only MD component is considered). The same relationships are found for Groups B, C, and D. In other words, the dispersion of the uncertainty band increases dramatically with the growth of the RRU model's complexity. For ensemble bias, we cannot find considerable changes among the scenarios in Groups A and B. For Groups C and D, the three bias indicators of Scenario 7 are slightly larger than the ones in Scenarios 8 and 9.

In summary, we can conclude that the width of the uncertainty band grows considerably with the complexity of the RRU model. This means that the estimated uncertainty of the model output will grow when more factors are taken into account. After performing the RRU model, the quality of the simulated ensemble flow improves dramatically. Only a slight improvement of the simulated flow is observed when SD (and TD) for GEV MD is considered. Nevertheless, this does not mean that there is no need to model the SD and TD. As the uncertainty band width of the ensemble flow grows significantly when considering spatiotemporal dependence of RRU, it should have a remarkable impact on flow simulation.

5. CONCLUSIONS

In this chapter, we study the sensitivity of flow simulation to key features of RRU. An ensemble generator named MDEG is implemented to generate ensemble rainfall values, which are used as input into a hydrological model (XAJ). A total of nine scenarios are designed to demonstrate the importance of SD and TD, MD, and joint distribution of MDEG on flow simulation. As mentioned earlier, we expected an extensible and simple RRU model for integrating it with other components of the hydrological model uncertainty. It is found that the MDEG model with Gaussian MD and spatiotemporal dependence components simulated using Gaussian copula is preferred.

The final goal of the flow simulation is to aid users to make decisions. Even if some features of input uncertainty do significantly affect the flow simulation, do they also play a key role in decision making? Most hydrological models are low-pass filters, so the high-frequency part of input rainfall is smoothed out in model output. On the other hand, the output of a decision system is generally of limited choices. In other words, a larger number of possible outcomes of the hydrological model are scaled or filtered to a couple of values. For this reason, we need to analyze how the features of RRU behave in the decision system and discover which ones are more sensitive for the final decision making.

REFERENCES

AghaKouchak, A., et al., 2010a. Conditional simulation of remotely sensed rainfall data using a non-Gaussian v-transformed copula. Advances in Water Resources 33 (6), 624–634.

AghaKouchak, A., et al., 2010b. Copula-based uncertainty modelling: application to multisensor precipitation estimates. Hydrological Processes 24 (15), 2111–2124.

Anagnostou, E.N., et al., 1998. Mean-field rainfall bias studies for WSR-88D. Journal of Hydrologic Engineering 3 (3), 149–159.

Barnston, A.G., 1991. An empirical method of estimating raingage and radar rainfall measurement bias and resolution. Journal of Applied Meteorology 30 (3), 282–296.

Blasone, R.-S., et al., 2007. Parameter Estimation and Uncertainty Assessment in Hydrological Modelling. Technical University of Denmark Danmarks Tekniske Universitet, Department of Hydrodynamics and Water ResocurcesStrømningsmekanik og Vandressourcer.

Borga, M., Tonelli, F., 2000. Adjustment of range-dependent bias in radar rainfall estimates. Physics and Chemistry of the Earth, Part B: Hydrology, Oceans and Atmosphere 25 (10), 909–914.

Ciach, G.J., Krajewski, W.F., 1999. On the estimation of radar rainfall error variance. Advances in Water Resources 22 (6), 585–595.

Ciach, G.J., et al., 2000. Conditional bias in radar rainfall estimation. Journal of Applied Meteorology 39 (11), 1941–1946.

Ciach, G.J., 2003. Local random errors in tipping-bucket rain gauge measurements. Journal of Atmospheric and Oceanic Technology 20 (5), 752–759.

Ciach, G.J., et al., 2007. Product-error-driven uncertainty model for probabilistic quantitative precipitation estimation with NEXRAD data. Journal of Hydrometeorology 8 (6), 1325–1347.

Collier, C., 1986. Accuracy of rainfall estimates by radar, Part I: calibration by telemetering raingauges. Journal of Hydrology 83 (3), 207–223.

Dai, Q., Han, D., Rico-Ramirez, M.A., Islam, T., 2014a. Modeling radar-rainfall estimation uncertainties using elliptical and Archimedean copulas with different marginal distributions. Hydrological Sciences Journal 59, 1992–2008.

Dai, Q., Han, D., Rico-Ramirez, M.A., Srivastava, P.K., 2014b. Multivariate distributed ensemble generator: a new scheme for ensemble radar precipitation estimation over temperate maritime climate. Journal of Hydrology 511, 17–27.

Dai, Q., Han, D., Rico-Ramirez, M.A., Islam, T., 2014c. Radar rainfall uncertainty modelling influenced by wind. Hydrological Processes 29, 1704–1716.

Germann, U., et al., 2009. REAL—Ensemble radar precipitation estimation for hydrology in a mountainous region. Quarterly Journal of the Royal Meteorological Society 135 (639), 445–456.

Habib, E., et al., 2004. A method for filtering out raingauge representativeness errors from the verification distributions of radar and raingauge rainfall. Advances in Water Resources 27 (10), 967–980.

Habib, E., et al., 2008. Analysis of radar-rainfall error characteristics and implications for streamflow simulation uncertainty. Hydrological Sciences Journal 53 (3), 568–587.

Habib, E., Qin, L., 2013. Application of a radar-rainfall uncertainty model to the NWS multi-sensor precipitation estimator products. Meteorological Applications 20 (3), 276–286.

Harrold, T., et al., 1974. The accuracy of radar-derived rainfall measurements in hilly terrain. Quarterly Journal of the Royal Meteorological Society 100 (425), 331–350.

He, X., et al., 2011. Statistical analysis of the impact of radar rainfall uncertainties on water resources modeling. Water Resources Research 47 (9).

Helton, J.C., et al., 2006. Survey of sampling-based methods for uncertainty and sensitivity analysis. Reliability Engineering & System Safety 91 (10), 1175–1209.

Kitchen, M., et al., 1994. Real-time correction of weather radar data for the effects of bright band, range and orographic growth in widespread precipitation. Quarterly Journal of the Royal Meteorological Society 120 (519), 1231–1254.

Saltelli, A., et al., 2000. Sensitivity Analysis. Wiley, New York.

Seo, D.-J., et al., 1999. Real-time estimation of mean field bias in radar rainfall data. Journal of Hydrology 223 (3), 131–147.

Smith, J.A., Krajewski, W.F., 1991. Estimation of the mean field bias of radar rainfall estimates. Journal of Applied Meteorology 30 (4), 397–412.

Villarini, G., et al., 2009. Product-error-driven generator of probable rainfall conditioned on WSR-88D precipitation estimates. Water Resources Research 45 (1).

Villarini, G., Krajewski, W.F., 2010. Review of the different sources of uncertainty in single polarization radar-based estimates of rainfall. Surveys in Geophysics 31 (1), 107–129.

Villarini, G., et al., 2010. Towards probabilistic forecasting of flash floods: the combined effects of uncertainty in radar-rainfall and flash flood guidance. Journal of Hydrology 394 (1), 275–284.

Zhao, R.-J., et al., 1980. The Xinanjiang model. In: Hydrological Forecasting Proceedings Oxford Symposium. Hydrological Forecasting Proceedings Oxford Symposium. IAHS, Oxford, pp. 351–356.

Zhao, R.-J., 1992. The Xinanjiang model applied in China. Journal of Hydrology 135 (1), 371–381.

FIELD-SCALE SENSITIVITY OF VEGETATION DISCRIMINATION TO HYPERSPECTRAL REFLECTANCE AND COUPLED STATISTICS

6

K. Manevski[1], M. Jabloun[1], M. Gupta[2], C. Kalaitzidis[3]

Aarhus University, Tjele, Denmark[1]; USRA/NASA Goddard Space Flight Center, Greenbelt, MD, United States[2]; MAICH/CIHEAM, Chania, Greece[3]

CHAPTER OUTLINE

1. INTRODUCTION

Over the past few decades, hyperspectral remote sensing has made breakthrough by recording the reflectance of various Earth features in numerous narrow virtually continuous spectral bands from the visible to the shortwave infrared parts of the electromagnetic spectrum. These remotely sensed data are helpful in discriminating features having similar spectra in the "traditional" multispectral domain. Spectroradiometers operated on the field, or on air- and space-borne platforms, provide spectral reflectance values of features that are of interest, within their field of view on the ground. Hence, spectral libraries are generated for important agroeconomic features, e.g., vegetation types and plant species, and are further utilized to answer the question whether there is a unique hyperspectral signature for a particular feature during its life cycle that can be used for large-scale land resource mapping. However, earlier studies have reported that hyperspectral signatures are not unique for vegetation species and have suggested that several species may have quantitatively similar spectra (Cochrane, 2000; Price, 1994). This is primarily due to intraspecies reflectance variation induced by canopy architecture and internal air, water, and biochemical concentrations (Manevski et al., 2011; Schmidt and Skidmore, 2003). To minimize the effect of spectral variability that is independent of the sources of variability, refined approaches that modify the geometrical shape of the spectral reflectance curve, standardize reflectance peaks, and enhance absorption features are being increasingly utilized. For example, continuum removal or derivative techniques are applied on the spectral reflectance for normalizing reflectance peaks and broadening absorption pits, thus making them potentially more "unique" for a given plant or plant status.

There are many different statistical approaches, with their prerequisites, strengths, and weaknesses, which may be coupled on the vegetation spectral reflectance for discrimination analysis, that is to say, to detect wavelengths at which the reflectance is statistically different between the vegetations being compared. The outcomes of these analyses may be further used by remote sensing studies, e.g., using satellite imagery, for discrimination of vegetation at higher scales (van Aardt and Wynne, 2001, 2007). All methods, nevertheless, aim to reduce the number of wavelengths to the most relevant for discrimination, without the loss of important information related to the study objectives. However, the vast majority of remote sensing studies, dealing with spectral discrimination of vegetation at field scale, first utilize the outcome of the analysis of variance (ANOVA) for the reflectance of vegetation being compared, as a basis for further reduction of the number of significant wavelengths. ANOVA may underlay two fundamental but different approaches: parametric and nonparametric, which makes the final discrimination result potentially sensitive to the choice of statistical test. Moreover, the final results may also depend on whether the ANOVA has been conducted on unaltered or processed spectral reflectance. The effects of the type of reflectance and statistical approach on the spectral discrimination has neither been studied nor summarized, so as to answer how much and to what extent the discrimination of vegetation is sensitive to unaltered or processed reflectance, and to parametric or nonparametric statistical tests.

Sensitivity analysis (SA) is a method of estimating the effect, i.e., the response in the output of a system to (systematic) change in the inputs. As such, it can be applied on a spectral library taken as a system. Specifically, the spectral reflectance can be varied as either unaltered or processed, given as input to either parametric or nonparametric model, for better understanding the discriminatory power of specific wavelengths or wavebands that are sensitive to the previously mentioned variations. Knowledge about the sensitivity of the discriminatory power to the type of reflectance and the associated statistical model is lacking, but it is very important for mapping, monitoring, and overall investigating vegetation condition in space and time using field-, air- and space-borne hyperspectral sensors.

This chapter aims at presenting the sensitivity of field-scale spectral discrimination to the type of reflectance and statistical test used, ultimately aiming to facilitate the vegetation mapping process at higher scales. The chapter is divided into two major parts. The first part provides an overview of the vegetation spectral signature and its important features, followed by an outline of the most common statistical tests used by the majority of studies, published so far, for spectral discrimination of vegetation at field scale. The different types of reflectance, in terms of unaltered and processed, are presented next, as these are one of the main determinants of the final discrimination results—the number and location of wavelengths relevant for discrimination between vegetation. The second part of the chapter provides the results of the first SA of spectral discrimination of vegetation to the type of reflectance and statistical test, across three major climatic zones. This part of the chapter does not interpret in detail the spectral discrimination results or link them with biochemical and biophysical properties of the investigated plants. Instead, it aims to discover if there is a general trend for single preferred statistical test and reflectance type that eases vegetation spectral discrimination, or if discrimination analysis is always case sensitive and plant dependent. Finally, the main conclusions are summarized in relation to the sensitivity of spectral discrimination of vegetation to the type of reflectance and statistical methods and the main challenges aimed at achieving a more precise estimation of the spectral discrimination of vegetation are highlighted.

2. BACKGROUND ON SPECTRAL DISCRIMINATION OF VEGETATION

The hyperspectral reflectance signature of vegetation is the most complex and inconsistent shapewise, when compared with other common land covers such as water, soil, or man-made impervious surfaces, and generally looks similar, irrespective of the vegetation type and health status (Hadjimitsis et al., 2009). The typical signature of vegetation in the visible (VIS) spectrum from 400 to 700 nm is composed of the maximum reflectance (minimum absorption) at 550 nm in the green portion and increased absorption in the red and blue portions, due to the selective reflection/absorption properties of chlorophyll and the auxiliary leaf pigments, such as carotenoids. Furthermore, the points of inflection at about 600 and 630 nm and minimum reflectance (maximum absorption) in the neighborhood of 680 nm is attributed to the presence of specific accessory pigments, whereas beyond 680 nm, the vegetation becomes highly reflective. The red edge (680–800 nm) is the region of rapid change in reflectance before the near-infrared (NIR) spectrum (700–1300 nm). Maximum reflectance is achieved in the NIR plateau from 800 to 1300 nm due to the multiple scattered reflectances in the air spaces—in both the mesophyll cells of the leaves and within the canopy. This is followed by one of the two major atmospheric blinds, i.e., from 1450 to 1530 nm where water vapor strongly absorbs the energy. Hence, the shortwave infrared (SWIR) spectrum (1300–2500 nm) is characterized by much lower reflectance compared with the NIR, and it contains a second major water vapor absorption zone from 1900 to 2000 nm. Other atmospheric gases such as CO_2, O_2, O_3, CH_4, and N_2O also display absorption blinds, but to a much lesser extent (Avery and Berlin, 1992). In addition, the natural anisotropy (direction dependency) of the reflectance introduces the Bidirectional Reflectance Distribution Function (Schaaf, 2009), which describes the change of the reflectance with the solar position geometry.

The remaining part of this section focuses on the statistical methods applied to the vegetation spectral signatures for discrimination purposes. First, an overview is provided of the two main statistical approaches employed for vegetation species discrimination using hyperspectral reflectance data. Then, the hyperspectral reflectance and common mathematical manipulations are outlined. The section closes with case studies that compare, directly or indirectly, the outcome of the spectral discrimination of various vegetation types as affected by the type of reflectance and the statistical tests employed.

2.1 PARAMETRIC VERSUS NONPARAMETRIC STATISTICAL TESTS

Vegetation reflectance is naturally variable, and its distribution may or may not cluster around the mean in a bell shape, if graphically presented. That is to say, the vegetation reflectance in each wavelength may or may not exhibit a normal, i.e., Gaussian distribution (Manevski et al., 2012; Noble and Brown, 2009). As a consequence, statistical techniques coupled with reflectance data for spectral discrimination can be overall grouped into parametric and nonparametric tests. Parametric tests assume normal data distribution and homogeneous variance as a measure of the amount of variation in the data being compared. As violations of some of the assumptions, primarily the data distribution, do not significantly affect the results when large number of samples are used (Robson, 1994), parametric tests are often considered as robust. The single-factor ANOVA (commonly known as one-way ANOVA) is used in the spectral discrimination of vegetation, as plant species or vegetation type is the sole factor to be studied for each wavelength. ANOVA explores the variability between the groups as a deviation of each group's mean from the "grand mean"—the mean of the means of all groups. Posthoc tests are further implemented when significant test results have already been detected by ANOVA, and additional pairwise exploration is needed on which means are significantly different from each other.

When distribution is obviously not bell-shaped, and neither is the variance homogeneous, for example, when one or more reflectance values are off scale, i.e., too high or too low, which is often true, causing variance homogeneity—the most important requirement in parametric ANOVA not to be achieved—alternative approaches like the nonparametric techniques are recommended. Nonparametric analysis ranks the reflectance values from low to high, and ANOVA is performed on the ranks. To this end, the Kruskal−Wallis nonparametric ANOVA compares the medians of the groups being compared. In vegetation spectral discrimination studies, it is most commonly followed by the Mann−Whitney U-test for further pairwise comparisons.

In both parametric and nonparametric ANOVA, the resulting p-value demonstrates the probability of making an error in concluding a difference between the groups being compared, when none really exists. Very few studies have compared the spectral discrimination outcome when both statistical approaches are coupled on the same vegetation reflectance data. More details about the sensitivity of the spectral discrimination outcome on the statistical method are discussed later in this section.

2.1.1 Other Discrimination Methods

The previous ANOVA methods were single factor, whereas other statistical methods may explain spectral differences between vegetation using multiple factors. For instance, principal component analysis (PCA) and canonical discriminant analysis are nonparametric multivariate data analysis techniques that take a cloud of data points analyzed and rotate/cluster it in such a way that the maximum variability is visible. The observations in the same principal component, i.e., cluster are similar in some sense, that the degree of association between two objects is maximal if they belong to the same group and minimal otherwise (Castro-Esau et al., 2004; Holden and LeDrew, 1998). Other quantitative methods attempt not only to determine spectral regions, where different covers are most likely to be statistically discriminated, but also to measure how well different covers can be separated. Such separability measures look at the distance between the class means (e.g., the Euclidean distance), or look at both the differences between the class means and the distribution of the values around those means (e.g., Jeffreys-Matusita or Bhattacharyya distance), at one or more wavelengths at a time.

2.2 UNALTERED VERSUS PROCESSED HYPERSPECTRAL REFLECTANCE

In case no quantitative manipulations of the measured spectral reflectance have been performed, the spectral reflectance is unaltered. Unaltered reflectance of vegetation has been widely studied because it is commonly viewed in a holistic sense, as an indicator for the properties of the landscape and the interaction between the local climate, rocks and soil, landforms, fauna, water and humans, as well as landscape factors, integrated within digital imagery (Schmidt and Skidmore, 2003). This implies that the value of the unaltered spectral reflectance takes into account the aforementioned variation sources (if present, some or all) inherent in the measured data. Therefore, unaltered spectral libraries are most often used as sets of reference spectra to delineate different land covers and mixed communities within a certain spatial extent (Nidamanuri and Zbell, 2011). Unaltered spectral libraries are also easier to link with air or satellite imagery, corresponding to the same or similar time of acquisition. A few publicly available spectral libraries containing unaltered spectral reflectance exist and are available today, such as the US Geological Survey (USGS) spectral library (http://speclab.cr.usgs.gov/), offering a comprehensive collection of field measured and Advanced Spaceborne Thermal Emission and Reflection Radiometer (ASTER)-based spectral signatures of various land targets, primarily soils and minerals (Clark et al., 2007), or SPECCHIO online spectral database system from the Remote Sensing Laboratories of the University of Zurich (http://www.specchio.ch/), including rich metadata sets enclosed within the reference spectra and spectral campaign data to ensure longevity and shareablity of spectral data between research groups. It is worth mentioning that a vast and valuable pool of unaltered spectral libraries of various vegetation across the world exists, as seen through published research articles. However, authors are reluctant to share their spectral libraries, presumably due to project funders' copyrights.

The continuum removal is one of the most common mathematical manipulations of the unaltered spectral reflectance in vegetation remote sensing studies (Petropoulos et al., 2014). It is a method that involves the use of a continuum line as convex hull, fitted over the top of a spectral signature using straight line segments that connect the local maxima. Since the first and last spectral data values are on the hull, the first and last bands in the output continuum-removed data file are equal to 1. Fig. 6.1 gives an overview of an unaltered and continuum-removed spectral reflectance of vegetation. It can be seen on the figure that the continuum removal enhances differences in absorption strength, but that it normalizes the absolute differences in reflectance peaks. Thus, continuum-removed spectral libraries are primarily designed for geomaterials to isolate absorption pits and falls for further spectral analysis, which would otherwise be difficult to detect on unaltered reflectance. For vegetation spectral analysis, this type of reflectance has been used either as a method to standardize spectral libraries obtained at different measurement setups or to potentially improve the spectral discrimination. However, vegetation spectral signature contains several distinct reflectance peaks in each major part of the spectrum, as presented earlier in this section. Hence, continuum removal modifies the spectral signature of vegetation in a way that may or may not improve the discrimination, depending on several factors (Manevski et al., 2011; Petropoulos et al., 2014). In addition, applying the continuum removal on wide spectral range such as the one commonly used in remote sensing from 350 to 2500 nm might affect the emphasis of the absorption features, i.e., some may be missed when slightly low local reflectance maxima do not fall on the continuum curve. Moreover, placing peak in more noisy or poorly defined spectral signature on the continuum curve can result in artificial representation of the absorption features. Some algorithms (ENVI, 2009) are able to define local maxima to increase the likelihood of identifying real absorption features.

Other mathematical processing of unaltered spectral reflectance include various derivatives, used primarily to eliminate the interference from changes in illumination or soil background in the spectral signature or for resolving complex spectra of several target species within individual pixels on digital imagery (Asner et al., 2000). Two derivatives of the unaltered reflectance are commonly used. The first

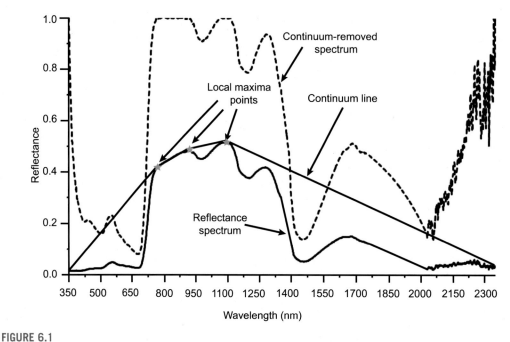

FIGURE 6.1

Vegetation spectral reflectance signature (*curved full line*), its continuum (*straight full line*) with the local maxima (*stars*), and the continuum-removed spectral reflectance signature (*dashed line*).

derivative is calculated for a wavelength intermediate between two wavelengths, whereas the second-derivative determination involves three closely spaced wavelength reflectance values. First and second derivatives emphasize sudden spectrum changes, for example, from a negative slope to a positive slope, in the case of the red edge feature of vegetation reflectance spectrum (Fig. 6.3). Such distinguishable features are especially useful for discriminating between peaks of overlapping bands (Morrey, 1968). Well-described details on derivatives of spectral reflectance can be found in, e.g., Morrey (1968).

2.3 CASE STUDIES FOR EFFECTS OF TYPE OF REFLECTANCE AND STATISTICAL TEST ON THE VEGETATION DISCRIMINATION RESULTS

Information on the spectral discrimination of vegetation at field scale facilitates the mapping process at higher scales, which may be further used by policy decision makers concerning, e.g., environmentally or ecologically protected areas, native habitat mapping and restoration, monitoring of desertification, and land degradation (Adam and Mutanga, 2009; Clark et al., 2005). However, the high variability and similarity of the vegetation spectral signatures imposed difficulties in discriminating vegetation types and species. Scientists realized that altering the geometrical structure of the vegetation spectral signature, i.e., the type of reflectance affects the discrimination results to a variable extent. Moreover, different statistical methods can be applied to the spectral signatures for discrimination analysis, each yielding potentially different results. The sensitivity of vegetation discrimination results to the type of reflectance has been realized since the 1980s (e.g., Clark and Roush, 1984), but more explicitly it has been investigated during 2000s, especially the sensitivity to type of statistical test (Manevski et al., 2011; Prasad and Gnanappazham, 2015). Schmidt and Skidmore (2003) applied nonparametric ANOVA and Mann—Whitney U tests to discriminate between 27 different salt marsh plants in the Netherlands (humid temperate climate) using field spectra in the range 350—2500 nm. Based on the number of statistically significant wavelengths for the plants altogether, the authors concluded that the use of unaltered reflectance increased the discrimination throughout the majority of the spectrum, whereas the use of continuum-removed reflectance narrowed it down as it weakened the statistical differences in the NIR and SWIR spectra. In another study, Manevski et al. (2012) also applied nonparametric ANOVA and Mann—Whitney U tests to discriminate between the hyperspectral reflectance of three Mediterranean tree species in Greece (semiarid continental climate) using field spectra in the range 350—2500 nm. Their study reported that the use of unaltered reflectance narrows the statistical difference between the plants to bands in the VIS and the SWIR spectrum, and weakens the difference in the NIR spectrum, compared with the use of continuum-removed reflectance analysis of the same plants already published (Manevski et al., 2011).

It appears that the differences due to pigment content between vegetation types and plant species are detected in the VIS, and the continuum removal is a useful method to enhance these differences for discrimination purposes. On the other hand, the differences in NIR reflectance between plants caused by canopy structure may be reduced with continuum removal, especially under more humid environments, where canopy structure is more important for discrimination than the internal water content. Variations of internal water of the plants are especially evident in the SWIR, and continuum removal probably will also emphasize detectable differences in this spectrum, especially for plants with highly variable water content in relation to drought adaptations (e.g., Manevski et al., 2011).

Recently, Prasad and Gnanappazham (2015) have compensated the eliminating effect of continuum removal on canopy structure variation by applying continuum removal on additive inverse spectra as a "pseudoabsorption" enhancement. In their study in India, the authors applied parametric and nonparametric statistical analyses on unaltered, continuum-removed, additive inverse and continuum-removed additive inverse reflectance to discriminate between the hyperspectral reflectance of eight mangrove species using field and laboratory spectra in the range 350—2500 nm. The authors indeed concluded that continuum removal of additive inverse spectra results in better discrimination than continuum removed spectra. These authors also concluded that the nonparametric test gave better discrimination between the plants than the parametric test. In fact, the large spectral variability, spatially and temporally, of vegetation seems to be more recognized as being nonparametric, as the majority of studies apply nonparametric ANOVA, even though some researchers apply parametric ANOVA in relation to their data.

Even if normality is assumed, especially when a large number of sampled spectra are available (the central limit theorem), such analysis needs to be reinforced by fulfilled assumption of equal reflectance variances between the vegetation being compared. Manevski et al. (2011) applied both parametric and nonparametric statistical test on a single spectral library, composed of continuum-removed reflectance from 350 to 2400 nm for five Mediterranean trees. Their results showed small effect of the type of statistical test on the discrimination results, although the use of nonparametric tests resulted in better discrimination, as implied by the higher number of wavelengths in the NIR and SWIR spectra.

From this discussion, it is evident that the outcome of a vegetation spectral discrimination depends on the type of reflectance given as input to the statistical test, i.e., whether the reflectance is unaltered or processed, and it also depends further on the type of the statistical test, i.e., parametric or nonparametric. It can also be noticed from these studies that ANOVA detects numerous wavelengths within the 350—2500 nm spectrum relevant for discrimination of various vegetation. Hence, the dimensionality of the results is still high and requires further processing to narrow down the number of wavelengths to the ones most relevant for discrimination. It is not intended in the present study to further select bands or wavelengths that discriminate between the plants the most, because there is no single band selector and the final results would be affected by the choice of band selector. For instance, Prasad and Gnanappazham (2015) used PCA, Manevski et al. (2011, 2012) used the lowest *p*-value, Schmidt and Skidmore (2003) used the highest frequency of statistical significance, and Adam and Mutanga (2009) used classification and regression tree with the purpose of reducing the dimensionality of results. Moreover, none of these techniques maximizes the discrimination accuracy. Therefore, systematic implementation of the multiple types of reflectance and statistical tests on spectral libraries may offer a clear idea about their effect on the vegetation discrimination results, as presented in the next section.

3. SENSITIVITY OF SPECTRAL DISCRIMINATION OF VEGETATION TO THE TYPE OF REFLECTANCE AND STATISTICAL TEST

This section demonstrates quantitatively the importance of reflectance type and statistical method in spectral discrimination using SA of hyperspectral remote sensing data. Three spectral libraries differing in plant species, spectral range, and number of observations were utilized in an extensive discrimination analysis, and the results were summarized and compared. The analysis was implemented as a monofactor SA, where the input, i.e., the reflectance was varied as either unaltered or continuum removed at each spectral wavelength, and the statistical test used to process the reflectance was also varied as either parametric or nonparametric ANOVA.

3.1 HYPERSPECTRAL DATA AND METHOD DESCRIPTION

A brief description of the spectral libraries used in the analysis is presented next.

(1) The first spectral library, named hereafter GREECE, is composed of hyperspectral reflectance of three trees growing in a typical semiarid Mediterranean climate. Details about the hyperspectral reflectance and the method of acquisition can be found in Manevski et al. (2012). Briefly, spectra from *Olea europaea* (old olive tree), *Ceratonia siliqua* (carob tree), and *Pistacia lentiscus* (mastic tree) were acquired by ASD FieldSpec Pro spectroradiometer (Analytical Spectral Devices, Boulder, CO, USA) operating in the range between 350 and 2500 nm during spring 2010 field campaign in Crete, Greece. Data were collected in 5 days in March 2010, at high Sun angle between 10:00 a.m. and 2:00 p.m. and Sun's azimuth angle ranging from 283 to 300 degrees. The spectroradiometer was mounted on the novel carrier lift system MUFSPEM@MED (Mobile Unit for Field SPEctral Measurements at the MEDiterranean; Manakos et al., 2010). Measurements were performed with the sensor positioned approximately 150 cm over the canopy of the trees. The instrument was configured to collect and automatically average 20 samples into a single spectral measurement. At least five such spectral measurements per plant, per day, were recorded over the canopies, after a calibration with a white reference reading from a Spectralon panel.

(2) The second spectral library, named hereafter INDIA, is composed of hyperspectral reflectance of winter wheat grown in typical humid tropical environment. *Triticum aestivum* L. (winter wheat), was sown on December 2, 2008, on silty loam soil and harvested on April 16, 2009. The crop was fertilized and sprayed against pests and diseases. The crop canopy spectral reflectance was measured with ASD Field Spec Pro spectroradiometer (Analytical Spectral Devices) operating in the range between 350 and 1050 nm on several occasions throughout the growth season. For the present analysis, data recorded at three major periods were compared, i.e., 23 January (juvenile), 09 February (early flowering), and 05 March 2009 (early grain filling).

(3) The third data set, named hereafter USA, is composed of laboratory spectral data from leaves of three trees growing in a typical continental climate. These data were downloaded from the United States Geological Survey (USGS) Digital Spectral Library available online (http://speclab.cr.usgs.gov/spectral.lib06/ds231/datatable.html), where description about their hyperspectral reflectance and the method of acquisition can be found. Briefly, laboratory spectra of *Quercus robur* (oak trea), *Populus tremuloides* (aspen tree), and *Pinus contorta* (lodgepole pine) measured with ASD FR in the range between 414 and 2440 nm were downloaded from the online library. The spectral signatures for the oak and the aspen trees were acquired by measuring single leaf, either fresh/green or yellow/dry, lying flat on a strongly absorbing dark surface, whereas for the lodgepole pine, green/dry needles were stacked in a sample cup and the head of the spectrometer fiber-optic cable was passed over the needles.

To investigate the sensitivity of the discrimination results on reflectance type and statistical method, either unaltered or continuum-removed reflectance for each spectral library was given as an input to parametric- and nonparametric ANOVA (Fig. 6.2). For each spectral library and each wavelength, parametric one-way ANOVA was followed by Tukey posthoc test for pairwise comparisons, whereas nonparametric ANOVA was followed by Wilcoxon Rank Sum (which is equivalent to Mann−Whitney U-test) posthoc test for pairwise comparisons. To cope with the presence of ties, a situation often possible to observe when unaltered reflectance is processed to continuum-removed reflectance, a Wilcoxon Rank Sum test using the Shift Algorithm by Streitberg and Röhmel (1986) was used to compute the exact p-values for both tied and untied samples. The outputs, i.e., the p-values, of all statistical tests were indexed 1 when significant at 99% confidence level ($p < .01$) or 0 when insignificant, thus resulting in 0-1 matrix for each spectral library/site. All data processing and analysis was conducted in R statistical platform (CRAN, 2016). The unaltered spectral signatures were transformed into continuum-removed ones using the function "continuumRemoval" from prospectr package (Stevens and Ramirez-Lopez, 2013). The parametric ANOVA was conducted using "aov" and "TukeyHSD" from the stats package (R Core Team, 2016). The nonparametric ANOVA was conducted with kruskal.test from the stats package (R Core Team, 2016) and "wilcox.exact" from exactRankTests package (Hothorn and Hornik, 2000).

The number of observations is an important factor affecting the results of ANOVA-based studies in relation to variability, and it differed markedly between the three data sets, with GREECE covering a large number of observations per plant, USA being the sparsest, and INDIA in between. Also, the number of observations differed within GREECE, being the largest for olive tree and the sparsest for mastic tree (Table 6.1). Hence, the analysis went one step further by investigating the effect of number of observations/replicates on the discrimination results. For this SA the same statistical tests as described earlier were conducted and the results were compared for GREECE data sets differing in their number of observations: all (note that the number of observations is not the same for each tree); 38 observations (equal number for all trees); 10, 20, and 6 (as is the case for INDIA); and 3 (as is the

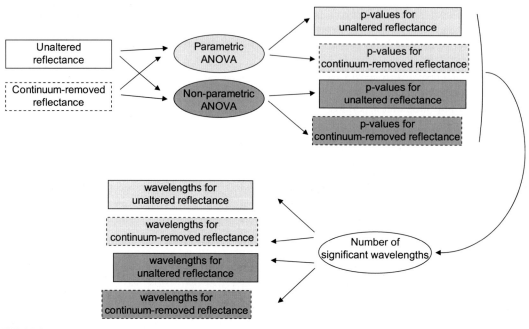

FIGURE 6.2

Overall methodology for the sensitivity analysis of the vegetation discrimination to type of reflectance and associated statistical test. *ANOVA*, analysis of variance.

Table 6.1 Overview of the spectral libraries used in the sensitivity analysis

Location	Site	Plant Species Latin Name	Common Name	Time of Measure-ments	Number of Obser-vations	Spectral Range
GREECE	Akrotiri, Crete	*Ceratonia siliqua*	Carob tree	5−30 Mar 2010	50	350−2350 nm
		Olea europaea	Olive tree		101	
		Pistacia lentiscus	Mastic tree		38	
INDIA	Saliyar village, Roorkee, Uttaranchal	*Triticum aestivum* L.	Winter wheat	23 Jan 2009 09 Feb 2009 05 Mar 2009	6 6 6	350−1050 nm
USA[a]	Lakewood, Colorado	*Quercus robur*	Oak tree	Unknown	2	414−2440 nm
	Boulder, Colorado	*Populus tremuloides*	Aspen tree		3	
	Yellowstone, Wyoming	*Pinus contorta*	Lodgepole pine		2	

[a]*Downloaded files from USGS Spectroscopy Lab library were "Oak Oak-Leaf-1 fresh W1R1Fa AREF" and "Oak Oak-Leaf-2 dried W1R1Fa AREF" for oak, "Aspen Aspen-1 Green-Top W1R1Fa AREF," "Aspen Aspen-3 YellowGreenTop W1R1Fa AREF" and "Aspen Aspen-4 Yellow-Top W1R1Fa AREF" for aspen, and "Lodgepole-Pine LP-Needles-1 W1R1Fa AREF" and "Lodgepole-Pine LP-Needles-3 W1R1Fa AREF" for lodgepole.*

case for USA). The observations were selected randomly for each tree in this data set. The results were interpreted and discussed within spectral library/site and between spectral library/site.

3.2 SENSITIVITY OF VEGETATION SPECTRAL DISCRIMINATION TO REFLECTANCE TYPE AND STATISTICAL METHOD

The sensitivity of the vegetation discrimination in the three spectral libraries to change in spectral reflectance and statistical model for each investigated wavelength is presented in Fig. 6.3. The figure is interpreted as follows: wavelengths (or wavebands) whose color varies within and between spectral libraries have higher sensitivity in their discrimination power, and vice versa. Overall, for the three spectral libraries, the spectral discrimination was more sensitive to the type of reflectance, than to the

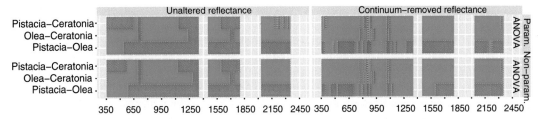

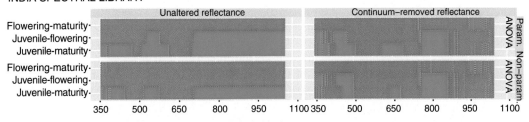

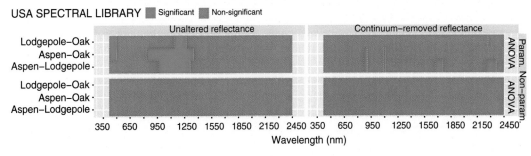

FIGURE 6.3

Sensitivity matrix of vegetation spectral discrimination on type of reflectance (unaltered versus continuum removed) and statistical test [parametric versus nonparametric analysis of variance (ANOVA)] for the three spectral libraries. Description of the spectral libraries can be found in Table 6.1.

type of statistical test. The use of unaltered reflectance resulted in lower number of wavelengths statistically significant for discrimination between the plant species in GREECE, and this was the case irrespective of the type of statistical test, i.e., parametric or nonparametric ANOVA (Fig. 6.3). For USA, however, the situation was the opposite and much more significant wavelengths were detected by the use of unaltered reflectance compared with continuum-removed reflectance. This is because the continuum removal has probably normalized the distinct reflectance peaks present between the spectral signatures of the three trees in this data set, whereas absorption pits were not very distinct between the signatures of the plants. For INDIA, the discrimination results were in between the other two data sets and showed similar response to the type of reflectance, yielding overall 500—980 significant wavelengths for the unaltered reflectance, and 57—660 for the continuum-removed reflectance.

From this, it can be concluded that the spectral discrimination results are case sensitive to the type of reflectance. Yet, the use of continuum-removed reflectance clearly enhances the discrimination in the NIR spectrum, and also in the SWIR spectrum, especially for the spectrally most similar plants/stages in the spectral libraries (e.g., olive tree and mastic tree in GREECE, and aspen and lodgepole in USA; Fig. 6.3).

For INDIA, winter wheat was spectrally discriminated in temporal context, i.e., between three major phenological phases. The use of unaltered reflectance was able to spectrally discriminate between the hyperspectral signature of the same plant measured at juvenile and grain filling/maturity to a higher degree compared with the continuum-removed reflectance. This is somewhat expected since the physiology of winter wheat starts dramatically during grain filling due to assimilate and water translocation from leaves and stem to the grain (Gaju et al., 2014), whereas at the juvenile stage the crop canopy is simply green and mostly dominated by leaves and tillers. During grain filling, wheat canopy gradually dries and turns yellowish, accompanied with defoliation and the developed flower, i.e., the grains are clearly the top of the canopy. Hence, spectral differences between juvenile and maturity stages were very pronounced by the use of unaltered reflectance. Likewise, the spectral signatures of winter wheat compared at juvenile and lowering were also different in the majority of the spectrum, i.e., NIR and SWIR. Hence, for temporal following of cereal crops using hyperspectral remote sensing at field scale, the use of unaltered reflectance may provide satisfactory results, which has also been demonstrated by other studies (e.g., Li et al., 2014, 2015).

In relation to the type of statistical test, even though results of both parametric and nonparametric statistical tests show that the plants in the three spectral libraries considered in this study are spectrally discriminant in many numbers of wavelengths, the results vary among them (Fig. 6.4). This is due to the fact that assumptions made are different in both types of analysis. The nonparametric Wilcoxon rank sum test is found to give a relatively higher number of significant wavelengths compared with the parametric Tukey test. As outlined in the previous section of this chapter, this is because parametric ANOVA requires homogeneity of variance, which is very hard to achieve in the case of vegetation spectra due to high intra- and interspecies variation in reflectance. Nonparametric ANOVA seems comparatively flexible, and it depends on ordinals of values rather than the actual value, which makes it easy to analyze.

The effect of the statistical test was not so evident for the discrimination between different plants in GREECE, because both parametric and nonparametric statistical tests resulted in similar number of bands, with small variations (Fig. 6.4). For the other two data sets, however, the type of statistical test had a pronounced effect on either the temporal discrimination of the same plant (INDIA data set) or the

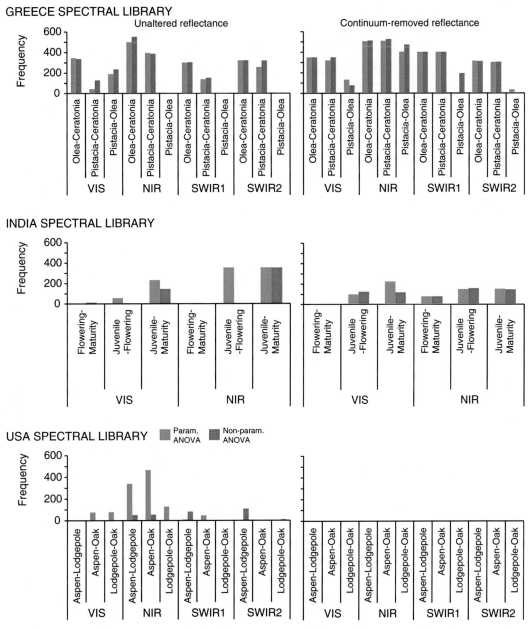

FIGURE 6.4

Frequency of the statistically significant wavelengths detected by the sensitivity analysis when using unaltered and continuum-removed reflectance coupled with parametric and nonparametric analysis of variance (ANOVA) within the visible (VIS, 400–700 nm), near infrared (NIR, 700–1300 nm), short-wave infrared 1 (SWIR1, 1300–1800 nm), and SWIR2 (1800–2500 nm) spectrum. Lower-right plot is empty due to nonexistence of statistically significant bands—see Fig. 6.3. Description of the spectral libraries can be found in Table 6.1.

discrimination between three different plants (USA data set). This is most probably in relation to the coverage of the spectral variability, i.e., these two data sets contained much less observations per plant (Table 6.1), which is discussed later in this section.

It should be mentioned that the implementation of the SA is robust, i.e., it is based on existing statistical functions within the available packages in R, as presented previously (Section 3.1). Thus, users can easily create their own R code for their spectral libraries. Moreover, in terms of computational time, the larger the dimensionality of the spectral library (e.g., GREECE; Table 6.1), the longer the SA time. This is due to the iterative pairwise calculations within a loop. Nevertheless, computational time was reasonable and one complete sensitivity run for the most detailed spectral library (GREECE) took about 10−12 min in real time on standard computer; users may further shorten the computational time by using a vectorized version of the "for" loop in R.

The results of the SA clearly imply that irrespective of the type of reflectance and statistical test, the NIR and the SWIR are the most important wavelength region in the spectra for discriminating different vegetation plants, agreeing with the fundamental knowledge for delineating vegetation by remote sensing. Spectral discrimination in the VIS is important for vegetation types and plants that are "obviously" different in color in their canopies, or if the same plant is followed throughout its phenological development. So the VIS spectrum is mostly relevant for discriminating between the most distant phenological stages—when the plant morphology also differs the most or when biotic/abiotic factors impact its morphological characteristics.

3.3 SENSITIVITY OF VEGETATION SPECTRAL DISCRIMINATION TO THE NUMBER OF OBSERVATIONS

The effect of number of observations on the spectral discrimination results using the most detailed spectral library GREECE is presented in Fig. 6.5. Analogous to Fig. 6.3, wavelengths in the figure whose color varies within and between spectral libraries have higher sensitivity in their discrimination power. It can be clearly seen on the figure that decreasing the number of observations drastically decreases the ability to discriminate between the plants when using unaltered reflectance, and six observations or less resulted in no significant results according to both parametric and nonparametric ANOVA. This also agrees with the low discrimination between the plants in the USA data set, which was characterized with the lowest number of observations as well. The discrimination results were less sensitive to the number of observations when using the continuum-removed reflectance, although the same trend of decreased discrimination with decreasing number of observations was also seen (Fig. 6.5). This situation implies on an important aspect of power analysis that has been mentioned too little in field spectroradiometry and vegetation remote sensing at field scale. Although many research situations are complex and may defy rational power analysis, field spectroradiometry and creation of vegetation spectral libraries must include well-designed field campaigns (Petropoulos et al., 2014; Rasaiah et al., 2014). Hence power analysis should be incorporated to minimize the problem of numbers of observations, i.e., sample size. It would allow determination of (1) the number of observations required to detect differences with a given degree of confidence and (2) the probability of detecting the differences for the given size with the given level of confidence. If the probability is unacceptably low, it would be probably wiser to alter, i.e., increase the number of observations, or simply to abandon the experiment.

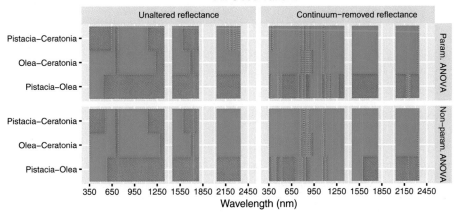

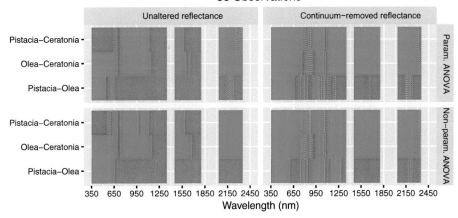

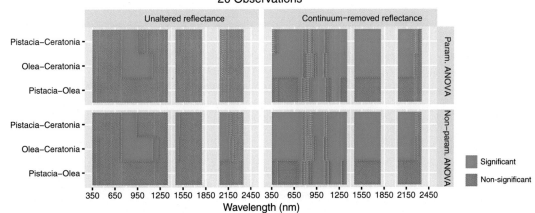

FIGURE 6.5

Effect of sample size on the spectral discrimination results when using unaltered and continuum-removed reflectance coupled with parametric and nonparametric analysis of variance (ANOVA). Description of the spectral libraries can be found in Table 6.1.

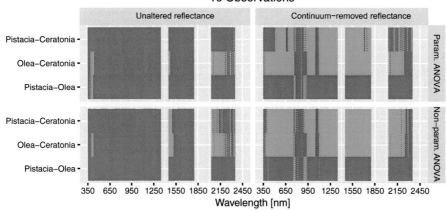

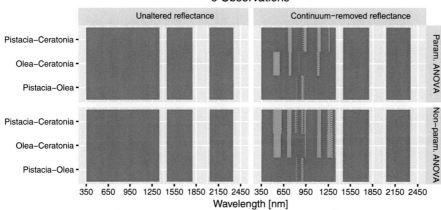

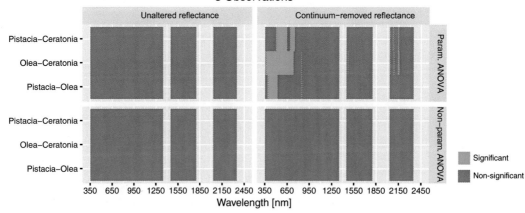

FIGURE 6.5 Cont'd

The implications of this SA are that the type of reflectance matters for vegetation discrimination, with the use of continuum-removed reflectance resulting in generally better discrimination, i.e., higher number of bands. Vegetation discrimination is less sensitive to the type of statistical test, i.e., parametric or nonparametric ANOVA. However, when the number of observations is involved, the situation becomes even more complicated and more difficult to generalize.

Summing up, a monofactor SA was conducted on the hyperspectral signatures of various vegetation types and plants, and it was used to determine the specific wavelengths and regions ideal for their remote identification. The results showed clearly the response of the spectral discrimination to the type of reflectance, and also to the statistical test used to conduct the analysis. The results are helpful for future discrimination studies at field scale in determining the need for processing the spectral reflectance, the choice of statistical test, and also the number of observations needed for relevant discrimination analyses. The results are also encouraging for the utilization of the knowledge derived from the field measurements to remote sensing analyses of air- and space-borne acquired data for improving sensor's capability to discriminate among vegetation species. In addition, further work should also be directed toward upscaling the results of an already large amount of field-scale studies in order to confirm, utilize or improve the discrimination at field scale and the synergy and complementarity of field spectroradiometry and air- and satellite-based remote sensing of land cover and management.

4. FINAL REMARKS

Contrary to earlier studies implying on a common nature of the vegetation spectral signature, an increasing number of recent studies show that many plants do have their distinct spectral signature. This is because the research dealing with remote sensing of vegetation utilizes an increasing number of methods for reflectance processing and accompanying statistics aiming to discriminate between the signatures of the vegetation. To this end, the present chapter shows the sensitivity of the discrimination results to the most common types of reflectance and statistical tests. The settings for the chapter have also included the first systematic monofactor SA of three spectral libraries, each representing different spectral settings and biogeographical regions.

The outcome of this work shows a distinct effect of the type of reflectance and the statistical test on the vegetation discrimination results, which further interacts with the number of observations. Overall, continuum-removed reflectance processed with nonparametric analysis, especially when a large number of observations are available, is preferable given the high variability and instinct complexity of vegetation spectral response at field scale. When the number of observations is low, however, the discrimination analysis should either be avoided, or an unaltered library should be processed with parametric/nonparametric analysis. The work also bares important implications for the design of field campaigns aiming to collect spectral reflectance of vegetation and create spectral libraries: power of analysis determining the optimum number of observations should be integrated in the field campaigns to determine the optimum number of observations for covering spectral variability and enhancing spectral discrimination. Finally, this chapter is also relevant for vegetation hyperspectral studies at multiple scales, including air- and space-borne hyperspectral image analyses and methodologies that involve field-scale spectral data in training classification algorithms.

REFERENCES

Adam, E., Mutanga, O., 2009. Spectral discrimination of papyrus vegetation (*Cyperus papyrus* L.) in swamp wetlands using field spectrometry. ISPRS Journal of Photogrammetry and Remote Sensing 64 (6), 612–620.

Asner, G.P., Wessman, C.A., Bateson, C.A., Privette, J.L., 2000. Impact of tissue, canopy, and landscape factors on the hyperspectral reflectance variability of arid ecosystems. Remote Sensing of Environment 74 (1), 69–84.

Avery, T.E., Berlin, G.L., 1992. Fundamentals of Remote Sensing and Airphoto Interpretation. NY MacMillan Publishing Company, New York.

Castro-Esau, K.L., Sanchez-Azofeifa, G.A., Caelli, T., 2004. Discrimination of lianas and trees with leaf-level hyperspectral data. Remote Sensing of Environment 90 (3), 353–372.

Clark, M.L., Roberts, D.A., Clark, D.B., 2005. Hyperspectral discrimination of tropical rain forest tree species at leaf to crown scales. Remote Sensing of Environment 96 (3–4), 375–398.

Clark, R.N., Roush, T.L., 1984. Reflectance spectroscopy – quantitative-analysis techniques for remote-sensing applications. Journal of Geophysical Research 89 (Nb7), 6329–6340.

Clark, R.N., et al., 2007. USGS Digital Spectral Library Splib06a, U.S. Geological Survey, Data Series 231.

Cochrane, M.A., 2000. Using vegetation reflectance variability for species level classification of hyperspectral data. International Journal of Remote Sensing 21 (10), 2075–2087.

CRAN, R., 2016. R: A Language and Environment for Statistical Computing. R Foundation for Statistical Computing, Vienna, Austria, ISBN 3-900051-07-0. URL: http://www.R-project.org/.

ENVI, 2009. ENVI User's Guide, ENVI Software.

Gaju, O., et al., 2014. Nitrogen partitioning and remobilization in relation to leaf senescence, grain yield and grain nitrogen concentration in wheat cultivars. Field Crops Research 155, 213–223.

Hadjimitsis, D.G., Padavid, G., Agapiou, A., 2009. Surface Reflectance Retrieval from Landsat TM/ETM+ Images for Monitoring Irrigation Demand in Cyprus, EARSeL Symposium "Imagine Europe". EARSeL, Chania, Greece.

Holden, H., LeDrew, E., 1998. Spectral discrimination of healthy and non-healthy corals based on cluster analysis, principal components analysis, and derivative spectroscopy. Remote Sensing of Environment 65 (2), 217–224.

Hothorn, T., Hornik, K., 2000. Exact Distributions for Rank and Permutation Tests. R package Vignette R package version 0.8-28.

Li, F., Mistele, B., Hu, Y.C., Chen, X.P., Schmidhalter, U., 2014. Reflectance estimation of canopy nitrogen content in winter wheat using optimised hyperspectral spectral indices and partial least squares regression. European Journal of Agronomy 52, 198–209.

Li, H., Zhao, C.J., Yang, G.J., Feng, H.K., 2015. Variations in crop variables within wheat canopies and responses of canopy spectral characteristics and derived vegetation indices to different vertical leaf layers and spikes. Remote Sensing of Environment 169, 358–374.

Manakos, I., Manevski, K., Petropoulos, G.P., Elhag, M., Kalaitzidis, C., 2010. Development of a spectral library for Mediterranean land cover types. In: Reuter, R. (Ed.), 30th EARSeL Symp.: Remote Sensing for Science, Education and Natural and Cultural Heritage. EARSeL, Chania, Greece, pp. 663–668.

Manevski, K., Manakos, I., Petropoulos, G.P., Kalaitzidis, C., 2011. Discrimination of common Mediterranean plant species using field spectroradiometry. International Journal of Applied Earth Observation and Geoinformation 13 (6), 922–933.

Manevski, K., Manakos, I., Petropoulos, G.P., Kalaitzidis, C., 2012. Spectral discrimination of Mediterranean Maquis and Phrygana vegetation: results from a case study in Greece. IEEE Journal of Selected Topics in Applied Earth Observations and Remote Sensing 5 (2), 604–616.

Morrey, J.R., 1968. On determining spectral peak positions from composite spectra with a digital computer. Analytical Chemistry 40 (6), 905–914.

Nidamanuri, R.R., Zbell, B., 2011. Transferring spectral libraries of canopy reflectance for crop classification using hyperspectral remote sensing data. Biosystems Engineering 110 (3), 231−246.

Noble, S.D., Brown, R.B., 2009. Plant species discrimination using spectral/spatial descriptive statistics. In: Proceedings of the 1st International Workshop on Computer Image Analysis in Agriculture, pp. 82−92. Potsdam, Germany.

Petropoulos, G.P., Manevski, K., Carlson, T.N., 2014. Hyperspectral remote sensing with emphasis on land cover mapping: from ground to satellite observations. In: Scale Issues in Remote Sensing. John Wiley & Sons, Inc, pp. 285−320.

Prasad, K.A., Gnanappazham, L., 2015. Multiple statistical approaches for the discrimination of mangrove species of *Rhizophoraceae* using transformed field and laboratory hyperspectral data. Geocarto International 1−22.

Price, J.C., 1994. How unique are spectral signatures? Remote Sensing of Environment 49 (3), 181−186.

Rasaiah, B., Jones, S., Bellman, C., Malthus, T., 2014. Critical metadata for spectroscopy field campaigns. Remote Sensing 6 (5), 3662−3680.

R Core Team, 2016. R Core Team and contributors worldwide - statistical functions version 3.3.0.

Robson, C., 1994. Experiment, Design and Statistics in Psychology Penguin Books.

Schaaf, C.B., 2009. Albedo and Reflectance Anisotropy: Assessment of the Status of the Development of the Standards for the Terrestrial Essential Climate Variables, Global Terrestrial Observing System.

Schmidt, K.S., Skidmore, A.K., 2003. Spectral discrimination of vegetation types in a coastal wetland. Remote Sensing of Environment 85 (1), 92−108.

Stevens, A., Ramirez-Lopez, L., 2013. An introduction to the prospectr package. R package Vignette R package version 0.1.3.

Streitberg, B., Röhmel, J., 1986. Exact calculations for permutation and rank tests: an introduction to some recently published algorithms. Statistical Software Newsletter 12, 10−17.

van Aardt, J.A.N., Wynne, R.H., 2001. Spectral separability among six southern tree species. Photogrammetric Engineering & Remote Sensing 67 (12), 1367−1375.

van Aardt, J.A.N., Wynne, R.H., 2007. Examining pine spectral separability using hyperspectral data from an airborne sensor: an extension of field-based results. International Journal of Remote Sensing 28 (1−2), 431−436.

GLOBAL (OR VARIANCE)-BASED SA METHODS: CASE STUDIES

A MULTIMETHOD GLOBAL SENSITIVITY ANALYSIS APPROACH TO SUPPORT THE CALIBRATION AND EVALUATION OF LAND SURFACE MODELS

7

F. Pianosi, J. Iwema, R. Rosolem, T. Wagener

University of Bristol, Bristol, United Kingdom

CHAPTER OUTLINE

1. INTRODUCTION

Global sensitivity analysis (GSA) is a set of mathematical techniques to investigate the behavior of numerical models, and in particular, how variations of the model's input factors impact the variability of its outputs. *Input factors* may include any aspect of the model that can be changed before its evaluation. Input factors typically considered in GSA are the parameters of the model, but they might also include the initial conditions, boundary conditions, time series of driving forcing inputs, etc., used to force the model simulation. Model *outputs* can be any scalar variable that is obtained after the model evaluation, for instance, an aggregate of the time series produced by the model simulation. When observations are available, the model output can be a performance metric (or *objective function*) measuring the model's fit to data. GSA can then be used to support the model calibration and diagnostic evaluation. Questions typically addressed by GSA are: what parameters are mostly influential on the model accuracy (the *ranking* question) and therefore should be the main focus of model calibration; which parameters have instead negligible influence (*screening*) and therefore can be fixed to default values; whether the model accuracy is also influenced by *interactions* among parameters and between parameters and uncertain initial conditions; and whether uncertainty ranges around parameter estimates can be narrowed down based on model performances (*mapping*). In addressing these questions, GSA gives the opportunity to verify that the model behavior is consistent with our expectations and to evaluate the information content in the available observations. By providing a structured and automatic approach to investigate the propagation of uncertainties through the model, GSA is particularly valuable when dealing with complex environmental models like land surface models (LSMs). Such models are commonly used to investigate multiple outputs such as the energy and water balance components and biogeochemistry (e.g., net carbon flux from plants and soils). They typically encompass many parameters, which are often calibrated and validated using Earth observations (EOs) such as soil moisture content and surface energy fluxes (e.g., Abramowitz et al., 2008; Blyth et al., 2010, 2011). Due to their complexity, manual approaches to investigate the response of LSMs to changes in their input factors, namely, parameters, are impracticable. GSA can then be used to determine which parameters are more influential and therefore should be the focus of computationally intensive calibration procedures. Such studies have become increasingly relevant especially as LSMs are adopting the so-called hyperresolution (Wood et al., 2011; Beven and Cloke, 2012; Rosolem et al., 2012) and novel ground-based techniques to inform model accuracy are being developed (Robinson et al., 2008; Romano et al., 2012; Romano, 2014).

The implementation and interpretation of GSA is complicated by a number of choices that GSA users have to make and for which multiple, equally sensible options might be available. These choices are discussed in details in Pianosi et al. (2016). Here we focus on a subset of them, which we consider particularly relevant for LSMs applications, and demonstrate their implications through an application example using the Joint UK Land Environment Simulator (JULES) (Best et al., 2011). We investigate the influence of nine parameters and three initial conditions on four measures of model accuracy in reproducing heat fluxes and soil moisture in an Ameriflux semiarid site located in Arizona (Santa Rita Creosote), *with the ultimate goal of gaining insights to support the model calibration.*

The first choice here investigated is that of the most appropriate GSA method for the problem at hand. The choice of the method is typically driven by the purpose of the analysis (i.e., screening, ranking, interactions, mapping) and by the available computing resource. The choice can be further driven by specific characteristics of the problem at hand, for instance, linearity or nonlinearity of the input—output relationship or statistical characteristics of the output distribution (e.g., skew). These are usually handled more or less efficiently by the different methods (Pianosi et al., 2016). The analyst's confidence with a specific GSA method or the availability of software tools might also condition the choice. For LSMs, a variety of methods have been applied, including derivatives-based approach (Alton et al., 2007), regional sensitivity analysis (RSA) (Bastidas et al., 1999; Bakopoulou et al., 2012), regression analysis (Hou et al., 2012; Huang et al., 2013), variance-based (Sobol') sensitivity analysis (Rosero et al., 2010; Rosolem et al., 2012), screening methods (Li et al., 2013), or a combination of screening and variance-based methods within a two-step approach (Gan et al., 2015). In this work, we establish and demonstrate a multimethod approach wherein different GSA methods are applied simultaneously (without any additional computational burden) to gather complementary information about the model behavior from different angles. Our multimethod approach allows users to validate the results produced by each individual method and therefore to reinforce the general conclusions drawn from GSA.

In addition, other technical choices that will be discussed in this chapter are the definition of the range of variations of the input factors and how to account for model accuracy within GSA. The definition of variability ranges is often uncertain and very critical. When the input factors are the model parameters, ranges are defined based on the parameter's physical meaning, from the previous literature, or using a priori knowledge about the characteristics of the study site. However, a number of studies demonstrate that using even slightly different range definitions, each considered equally plausible by the analyst, can significantly change the values of sensitivity measures (e.g., Wang et al., 2013; Shin et al., 2013). Here, we start from the largest possible definition of the ranges and show how GSA itself can be used to find a more appropriate definition of the ranges based on comparison between model simulations and observations and compare the range constraining so obtained with the one suggested by prior knowledge of the characteristics of the study site. Another related aspect is that of how to account for model accuracy within GSA. In fact, when dealing with complex models like LSMs, it may happen that a sampled combination of input factors generates a model's response that the modeler would reject as unacceptable, for instance, because the mismatch with observations is too large. When this happens, the question arises whether these simulations should contribute to estimate model sensitivities or not. Here, we show how model accuracy can be accounted for in estimating sensitivity indices in different GSA methods and compare sensitivity estimates obtained when using all model evaluations or when discarding unacceptable ones.

Although our application results refer to a specific LSM (i.e., JULES) and to a subset of its input factors (nine parameters and three initial conditions), the methodology presented is generic and could easily be applied to different models and GSA setups. However, it should be noted that even for the same model and GSA setup, the results might vary with the study location. For instance, Rosero et al. (2010) show that sensitivity to parameters of the Noah LSM changes when the model is applied to different Fluxnet sites along a precipitation gradient. Hence sensitivity results can be difficult to generalize, and whenever possible, GSA should be reapplied for new sites or conditions. In this chapter, we therefore focus on the methodological and procedural aspects and aim at providing good practice examples for the application of GSA, by showing how critical implementation choices can be made and revised to reinforce the conclusions drawn from GSA.

2. MODEL AND METHODS

GSA methods differ from each other in the way they define and measure sensitivity. However, their numerical implementation always encompasses the same three basic steps: (1) sampling the variability space of the input factors, (2) evaluating the model, and (3) postprocessing the input and output samples to compute the sensitivity indices. In this chapter, we show how a multimethod approach is made possible by the reiterated application of different postprocessing methods to the same data set of

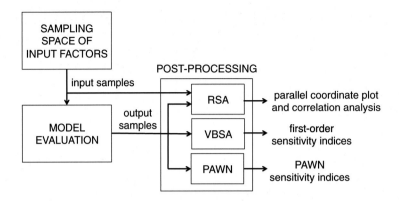

FIGURE 7.1

Schematic of the multimethod approach for GSA. The three GSA methods applied in this study are regional sensitivity analysis (RSA), variance-based sensitivity analysis (VBSA), and the density-based PAWN method.

input—output samples. A schematic of the multimethod approach is given in Fig. 7.1. Since the computational burden of GSA mainly derives from the model evaluations used to generate the input—output samples, rather than their postprocessing by the individual GSA methods, our approach comes at no additional computing cost with respect to the application of one specific GSA method.

In the next paragraphs, we describe the three GSA methods applied in our approach, RSA, variance-based (Sobol'), and a novel density-based approach called PAWN. We also discuss how we adjusted the approximation techniques of the variance-based and density-based indices to apply them to a generic input—output data set, rather than to a data sample obtained via a tailored sampling strategy, as commonly done in the literature. Then, we describe the JULES model here analyzed, the case study site, and the experimental setup for our GSA.

2.1 REGIONAL SENSITIVITY ANALYSIS

RSA, first proposed by Young et al. (1978), and Spear and Hornberger (1980), looks for regions of the input space that map into output values of particular interest. Typically (but not exclusively, see Wagener et al., 2001), two binary regions are created: "behavioral" and "nonbehavioural," depending on whether the model outputs are above or below a prescribed threshold. The divergence between the input distributions in the behavioral and nonbehavioural sets is taken as a measure of sensitivity.

In our application, the input factors are the model parameters and the initial conditions of the model simulation, whereas the outputs are four objective functions measuring the model's fit to soil moisture and heat fluxes data. It thus makes sense to define as behavioral those combinations of parameters and initial conditions that produce model performances better than the default setup of JULES. Formally, the set X_b of behavioral inputs is

$$X_b = \left\{ x \middle| y_j = f_j(x) \leq \bar{y}_j \text{ for all } j \right\} \tag{7.1}$$

where $\mathbf{x} = [x_1, \ldots, x_M]$ is the vector of the M input factors subject to GSA, y_j is one of the objective functions measuring the model's fit to data, and $\bar{y}_j$ are the values of the objective functions when using the default setup (we assume here that the objective function expresses the distance between simulated and observed variables, and therefore should be minimized). Notice that the criterion in Eq. (7.1) is very strict since it considers as behavioral only those setups that improve the model's accuracy with respect to all objective functions simultaneously. A more relaxed criterion might be a Pareto efficiency criterion, where a setup is accepted when at least one objective function is improved. Application examples are given in Bastidas et al. (1999) and Rosolem et al. (2012).

Once the behavioral and nonbehavioural sets are built, we compare them from two different angles: (1) the parallel coordinate plot (PCP) is used to assess whether it is possible to narrow down the variability range of the input factors and (2) correlation analysis is used to gather information about the interactions between input factors. Further details about this visualization and analytical tools are given in the results section.

2.2 VARIANCE-BASED SENSITIVITY ANALYSIS

Variance-based sensitivity analysis (VBSA) measures the sensitivity to a given input by its contribution to the variance of the model output (Sobol', 1990; Saltelli et al., 2008). Several variance-based indices can be defined. In this application, we focus on the first-order index (or "main effect"), which measures the direct contribution of the variation of an input factor to the output variance, i.e., without considering interactions with other inputs. The definition of the first-order index for the ith input factor is

$$S_i = \frac{V_{x_i}[E_{x_{\sim i}}(y|x_i)]}{V(y)} \tag{7.2}$$

where E denotes the expected value, V denotes the variance, and $\mathbf{x}_{\sim i} = [x_1, ..., x_{i-1}, x_{i+1}, ..., x_M]$ is a vector that includes all input factors but the ith.

Since the analytical solution of Eq. (7.2) is usually impossible, approximate techniques are used. The most established ones are the Fourier Amplitude Sensitivity Test [FAST (Cukier et al., 1973)], and the use of the estimators reviewed in Saltelli et al., 2010. However, both these approximation techniques require a tailored sampling strategy, and therefore they cannot be included in our multimethod approach. Hence, we need to develop a numerical approach to approximate Eq. (7.2) that can be applied to a generic input–output data set. The approach here proposed is similar to the one presented by Strong et al., 2014. For each input factor, we approximate the term $E_{x_{\sim i}}(y|x_i)$ in Eq. (7.2) via a regression function. For instance, in our application we use a linear combination of radial basis functions, i.e.,

$$\widehat{E}_i = \sum_{j=1}^{n} \left[a_j \exp\left(-(x_i - w_j)^2 \right) \right] \tag{7.3}$$

Then, the variance $V_{x_i}[E_{x_{\sim i}}(y|x_i)]$ in Eq. (7.2) can be approximated by the sample variance of $\widehat{E}_i$. The unconditional variance $V(y)$ in Eq. (7.2) is easily approximated by the sample variance of the output y. Operationally, for each input factor the steps are (1) to calibrate the regression function of Eq. (7.3), i.e., to estimate the parameter a_j, w_j that optimally fit the samples $< x_i, y >$; (2) to evaluate the optimally calibrated regression function against the available samples of x_i; and (3) to compute the sample variance of the evaluated E_i.

This approximation strategy is quite straightforward to apply, and it provides robust estimates of first-order indices (as shown in the VBSA results section by using bootstrapping). The drawback is that it cannot be applied to estimate total-order indices. Total-order indices measure the contribution to output variance from both individual variations of an input factor and its interactions with other inputs (Saltelli et al., 2008) and provide complementary information to first-order indices. For example, the difference between total-order and first-order indices measures the level of interactions of the input factors. When using the estimators reviewed in Saltelli et al. (2010), total-order indices can be estimated from the same input–output data set as first-order indices. However, as already noted, such an input-output data set is obtained by a tailored sampling strategy and thus cannot be integrated in our multimethod approach without adding up to computational complexity. To compensate for this limitation and gain some insights about the interactions between input factors, we will use a different approach that combines RSA with correlation analysis. This is further explained later on in the RSA results section.

2.3 THE PAWN DENSITY-BASED METHOD

Density-based methods investigate the entire distribution of model output rather than its variance only. Sensitivity is measured by the divergence between the unconditional output distribution that is induced when varying all input factors, and the conditional distribution that is obtained when conditioning the input under study. Several density-based methods are reviewed in Pianosi and Wagener (2015). The article also presents a novel density-based approach, called PAWN, which will be used within our multimethod approach. The PAWN sensitivity index for the ith input factor is defined as

$$S_i = \max_{x_i} \max_{y} \left| F_y(y) - F_{y|x_i}(y|x_i) \right| \tag{7.4}$$

where $F_y(y)$ and $F_{y|x_i}(y|x_i)$ are the unconditional and conditional cumulative distribution functions (CDFs) of the output.

In the operational implementation of the method, the outer maximum in Eq. (7.4), i.e., the one with respect to the conditioning value of x_i, is approximated by the maximum over a prescribed number of conditioning values. For each of these, the inner maximum, i.e., the maximum absolute difference between CDFs, is approximated by using empirical distribution functions. These are obtained by evaluating the model against input samples where all input factors vary (unconditional distribution), and against samples where the ith input is fixed to the conditioning value and the others vary (conditional distribution). When applying PAWN within our multimethod approach, however, the conditional distributions must be computed over the already available input–output data set. Since this data set does not contain multiple samples with exactly the same value of x_i, conditional distributions must be conditioned on "similar" values of x_i. In other terms, the sensitivity index of Eq. (7.4) is approximated as

$$S_i = \max_{k=1,\ldots,n} \max_{y} \left| F_y(y) - F_{y|x_i}(y|x_i \in J_k) \right| \tag{7.5}$$

where $J_k(k = 1, \ldots, n)$ are n (e.g., 10) equally spaced intervals over the range of variation of x_i. One interesting feature of density-based methods like PAWN is that sensitivity indices can be easily tailored to focus on a range of the output distribution of particular interest. For instance, in our case we can focus on the lower tail of the output distribution where the objective function is below the threshold value produced by the default setup. Sensitivity indices will be then computed as

$$S_i = \max_{k=1,\ldots,n} \ \max_{y \leq \bar{y}_j} \left| F_y(y) - F_{y|x_i}(y|x_i \in J_k) \right| \qquad (7.6)$$

2.4 THE JULES MODEL

The JULES is a community LSM whose development was initiated at the UK Met Office. It serves as the lower boundary condition to the Met Office Unified Model for weather and climate forecasting. Best et al. (2011) provide a detailed description of the model structure and underlying assumptions, whereas Blyth et al., 2011 review the application of JULES across a set of study sites. JULES has a modular structure, which makes it easy to remove, add, or replace modules. The modules are coupled through energy, water, and carbon fluxes. The complete coupled modules, with their processes and coupling, result in evapotranspiration, heat, momentum, and carbon fluxes going back to the atmosphere. Water, heat, and carbon fluxes leave JULES through the subsurface. Fig. 7.2 illustrates the key water and energy processes simulated in JULES. The model's forcing inputs are hourly downward shortwave radiation, downward longwave radiation, air temperature, atmospheric pressure, precipitation, wind speed, and atmospheric humidity. The main simulation outputs are time series of soil moisture, water fluxes (e.g., surface and subsurface runoff, drainage, evaporation), and energy fluxes

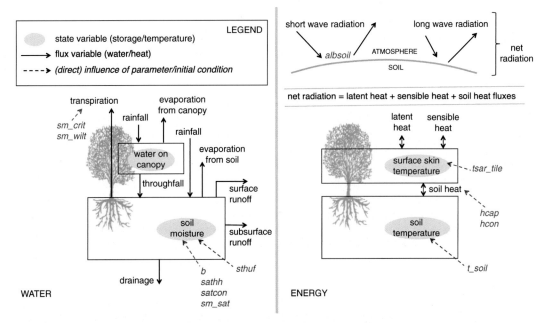

FIGURE 7.2

Schematic of the main state and flux variables simulated in JULES. Left: water fluxes, right: energy fluxes. The model parameters (red) and initial conditions (blue) that will be investigated by GSA are also reported.

(i.e., latent and sensible heat), in our case simulated at daily resolution. JULES forcing inputs usually come from meteorological observations, while soil moisture observations and surface energy flux data can be used to evaluate JULES simulation outputs and thus investigate the model structure and performance. Hence, EO data has played an important role in the development and increasing understanding of LSMs like JULES (Dirmeyer et al., 2006; Blyth et al., 2010, 2011). EO data that are used in this study to force the JULES model and to assess simulated outputs are described in the next paragraph. The model parameters and initial conditions that will be investigated by GSA are further described in Section 2.6, and they are reported in Fig. 7.2 next to the variable that they directly influence. In this study, the three base modules of JULES (soil dynamics, surface exchange, and plant physiology) were used, whereas any additional module (e.g., vegetation dynamics) was switched off. Model settings were similar to the United Kingdom Variable resolution science configuration with the only difference that a single point was used instead of a grid.

2.5 THE SANTA RITA CREOSOTE STUDY SITE

The Santa Rita Creosote is a semiarid site in Arizona (latitude 31.9085 N, longitude 110.839 W, altitude 989 masl) (Skott et al., 1990), which is sparsely vegetated ($\sim$24% of surface area) with Creosote bush ($\sim$14% of surface area) and other species of bushes, grasses, and cacti (Cavanaugh et al., 2011). The average temperature is 17.8°C, with daytime temperatures exceeding 35°C in summer. The average yearly precipitation is 415 mm (Skott et al., 1990). According to Cavanaugh et al. (2011), the soil texture can be characterized as sandy loam with 5–15% gravel. However, the Harmonized World Soil Database (HWSD) classifies the soils at Santa Rita Creosote as loam and clay loam. This is a typical example of uncertainty that land surface modelers face when defining soil properties in such applications.

Available data for the Santa Rita Creosote site include meteorological data (precipitation, air temperature, wind speed, air pressure, relative humidity, and downward shortwave and longwave radiations), land surface energy fluxes, and point-scale soil moisture data from Time Domain Transmission (TDT) sensors, coming from Ameriflux Level2 (ORNL-Ameriflux web page, 2015; Papuga, 2009). Precipitation was measured with a tipping bucket rain gauge at 1 m above ground. The wind speed (RM Young Wind Sensor), air temperature (HMP sensor), and relative humidity (HMP sensor) were measured at 3.3 m. Atmospheric pressure was measured at 3.75 m with an LI17500. Downward shortwave and longwave radiations were measured by a radiometer (CNR1) installed 2.75 m above ground. Latent and sensible heat density fluxes were determined from an (open path) LiCor7500 and CSAT 3D Sonic eddy covariance system located at 3.75 m. Two soil temperature sensors are located at 1 cm depth and two at 4 cm depth. Data used in our application are the mean of these sensors for each measurement time step. Point-scale soil moisture data come from three TDT sensors located at −2.5 cm and three sensors located at −12.5 cm. Observations here used refer to the average values for each layer in each hour. Additionally, soil moisture data from Cosmic-Ray Neutron Sensor (CRNS) are also available (Zreda et al., 2012). The CRNS was installed on June 2, 2010, and CRNS neutron counts were corrected as described in Iwema et al. (2015).

2.6 EXPERIMENTAL SETUP: DEFINITION OF INPUT FACTORS AND OUTPUTS

In our application, the input factors subject to GSA are nine parameters and three initial conditions, listed in Table 7.1. The nine selected parameters are related to the soil hydraulic characteristics and the

Table 7.1 The nine parameters and three initial conditions subject to GSA, their default value, and range of variation (*sau = same as unconstrained*).

Name	Description	UoM	Default Value	Unconstrained Range		Site Range	
1. *b*	Parameter of van Genuchten–Mualem function	–	4.6253	0.2801	11.1111	2	10
2. *sathh*	Parameter of van Genuchten–Mualem function	M	0.2110	0.0968	833.33	0.0971	0.6667
3. *satcon*	Saturated hydraulic conductivity	mm/s	0.0061	0.0001	0.14	0.0018	0.07
4. *sm_sat*	Saturated soil moisture content	cm^3/cm^3	0.4188	0.36	0.74	0.36	0.51
5. *sm_crit*	Critical point soil moisture	cm^3/cm^3	0.2280	0.0036	0.74	sau	sau
6. *sm_wilt*	Wilting point soil moisture	cm^3/cm^3	0.1012	0.000036	0.74	sau	sau
7. *hcap*	Soil heat capacity	$J/m^3/K$	1,185,676	1,000,000	3,000,000	sau	sau
8. *hcon*	Soil heat conductivity	W/m/K	0.2269	0.1	3	0.2	3
9. *albsoil*	Soil albedo	–	0.1600	0.1	0.5	sau	sau
10. *tsar_tile*	Initial surface skin temperature	K	277.67	260	320	sau	sau
11. *sthuf*	Initial soil moisture content as fraction of saturation	–	0.2410	0.01	1	sau	sau
12. *t_soil*	Initial soil temperature	K	279.34	260	320	sau	sau

soil—plant and soil—heat processes and therefore are expected to have an influence on the soil dynamics, which is the main focus of this study. However, the JULES model includes other 57 parameters to simulate biophysical processes, which will not be considered in the GSA here, and are set to their default values. A complete list of the model parameters and equations can be found in Best et al. (2011). For each parameter, the same value was applied for each vertical layer of the model. Similarly, the same temperature and soil moisture initialization was used for all layers.

The outputs analyzed by GSA are four objective functions based on comparison between simulated variables and daily average observations available over a 2-year time period from 2012 to 2013. Specifically, we compute:

1. The root mean squared error (RMSE) of sensible heat flux density.
2. The RMSE of latent heat flux density.
3. The RMSE of soil moisture simulations with respect to TDT sensor (i.e., point-scale) observations at depths 2.5 and 12.5 cm. A depth-weighted mean was calculated from the observations at these two depths and then compared with JULES simulations for the first soil layer (0—10 cm).
4. The RMSE of soil moisture simulations with respect to CRNS observations. Soil moisture of the JULES profile is postprocessed by the Cosmic-Ray Soil Moisture Interaction Code (Shuttleworth et al., 2013) to estimate the equivalent neutron count, which is then compared with the observed neutron count. This minimizes the propagation of error due to the nature of the CRNS sensor, which presents a variable effective depth based on the soil wetness conditions (as discussed in Shuttleworth et al., 2013 and Rosolem et al., 2014).

All comparisons between simulated and observed variables were done using daily average values. The objective functions are computed over a 2-year period from January 2012 to December 2013, but each simulation started on July 1, 2011. In the Results section, we will investigate through GSA whether this 6-month warm-up period is sufficient to absorb the influence of the uncertain initial conditions.

2.7 DEFINITION OF THE RANGE OF VARIATION OF THE INPUT FACTORS

In our analysis, we assume uniform, independent distributions for all parameters and initial conditions. The definition of the input variability space then only requires defining the range of variation of each input factor. In this study, we decided to use ranges as wide as possible. For the three initial conditions, ranges are confined by the minimum and maximum observed values of that variable over the entire data set. For the nine parameters, we consider the entire feasible range according to the physical meaning of the parameters and with no consideration of the specific characteristics of the Santa Rita site. These range values are reported in Table 7.1 under the name "unconstrained ranges." For the nine parameters, Table 7.1 also reports a set of narrower ranges, called "site ranges," which are more realistic for the mixture of loam, clay loam, and sandy loam that is found in the Santa Rita site, according to the Wösten pedotransfer function (Wösten et al., 2001). However, the latter ranges are only reported for the sake of comparison and are not used in computations.

The first three parameters in Table 7.1 (b, *sathh*, and *satcon*) are related to the Mualem—Van Genuchten soil hydraulic functions [Eqs. (21), (60), and (61) in Best et al., 2011]. Parameter *satcon* is the hydraulic conductivity for saturated soil [called K_{hs} in Eq. (61) in Best et al., 2011]. Parameter *b* is related to the *n* parameter in the Van Genuchten function by the equation $b = 1/(n + 1)$. Parameter *sathh* is related to the Mualem parameter α by the equation $sathh = 1/a$. The other two parameters appearing in the Mualem−Van Genuchten functions, ξ and *m*, are set to the recommended values of $\xi = 0.5$ and $m = 1/(n + 1)$ (Best et al., 2011). The ranges of the soil hydraulic parameters *b*, *sathh*, *satcon*, and *sm_crit* are derived from soil textural (sand, silt, and clay) and organic matter content values via the Wösten pedotransfer function (Wösten et al., 2001). In the "unconstrained ranges" case, possible values of soil textural composition were determined by taking the entire United States Department of Agriculture soil texture triangle. Three soil organic carbon (OC) content classes were used in this study (DeLannoy et al., 2014), which were represented by class nominal values: 0.26%, 0.46% and 1.12% OC. OC was converted to organic matter (OM) via the equation $OM = 1.72\ OC$ for use in the Wösten pedotransfer function. The Wösten pedotransfer function also takes into account whether the topsoil or subsoil is concerned. We considered both options, for a total of six soil texture triangles. First, bulk density was calculated following the procedure described in DeLannoy et al. (2014). We took the minimum and maximum values found for each of the parameters mentioned. In the "site ranges" case, instead, we only consider soil textural composition values of loam, clay loam, and sandy loam, which are the soil types found in the Santa Rita site, thus obtaining much narrower ranges.

The critical soil moisture content *sm_crit* is defined as a fraction of the saturated soil moisture content *sm_sat*, and the wilting point soil moisture content *sm_wilt* is defined as a fraction of the critical soil moisture content *sm_crit*. This is done to better accommodate the related constraints on such quantities (i.e., $sm\_wilt < sm\_crit < sm\_sat$). The ranges of *sm_crit* and *sm_wilt* are thus obtained from the range of *sm_sat* by applying a multiplier varying between 0 and 1.

The ranges of the soil heat capacity (*hcap*), soil heat conductivity (*hcon*), and bare soil albedo (*albsoil*) were taken as margins around values reported by De Vries (1963) and Smits et al. (2010) in Moene and Van Dam (2014) for clay to sand.

3. RESULTS

The multimethods GSA approach is here applied to an input−output data set of 20,000 samples. Input combinations were sampled via a maximin Latin hypercube design (Forrester et al., 2008; Section 1.4) of the input variability space defined by the unconstrained ranges (see Table 7.1). Model simulations required about 72 h on a 64-bit (with eight CPUs) Linux desktop computer. Sensitivity analysis was performed using the SAFE Toolbox (Pianosi et al., 2015).

3.1 RESULTS OF REGIONAL SENSITIVITY ANALYSIS

Results of RSA are visualized in the PCP shown in Fig. 7.3. Here, each line represents a combination of (normalized) input factors (parameters and initial conditions). Red lines are behavioral samples, i.e., those improving the model's accuracy with respect to the default setup for all four objective functions simultaneously, whereas gray lines are nonbehavioral samples. Values of parameters and initial conditions in the default setup are represented by black dots. Vertical black lines represent the "site ranges" of the model parameters, i.e., the reduced ranges recommended in the literature (Ameriflux, COSMOS websites, and the HWSD) for the soil type (a mixture of loam, clay loam, and sandy loam) of the Santa Rita Creosote site.

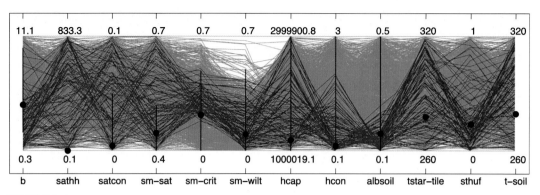

FIGURE 7.3

Parallel coordinate plot. *Gray lines* are nonbehavioural input combinations, *red lines* are behavioral ones, *black dots* are default inputs, and *black lines* are "site ranges" (parameter ranges for the soil type of the case study area).

- The number of behavioral samples is quite small (111 of 20,000), which suggests that improving the default setup by model calibration is possible but not straightforward.
- For the majority of the parameters (all but *sm_crit*, *sm_wilt*, *hcon*, and *albsoil*), behavioral values are scattered all along the unconstrained range, and thus constraining performances does not help to narrow down those ranges. For parameters *sm_crit*, *hcon*, and *albsoil*, ranges might be reduced toward the higher end (*sm_crit*) or the lower end (*hcon* and *albsoil*).
- For parameters *b*, *sathh*, *sat con*, and *sm_sat*, the majority of behavioral samples fall outside the "site ranges." In other words, for our case study, constraining model performances would not constrain those parameters to recommended ranges for the soil types found in the site area. This might be an evidence of a limited information content in the available observations or of the overparameterization of the model.
- Interestingly, we find no support for reducing the range of *sathh*, despite the fact that the unconstrained range is extremely large (its upper limit is 833 m, whereas the upper limit of the "site range" is 0.6667). One possible explanation is that all four objective functions are insensitive to this parameter, as it will be later shown by the variance-based and density-based sensitivity (PAWN) analyses.

To investigate the interactions between input factors, we combined the RSA approach with correlation analysis. For each objective function, we redefine as behavioral the samples whose objective value is below the threshold value produced by the default setup. In our case study, we find 486 behavioral samples for objective function 1 (i.e., sensible heat RMSE <27 W/m^2), 10,145 for objective function 2 (latent heat RMSE <26 W/m^2), 7374 for objective function 3 (TDT soil moisture RMSE <0.05 m^3/m^3), and 7158 for objective function 4 (CRNS soil moisture RMSE <347 cph). For each behavioral set, we then compute the Spearman cross-correlation between all possible pairs of input factors. Results are shown in Fig. 7.4. The underlying idea is that high cross-correlation within the behavioral set suggests an interaction between the two input factors.

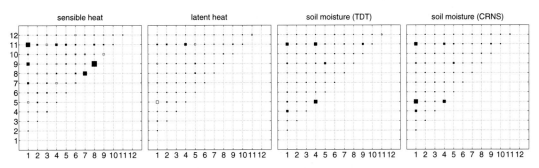

FIGURE 7.4

Spearman cross-correlation between pairs of input factors within the behavioral set of each objective function. *Marker color* indicates the sign (white, positive; black, negative); marker size is proportional to the absolute value of the cross-correlation. Absolute ranges vary from a minimum of 10^{-5} to a maximum of 0.3593. Input factors are: 1, *b*; 2, *sathh*; 3, *satcon*; 4, *sm_sat*; 5, *sm_crit*; 6, *sm_wilt*; 7, *hcap*; 8, *hcon*; 9, *albsoil*; 10, *tsar_tile*; 11, *sthuf*; 12, *t_soil*.

By defining an absolute correlation greater than 0.20 as "significant," we find that:

- For objective function 1 (RMSE of sensible heat simulations), cross-correlation (and thus interactions) is significant between input factors 1 and 9 (parameter *b* and *albsoil*; correlation is −0.24), 1 and 11 (*b* and *sthuf*; −0.39), 7 and 8 (*hcap* and *hcon*; −0.35), and 8 and 9 (*hcon* and *albsoil* −0.48).
- For objective function 2 (RMSE of latent heat simulations), cross-correlation is significant only between 1 and 5 (*b* and *sm_crit*; 0.22).
- For objective functions 3 and 4 (RMSE of soil moisture simulations against TDT observations and CRNS observations) there is a significant negative correlation between 4 (*sm_sat*) and 5 (*sm_crit*) (−0.30 for TDT and −0.24 for CRNS), and between 1 and 11 (parameter *b* and initial soil moisture *sthuf*) (−0.22 for TDT and −0.27 for CRNS). Objective function 3 is also sensitive to interactions between 4 and 11 *(sm_sat* and *sthuf*; −0.25). Objective function 4 shows interactions between 1 and 5 (*b* and *sm_crit* −0.36).

Interactions between *sm_crit* and *sm_sat* are expected since the latter is obtained as a fraction of the former through a multiplier (see Section 2.7). A more interesting interaction is the one between the initial condition *sthuf* and the soil parameters *b* and *sm_sat*. It means that the initial condition has an influence on the simulated soil moisture, at least through interactions, even after the 6-month warm-up period, and that a robust estimate of parameters *b* and *sm_sat* can only be obtained by reducing the uncertainty on the initial condition.

3.2 RESULTS OF VARIANCE-BASED SENSITIVITY ANALYSIS

Based on the same sample of 20,000 model evaluations, we can approximate the variance-based first-order sensitivity indices using the approach described in the previous section. Fig. 7.5 reports such a set of indices, one per objective function. Indices were computed using all the 20,000 output samples (white circles in Fig. 7.5) or using only the behavioral samples (i.e., those below the threshold value

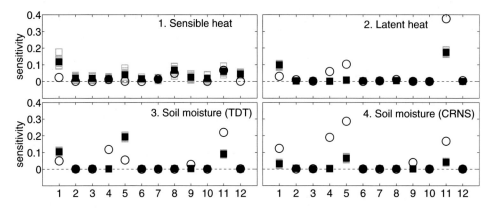

FIGURE 7.5

Variance-based first-order sensitivity indices for the four objective functions. *White circles*, computed using all samples; *gray squares*, computed using one bootstrap resample of the output samples that are below the threshold value in the default setup; *black squares*, average of 10 bootstrap resamples. Input factors are 1, *b*; 2, *sathh*; 3, *satcon*; 4, *sm_sat*; 5, *sm_crit*; 6, *sm_wilt*; 7, *hcap*; 8, *hcon*; 9, *albsoil*; 10, *tsar_tile*; 11, *sthuf*; 12, *t_soil*. TDT, Time Domain Transmission.

produced by the default setup for that objective function). Since in the latter case the sample size is reduced, the approximation of the indices was repeated 10 times using bootstrap resamples (Efron and Tibshirani, 1993) of the original output sample. Fig. 7.5 reports the indices estimated at each resample (gray) as well as the average values of indices across resamples (black). This figure shows that

- For objective function 1 (RMSE of sensible heat), the spread of gray markers indicates that there is some uncertainty in sensitivity estimates computed over behavioral samples. The reason is that the sample size after discarding nonbehavioural samples is very small (only 486 samples left). Still, the ranking of input factors can be clearly assessed. The uncertainty of sensitivity indices for the other objective functions is rather low.
- For all objective functions, sensitivity estimates are very different if computed over the entire sample or after filtering out samples associated with poor model performance (compare white circles and black squares). This highlights the importance of clearly defining the purpose of GSA, i.e., to assess the relative influence of input factors over the overall model's response or over a subregion of that response (for instance, the subregion of behavioral models).
- Since the ultimate goal of our GSA is to support the model calibration, we focus on the results obtained using behavioral samples only (black squares). These results show that all objective functions are sensitive to input 1 and 11 (*b* and *sthuf*). The initial condition *sthuf* thus has a direct influence on model accuracy, besides the indirect effect through interactions that was detected by the previous correlation analysis. Objective functions 1, 3, and 4 are also sensitive to input 5 (*sm_sat*). Objective function 1 is also sensitive to inputs 8 and 9 (parameters *hcon* and *albsoil*). According to VBSA, all other input factors have no influence on the objective functions. Before giving an interpretation of these results and their implications, in the next section, they will be verified and complemented by application of the PAWN approach.

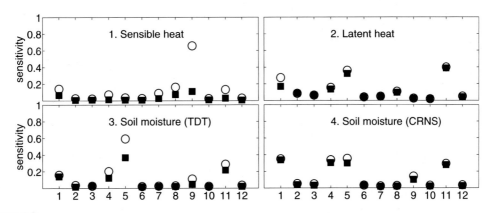

FIGURE 7.6

PAWN sensitivity indices for the four objective functions. *White circles,* computed over the entire output range; *black square,* computed over the subrange where output is below the threshold value of the default setup. Input factors are 1, b; 2, *sathh*; 3, *satcon*; 4, *sm_sat*; 5, *sm_crit*; 6, *sm-wilt*; 7, *hcap*; 8, hcon; 9, *albsoil*; 10, *tsar_tile*; 11, *sthuf*; 12, *t_soil*. *TDT,* Time Domain Transmission.

3.3 RESULTS OF PAWN

Fig. 7.6 reports the PAWN sensitivity indices for the four objective functions. Indices were computed using the entire range of variation of the output as in Eq. (7.5) or only the subrange below the threshold value produced in the default setup, as in Eq. (7.6). Using Eq. (7.6) is indeed a way to focus GSA on "behavioral" models. Fig. 7.6 shows that, just as in previous VBSA, this choice clearly changes sensitivity estimates. Since the ultimate goal of our analysis is to provide background to inform model calibration, again we analyze only the results relative to the subrange of behavioral models (black squares in Fig. 7.6). We notice that:

- Sensitivities detected by previous VBSA (all objective functions sensitive to input 1 and 11; objective function 1 sensitive to inputs 8 and 9; objective functions 3 and 4 sensitive to input 5) are all confirmed. The only exception is input 5 *(sm_crit)*, which appeared to be influential on objective function 1 in VBSA, whereas it is not according to the PAWN method.
- Additionally, PAWN rates influential several other input factors. These are inputs 4, 5, and 8 for objective function 2 and inputs 4 and 9 for objective functions 3 and 4. PAWN indices thus seem to be a valuable complement of first-order variance-based sensitivity indices.

3.4 OVERALL SENSITIVITY ASSESSMENT FROM THE MULTIMETHOD APPROACH

Fig. 7.7 provides a summary of the results produced by correlation analysis, VBSA, and PAWN. It shows that overall only a limited number of input factors affect the model's response in the subregion of behavioral models.

These influences are all in line with our expectation about the model's behavior. Heat capacity (*hcap*), conductivity *hcon*, and albedo *albsoil* clearly have an influence on simulations of sensible heat fluxes (objective function 1). Bulk heat properties are also influenced by the wetness of the soil and

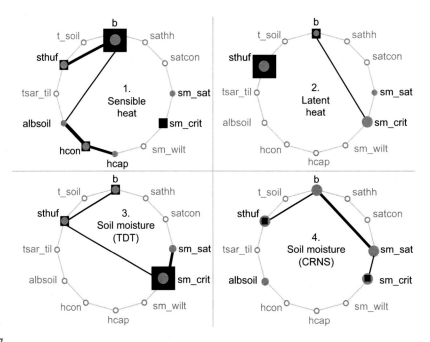

FIGURE 7.7

Summary of the multimethod approach for each objective function. *Black squares* denote influential input factors according to VBSA (the larger the marker, the higher the variance-based first-order sensitivity index). *Gray circles* denote influential input factors according to PAWN (again, marker size is proportional to PAWN sensitivity index). For both methods, sensitivity estimates are obtained focusing on behavioral models only. *Connecting lines* denote interactions as detected by cross-correlation.

therefore parameters *b*, *sm_sat*, and *sm_crit*. Latent heat simulations (objective function 2) are mostly influenced by the soil–plant parameter *sm_crit* and the initial wetness *sthuf* since latent heat fluxes are controlled by water losses through evapotranspiration, particularly in a semiarid site like Santa Rita Creosote. Similarly to latent heat, soil moisture simulations (objective functions 3 and 4) are also affected by *b*, *sm_sat*, *sm_crit*, and *sthuf*. Parameters *sathh*, *satcon*, and *sm-wilt*, instead, are not considered influential either directly or through interactions by any of the applied methods. This constitutes a strong indication that these parameters could be fixed to default values and excluded from model calibration.

As already highlighted, the initial value of soil moisture *(sthuf)* has a large influence on all objective functions both directly and through interactions. This means that either the warm-up period can be extended or the uncertainty about this initial condition needs to be reduced as a prerequisite for a robust calibration of the model. None of the objective functions instead was found sensitive to the initial soil temperature *t_soil* by any of the applied methods. An explanation could be that, at this site, the heat fluxes and soil moisture observations are predominantly related to the upper soil layers, where temperature changes relatively rapidly, and therefore the initial soil temperature *t_soil* has negligible effect after the 6-month warm-up period.

Finally, the input factors that influence the accuracy of soil moisture simulations compared with TDT observations (objective function 3) are the same as when comparing with CRNS counts (objective function 4) with the only exception of parameter *albsoil*, which is influential only when using CRNS counts. One possible explanation is that the CRNS footprint is larger and reduces the mismatch in supporting scale with respect to point measurements from TDT, thus allowing for identifying stronger links in the overall energy balance of JULES and therefore enabling calibration of the associated parameters.

4. CONCLUSIONS

In this chapter, we illustrated a multimethod approach for GSA of the JULES LSM. The ultimate aim is to support the calibration of the model for the Santa Rita Creosote site, where soil moisture and heat flux observations are available. In our application, we combine three GSA approaches (RSA, VBSA, and PAWN) to explore the relative influence and interactions of nine model parameters and three initial conditions on the model performance, as measured by four objective functions related to simulation of soil moisture, sensible heat, and latent heat fluxes.

RSA is first used to narrow down the uncertainty ranges of the input factors subject to GSA. Results show that, based on the available observations of soil moisture and heat fluxes, the uncertainty ranges of three parameters can be significantly reduced. RSA combined with correlation analysis is used to detect interactions. Results show strong interactions between one of the initial conditions (the initial soil moisture content) and several hydraulic parameters, which indicates that a robust estimation of these parameters will require extending the warm-up period or reducing the uncertainty about that initial condition.

The direct influence of input factors on the model's response is assessed by two different approaches, VBSA and PAWN. Since the two methods capture different aspects of the model's response, they measure a different level of sensitivity for the same input factors and therefore complement each other. Three parameters are not detected as influential by any of the applied methods, and therefore they can be reasonably determined as uninfluential. Hence, they can be fixed to default values and excluded from further calibration, thus reducing its computing burden at no significant loss of information.

In our application, all detected sensitivities are consistent with the links between the physical processes represented in the model, which constitutes an additional validation of the model's ability to reproduce the system behavior. Since two types of soil moisture observations were available, i.e., from point-scale sensors and from CRNS, we also used GSA to assess the information content of the two data sets. In this application, sensitivities are quite similar, although using CRNS seems to increase the sensitivity to the bare soil albedo, which might allow for a better estimation of that parameter during model calibration.

In conclusion, the multimethod approach here presented provides a picture of model's sensitivities from different angles and allows for addressing different questions (mapping, ranking, interactions) simultaneously. Also, it avoids the (subjective) choice of one specific method. On the other hand, since the different GSA methods are applied to the same sample of model evaluations, it does not increase the computational burden with respect to the application of an individual method. This is particularly important when each model evaluation is time consuming, as often is the case with complex LSMs. As such, we hope our approach will contribute to the uptake of GSA in the LSM community and hence contribute to a more effective use of increasingly available EO data sets for understanding, calibrating, and validating LSMs.

ACKNOWLEDGMENTS

F. Pianosi and T. Wagener were supported by the Natural Environment Research Council (Consortium on Risk in the Environment: Diagnostics, Integration, Benchmarking, Learning and Elicitation (CREDIBLE); grant number NE/J017450/1). R. Rosolem was supported by the Natural Environment Research Council (A MUlti-scale Soil moisture Evapotranspiration Dynamics study (AMUSED); grant number NE/M003086/1). J. Iwema was supported by the Queen's School of Engineering (University of Bristol) PhD Scholarship. We thank the Principal Investigator of the Ameriflux Santa Rita Creosote site, Shirley Papuga (University of Arizona), for providing data. Funding for AmeriFlux data resources was provided by the US. Department of Energy's Office of Science. The Matlab/Octave SAFE Toolbox for GSA is freely available for noncommercial purposes).

REFERENCES

Abramowitz, G., Leuning, R., Clark, M., Pitman, A., 2008. Evaluating the performance of land surface models. Journal of Climate 21, 5468–5481.

Alton, P., Mercado, L., North, P., 2007. A sensitivity analysis of the land-surface scheme JULES conducted for three forest biomes: biophysical parameters, model processes, and meteorological driving data. Global Biogeochemical Cycles 20 (GB1008).

Bakopoulou, C., Bulygina, N., Butler, A., McIntyre, N., 2012. Sensitivity analysis and parameter identifiability of the land surface model JULES at the point scale in permeable catchments. In: BHS Eleventh National Symposium, Hydrology for a Changing World, Dundee, pp. 1–7.

Bastidas, L.A., Gupta, H.V., Sorooshian, S., Shuttleworth, W.J., Yang, Z.L., 1999. Sensitivity analysis of a land surface scheme using multi-criteria methods. Journal of Geophysical Research 104 (D16), 19481–19490.

Best, M.J., Pryor, M., Clark, D.B., Rooney, G.G., Essery, R.L.H., Menard, C.B., Edwards, J.M., Hendry, M.A., Porson, A., Gedney, N., Mercado, L.M., Sitch, S., Blyth, E., Boucher, O., Cox, P.M., Grimmond, C.S.B., Harding, R.J., 2011. The Joint UK land environment simulator (JULES), model description – Part 1: energy and water fluxes. Geoscientific Model Development 4 (3), 677–699.

Beven, K.J., Cloke, H.L., 2012. Comment on "Hyperresolution global land surface modeling: meeting a grand challenge for monitoring Earth's terrestrial water" by Eric F. Wood et al. Water Resources Research 48 (W01801).

Blyth, E., Gash, J., Lloyd, A., Pryor, M., Weedon, G.P., Shuttleworth, J., 2010. Evaluating JULES land surface model energy fluxes using FLUXNET data. Journal of Hydrometeorology 11, 509–519.

Blyth, D., Clark, E., Ellis, R., Huntingford, C., Los, S., Pryor, M., Best, M., Sitch, S., 2011. A comprehensive set of benchmark tests for a land surface model of simultaneous fluxes of water and carbon at both the global and seasonal scale. Geoscientific Model Development 4, 255–269.

Cavanaugh, M.L., Kurc, S.A., Scott, R.L., 2011. Evapotranspiration partitioning in semiarid shrubland ecosystems: a two-site evaluation of soil moisture control on transpiration. Ecohydrology 4 (5), 671–681.

Cukier, R.I., Fortuin, C.M., Shuler, K.E., Petschek, A.G., Schaibly, J.H., 1973. Study of the sensitivity of coupled reaction systems to uncertainties in rate coefficients. I Theory. The Journal of Chemical Physics 59 (8), 3873–3878.

DeLannoy, G.J.M., Koster, R.D., Reichle, R.H., Mahanama, S.P.P., Liu, Q., 2014. An updated treatment of soil texture and associated hydraulic properties in a global land modeling system. Journal of Advances in Modeling Earth Systems 6, 957−979.

De Vries, D.A., 1963. Physics of Plant Environment, Chapter Thermal Properties of Soils, pp. 210−235.

Dirmeyer, P.A., Koster, R.D., Guo, Z., 2006. Do global models properly represent the feedback between land and atmosphere? Journal of Hydrometeorology 7 (6), 1177−1198.

Efron, B., Tibshirani, R., 1993. An introduction to the bootstrap. Chapman & Hall/CRC.

Forrester, A., Sobester, A., Keane, A., 2008. Engineering Design via Surrogate Modelling: A Practical Guide. John Wiley & Sons.

Gan, Y., Liang, X.-Z., Duan, Q., Choi, H.I., Dai, Y., Wu, H., 2015. Stepwise sensitivity analysis from qualitative to quantitative: application to the terrestrial hydrological modeling of a conjunctive surface-subsurface process (CSSP) land surface model. Journal of Advances in Modeling Earth Systems 7 (2), 648−669.

Hou, Z., Huang, M., Leung, L.R., Lin, G., Ricciuto, D.M., 2012. Sensitivity of surface flux simulations to hydrologic parameters based on an uncertainty quantification framework applied to the Community Land Model. Journal of Geophysical Research: Atmospheres 117 (D15).

Huang, M., Hou, Z., Leung, L.R., Ke, Y., Liu, Y., Fang, Z., Sun, Y., 2013. Uncertainty analysis of runoff simulations and parameter identifiability in the Community Land Model: evidence from MOPEX basins. Journal of Hydrometeorology 14, 1754−1772.

Iwema, J., Rosolem, R., Baatz, R., Wagener, T., Bogena, H.R., 2015. Investigating temporal field sampling strategies for site-specific calibration of three soil moisture − neutron flux interaction models. Hydrology and Earth System Sciences Discussion 19, 3203−33216.

Li, J., Duan, Q.Y., Gong, W., Ye, A., Dai, Y., Miao, C., Di, Z., Tong, C., Sun, Y., 2013. Assessing parameter importance of the common land model based on qualitative and quantitative sensitivity analysis. Hydrology and Earth System Sciences 17 (8), 3279−3293.

Moene, A., Van Dam, J., 2014. Transport in the Atmosphere-vegetation-Soil Continuum, first ed. Cambridge University Press, NY, USA.

ORNL-DAAC, 2015. Ameriflux Web Page. http://ameriflux.ornl.gov.

Papuga, S., 2009. Highlight fluxnet site santa rita creosotebush. FluxLetter: The Newsletter of FLUXNET 2 (4), 1−4.

Pianosi, F., Sarrazin, F., Wagener, T., 2015. A matlab toolbox for global sensitivity analysis. Environmental Modelling & Software 70, 80−85.

Pianosi, F., Wagener, T., 2015. A simple and efficient method for global sensitivity analysis based on cumulative distribution functions. Environmental Modelling & Software 67, 1−11.

Pianosi, F., Wagener, T., Beven, K., Freer, J., Hall, J., Rougier, J., Stephenson, D., 2016. Sensitivity analysis of environmental models: a systematic review with practical workflow. Environmental Modelling & Software 79, 214−232.

Robinson, D., Campbell, C., Hopmans, J., Hornbuckle, B., Jones, S., Knight, R., Ogden, F., Selker, J., Wendroth, O., 2008. Soil moisture measurement for ecological and hydrological watershed-scale observatories: a review. Vadose Zone Journal 7, 358−389.

Romano, N., 2014. Soil moisture at local scale: measurements and simulations. Journal of Hydrology 516, 6−20.

Romano, N., Angulo-Jaramillo, R., Javaux, M., Van der Ploeg, M.J., 2012. Interweaving monitoring activities and model development towards enhancing knowledge of the soil-plant-atmosphere continuum. Vadose Zone Journal 11 (3).

Rosero, E., Yang, Z.-L., Wagener, T., Gulden, L., Yatheendradas, S., Niu, G., 2010. Quantifying parameter sensitivity, interaction, and transfer- ability in hydrologically enhanced versions of the Noah land surface model over transition zones during the warm season. Journal of Geophysical Research 115 (D03106).

Rosolem, R., Gupta, H., Shuttleworth, W., Zeng, X., deGoncalves, L., 2012. A fully multiple-criteria implementation of the Sobol' method for parameter sensitivity analysis. Journal of Geophysical Research 117 (D07103).

Rosolem, R., Hoar, T., Arellano, A., Anderson, J.L., Shuttleworth, W.J., Zeng, X., Franz, T.E., 2014. Translating aboveground cosmic-ray neutron intensity to high-frequency soil moisture profiles at sub-kilometer scale. Hydrology and Earth System Sciences 18, 4363—4379.

Saltelli, A., Annoni, P., Azzini, I., Campolongo, F., Ratto, M., Tarantola, S., 2010. Variance based sensitivity analysis of model output. Design and estimator for the total sensitivity index. Computer Physics Communications 181 (2), 259—270.

Saltelli, A., Ratto, M., Andres, T., Campolongo, F., Cariboni, J., Gatelli, D., Saisana, M., Tarantola, S., 2008. Global Sensitivity Analysis, The Primer. Wiley.

Shin, M.-J., Guillaume, J.H., Croke, B.F., Jakeman, A.J., 2013. Addressing ten questions about conceptual rainfall-runoff models with global sensitivity analyses in R. Journal of Hydrology 503, 135—152.

Shuttleworth, W.J., Rosolem, R., Zreda, M., Franz, T., 2013. The cosmic-ray soil moisture interaction code (cosmic) for use in data assimilation. Hydrology and Earth System Sciences 17, 3205—3217.

Skott, R.L., Cable, W.L., Hultine, K.R., 1990. The ecohydrologic significance of hydraulic redistribution in a semiarid savanna. Water Resources Research 44 (2).

Smits, K.M., Sakaki, T., Limsuwat, A., Illangasekare, T.H., 2010. Thermal conductivity of sands under varying moisture and porosity in drainage-wetting cycles. Vadose Zone Journal 9, 172—180.

Sobol', I., 1990. Sensitivity analysis for non-linear mathematical models. Mathematical Modelling and Computational Experiment 1, 112—118. Translated from Russian: Sobol', I.M., 1993. Sensitivity estimates for nonlinear mathematical models. Matematicheskoe Modelirovanie 2, 407—414.

Spear, R., Hornberger, G., 1980. Eutrophication in peel inlet. II. Identification of critical uncertainties via generalized sensitivity analysis. Water Research 14 (1), 43—49.

Strong, M., Oakley, J.E., Brennan, A., 2014. Estimating multiparameter partial expected value of perfect information from a probabilistic sensitivity analysis sample: a nonparametric regression approach. Medical Decision Making 34 (3), 311—326.

Wagener, T., Boyle, D., Lees, M., Wheater, H., Gupta, H., Sorooshian, S., 2001. A framework for development and application of hydrological models. Hydrology and Earth System Sciences 5, 13—26.

Wang, J., Li, X., Lu, L., Fang, F., 2013. Parameter sensitivity analysis of crop growth models based on the extended Fourier Amplitude Sensitivity Test method. Environmental Modelling & Software 48, 171—182.

Wood, E.F., Roundy, J.K., Troy, T.J., van Beek, L.P.H., Bierkens, M.F.P., Blyth, E., de Roo, A., Döll, P., Ek, M., Famiglietti, J., Gochis, D., van de Giesen, N., Houser, P., Jaff, P.R., Kollet, S., Lehner, B., Lettenmaier, D.P., Peters-Lidard, C., Sivapalan, M., Sheffield, J., Wade, A., Whitehead, P., 2011. Hyperresolution global land surface modeling: meeting a grand challenge for monitoring earth's terrestrial water. Water Resources Research 47 (5).

Wösten, J., Pachepsky, Y.A., Rawls, W., 2001. Pedotransfer functions: bridging the gap between available basic soil data and missing soil hydraulic characteristics. Journal of Hydrology 251, 123—150.

Young, P.C., Spear, R.C., Hornberger, G.M., 1978. Modeling badly defined systems: some further thoughts. In: Proceedings SIMSIG Conference, Canberra, pp. 24—32.

Zreda, M., Shuttleworth, W., Zeng, X., Zweck, C., Desilets, D., Franz, T., Rosolem, R., 2012. Cosmos: the cosmic-ray soil moisture observing system. Hydrology and Earth System Sciences 16, 4079—4099.

GLOBAL SENSITIVITY ANALYSIS FOR SUPPORTING HISTORY MATCHING OF GEOMECHANICAL RESERVOIR MODELS USING SATELLITE INSAR DATA: A CASE STUDY AT THE CO_2 STORAGE SITE OF IN SALAH, ALGERIA

8

J. Rohmer, A. Loschetter, D. Raucoules
BRGM, Orléans, France

CHAPTER OUTLINE

1. INTRODUCTION

The space-borne differential Synthetic Aperture Radar (SAR) interferometry is a powerful monitoring technique to reveal surface deformation response induced by volumetric changes (through time and

Sensitivity Analysis in Earth Observation Modelling. http://dx.doi.org/10.1016/B978-0-12-803011-0.00008-2

space) resulting from subsurface fluid withdrawal/injection operations into porous reservoirs (Bell et al., 2008). In the particular case of CO_2 geological storage, this technique has successfully been implemented to measure the surface deformation at In Salah Gas Project in Algeria, where 0.5–1 million tons of CO_2 per year have been injected (e.g., see site context described by Mathieson et al., 2009). This analysis revealed an average ground uplift of 5 mm/year.

In the CO_2 storage domain, available characterization data on the underground is scarcer than in traditional oil and gas exploration (with few wells drilled) (see, e.g., Bouc et al., 2010). An efficient approach to constrain such uncertainty is to rely on a history-matching exercise, which basically aims at calibrating the reservoir model to "fit" the observations. This problem can be solved from a Bayesian perspective; from a data assimilation perspective using, for instance, Ensemble Kalman Filter; or from an optimization-based perspective (see an overview by Oliver and Chen, 2011). To date, these exercises applied on the InSAR data at In Salah have mainly been used to get better insight into the flow behavior of the reservoir and caprock formations (e.g., Mathieson et al., 2009; Vasco et al., 2008; de la Torre Guzman et al., 2014). Therefore, we focused in the present study on the geomechanical properties of the main rock formations. Yet, such a task is hindered by the large number of sources of uncertainty, which is typically of the order of 10 or more. The problem of calibration is generally ill-posed, because there are too many model parameters to take into account. To set priorities on the model parameters, a preliminary step can rely on global sensitivity analysis (GSA), and more specifically on variance-based techniques (Saltelli et al., 2008), which allows handling such a problem by accounting for the whole range of variations of the unknown parameters and without introducing assumptions on the relationship between model inputs and outputs (the analysis is said to be "model free"). However, such type of analysis has a "price," namely, the large number of different simulations to run (typically over 1000). This requirement is generally incompatible with the huge computational cost of the reservoir simulation code (>several hours) so that GSA is usually combined with meta-modeling (also known as surrogate model, response surface) techniques (e.g., Rohmer and Foerster, 2011). This procedure basically consists in replacing the reservoir simulation code (simulator) by a mathematical approximation, which corresponds to a function constructed using a limited number of time-consuming simulations (typically 50–100), and aims at reproducing the behavior of the "true" model in the domain of model input parameters (here the properties of the subsurface) and at predicting the model responses (here the ground displacements) with a negligible computation time (further details are provided in Section 3.2).

The chapter is organized as follows. Section 2 describes the surface deformation measured by processing satellite images during CO_2 injection at the KB-501 well of In Salah (Algeria) over the time period between 2004 and 2009. Section 3 describes in more details the proposed workflow and the associated methods (GSA and metamodels) to overcome the aforedescribed difficulties. Finally, Section 4 describes the history-matching exercise conducted using the geomechanical reservoir model and discusses the results regarding storage site characterization.

2. CASE STUDY

In this section, we describe the spatiotemporal ground displacements measured during CO_2 injection at the KB-501 well, In Salah (Section 2.1), and the setup of a geomechanical reservoir model used to reproduce these observations (Section 2.2). In the following, the observed and numerically simulated ground displacements, U_z and $\tilde{U}_z$, at a given spatial location of coordinates' vector s and at a given time t are all expressed in the line of sight of the satellite.

2.1 SURFACE DEFORMATION AT THE KB-501 WELL OF IN SALAH SITE

The surface deformation during CO_2 injection at the KB-501 well of In Salah (Algeria, see Fig. 8.1) was analyzed. A total of 31 satellite images over the period 2003–09 from the descending orbit (ASAR-sensor of ENVISAT mission) were processed for the Persistent Scatterers Interferometry (PSI) analysis (e.g., Ferretti et al., 2000) using the GAMMA/IPTA software (Wegmuller et al., 2013). Note

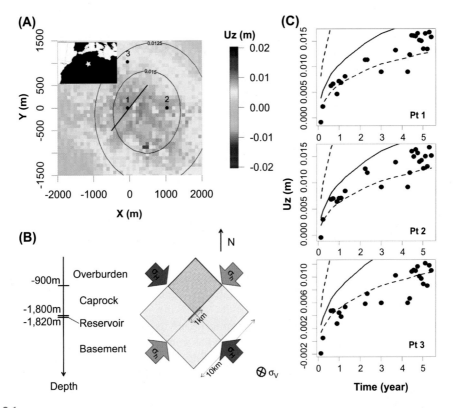

FIGURE 8.1

(A) Spatial distribution of the ground displacements (measured in the line of sight of the sensor) at In Salah, Algeria (well KB-501), after 5 years of injection. The *inset* indicates the location in Algeria. The *horizontal line* marks the horizontal injection well, and the *dots* indicate the observation points. The *black isocontours* indicate the median values calculated using a set of 100 model runs (with different input parameters). (B) Distribution over depth of the rock formations considered in the study (left) and upper view of the reservoir model geometry and boundary conditions (right). (C) Time series of the measured ground displacements (*dots*) when compared with the median (*straight line*), lower, and upper bounds (*dashed lines*) for a set of 100 model runs at three observation points.

that the analysis of surface deformation is restricted to the period ranging from mid-2004 (start of injection) to end of 2009 (covered by only 25 images of the set).

Due to the ideal surface conditions (arid rocky desert), the density of Persistent Scatterers is very large (density > 1000/km^2). Fig. 8.1A presents the spatial distribution of the bilinearly interpolated (100 × 100 grid cells of ~90 × 80 m^2) surface deformations (denoted U_z) measured in the line of sight of the satellite after 5 years of injection. In the following, we focus on the 4 by 3 km^2 area around the horizontal injection well referred to as the "region of interest". The deformation pattern associated with the CO_2 injection corresponds to an anisotropic ground heave with a principal axis oriented SE-NW. The bottom left-hand corner corresponds to a zone of subsidence related to gas production, whereas the top right-hand corner corresponds to a "No Data" zone. Fig. 8.1C presents the time series of surface displacement at three observation points.

2.2 THREE-DIMENSIONAL HYDROMECHANICAL MODEL OF KB-501

The modeling of ground uplift (surface heave) during CO_2 injection at In Salah has been addressed by several past studies, either through semianalytical (Vasco et al., 2008) or through full-field-coupled flow geomechanical numerical simulations (Rutqvist et al., 2010; Morris et al., 2011; Rinaldi and Rutqvist, 2013). In this study, we concentrated on the second approach and a reservoir geomechanical model was setup following the assumptions made by Rutqvist et al. (2010). The hydromechanical processes related to the CO_2 injection at KB-501 well at In Salah were solved using a one-way coupling between a multiphase flow transport simulator, TOUGH2 (Pruess, 1999), and a geo-mechanical code, Code_Aster (details on the numerical implementation are available in EdF R&D, 2012), as described by Rohmer and Seyedi (2010). The multiphase fluid-flow simulation was con-ducted with the fluid property module ECO2N (Pruess, 2005), which contains a comprehensive description of the thermodynamic and thermophysical properties of water—NaCl—CO_2 mixtures needed for the analysis of CO_2 storage in brine-saturated formations. The injection was carried out through a 1815-m-deep, ~1-km-long horizontal injection well (aligned in a NE-SW direction, see Fig. 8.1) considering a constant injection rate (taken as the average over time using the data by Bissell et al., 2011) of 10.3 kg/s (~0.32 Mt/y). Four spatially homogeneous rock formations were considered: overburden (depth from 0 to 900 m), caprock (900—1,800 m), reservoir (1800—1820 m), and base-ment (>1820 m), see Fig. 8.1B. Due to the symmetry of the problem, one-quarter of the three-dimensional system (10 km by 10 km by 2.5 km) was considered. A no-flow boundary condition was considered on the bottom side and on each of the lateral sides, and nil lateral displacements were imposed on the SW and SE lateral sides (for symmetry reasons), whereas mechanical loads were imposed on the NW and NE lateral sides (Fig. 8.1B). Model assumptions were primarily based on Rutqvist et al. (2010). Elastic properties of the different rock formations are pervaded with uncertainty: best estimates of the lower and upper bounds on those model parameters are reported in Table 8.1. Besides, the initial stress state defined by a ratio of minimum horizontal to vertical total stress K_h and by the ratio of maximum horizontal to vertical total stress K_H is also considered uncertain with K_h and K_H, assumed to, respectively, range from 0.6 to 0.8 and from 1.0 to 1.5. Note that we assumed that the flow parameters used by Rutqvist et al. (2010), namely, permeability, porosity, and multiphase flow parameters, are affected by lower levels of uncertainty. Thus we restricted the history-matching ex-ercise to the elastic parameters of the four formations and to the two parameters defining the initial stress state. In total, 10 uncertain parameters were considered.

Table 8.1 Lower and upper bounds assumed for the rock formations' elastic properties						
Property	**Symbol**	**Unit**	**Overburden (o)**	**Caprock (c)**	**Reservoir (r)**	**Basement (b)**
Young's modulus	E	GPa	0.5–2.5	10–20	2.0–20	10–20
Poisson's ratio	ν	–	0.15–0.30	0.10–0.20	0.15–0.35	0.10–0.20

Fig. 8.1A provides the isocontour at 0.01 and 0.015 m derived from the median value of a set of 100 numerical simulations, whose input parameters were randomly and uniformly chosen within the intervals defined by the bounds reported in Table 8.1. In Fig. 8.1C, we also report the numerically simulated time series (straight line: median, dashed lines: lower and upper bounds). We can notice that (1) the spatial distribution of numerically calculated $\tilde{U}_z$ presents a quasicircular shape, which is decentered regarding the injection well (recall that the model results were converted so that they are in the line of sight of the satellite), whereas the observed U_z presents a more complex spatial pattern; (2) the observations lie to a large extent within the intervals between the median and the lower bound of the numerical simulations: this indicates that the reservoir model tends to overestimate the observations.

3. METHODS

In this section, we first describe the key ingredients of the proposed history-matching workflow, namely GSA (Section 3.1) and metamodeling techniques (Section 3.2). We then describe in more details the different steps of the overall procedure (Section 3.3).

3.1 VARIANCE-BASED GLOBAL SENSITIVITY ANALYSIS

Different methods of GSA exist in the literature (see, e.g., Iooss and Lemaître, 2015; Ireland et al., 2015; Petropoulos et al., 2014). Here, it is chosen to focus on variance-based sensitivity analysis (VBSA), which can be considered among the most advanced methods to fulfill the requirements of GSA. The basic concepts of VBSA are first briefly introduced in the present section. For a more complete introduction, the interested reader can refer to Saltelli et al. (2008) and the references therein.

Let us define f as the geomechanical reservoir model used to support the history-matching study. Consider the m-dimensional vector X as a random vector of independent random variable X_i (here the mechanical properties of the rock formations), with $i = 1,2,...,m$, then the output (here the differences between the observed and the numerically calculated surface displacements) $Y = f(X)$ is also a random variable (as a function of a random vector). VBSA aims at determining the part of the total unconditional variance Var(Y) of the output Y of f resulting from the variation of each input random variable X_i. Assuming that the variance can adequately capture the uncertainty, this analysis relies on the functional analysis of variance decomposition of f based on which the Sobol' indices (ranging between 0 and 1) can be defined:

$$S_i = \frac{\text{Var}[\text{E}(Y|X_i)]}{\text{Var}(Y)}, \quad S_{ij} = \frac{\text{Var}[\text{E}(Y|X_i, X_j)]}{\text{Var}(Y)} - S_i - S_j \tag{8.1}$$

The first-order index S_i is referred to as "the main effect of X_i" and can be interpreted as the expected amount of $Var(Y)$ (i.e., representing the uncertainty in Y) that would be reduced if it was possible to learn the true value of X_i. This index provides a measure of importance useful to rank in terms of importance the different input parameters within a "factors' prioritizing setting" (Saltelli et al., 2008). The second-order term S_{ij} measures the combined effect of both parameters X_i and X_j. Higher order terms can be defined in a similar fashion. The total number of sensitivity indices reaches $2^m - 1$. In practice, the sensitivity analysis is generally limited to the pairs of indicators corresponding to the main effect S_i and to the total effect S_{Ti} of X_i (Saltelli et al., 2008). The latter is defined as follows:

$$S_{Ti} = 1 - \frac{Var[E(Y|X_{-i})]}{Var(Y)} \tag{8.2}$$

where $X_{-i} = (X_1,\ldots, X_{i-1}, X_{i+1},\ldots X_m)$. The total index corresponds to the fraction of the uncertainty in Y that can be attributed to X_i and its interactions with all other input parameters. $S_{Ti} = 0$ means that the input factor X_i has no effect so that X_i can be fixed at any value over its uncertainty range within a "factors' fixing" setting (as described in Saltelli et al., 2008).

Different algorithms are available for the estimation of the Sobol' indices like (extended) Fourier Amplitude Sensitivity Test (Saltelli et al., 1999) or the Sobol' algorithm (Sobol', 1993). A more extensive introduction is provided by Saltelli et al. (2008: Chapter 4). In the following, we relied on the algorithm described by Saltelli et al. (2010), which requires running $N \times (m + 2)$ model simulations, with N being the number of Monte Carlo samples and m the number of uncertain parameters. Considering N of the order of 1000 and $m = 10$, the total cost of GSA should reach 12,000 numerical simulations, which is obviously hardly achievable by using a large-scale reservoir model as the one described in Section 2.2: this has a computation time cost of several hours for a single model run.

3.2 PRINCIPLES OF METAMODELING

Metamodeling consists in replacing f by a mathematical approximation referred to as "metamodel" (also named "response surface," or "surrogate model"). This corresponds to a function constructed using a few computer experiments (typically 50–100), i.e., a limited number of time-consuming simulations (training data), whose input parameters can be randomly chosen depending on prior pieces of information (like the lower and upper bounds, a best estimate, etc.). Once constructed, the predictive quality should be assessed, i.e., the expected level of fit of the metamodel to a data set that is independent of the original training data that were used to construct the metamodel, i.e., to "yet-unseen" data. This can be achieved using a cross-validation procedure (e.g., Hastie et al., 2009; Petropoulos et al., 2015), which involves (1) randomly splitting the initial training data into q equal subsets (q is typically between 1 and 10), (2) removing each of these subsets in turn from the initial set and fitting a new metamodel using the remaining $q - 1$ subsets, and (3) using the subset removed from the initial set as a validation set and estimating it using the new metamodel. Using the residuals computed at each iteration of this procedure, a coefficient of determination (denoted R^2) can be computed using a formula similar to Eq. (8.3). For small training sets, the cross-validation procedure with $q = 1$ is usually used corresponding to the so-called leave-one-out cross-validation procedure. In this case, R^2 yields as follows

$$R^2 = 1 - \frac{\sum_{i=1}^{n} (y_i - \hat{y}_i)^2}{\sum_{i=1}^{n} (y_i - \bar{y})^2} \tag{8.3}$$

where y_i corresponds to the observations (i.e., the model results) in the validation set (for i from 1 to n), $\bar{y}$ to the corresponding mean and $\hat{y}$ to estimated values using the metamodel. The coefficient R^2 provides a metric of the quality of prediction so that a coefficient close to 1 indicates that the metamodel is successful in matching the validation data.

After this validation, the metamodel can be used to reproduce the behavior of the "true" model in the domain of model input parameters (here the properties of the rock formations and the initial stress state) and to predict the model responses with a negligible computation time, hence making GSA feasible. In the following, we focus on kriging metamodeling (Forrester et al., 2008).

3.3 INTRODUCTION TO KRIGING METAMODEL

We introduce here the basic concepts of kriging metamodeling, which can be viewed as an extension to computer experiments of the kriging method used for spatial data interpolation and originally developed by Krige (1951) for mining applications. This nonparametric technique presents several attractive features. It is flexible to any kind of functional form of the simulator. In particular, it introduces less restrictive assumptions on the functional form of the simulator than a polynomial model would imply. It is an exact interpolator, which is an important feature when the simulator is deterministic; it provides a variance estimate of the prediction, the latter being very useful to guide the selection of future training samples according to the target of the optimization problem (see, e.g., Jones, 2001). For a more complete introduction to kriging metamodeling and full derivation of equations, the interested reader can refer to Forrester et al. (2008).

The kriging model considers the deterministic (i.e., not random) response of the simulator $y = f(x)$ as a realization of a Gaussian stochastic process F so that $f(x) = F(x, \omega)$ where ω belongs to the underlying probability space Ω. In the following, we use the notation $F(x)$ for the process and $F(x, \omega)$ for one realization. The process F results from the summation of two terms:

- $f_0(x)$, the deterministic mean function, which is usually modeled by a constant or a linear model and represents the trend of f;
- $Z(x)$, the Gaussian centered stationary stochastic process (with zero mean and covariance described later), which describes the deviation (i.e., departure) of the model from its underlying trend f_0.

The stochastic process Z is characterized by the covariance matrix C, which depends on the variance σ_Z^2 and on the correlation function R, which governs the degree of correlation through the use of the vector of length-scale parameters ω between any input vectors. The covariance between $Z(u)$ and $Z(v)$ is then expressed as $C(u; v) = \sigma_Z^2 \cdot R(u; v)$, where $u = (u_1; u_2; \ldots; u_m)$ and $v = (v_1; v_2; \ldots; v_m)$ are two input vectors of dimension m. A variety of correlation (and covariance) functions have been proposed in the literature (see, e.g., Stein, 1999), and we used the Matérn correlation function (with parameter $v = 5/2$), which presents more flexibility than the commonly used Gaussian correlation function, since it allows regulating the degree of differentiability of the underlying random process in addition to the length-scale parameters. Let us define X_D the design matrix composed of the vectors of input parameters x (i.e., the training samples) to be simulated so that $X_D = (x^{(1)}; .x^{(2)}; \ldots; x^{(n)})$ and y_D is the vector of model response associated with each selected training samples so that $y_D = (y^{(1)} = f(x^{(1)}); y^{(2)} = f(x^{(2)}); \ldots; y^{(n)} = f(x^{(n)}))$.

Under the aforedescribed assumptions, the distribution of the model response for a new input vector of input parameters x^* follows a Gaussian distribution conditional on the design matrix X_D and on the corresponding model results y_D with expected value $\widetilde{y}_m$ for the new configuration x^* defined by Eq. (8.4).

$$\widetilde{y}(x^*) = E(F(x^*)|X_D; \; \xi_D) = \widehat{y}_m + r(x^*)^T \cdot R_D^{-1} \cdot (\xi_D - I \cdot \widehat{y}_m) \tag{8.4}$$

with the constant $\widehat{y}_m = \left(I^T \cdot R_D^{-1} \cdot I\right)^{-1} \cdot \left(I^T \cdot R_D^{-1} \cdot y_D\right)$ and R_D the correlation matrix calculated for the training data.

The conditional mean in Eq. (8.4) is used as a predictor, i.e., the "best estimate." This formulation is categorized as "ordinary kriging" and is the most common version of kriging used in engineering (Forrester et al., 2008). A more general form of kriging equations exists, known as "universal kriging," and allows computing the deterministic mean function f_0 as a polynomial regression (generally of low order) with unknown coefficients (see Forrester et al., 2008 for details).

3.4 DESCRIPTION OF THE WORKFLOW

As described in the introduction, there are several strategies for conducting a history-matching exercise. We focus in the following discussion on an optimization-based strategy and aim at identifying the values of the reservoir model parameters (Table 8.1), which minimize the root mean square error (*RMSE*) at any spatial location s of the region of interest as displayed in Fig. 8.1 over the time period of injection from 2004 to 2009 as follows:

$$RMSE(s) = \sqrt{1/n_t \sum_{t=1}^{n_t} \left(U_z(s, \, t) - \widetilde{U}_z(s, \, t)\right)^2} \tag{8.5}$$

where n_t is the number of time instants (here 25), $U_z(s, t)$ corresponds to the PSI-derived (observed) ground displacement at the given location s and at the given time instant t, and $\widetilde{U}_z$ is the ground displacement estimated by the reservoir numerical model.

Since there are a large number of uncertainty sources, the objective is then to identify which input parameters should be characterized in priority to minimize *RMSE*. The main steps of the proposed workflow are as follows:

1. We first simulate $\widetilde{U}_z$ with values of the model input parameters randomly and uniformly chosen within the interval defined by the bounds reported in Table 8.1. On this basis, a kriging-type metamodel is constructed at each spatial location s to approximate *RMSE(s)*.
2. The predictive quality of the surface response can then be validated through a leave-one-out cross-validation.
3. At each spatial location, the validated metamodel can be used to compute the main and total effects assigned to each uncertain parameters using the algorithm of Saltelli et al. (2010). The parameter with the largest main effect is the one influencing the most the variability of *RMSE(s)*, whereas the parameters with low values of total effects can be considered noninfluential.
4. Using the results of GSA, the number of uncertain parameters can be reduced. The relationship between *RMSE(s)* and this limited number of input parameters can be approximated by a new kriging-type metamodel constructed with a new set of numerical simulations. Once validated, the global minimum of this approximated relationship can be sought by solving the associated

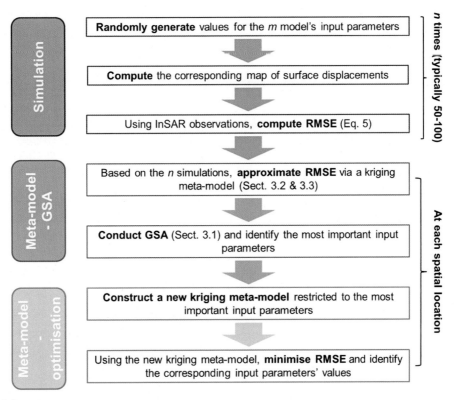

Simulation

| Randomly generate values for the m model's input parameters |

| Compute the corresponding map of surface displacements |

| Using InSAR observations, compute RMSE (Eq. 5) |

n times (typically 50-100)

Meta-model - GSA

| Based on the n simulations, approximate RMSE via a kriging meta-model (Sect. 3.2 & 3.3) |

| Conduct GSA (Sect. 3.1) and identify the most important input parameters |

Meta-model - optimisation

| Construct a new kriging meta-model restricted to the most important input parameters |

| Using the new kriging meta-model, minimise RMSE and identify the corresponding input parameters' values |

At each spatial location

FIGURE 8.2

Flowchart of the whole procedure for conducting the GSA meta model optimization-based history matching. *GSA*, global sensitivity analysis; *RMSE*, root mean square error.

optimization problem using the metamodel: this corresponds to the values of the most important rock properties for which the differences between the observations and the model results are minimized (i.e., *RMSE*).

The flowchart in Fig. 8.2 summarized the whole procedure.

4. APPLICATION

In this section, we apply the workflow described in Section 3.4 (Fig. 8.2). The most influential input parameters regarding the discrepancy between the observations and the model responses are first identified (Section 4.1). This first stage eases the optimization-like history-matching task (Section 4.2). This two-stage procedure was applied at each spatial location of the region of interest discretized over an evenly spaced grid with each grid cell being of approximately 90 by 80 m^2. Statistical analysis was conducted using packages named "sensitivity" (http://cran.r-project.org/web/packages/sensitivity/index.html) and "DiceKriging" developed by Roustant et al. (2012) of the R software (R Development Core Team, 2014).

4.1 REDUCING THE NUMBER OF UNCERTAINTY INPUT PARAMETERS

The preliminary "metamodel—global sensitivity analysis" was conducted using 100 simulations. Note that preliminary tests showed us that this minimum number of simulations was necessary to achieve a satisfactory level of predictive quality (see later). Based on these training data, a kriging-type of 10 dimensions (Matérn-type covariance model with constant trend and without nugget effect) was constructed to approximate $RMSE$ at each spatial location of the region of interest. The predictive quality of the surface response was validated through a leave-one-out cross-validation procedure, which yielded a high coefficient of determination greater than 95% for all kriging metamodels (Fig. 8.3B). Fig. 8.3A illustrates this validation step by depicting the observed (derived from the reservoir simulation) versus the kriging-based approximated $RMSE$ considering the three observation points

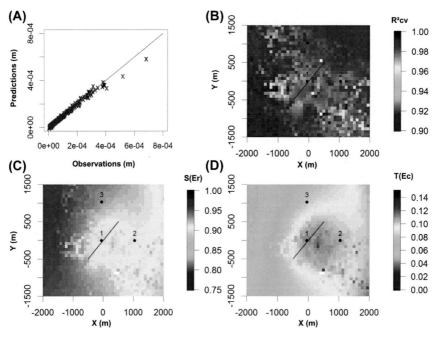

FIGURE 8.3

(A) Relationship between the simulated root mean square error (named "observations") and the ones approximated (named "predictions") by a kriging metamodel constructed at three observation points (black, observation point 1; blue, N°2; red, N°3). The closer the cross-type markers from the straight line, the better the approximation. (B) Coefficient of determination resulting from the cross-validation (the closer to 1, the better the approximation). (C) Main effect denoted S associated with the reservoir Young's modulus E_r. (D) Total effect denoted T associated with the caprock Young's modulus E_c.

(see locations in Fig. 8.1A). The closer the dots from the diagonal (marked by a straight black line), the higher the predictive quality of the metamodel. For all three metamodels, the coefficients of determination all exceeded 98%.

On this basis, the kriging models were used to compute the sensitivity measures (Sobol' indices of first order) at each spatial location through the approach of Saltelli et al. (2010) using 5000 Monte Carlo samples. Preliminary tests indicated that such a number of random samples yielded satisfactory convergence of the sensitivity measures to two decimal places. Hence, a total of $5000 \times (10 + 2) = 60,000$ model runs were necessary for computing the sensitivity indices. Recalling the computation time of several hours for a single simulation, it is worth noticing that GSA would not have been possible by directly using the reservoir model. To reach a computation time of 1 day, an appropriate grid-computing architecture composed of thousands of computer units would have been necessary.

Fig. 8.3C shows the map of the main effects over the whole region of interest: values all exceed 80% and are all associated with the reservoir Young's modulus E_r. The second most important parameter corresponds to the caprock Young's modulus E_c, but with limited influence (<10%). Regarding the total effects, the analysis of total effects revealed that all parameters have very low values of the order of <2% except for E_r (total effects >90%) and E_c ($\sim 11\%$ in average over space as shown in Fig. 8.3D). Table 8.2 provides the numerical values of the total and main effects for the three observation points considering both Young's moduli.

4.2 CALIBRATION OF THE RESERVOIR YOUNG'S MODULUS

Recall that a total effect close to zero indicates that the corresponding parameter has negligible influence, so that the error made by fixing them at any nominal value is twice the value of the total index (provided that the corresponding random parameter associated to the total effect is uniformly distributed, see Sobol' et al., 2007). Here, GSA can be used to restrict the optimization problem to one single property, namely E_r. Yet, at some locations, E_c presents some low-to-moderate values as shown in Fig. 8.3D: we further investigated the validity of reducing the analysis to one single parameter (instead of both E_r and E_c) by conducting the following procedure: (1) construct a one-dimensional (1D) kriging metamodel (instead of a kriging of 10 dimensions as for Section 4.1) using the same set of training data (100 simulations) as for the "metamodel-GSA" approach as aforedescribed, (2) use the 1D kriging metamodel to approximate *RMSE* at any spatial location, and (3) compute the error and summarize the approximation quality by a coefficient of determination. This procedure was applied at

Table 8.2 Main (S) and total (T) effects (%) at the three observation points		
Observation Point	Caprock Young's Modulus E_c (GPa)	Reservoir Young's Modulus E_r (GPa)
N°1	S = 9.0	S = 79.2
	T = 11.3	T = 89.5
N°2	S = 7.2	S = 79.3
	T = 10.1	T = 90.9
N°3	S = 8.5	S = 79.0
	T = 10.5	T = 89.5

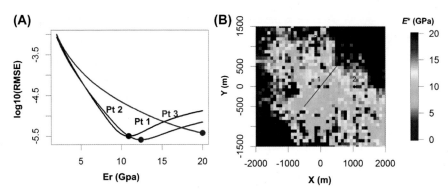

FIGURE 8.4

(A) Relationship between the logarithm (base 10) of root mean square error (*RMSE*) and E_r at three different observation points (see location in Fig. 8.1A). The optimal value E^* corresponds to 10.9, 12.4, and 20 GPa for observation points N°1 (black), N°2 (blue), and N°3(red), respectively. (B) Spatial distribution of E^* estimated at each measurement point.

any spatial location of the region of interest and showed that on average the coefficient of determination reached values of the order of 85%, which was considered satisfactory.

On this basis, we chose to run a series of 40 additional reservoir simulations by randomly varying E_r within the interval [2.0 GPa, 20 GPa] and fixing the nine other parameters (noninfluential) at their nominal values (using the best estimates provided by Rutqvist et al., 2010). A new 1D kriging metamodel was constructed at each spatial location and was used to estimate the relationship between the logarithm (base 10) of *RMSE* and E_r (see Fig. 8.4A for the three observation points). In this manner, the optimal value E^*, which minimizes *RMSE* was selected. At some location, the solution to the optimization problem corresponds to the upper bound assumed for E_r, whereas in the central area of the region of interest, the solution reaches values of the order of 10–15 GPa. Fig. 8.4B depicts the spatial distribution of E^*.

Fig. 8.4B shows that the differences with satellite-based observations are better explained (i.e., by minimizing *RMSE*) using the reservoir model considering a reservoir rock formation with a heterogeneous spatial distribution of E_r ranging from 10 to 15 GPa in the vicinity of the horizontal injection well and with more rigid surroundings (E_r reaches the maximum considered values of 20 GPa). The spatial pattern of E^* approximately follows the anisotropic ground heave with the principal axis oriented SE-NW. A possible explanation may be related to the presence of fractures' networks within the reservoir (fractured rocks are known to present reduced stiffness), which greatly influence the flow and the induced volumetric deformation within the reservoir as outlined by the study by Bond et al. (2013). This suggests that interferometry analysis of satellite images may be useful for detecting such deformation features at reservoir's depth: this should be verified via further work by confronting with other geophysical surveys.

SUMMARY AND FUTURE WORK

In the present study, we have shown how useful GSA can be useful to ease the calibration exercise by allowing to reduce the number of uncertain properties, i.e., by highlighting which property should be calibrated in priority to better explain the observations. By analyzing the main and the total effects, we highlighted that the number of uncertain parameters could be dropped from 10 to only 1. Yet, the main challenge was the number of different model runs necessary to conduct GSA (here it exceeded several thousands). We tackled this issue by combining the analysis with a metamodel of type kriging and showed that the model response could be approximated using only 100 of different numerical simulations, which made GSA achievable at a reasonable computation time cost (i.e., less than a couple of hours). Besides, an attractive feature of kriging is to be associated to variance estimates, which is useful to adaptively guide the sampling for solving the optimization problem posed by the history-matching exercise (see, e.g., Jones, 2001). However, the use of kriging metamodels may show some limitations in practice. First, a "basic" kriging metamodel assumes that the simulator responds smoothly to its inputs, which is often not the case in practice: the numerical code may show discontinuities or sharp changes leading to poor local approximation. Advanced nonstationary, semi-parametric, nonlinear regression techniques like treed Gaussian processes by Gramacy and Lee (2008) may bring useful answers to this problem. Second, noise may perturb the observations, which might make the optimization problem more difficult to solve: possible approaches can rely on more sophisticated infill criteria (Picheny et al., 2013) than the one proposed by Jones (2001).

An additional major difficulty for the history-matching exercise can be related to the nature of the observations and model results, namely, they are complex functions of time and space. In the present study, we chose a pragmatic approach by setting up a metamodel at each spatial location (in practice 10,000 metamodels were constructed). Although simple, this procedure can be time consuming (the metamodel construction and validation had to be reconducted 10,000 times) and remains redundant in the sense that by doing so, we neglected the spatial structure of surface displacements (e.g., it can be expected that displacement measured close to each other should present some similarities). In this situation, advanced techniques could be envisaged by using, for instance, feature extraction procedures to summarize this spatial information (see, for instance, an example using time series by Rohmer, 2014).

Finally, it should be acknowledged that, although allowing capturing the main features of the deformation processes, the reservoir model used here does not capture the whole complexity of the coupled hydraulic−mechanical processes. In the future, applying the GSA metamodel calibration procedure on a more complete reservoir model (for instance, Bissell et al., 2011) should be done to better capture the spatial structure of the deformation pattern.

ACKNOWLEDGMENTS

This work has been supported by the French Environment and Energy Management Agency (ADEME) through the STOCK-CO2 program (project AMIRAL, no. 1294C0025).

REFERENCES

R&D, E.D.F., 2012. Code_Aster, Analysis of Structures and Thermomechanics for Studies & Research, Description on Numerical Implementation of Thermo-hydraumechanical Processes. Available at: http://www.code-aster.org/V2/doc/default/en/man_r/r7/r7.01.10.pdf.

Bell, J.W., Amelung, F., Ferretti, A., Bianchi, M., Novali, F., 2008. Permanent scatterer InSAR reveals seasonal and long-term aquifer-system response to groundwater pumping and artificial recharge. Water Resources Research 44. http://dx.doi.org/10.1029/2007WR006152. W02407.

Bissell, R.C., Vasco, D.W., Atbi, M., Hamdani, M., Okwelegbe, M., Goldwater, M.H., 2011. A full field simulation of the In Salah gas production and CO_2 storage project using a coupled geo-mechanical and thermal fluid flow simulator. GHGT-10. Energy Procedia 4, 3290–3297.

Bond, C.E., Wightman, R., Ringrose, P.S., 2013. The influence of fracture anisotropy on CO_2 flow. Geophysical Research Letters 40 (7), 1284–1289.

Bouc, O., Bellenfant, G., Dubois, D., Guyonnet, D., Rohmer, J., Gastine, M., Wertz, F., Fabri, H., 2010. Geological storage safety assessment: methodological developments. In: PSAM 10–10th International Probabilistic Safety Assessment & Management Conference (Seattle, United States).

Ferretti, A., Prati, C., Rocca, F., 2000. Nonlinear subsidence rate estimation using permanent scatterers in differential SAR interferometry. IEEE Transactions on Geoscience and Remote Sensing 38 (5), 2202–2212.

Forrester, A., Sobester, A., Keane, A., 2008. Engineering Design via Surrogate Modelling: A Practical Guide. John Wiley & Sons.

Gramacy, R.B., Lee, H.K., 2008. Bayesian treed Gaussian process models with an application to computer modeling. Journal of the American Statistical Association 103 (483), 1119–1130.

Hastie, T., Tibshirani, R., Friedman, J., 2009. The Elements of Statistical Learning: Data Mining, Inference, and Prediction. Springer-Verlag, New York.

Iooss, B., Lemaître, P., 2015. Uncertainty Management in Simulation-optimization of Complex Systems: Algorithms and Applications. In: Meloni, C., Dellino, G. (Eds.), A Review on Global Sensitivity Analysis Methods, Springer.

Ireland, G., Petropoulos, T.N., Carlson, G.P., Purdy, S., 2015. Addressing the ability of a land biosphere model to predict key biophysical vegetation characterisation parameters with Global Sensitivity Analysis Environmental Modelling & Software, 65, 94–107. http://dx.doi.org/10.1016/j.envsoft.2014.11.010.

Jones, D.R., 2001. A taxonomy of global optimization methods based on response surfaces. Journal of Global Optimization 21, 345–383.

Krige, D., 1951. A statistical approach to some basic mine valuation problems on the Witwatersrand. Journal of the Chemical, Metallurgical and Mining Society of South Africa 52 (6), 119–139.

Mathieson, A., Wright, I., Roberts, D., Ringrose, P., 2009. Satellite imaging to monitor CO_2 movement at Krechba, Algeria. Energy Procedia 1 (1), 2201–2209.

Morris, J.P., Hao, Y., Foxall, W., McNab, W., 2011. A study of injection-induced mechanical deformation at the In Salah CO_2 storage project. International Journal of Greenhouse Gas Control 5 (2), 270–280.

Oliver, D.S., Chen, Y., 2011. Recent progress on reservoir history matching: a review. Computers & Geosciences 15 (1), 185–221.

Petropoulos, G.P., Griffiths, H.M., Carlson, T.N., Ioannou-Katidis, P., Holt, T., 2014. SimSphere Model Sensitivity Analysis Towards Establishing its Use for Deriving Key Parameters Characterising Land Surface Interactions. Geoscientific Model Development 7, 1873–1887. http://dx.doi.org/10.5194/gmd-7-1873-2014.

Petropoulos, G.P., Ireland, G., Griffiths, H., Kennedy, M.C., Ioannou-Katidis, P., Kalivas, D.K.P, 2015. Extending the Global Sensitivity Analysis of the SimSphere model in the Context of its Future Exploitation by the Scientific Community. Water MDPI 7, 2101–2141. http://dx.doi.org/10.3390/w7052101.

Picheny, V., Wagner, T., Ginsbourger, D., 2013. A benchmark of kriging-based infill criteria for noisy optimization. Structural and Multidisciplinary Optimization 48 (3), 607–626.

Pruess, K., Oldenburg, C.M., Moridis, G.J., 1999. TOUGH2 User's Guide, Version 2.0. Lawrence Berkeley National Laboratory Report LBNL-43134 (Berkeley, CA, USA).

Pruess, K., 2005. ECO2N—a TOUGH2 Fluid Property Module for Mixtures of Water, NaCl, and CO2, Lawrence Berkeley National Laboratory Report LBNL-57952.

R Core Team, 2014. R: A Language and Environment for Statistical Computing. R Foundation for Statistical Computing, Vienna, Austria. http://www.R-project.org/.

Rinaldi, A.P., Rutqvist, J., 2013. Modeling of deep fracture zone opening and transient ground surface uplift at KB-502 CO_2 injection well, In Salah, Algeria. International Journal of Greenhouse Gas Control 12, 155–167.

Rohmer, J., Seyedi, D., 2010. Coupled large scale hydromechanical modelling for caprock failure risk assessment of CO_2 storage in deep saline aquifers. Oil & Gas Science and Technology — Rev. IFP 65 (3), 503—517.

Rohmer, J., Foerster, E., 2011. Global sensitivity analysis of large-scale numerical landslide models based on Gaussian-process meta-modeling. Computers & Geosciences 37 (7), 917—927.

Rohmer, J., 2014. Dynamic sensitivity analysis of long-running landslide models through basis set expansion and meta-modelling. Natural Hazards 73 (1), 5—22.

Roustant, O., Ginsbourger, D., Deville, Y., 2012. DiceKriging, DiceOptim: two R packages for the analysis of computer experiments by kriging-based metamodeling and optimization. Journal of Statistical Software 51 (1), 1—55.

Rutqvist, J., Vasco, D.W., Myer, L., 2010. Coupled reservoir-geomechanical analysis of CO_2 injection and ground deformations at In Salah, Algeria. International Journal of Greenhouse Gas Control 4 (2), 225—230. http://dx.doi.org/10.1016/j.ijggc.2009.10.017.

Saltelli, A., Tarantola, S., Chan, K.P.S., 1999. A quantitative model-independent method for global sensitivity analysis of model output. Technometrics 41 (1), 39—56.

Saltelli, A., Ratto, M., Andres, T., Campolongo, F., Cariboni, J., Gatelli, D., Tarantola, S., 2008. Global Sensitivity Analysis: The Primer. John Wiley & Sons.

Saltelli, A., Annoni, P., Azzini, I., Campolongo, F., Ratto, M., Tarantola, S., 2010. Variance based sensitivity analysis of model output. Design and estimator for the total sensitivity index. Computer Physics Communications 181, 259—270.

Sobol', I.M., 1993. Sensitivity estimates for non linear mathematical models. Mathematical Modelling and Computational Experiments 1, 407—414.

Sobol', I., Tarantola, S., Gatelli, D., Kucherenko, S., Mauntz, W., 2007. Estimating the approximation error when fixing unessential factors in global sensitivity analysis. Reliability Engineering & Systems Safety 92 (7), 957—960.

Stein, M.L., 1999. Statistical Interpolation of Spatial Data: Some Theory for Kriging. Springer, New York.

De la Torre Guzman, J., Babaei, M., Shi, J.-Q., Korre, A., Durucan, S., 2014. Coupled flow-geomechanical performance assessment of CO_2 storage sites using the Ensemble Kalman Filter. Energy Procedia 63, 3475—3482.

Vasco, D.W., Ferretti, A., Novali, F., 2008. Reservoir monitoring and characterization using satellite geodetic data: interferometric synthetic aperture radar observations from the Krechba field, Algeria. Geophysics 73 (6), WA113—WA122.

Wegmuller, U., Werner, C., Strozzi, T., 2013. GAMMA Interferometric Point Target Analysis (IPTA) Documentation. Gamma Remote Sensing, Gumligen.

ARTIFICIAL NEURAL NETWORKS FOR SPECTRAL SENSITIVITY ANALYSIS TO OPTIMIZE INVERSION ALGORITHMS FOR SATELLITE-BASED EARTH OBSERVATION: SULFATE AEROSOL OBSERVATIONS WITH HIGH-RESOLUTION THERMAL INFRARED SOUNDERS

P. Sellitto

Laboratoire de Météorologie Dynamique/École Normale Supérieure, Paris, France

CHAPTER OUTLINE

1. INTRODUCTION

Satellite observations are an important source of information on the Earth and its subsystems (atmosphere, oceans, continental surfaces, cryosphere, and biosphere), due to the continuous operations and global coverage. Today, a huge amount of data are produced by satellite measurements. Practically, a substantial part of the information contained in these observations is not exploited, due to the computationally demanding inversion algorithms involved. Then, fast and precise inversion algorithms are required to deal with this increasing volume of data. One particular aspect of the increasing data flux is the increase in spectral resolution and spectral sampling of the new-generation satellite sensors (see, e.g., Crevoisier et al., 2014), producing data with greater dimensionality. Notably, the better spectral resolution is a target of the design of new-generation sensors for the observation of the atmosphere, due to the requirements in this sense to fully resolve the spectral absorption features of atmospheric gases and, to a lesser extent, to resolve features of atmospheric particles (aerosols) and hydrometeors (cloud droplets and precipitations) and their chemical composition and microphysics.

Neural networks (NNs) are a useful inversion tool for Earth observation, due to their fast computation once trained. One prerequisite of NN inversion schemes is that the complexity of the algorithm (the number of interconnections between neurons and the number of the neurons itself) is optimized to assure generalization capabilities of the trained NNs (more details on NN theory are given in Section 2.1). In particular, a careful spectral input (the radiation measurements performed by the satellite sensor) selection, thus avoiding redundancy from highly correlated spectral information, is beneficial for the generalized performances of the NNs. This is getting more important as the spectral resolution of the sensors is getting higher, giving the technological improvements in the satellite industry. Then, the design of reliable NN-based inversions schemes for the analysis of high-resolution satellite observations should be accompanied by careful spectral sensitivity analyses tailored to the specific geophysical problem, to select the radiation measurements at the most informative wavelengths, discarding the other less informative wavelengths. In this work, two NN-based approaches for spectral input vector reduction are tested: the extended pruning (EP) and the autoassociative neural network (AANN) methods, and these are applied to the retrieval of chemical and microphysical properties of secondary sulfate aerosols (SSAs) from high-resolution thermal infrared (TIR) satellite observations.

This chapter is organized as follows. In Section 2 the data and methods used in this work are introduced. Section 2.1 gives a brief overview of the NN theory. Section 2.2 presents the bases of the two input vector reduction techniques tested in the remainder of the chapter: the EP (Section 2.2.1) and the AANN (Section 2.2.2). The sample data set used to train and test the NNs are discussed in Section 2.3. Results are presented in Section 3, including the spectral sensitivity analysis (Section 3.2) and the comparison of performances of the maximum dimensionality NN (introduced in Section 3.1) and the input-reduced NNs (Section 3.3). Conclusions are given in Section 4.

2. DATA AND METHODS
2.1 ARTIFICIAL NEURAL NETWORKS: OVERVIEW

In the following, a brief overview of artificial NNs is given. Several exhaustive treatments of NN theory, and the more general context of soft computing algorithms, are available (e.g., Bishop, 1995; Haykin, 1999; Kecman, 2001); the interested reader can refer to these books for further details.

Artificial NNs are statistical models that perform nonlinear regression and function approximation, basing their functioning on the modeling of networks of biological neurons. As such, they are based on parallel and distributed computation. NNs belong to the class of soft computing algorithms. A hard computing algorithm is aimed at mapping an input onto an output space by a (more or less) complete knowledge of the system under investigation, e.g., by the knowledge of the physical processes governing it. On the contrary, in a soft computing algorithm a statistical model between input and output quantities is searched from sets of experimental or simulated data sets, by means of a learning phase.

NNs are structures of elementary processing units called neurons, interconnected by weighted synapses (of weight w) and organized in layers. A neuron processes its inputs (I) by means of an activation function (AF) to produce its output (O). The typical operation of a neuron is to accept a linear combination of the inputs (whose coefficients are the synapse's weights) and then to produce the output by means of a sigmoidal AF (σ):

$$O = \sigma \sum_{i=1}^{n} (w_i I_i) \tag{9.1}$$

Different kinds of NNs exist, depending on their structure, i.e., the organization and interconnection of the neurons and the layers. In the following, we will restrict our attention to feedforward NNs, i.e., NNs whose connections between neurons do not realize cycles; from another point of view, these are NNs in which the information only propagates forward (from a layer i to a layer j, with i closer to the input layer than j). On the contrary, NNs with directed cycles are called recurrent NNs and are not considered in the following. In the class of feedforward NNs, we will consider NN models called multilayer perceptrons (MLPs), i.e., NNs linking some input to output quantities (represented by the input and the output layers) by means of at least one connecting layer (these latter layers are called hidden layers), thus generating a multiple layer NN scheme. It must be considered that the input layer is not generally composed of processing units—it just transfers the input quantity to the following layers—and then, if hidden layers are not present, the only processing is performed by the output layer, thus generating a single-layer NN scheme. Fig. 9.1 shows the general scheme of an MLP with a single hidden layer structure. Schemes with more than one hidden layer are explored in this work (see Section 3), but their performances are found to be less good than that of a single hidden layer structure for our application.

The aim of an NN algorithm is to extract the underlying relationship between a set of input and output quantities by means of a *learning phase*. During the learning phase, a set of statistically representative input and output *samples* are presented to the NN (the outputs are also called *target* values, the vector T). Adjustment of internal parameters w of the NN, also called *weights* (with the statistical nature of these models in mind) or *synapses weights* (with the parallel with biological neurons in mind), is performed to fit the NN model, i.e., to minimize the distance, in a statistical sense, between the computed and the target outputs. This distance, and then the performances of the NN model, are monitored by means of a weight-dependent error function $E = E(w)$; the absolute minimum of E is searched by the optimization of the parameters w. The hypersurface E in the space defined by the set of w can be characterized by several relative minima, and the training an NN falls into the category of the nonlinear optimization problems.

Once trained, an MLP performs a mapping $R^{N_i} \rightarrow R^{N_O}$, approximating an unknown function f between N_i input (the N_i-dimensional parameters vector I) and N_O output variables (the N_O-dimensional parameters vector O) with a nonlinear model F, which is a function of the of the weights matrix W:

$$O(I, W) = F(I, W) \tag{9.2}$$

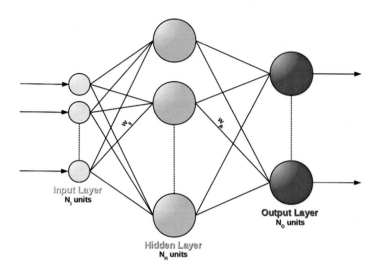

FIGURE 9.1

General scheme of a single hidden layer multilayer perceptron.

In Eqn (9.2), F is the NN model that approximates the actual relationship f. With reference to Fig. 9.1, and considering sigmoidal AFs σ_j and N_H hidden neurons, the NN model F for every kth output can be written as follows:

$$O_k = \sum_{j=1}^{N_H} w_{jk} \left[\sigma_j \sum_{i=1}^{N_i} (w_{ij} I_i) \right] \tag{9.3}$$

Two questions arise in the NN theory: (1) which are the functions that NN models can approximate and (2) how can one produce the approximation of the desired function f.

As for the first question, Cybenko (1989), Funahashi (1989) and Hornik et al. (1989) demonstrated that the Weierstrass approximation theorem ("every continuous function defined on a closed interval $[a, b]$ can be uniformly approximated as closely as desired by a combination of polynomial function") can be extended to combinations of sigmoidal functions (indeed, combinations of any bounded monotonically increasing continuous function)—the so-called Cybenko–Hornik–Funahashi approximation theorem—and then the NNs can approximate any function f with the desired accuracy. The order of the approximating function, so, with reference to Eq. (9.3), the number of sigmoidal functions in F and the number of hidden neurons N_H must be great enough for F to be sufficiently flexible to approximate the unknown function f but small enough to avoid overfitting during the learning stage. Overfitting occurs when the NN model specializes, during the training stage, to the learning samples, thus limiting the generalization capabilities of the model. The training samples can be limited in the statistical representativity of the actual function f and affected by noise.

As for the second, once a norm or error function **E** is defined [usually the mean square error, MSE] to measure the goodness of the approximation, the problem falls into the category of the nonlinear optimization problems: which set of parameters **W** (the weight matrix in our case) of the model minimizes the error? The error function is estimated using the P training samples during the learning phase and can assume the following form or similar:

$$E = \sum_{p=1}^{p} [T_p - O_p] = \sum_{p=1}^{p} [T_p - F(I, W)] \tag{9.4}$$

Generally, the minimization of E is performed by weights adaptation by learning, using descending gradient methods, i.e., by estimating the local direction of the decreasing gradient g of E. **G** is a function of the weights **W** and then contains information on how to modify **W** to minimize E. One common technique is to iteratively adjust the weights by backpropagating the gradient signal (error back-propagation methods).

2.1.1 Artificial Neural Networks for the Inversion of Satellite Measurements of Spectral Radiation for the Observation of the Earth's Atmosphere

We focus our attention on the use of NNs in atmospheric remote sensing. Since the 1990s, NN algorithms have been used in this context to retrieve temperatures, (e.g., Blackwell, 2005; Aires et al., 2013), for height-resolved ozone retrievals, (e.g., Del Frate et al., 2002; Turquety et al., 2004; Sellitto et al., 2011; 2012a,b), and to determine aerosol microphysical and optical properties, (e.g., Brajard et al., 2006; Taylor et al., 2014) from microwave, infrared, ultraviolet, and visible satellite observations, as well as in other applications.

In this context, an NN model is used to infer some (possibly) height-dependent atmospheric variable [a profile $P(z)$, sampled at some altitudes z_k to be determined depending on the band-dependent vertical sensitivity of the satellite observations] from radiometric measurements $R(\lambda)$ at a set of sensitive wavelengths and using additional inputs of relevant parameters Q_{ADD} (e.g., other known atmospheric variables, information on the observation geometry such as the satellite viewing angle, surface information if relevant) that can better constrain the mapping. The NN model, for a single hidden layer MLP with sigmoidal AFs, can be written as follows:

$$P(z_k) = a_k \left[\sum_{j=1}^{N_H} w_{jk}^{(2)} \sigma_j \left(\sum_{iSP=1}^{N_{SP}} w_{iSP}^{(1)} R(\lambda_{iSP}) + \sum_{iADD=1}^{N_{iADD}} w_{iADD}^{(1)} Q_{iADD} \right) \right] \qquad (9.5)$$

This expression is identical to that of Eq. (9.3), with an explicit indication of the different radiometric [$R(\lambda)$] and additional (Q_{ADD}) inputs and the weights of the synapses in the spectral (w_{iSP}) and additional (w_{iADD}) part of the NN structure. The factor a_k takes into account the possibility that the output layers AF may be linear. The two superscripts (1) and (2) identify the first (hidden) and the second (output) layers, respectively.

2.2 NEURAL NETWORK—BASED TECHNIQUES TO REDUCE THE INPUT VECTOR DIMENSIONALITY

In the following, techniques to reduce the complexity of inversion NNs, for atmospheric remote sensing application, are proposed. This complexity reduction is particularly aimed at reducing the spectral inputs, e.g., spectral radiances or brightness temperatures (BTs) acquired by a satellite sensor, which can be highly correlated in selected ranges and then bring more noise than information in many cases. In addition, as visible from Eq. (9.5), the reduction of spectral hidden units (possible if spectral input units are reduced) reduces the order of the NN model and subsequently enhances the generalization capability of the NN model, as discussed in Section 2.1. Two techniques are used in the present work, and discussed in the in the following paragraphs: the EP procedure (Section 2.2.1) and AANN schemes (Section 2.2.2).

2.2.1 Extended Pruning

The *pruning* techniques are aimed at reducing the complexity of the final NN by eliminating the least important connections between neurons, e.g., those with the smallest value of the synaptic weight.

The pruning of an NN algorithms can be *extended* to the inputs layer with the aim of removing the redundant inputs. If the input is a spectral quantity, e.g., the spectral radiances or BTs acquired by a satellite sensor as a function of the wavelength, the EP pruning procedure allows the selection of the most informative wavelengths and then a spectral sensitivity analysis of the input vector. In practice, as introduced by Del Frate et al. (2005), in an EP procedure the weights of a trained NN (after the initial training phase, considering the maximum dimensionality of the input vector) are scanned to assess their relative importance, and the least important synapses are removed. The criterion adopted in this work for the elimination of the weights is based on the absolute value of their weight: the synapse with the smallest weight is removed. The pruned network is then retrained in its new configuration (less synapses). Generally, the retraining is performed with a lower number of training cycles with respect to the initial training. The sequential pruning and training phases may be repeated for several cycles, until one neuron of the input layer is removed, i.e., all synapses connecting it to the hidden layer are removed. The elimination of input neurons continues until the performances of the reduced NN decrease significantly.

The EP scheme can be summarized as follows (with reference to Fig. 9.2):

1. an NN with a maximum input dimensionality is trained;
2. the synapse weights are scanned;
3. the synapse with the weight characterized by the lowest magnitude, or another criterion to estimate its importance in the overall NN, is removed (or pruned);
4. the reduced net is retrained without reinitialization;
5. the performances of the new NN are estimated by means of the training/test/validation error E;

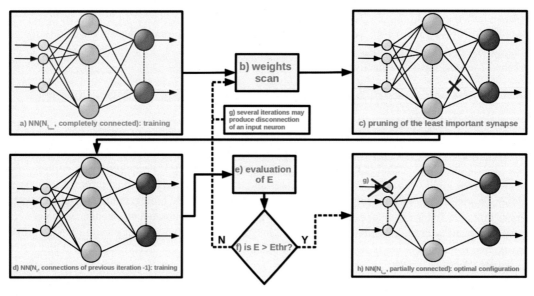

FIGURE 9.2

Extended pruning procedure scheme.

6. depending of the value of the error E with respect to a threshold value E_{thr}, the pruning is stopped or restarted from point c;
7. after several pruning phases, an input neuron may be disconnected from the NN;
8. at the end of the procedure, i.e., as for point f, the remaining input wavelengths are retained as the most informative for the given application.

Further examples of the application of EP procedures for spectral sensitivity analysis, input vector reduction, and NN complexity optimization are the works of Sellitto et al. (2011; 2012a,b) on NN algorithms for the retrieval of tropospheric ozone column, ozone and temperature profiles from UV/VIS (ultraviolet/visible) nadir satellite observations, NASA-Aura OMI (Ozone Monitoring Instrument) (Levelt et al., 2006) and ESA-EnviSat SCIAMACHY (SCanning Imaging Absorption SpectroMeter for Atmospheric ChartographY) (Bovensmann et al., 1999) in these cases.

2.2.2 Autoassociative Neural Networks

AANN are MLP models aimed at approximating the identity input—output mapping (Kramer, 1992). This is usually achieved with a scheme having the same number of input and output nodes, with a *bottleneck layer* having smaller dimensionality. The general scheme of AANN is shown in Fig. 9.3.

In general, an AANN can be conceptually divided into two subcomponents: the encoding (the input, one or more hidden, and the bottleneck layers) and the decoding NNs (the bottleneck, one or

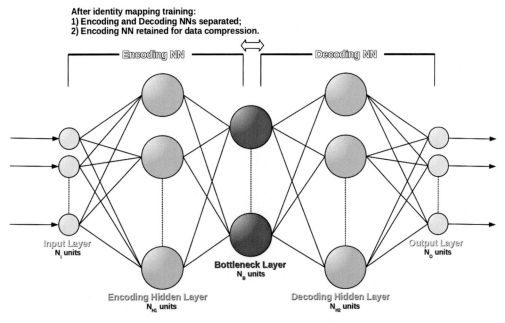

FIGURE 9.3

General scheme of a single hidden layer decoding and encoding autoassociative neural networks. After identity mapping training (1) encoding and decoding NNs separated and (2) encoding NNs were retained for data compression. *NN*, neural network.

more hidden, and the output layer). Once an AANN is trained, the encoding NN is retained. The smaller dimensionality of the bottleneck layer allows then a compression of the information, initially distributed in the N_i-dimensional input space, into the smaller N_B-dimensional bottleneck space. The nonlinear combination of inputs represented by the N_B bottleneck neurons are also called *endmembers*. Then, for our application, the encoding NN is connected to an inversion NN, which has smaller dimensionality than an inversion NN accepting as inputs all the spectral inputs from satellite observations. From another perspective, the bottleneck units carry nonlinear combinations of the inputs (this process is often called nonlinear principal component analysis), and analyzing these combinations and the relative importance of different wavelength allows a spectral sensitivity analysis of the input signal.

2.3 SAMPLE DATA SET

We test the two aforementioned spectral reduction techniques to the retrieval of chemical and microphysical parameters of upper troposphere-lower stratosphere (UTLS) SSA layers. Infrared Atmospheric Sounding Interferometer (IASI) pseudo-observations are generated with a radiative transfer model, considering idealized aerosol layers. The data set is thoroughly described in Sellitto and Legras (2016), and a brief overview is proposed herein.

2.3.1 Sulfate Aerosols and Their Extinction Coefficient

SSAs are produced from the conversion of gaseous precursors, like sulfur dioxide, emitted by natural (e.g., volcanoes) and anthropic sources (e.g., industrial pollution). Their importance resides in their climatic effects when present at higher altitudes, e.g., in the UTLS (SPARC, 2006).

In the present work, SSAs are modeled as spheric droplets of binary solutions of sulfuric acid (H_2SO_4) and water, with varying realistic (SPARC, 2006) UTLS mixing ratios, from 60% to 75% in weight of H_2SO_4. The optical parameters, including the extinction coefficient, of the layers are produced with the spectroscopic codes [solving the electromagnetic problem of the interaction of radiation and the distribution of homogeneous spheres using the Mie theory (van de Hulst, 1957)] of the Earth Observation Data Group of the Department of Physics of Oxford University. The inputs of the Mie code are the refractive index and the size distribution of the aerosol layer. The mixing ratio- and temperature-dependent spectral refractive indices for H_2SO_4/water binary systems, from laboratory measurements (Biermann et al., 2000), are extracted from the GEISA (Gestion et Etude des Informations Spectroscopiques Atmosphériques: Management and Study of Spectroscopic Information) database. Biermann et al. (2000) have measured the complex refractive index (the real and imaginary parts being linked to the scattering and absorption properties of the material) in the spectral range $500.0-5000.0 \text{ cm}^{-1}$, for a wide array of mixing ratios and temperatures of the sulfate-containing water solutions. For our purposes, we have considered four mixing ratios: 60, 64, 70, and 75%. The size distribution $n(r)$ [$n(r)$dr is the number of particles per unit volume with radius between r and r + dr] of the dispersions of aerosol droplets are described by means of log-normal distributions:

$$n(r) = \frac{N_0}{r \ln \sigma_r \sqrt{(2\pi)}} e^{-\frac{1}{2}\left(\frac{\ln(r/r_m)}{\ln \sigma_r}\right)^2} \tag{9.6}$$

with different number concentrations ($N_0 = 8, 9, 10, 12, 15, 20, 25,$ and 30 particles/cm^3) and mean radii ($r_m = 0.06, 0.07, 0.08, 0.1, 0.15, 0.2, 0.3,$ and 0.4 m). The standard deviation $\ln \sigma_r$ of the distribution is fixed at 1.86.

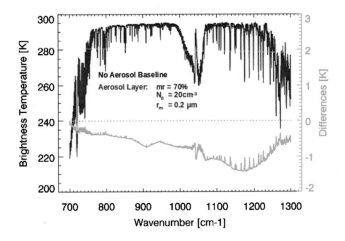

FIGURE 9.4

Example of IASI pseudo-observations (brightness temperatures spectra) for an atmosphere with a sulfate aerosol layer with 70% mixing ratio, 20 cm^{-3} number concentration, and 0.2 μm mean radius (in red). A no-aerosol baseline is also reported as reference (in blue), as well as brightness temperature difference baseline minus aerosol simulations (in green).

2.3.2 *Thermal Infrared Satellite Pseudo-Observations*

Pseudo-observations are derived with the TIR radiative transfer model 4A/OP (Automatized Atmospheric Absorption Atlas Operational) (Scott and Chedin, 1981), to simulate actual TIR sounder observations with various SSA layers. Sulfate aerosol layer extinctions for layers with different H_2SO_4 mixing ratios, number concentrations, and mean radii, derived as described in Section 2.3.1, are used. The aerosol layer has been placed at a fixed altitude (approximately 150 hPa, i.e., in the tropical tropopause) and with a fixed thickness of about 1 km. Vertical profiles of atmospheric gases and thermodynamic parameters (temperature and humidity) are taken for a standard tropical atmosphere. The IASI instrument (Clerbaux et al., 2009) has been simulated by means of its radiometric spectral response, operating wavelengths, spectral resolution, and radiometric noise. BTs in the range 700.0–1300.0 cm^{-1}, with an apodized spectral resolution of 0.50 cm^{-1} and a spectral sampling of 0.25 cm^{-1}, have been simulated. The full BT spectra dimensionality is then 2401. Finally, a set of 256 spectra are obtained by varying the chemical (4 mixing ratios) and microphysical (eight number concentrations and eight mean radii) properties of the sulfate aerosol layer, plus one baseline run with the same atmosphere and no aerosols. In Fig. 9.4 one example of these pseudo-observations, the baseline pseudo-observation and the difference between them (the spectral signature of the sulfate aerosol layer) are shown.

3. RESULTS

Different MLP NNs are trained and tested to investigate how performances vary as a function of different spectral input selection techniques. This allows the investigation of the relative importance and correlation of the BTs at different wavenumbers, and then a spectral sensitivity analysis. The tool used for simulating NNs is the SNNS (Stuttgart Neural Network Simulator) software (available at www.ra.cs.uni-tuebingen.de/SNNS/).

The available data set has been split into two: the training (200 samples, about 75% of the total) and the test data set (56 samples, about 25% of the total). This relative proportion between training and test samples is consistent with the proportion proposed by Hornik et al. (1989). The performances of the NNs are estimated by means of the MSE of the target versus estimated outputs, on the test data set.

In the following, the output vector dimensionality of the NNs is fixed to 3, considering the three varying sulfate aerosol parameters in our pseudo-observation generation: the H_2SO_4 mixing ratio, the number concentration, and the mean radius of the aerosol dispersion.

Another general issue to be set is the learning algorithm used to train the NN. The scaled conjugate gradient method (Møller, 1993) is chosen for its property to converge with a smaller number of iterations than the standard error backpropagation methods. The *early stopping* criterion has been applied to determine when to stop the training of the NNs. During the training stage, the MSE is calculated both on the training and the test data sets. In general, the MSE keep decreasing asymptotically on the training data set, whereas it reaches a minimum and then starts increasing on the test data set. To maximize the generalization properties of the trained NN, the training is stopped when the minimum value of the MSE is reached for the test data set.

3.1 TRAINING AND TESTING THE MAXIMUM DIMENSIONALITY NEURAL NETWORK

The first step is the training and testing of the maximum dimensionality NN, i.e., with the full spectral dimensionality input vectors described in Section 2.3.2. The full input vector has a dimensionality of 2401 units. Using 256 training samples for such a high dimensionality, NN can be less than desirable; nevertheless, the high computational cost of the generation of the training data set (radiative transfer calculations with aerosol scattering is very time consuming) limits the capability to produce bigger data sets. From a different point of view, this is another aspect suggesting a careful selection of the most informative wavelengths to limit the demands in terms of the direct and inverse radiative transfer calculations, necessary in physically-based inversion techniques.

Once the input is set by the maximum dimensionality of the BT spectra, i.e., the maximum spectral resolution of the IASI pseudo-observations (2401 wavenumbers), and the output is set to the varying SSA parameters in our simulations (3 parameters), the topology of the NN model is defined by the number and units of hidden layers. The optimality of the hidden layers has been searched empirically, i.e., by training and testing NNs with different configuration and then selecting of the best performances in terms of the MSE on the test data set. The best performances are found with one hidden layer having 2143 neurons. The final topology is then 2401-2143-3. The NN is trained with about 10^4 iterations.

The performances of the maximum dimensionality NN are shown in Table 9.1, in which the percent MSE and bias for the three output parameters are reported for the test data set.

3.2 SELECTION OF THE INPUT WAVELENGTHS AND SPECTRAL SENSITIVITY ANALYSIS

The spectral input vector dimensionality is reduced using the EP (see Section 2.2.1) and AANN (see Section 2.2.2) techniques. For the two methods, the whole training data set (256 samples) is used.

Table 9.1 Maximum input dimensionality (NN1) and reduced input (EP, NN2; AANN, NN3) NN performances

	NN1 (maximum dimensionality)			NN2 (EP)			NN3 (AANN)		
	mr	N_0	r_m	mr	N_0	r_m	mr	N_0	r_m
MSE(%)	84.5	95.1	60.4	42.1	54.9	22.3	39.5	60.0	25.2
Bias(%)	+23.3	+1.2	−2.1	+10.2	−0.5	−1.3	+11.7	+1.0	−1.1

AANN, *autoassociative neural network;* EP, *extended pruning;* MSE, *mean square error;* NN, *neural network.*

For the EP method, the scheme described in Section 2.2.1, applied to the highest dimensionality NN of Section 3.1 (topology: 2401-2143-3), has led to a substantial reduction of the inputs. The criterion on the error threshold E_{thr} is matched when only 35 inputs are surviving, i.e., about 85% of the input wavenumbers are eliminated. The final NN after the EP procedure has a 35-51-3 topology. The EP procedure generates an NN that is not fully connected (a number of input−hidden and hidden−output connecting synapses are pruned before the stopping criterion is matched); in our case, only 832 connections over 1938 [(35*51) + (51*3)] possible are present in the final NN. The reduced NN is trained with about 10^2 iterations, so a factor 10^2 less than for the maximum dimensionality/fully connected NN.

For the AANN method, the first step is to define the hidden structure of the encoding and decoding NNs. We have selected one hidden layer of 1500 neurons for both NNs and found that the identity mapping is not markedly sensitive to the number of hidden neurons, once the number of hidden layers has been set to 1 (on the contrary, using two hidden layers, for one between the encoding and decoding NN or both, significantly degrades the performances). In terms of the bottleneck layer, the minimum number of neurons assuring satisfactory performances is found as six. The encoding NN, resulting in the six endmembers, is then connected to an inversion NN and then retrained to retrieve the three aerosol parameters. The topology of the inversion NN is found again in an empirical manner (train and test with different topologies and then selection of the topology with the best performances on the test data set). The optimality is found with a 6-25-3 topology. In this case, the procedure produces a fully connected NN, so, in this case, with 225 [(6*25)+(25*3)] connections. The complexity of the AANN-reduced NN, in terms of the connections, is then about 3.5 times smaller than that of the EP-reduced NN. Nevertheless, it should be noticed that the number of inputs is about six times smaller than that of the EP procedure. Due to the partial connection of the EP-reduced NN, the complexity of the two NNs are comparable. The number of synapses for the two reduced NNs is greatly reduced with respect to the maximum dimensionality NN (more than 5 millions connections). The inversion NN, in the AANN model, is trained with about 10^2 iterations, as for the EP-reduced NN.

It is instructive to look at the selected wavenumbers for the two methodologies. The spectral input analysis obtained with these two independent methods allows the identification of the most informative spectral BT for this problem. The surviving wavenumbers of the EP can be considered the most important for the retrieval of the three aerosol properties, from the point of view of this selection procedure. The AANN endmembers can be studied by the analysis of the proportion of the initial inputs, by calculating the coefficient for each endmember (EM) using the weights values. Fig. 9.5 shows the first three EM in the function of the full input wavenumbers and the 35 selected

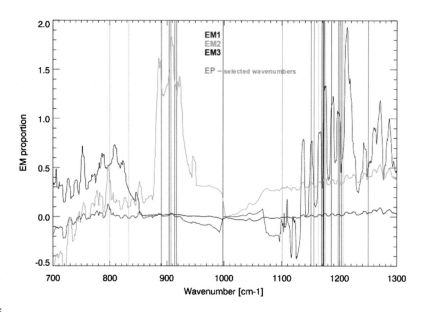

FIGURE 9.5

Spectral proportions of the first three (over five) endmembers from the AANN (EM1—3, *red, green,* and *blue lines*) and the 35 wavenumbers selected with the EP procedure (*gray vertical lines*). *AANN,* autoassociative neural network; *EM,* endmember; *EP,* extended pruning.

wavenumbers with the EP procedure. The other three EMs not shown here have a less interpretable spectral information and/or have similar behavior as those of the three shown.

The two methods select very similar spectral information. EM1 has higher values in the range $1150-1300$ cm^{-1}, where the maximum absorption of the undissociated sulfuric acid molecule can be found. EM2 has a marked peak in the range $890-940$ cm^{-1}, where a secondary maximum of the sulfuric acid absorption is located. These two spectral features are the dominating features in SSA absorption (along with smaller absorption features due to sulfate and bisulfate ions in the solution droplets), as shown, e.g., by Miller et al. (2005), Boer et al. (2007) and Sellitto and Legras (2016). EM3 has its maximum around 800 cm^{-1}, where sulfates have minimum absorption, and then probably is used by the NN model as a baseline. The EMs show wavy patterns, probably due to the many water vapor absorption lines present in the spectral range. The punctual wavenumbers selected by the EP are even more specifically centered around the sulfuric acid absorption peaks at 905 and 1170 cm^{-1}. A few wavenumbers are selected in the minimum absorption range around 800 cm^{-1}. One wavenumber is selected in the range of ozone absorption (around 1000 cm^{-1}) and one wavenumber is selected at wavenumbers higher than 1230 cm^{-1}.

3.3 COMPARING THE PERFORMANCES OF REDUCED DIMENSIONALITY NEURAL NETWORK

As it can be seen from Table 9.1, the performances of the maximum dimensionality NN are very poor in terms of the MSE, with values of 85—95% for the mixing ratio and number concentration and slightly smaller for the mean radius (about 60%). Both input dimensionality—reduced NNs have significantly better performances than the maximum dimensionality NN. The performances improve by a factor 1.5 (number concentration, AANN) to almost a factor 3 (mean radius, EP). In both cases, a specific improvement is found in the retrieval of the mean radius, with MSEs reaching values as low as 20—25%. The number concentration is the most noisy retrieval, with MSEs of 55—60%. These results are consistent with the ones shown by Sellitto and Legras (2016). In this work, it was shown that the spectral signature of the SSA is largely dominated by the mean (or effective) radius. The NNs with spectral inputs reduced by EP and AANN have similar performances, and it does not emerge one clear criterion to choose one of the two approaches. The retrievals are near zero-biased for all NNs, for the size distribution parameters, whereas about 20% and 10% positive bias of the mixing ratio is found for the maximum dimensionality and the input dimensionality—reduced NNs, respectively. The systematic positive underestimation of the sulfuric acid mixing ratio can be partially explained with the very small spectral signature of SSA with smaller mixing ratios.

4. CONCLUSIONS

Spectral input selection and structure optimization is beneficial when dealing with NN-based satellite data inversion schemes, to reduce the complexity of the NN models and then enhance the performances of these algorithms in terms of their generalization capabilities. This also provides useful spectral sensitivity analysis, i.e., the indication of the most informative wavelengths in the initial high-spectral-resolution input vector, singling out the useful from the redundant information. In this sense, these techniques can be seen as metamodeling-based spectral sensitivity analyses. In this work, two methods, the EP and the AANN, are used for the input wavenumber selection/reduction of a known inversion problem, the retrieval of chemical and microphysical properties of sulfate aerosol layers, building on the sensitivity analyses presented by Sellitto and Legras (2016). The two methods greatly reduced the complexity of the inversion NN schemes, from 2401 inputs (>5 million connections) of the full-resolution inputs to 35 (832 connections, EP) and 6 inputs (225 connections, AANN) of the reduced NNs. Then, the reduced neural schemes have the number of connections reduced by a factor 5×10^3 (EP) and 2×10^4 (AANN) with respect to the maximum dimensionality/fully connected NN. This reduction of complexity allows the reduction of the number of iterations needed to converge of a factor 10^2, with respect to the maximum dimensionality/fully connected NN. In addition, the reduced NNs have sensibly better performances when compared with the maximum dimensionality/fully connected NNs, during the retrieval phase. The aerosol properties are retrieved by the EP- and AANN-reduced NNs with a factor 1.5—3 better MSEs and smaller biases, on a test data set (data not used during the training stage). The sets of the 35 selected wavenumbers of the EP and of the 6 endmembers of the AANN are consistent between them and reflect the spectroscopy of the problem, with a stronger contribution in three spectral ranges: around 900, 1170 (the two main molecular absorption peaks of the undissociated sulfuric acid), and 800 cm^{-1} (region with the smallest absorption in the whole band, baseline).

The application of the reduced NN schemes to real IASI observations is ongoing. It is important to mention that, once trained, NN retrieval schemes are particularly effective from the computational point of view. Correspondingly, they are very well suited for the analysis of long series of archived spectral observations and for the generation of climatological series of climate-relevant variables. This is even more important when dealing with high-spectral-resolution observations. The use of physically based retrieval algorithms, like optimal estimation schemes, is impractical when dealing with inversion over long time periods on a global scale, and then efficient and accurate NN schemes, like the one discussed in the present study, qualify as useful alternative tools in climate studies.

ACKNOWLEDGMENTS

The optical parameters of sulfate aerosols layers used in this work are obtained with the IDL Mie scattering routines developed by the Earth Observation Data Group of the Department of Physics of Oxford University, and available via the following website: http://www.atm.ox.ac.uk/code/mie/. NOVELTIS is gratefully acknowledged for the support with the 4A/OP model. The neural networks used in this work have been simulated with the SNNS software, available via the following website: http:/ra.cs.uni-tuebingen.de/SNNS/.

REFERENCES

Aires, F., Bernardo, F., Prigent, C., 2013. Atmospheric water vapour profiling from passive microwave sounders over ocean and land. Part I: methodology for the Megha-Tropiques mission. Quarterly Journal of the Royal Meteorological Society 139, 852−864. http://dx.doi.org/10.1002/qj.1888.

Bishop, C.M., 1995. Neural networks for pattern recognition. Oxford University Press, New York.

Blackwell, W.J., 2005. A Neural-Network technique for the retrieval of atmospheric temperature and moisture profiles from high spectral resolution sounding data. IEEE Transactions on Geosciences and Remote Sensing 43, 2535−2546. http://dx.doi.org/10.1109/TGRS.2005.855071.

Brajard, J., Niang, A., Sawadogo, S., Fell, F., Santer, R., Thiria, S., 2006. Estimating Aerosol parameters from MERIS ocean colour sensor observations by using topological maps. International Journal of Remote Sensing 28 (3), 781−795. http://dx.doi.org/10.1080/01431160600821101.

Bovensmann, H., Burrows, J.P., Buchwitz, M., Frerick, J., Noel, S., Rozanov, V.V., Chance, K.V., Goede, A.H.P., 1999. SCIAMACHY—mission objectives and measurement modes. Journal of Atmospheric Sciences 56, 127−150.

Biermann, U.M., Luo, B.P., Peter, T., 2000. Absorption spectra and optical constants of binary and ternary solutions of H2SO4, HNO3, and H2O in the mid infrared at atmospheric temperatures. The Journal of Physical Chemistry A 104, 783−793. http://dx.doi.org/10.1021/jp992349i.

Boer, G.J., Sokolik, I.N., Martin, S.T., 2007. Infrared optical constants of aqueous sulfate/nitrate/ammonium multi-component tropospheric aerosols from attenuated total reflectance measurements: Part II. An examination of mixing rules. Journal of Quantitative Spectroscopy and Radiative Transfer 108, 39−53. http://dx.doi.org/10.1016/j.jqsrt.2007.02.018.

Crevoisier, C., Clerbaux, C., Guidard, V., Phulpin, T., Armante, R., Barret, B., Camy-Peyret, C., Chaboureau, J.-P., Coheur, P.-F., Crépeau, L., Dufour, G., Labonnote, L., Lavanant, L., Hadji- Lazaro, L., Herbin, H., Jacquinet-Husson, N., Payan, S., Péquignot, E., Pierangelo, C., Sellitto, P., Stubenrauch, C., 2014. Towards IASI- New Generation (IASI-NG): impact of improved spectral resolution and radiometric noise on the retrieval of thermodynamic, chemistry and climate variables. Atmospheric Measurement Techniques 7, 4367−4385.

Cybenko, G., 1989. Approximation by superpositions of a sigmoidal function. Math. Control Signal 2, 303−314. http://dx.doi.org/10.1007/BF02551274.

Clerbaux, C., Boynard, A., Clarisse, L., George, M., Hadji-Lazaro, J., Herbin, H., Hurtmans, D., Pommier, M., Razavi, A., Turquety, S., Wespes, C., Coheur, P.-F., 2009. Monitoring of atmospheric composition using the thermal infrared IASI/MetOp sounder. Atmospheric Chemistry and Physics 9 (16), 6041−6054. http://dx.doi.org/10.5194/acp-9-6041-2009.

Funahashi, K.-I., 1989. On the approximate realization of continuous mappings by neural networks. Neural Networks 2, 183−192. http://dx.doi.org/10.1016/0893-6080(89)90003-8.

Del Frate, F., Ortenzi, A., Casadio, S., Zehner, C., 2002. Application of neural algorithms for a real-time estimation of ozone profiles from GOME measurements. IEEE Transactions on Geosciences and Remote Sensing 40, 2263–2270. http://dx.doi.org/10.1109/TGRS.2002.803622.

Del Frate, F., Iapaolo, M.F., Casadio, S., Godin-Beekmann, S., Petitdidier, M., 2005. Neural networks for the dimensionality reduction of GOME measurement vector in the estimation of ozone profiles. Journal of Quantitative Spectroscopy and Radiative Transfer 92, 275–291.

Haykin, S., 1999. Neural Networks: A Comprehensive Foundation. Prentice Hall, Upper Saddle River.

Hornik, K., Stinchcombe, M., White, H., 1989. Multilayer feedforward networks are universal approximators. Neural Networks 2, 359–366. http://dx.doi.org/10.1016/0893-6080(89)90020-8.

van de Hulst, H., 1957. Light Scattering by Small Particles, Dover Books on Physics Series. Dover Publications. http://books.google.fr/books?id=PlHfPMVAFRcC.

Kecman, V., 2001. Learning and Soft Computing, Support Vector Machines, Neural Networks and Fuzzy Logic Models. MIT Press, Cambridge.

Kramer, M.A., 1992. Auto-associative neural networks. Computers & Chemical Engineering. ISSN: 0098-1354 16 (4), 313–328. http://dx.doi.org/10.1016/0098-1354(92)80051-A.

Levelt, P.F., van den Oord, G.H.J., Dobber, M.R., Mlkki, A., Visser, H., de Vries, J., Stammes, P., Lundell, J.O.V., Saari, H., 2006. The ozone monitoring instrument. IEEE Transactions on Geosciences and Remote Sensing 44, 1093–1101.

Møller, M.F., 1993. A scaled conjugate gradient algorithm for fast supervised learning. Neural Networks 6 (4), 525–533.

Miller, Y., Chaban, G.M., Gerber, R.B., 2005. Ab initio vibrational calculations for H_2SO_4 and H_2SO_4 x H_2O: spectroscopy and the nature of the anharmonic couplings. Journal of Physical Chemistry A 109, 6565–6574. http://dx.doi.org/10.1021/jp058110l.

Sellitto, P., Bojkov, B.R., Liu, X., Chance, K., Del Frate, F., 2011. Tropospheric ozone column retrieval at northern mid-latitudes from the Ozone Monitoring Instrument by means of a neural network algorithm. Atmospheric Measurement Techniques 4, 2375–2388. http://dx.doi.org/10.5194/amt-4-2375-2011.

Sellitto, P., Solimini, D., Del Frate, F., Casadio, S., 2012a. Tropospheric ozone column retrieval from ESA-Envisat SCIAMACHY nadir UV/VIS radiance measurements by means of a neural network algorithm. IEEE Transaction in Geosciences and Remote Sensing 50 (3), 998–1011. http://dx.doi.org/10.1109/TGRS.2011.2163198.

Sellitto, P., Di Noia, A., Del Frate, F., Burini, A., Casadio, S., Solimini, D., 2012b. On the role of visible radiation in ozone profile retrieval from nadir UV/VIS satellite measurements : an experiment with neural network algorithms inverting SCIAMACHY data. Journal of Quantitative Spectroscopy and Radiative Transfer 113 (3), 1429–1436.

Sellitto, P., Legras, B., 2016. Sensitivity of thermal infrared nadir instruments to the chemical and microphysical properties of UTLS secondary sulfate aerosols. Atmospheric Measurement Techniques 9, 115–132. http://dx.doi.org/10.5194/amt-9-115-2016.

SPARC, 2006. Assessment of stratospheric aerosol properties. In: Thomason, L., Peter, T. (Eds.), WCRP-124 WMO/TD- No.1295, 4.

Scott, N.A., Chedin, A., 1981. A fast Line-by-Line method for atmospheric absorption computations: the automatized atmospheric absorption atlas. Journal of Applied Meteorology 802–812. http://dx.doi.org/10.1175/1520-0450(1981)020h0802:AFLBLMi2.0.CO;2.

Turquety, S., Hadji-Lazaro, J., Clerbaux, C., Hauglustaine, D.A., Clough, S.A., Cassé, V., Shlüssel, P., Mégie, G., 2004. Trace gas retrieval algorithm for the infrared atmospheric sounding Interferometer. Journal of Geophysical Research 109, D21301. http://dx.doi.org/10.1029/2004JD00482.

Taylor, M., Kazadzis, S., Tsekeri, A., Gkikas, A., Amiridis, V., 2014. Satellite retrieval of aerosol microphysical and optical parameters using neural networks: a new methodology applied to the Sahara desert dust peak. Atmospheric Measurement Techniques 7, 3151–3175. http://dx.doi.org/10.5194/amt-7-3151-2014.

GLOBAL SENSITIVITY ANALYSIS FOR UNCERTAIN PARAMETERS, MODELS, AND SCENARIOS

10

M. Ye[1], M.C. Hill[2]

Florida State University, Tallahassee, FL, United States[1]; University of Kansas, Lawrence, KS, United States[2]

M. Ye[1], M.C. Hill[2]

Florida State University, Tallahassee, FL, United States[1]; University of Kansas, Lawrence, KS, United States[2]

CHAPTER OUTLINE

1. INTRODUCTION

Sensitivity analysis (SA) is a vital tool in Earth and environmental sciences for model development and improvement. Razavi and Gupta (2015) summarized the uses of SA in the literature, including (1) assessing similarity between model simulations and the underlying system, (2) identifying important model parameters and factors, (3) investigating the regions of important parameters and factors, (4) examining independence of model parameters and factors, (5) identifying unimportant model parameters and factors, and (6) quantifying attribution of the uncertainty in model parameters and factors to prediction uncertainty of the model. This chapter is focused on the function of SA for identifying important model parameters so that they can be built into a model and calibrated to match model simulations to corresponding field observations. Model parameters here are those used to characterize model properties and processes. Taking a system of groundwater reactive transport as an example, hydraulic conductivity is the parameter used to describe the system's hydraulic property related to groundwater flow and reaction coefficients are the parameters used to characterize the system's chemical processes related to solute transport. The number of model parameters can be significantly large, when a large-scale biogeochemical model is used for carbon and nitrogen modeling in climate change study (White et al., 2000). Parameter estimation for a large number of model parameters has been a long-lasting challenge. This challenge can be partly addressed by using large data sets for parameter estimation, which is made possible by the rapid development of techniques for acquiring multisensor remote sensitivity data [e.g., Moderate-Resolution Imaging Spectroradiometer (MODIS) and Landsat] (Running et al., 2004; Peng et al., 2011). Global sensitivity analysis (GSA) continues playing an important role in the data-rich research fields that use abundant satellite-derived measurements (e.g., Tang and Zhuang, 2009; Lu et al., 2013).

The SA methods for identifying important model parameters can be categorized into local and global methods. The basic idea of local methods is to evaluate $\partial y/\partial x|_{x=x_0}$, i.e., the change of model output, y, to the change of model parameter, x, around parameter value, x_0. For an important parameter, a small change may cause a big change in model output. Based on the derivative, a number of statistics have been defined for better measuring parameter influence on model outputs (Morgan and Henrion, 1990; Hill and Tiedeman, 2007). The parameter value, x_0, also called nominal values, can be obtained based on literature data, best engineering judgment, and/or model calibration. The requirement of knowing the nominal value poses a limitation to local sensitivity analysis (LSA), because the nominal value is largely unknown. For example, if the nominal value is taken as a calibrated value, it may only represent the mean or mode of the probability distribution of the parameter. Given that models used in Earth and environmental sciences are always nonlinear and that interaction between model parameters may play an important role to control model outputs, the LSA for the nominal value may not reflect the true sensitivity of model outputs to model parameters.

Given the parametric uncertainty, GSA methods have been developed to answer this question: can we identify important model parameters if we do not know the parameter values? For GSA, there is no need to choose nominal parameter values, and SA is conducted for parameter values in the entire parameter range. This can be done in multiple ways, and this chapter discusses two widely used methods: the Morris method and the Sobol' method. The Morris method extends the LSA from the single nominal value to multiple values in the parameter space; the Sobol' method is based on the variance-decomposition method and evaluates to what extent each uncertain parameter and its interaction with other parameters contribute to predictive uncertainty of model outputs. There has been a growing trend of using GSA; see the review article of Song et al. (2015) and references therein. However, most GSA methods are implemented using Monte Carlo (MC) approaches that are computationally expensive, and the computational cost has been a barrier to the use of GSA in real-world problems.

This chapter discusses two alternative methods for reducing the computational cost of GSA without using MC approaches.

Although existing methods of GSA consider parametric uncertainty, they are designed for a fixed model and modeling scenario without considering uncertainty in model structures and scenarios. The recognition of model and scenario uncertainties in Earth and Environmental sciences has raised new challenges to GSA. Model uncertainty is caused by multiple plausible interpretations of the system of interest based on available data and knowledge. In Earth and Environmental modeling, model uncertainty is often inevitable, because the system is open and complex and can be conceptually interpreted and mathematically described in multiple ways (Beven, 2002, 2006; Bredehoeft, 2003, 2005; Neuman, 2003). Scenario uncertainty is aleatory and an important source of predictive uncertainty. According to IPCC (2000, p. 62), "scenarios are images of the future, or alternative futures. They are neither predictions nor forecasts. Rather, each scenario is one alternative image of how the future might unfold. A set of scenarios assists in the understanding of possible future developments of complex systems." Following Meyer et al. (2014), a scenario is defined in this chapter as a future state or condition assumed for a system. A common way to address model and scenario uncertainties is to consider multiple models and modeling scenarios in the process of model development, calibration, and improvement. The consideration of multiple models and model scenarios immediately raises the following two questions to the current practice of GSA:

1. Are the parameters important to one model also important to other models?
2. Are the parameters important under one modeling scenario also important under other modeling scenarios?

Addressing the two questions requires developing new methods of GSA under model and scenario uncertainties. Table 10.1 illustrates the needs of the new development. The table lists fictitious numbers of total-effect sensitivity index used often in the Sobol' method for two models, M_1 and M_2. Model M_1 has four parameters (A, B, C, and D), whereas model M_2 has only three parameters (A, B, and C). Based on the table, the following questions can be asked:

1. Parameter A is the least important parameter to model M_1 but the most important one to model M_2. How important is this parameter to the system of interest, if we do not know which model is correct?
2. Parameter B has the same value of sensitivity index for the two models. Does this mean that the parameter is of the same importance to the two different models and to the system of interest?

Table 10.1 An example to illustrate the needs of developing new methods of global sensitivity analysis under model uncertainty

	S_A	S_B	S_C	S_D
Model M_1	10%	20%	30%	30%
Model M_2	50%	20%	10%	N/A

The numbers are total-effect sensitivity index of the Sobol' method for two models, M_1 and M_2. M_1 has four parameters A, B, C, and D, but M_2 has only three parameters A, B, and C.

3. Parameter D is only specific to model M_1. Is this parameter important to the system of interest, if we do not know which model is correct?

Dai and Ye (2015) developed a new method to tackle these questions by integrating the Sobol' method into a hierarchical framework of quantifying uncertainty in model parameters, structures, and scenarios. This new method will be introduced in this chapter.

The rest of this chapter is organized as follows. Section 2 introduces the Morris method to screen important parameters out of unimportant parameters, followed by a numerical example given by Pohlmann and Ye (2012). Section 3 discusses the sample-based Sobol' method and its numerical implementation using the MC approach of Saltelli et al. (2010) and two alternative methods for improving computational efficiency without using MC approaches. Section 4 first presents the hierarchical framework of uncertainty quantification, followed by the integration of the Sobol' method into the framework. Section 5 presents the numerical example of Dai and Ye (2015) to demonstrate the new method of GSA under the uncertainty of model parameters, structures, and scenarios. Section 6 discusses existing and potential applications of GSA to models using satellite data. Section 7 offers the conclusion and perspectives of GSA in future studies.

2. MORRIS METHOD

The Morris method extends the LSA at a single nominal parameter value to multiple parameter values in the entire parameter range. The extension also implicitly considers interaction between model parameters. Since the Morris method is conceptually straightforward and computationally efficient, it is one of the methods that is widely used for GSA. However, it should be noted that the Morris method is a screening method to separate (not in a theoretically rigorous manner though) important and unimportant parameters. In this sense, the Morris method is generally not used to quantitatively rank the relative importance of model parameters.

Following Saltelli et al. (2004), this chapter gives a brief description of the Morris method. For a vector of model parameter, $X = (x_1, x_2, \ldots, x_k)$, the method first evaluates the elementary effect of the ith parameter via

$$d_i(X) = \frac{y(x_1, \ldots, x_{i-1}, x_i + \Delta, x_{i+1}, \ldots, x_k) - y(X)}{\Delta} \tag{10.1}$$

where Δ is a value in $\{1/(p-1), \ldots, 1-1/(p-1)\}$ with p being the number of levels and $y(X)$ is model simulation for the parameter vector, X. Morris (1991) proposed two sensitivity measures, the mean (μ) and standard deviation (σ) of the elementary effects, which can be calculated as

$$\mu_i = \frac{1}{r} \sum_{j=1}^{r} d_i(j) \tag{10.2}$$

$$\sigma_i = \sqrt{\frac{1}{r-1} \sum_{j=1}^{r} \left[d_i(j) - \frac{1}{r} \sum_{j=1}^{r} d_i(j) \right]^2} \tag{10.3}$$

where $d_i(j)$ is the elementary effect evaluated for the ith parameter using the jth sample point, and r is the number of repeated sampling designs or trajectories of sample points in the parameter space. Generally speaking, μ is used to determine the overall effect of each parameter on the model output, and σ is used to detect the effects of nonlinearity of a parameter and/or interaction between parameters on the output. When evaluating the mean, μ, for nonmonotonic models, elementary effects with opposite signs may cancel out. This may lead to a small μ value, and thus affect the detection of important parameters using μ. To resolve this problem, Campolongo et al. (2007) proposed to use the absolute values of the elementary effect for evaluating the mean, i.e.,

$$\mu_i^* = \frac{1}{r} \sum_{j=1}^{r} d_i(j) \qquad (10.4)$$

Eqs. (10.1)–(10.4) indicate that implementing the Morris method is computationally inexpensive because the mean and standard deviation can be estimated using a small number of model executions (NMEs). This is especially true when using the Morris trajectory sampling method described in Saltelli et al. (2004).

The common practice of using the mean and standard deviation of elementary effect is to plot them in the μ–σ plane for examining the relative importance of the parameters. Fig. 10.1 is an example plot adopted from a report of Pohlmann and Ye (2012) for an exercise of groundwater modeling. Pohlmann and Ye (2012) used the Morris method to identify the parameters important to a simulated interbasin flow at the northern Yucca Flat area located in the Nevada National Security Site. The details of the numerical simulation are present in Pohlmann et al. (2007) and Ye et al. (2010, 2016). For a model considered by Pohlmann and Ye (2012), by using the Morris method, 9 important parameters were selected out of 58 parameters, and an MC simulation was conducted for these 9 parameters rather than

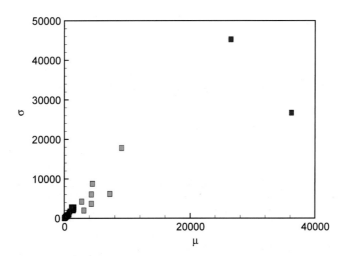

FIGURE 10.1

Mean and standard deviation of the elementary effects for a model considered in Pohlmann and Ye (2012). The *red and blue squares* represent the most and least important parameters, respectively.

for the 58 parameters. In Fig. 10.1, the two parameters represented by the red squares are the most important parameters, as they have the largest mean or the largest standard deviation. The parameters represented by the blue squares were considered to be unimportant due to their small values of the mean and standard deviation. Although the parameters in green squares were considered to be important, the distinction between them and the unimportant parameters (the blue squares) is not apparent. This is a disadvantage of the Morris method that it does not provide a measure for quantitatively ranking the relative importance of the parameters. For this reason, the Morris method can only be used as a screening method.

3. SOBOL' METHOD

The Sobol' method provides two sensitivity indices for quantitatively ranking the relative importance of model parameters, which makes the Sobol' method theoretically superior to the Morris method. This section first introduces the two indices defined in the Sobol' method and then discusses the numerical implementation of the method using the MC approach of quasirandom sampling developed by Saltelli et al. (2010). Subsequently, this chapter introduces two alternative methods that approximate the indices in a computationally efficient manner without using MC approaches.

3.1 FIRST-ORDER AND TOTAL-EFFECT SENSITIVITY INDICES

For a model with of form $\Delta = f(\boldsymbol{\theta}) = f(\theta_1, \ldots, \theta_d)$, where Δ is a scalar model output and $\boldsymbol{\theta} = \{\theta_1, \ldots, \theta_d\}$ denotes the vector of model parameters (Δ and $\boldsymbol{\theta}$ are, respectively, equivalent to y and X used above for the Morris method), the total output variance, V, of Δ can be decomposed as (Sobol', 1993)

$$V(\Delta) = \sum_{i=1}^{d} V_i + \sum_{i=1}^{d} \sum_{j>i}^{d} V_{i,j} + \ldots + V_{1,\ldots,d}, \tag{10.5}$$

where V_i represents the partial variances contributed by parameter θ_i, and $V_{i,j}$ to $V_{1,\ldots,k}$ represent the partial variances caused by parameter interaction. The Sobol' method defines the first-order sensitivity index of parameter θ_i as (Sobol', 1993)

$$S_i = \frac{V_i}{V(\Delta)}, \tag{10.6}$$

which measures the main or first-order effect of the parameter θ_i on the model output. In other words, this index is the portion of the variance of Δ caused by parameter θ_i. Parameters with larger index values are more important than those with smaller index values. The index can also be defined by using the law of total variance, which expresses the total variance as

$$V(\Delta) = V_{\theta_i}\left(E_{\boldsymbol{\theta}_{\sim i}}(\Delta|\theta_i)\right) + E_{\theta_i}\left(V_{\boldsymbol{\theta}_{\sim i}}(\Delta|\theta_i)\right), \tag{10.7}$$

The first-order sensitivity index is defined as

$$S_i = \frac{V_{\theta_i}\left(E_{\boldsymbol{\theta}_{\sim i}}(\Delta|\theta_i)\right)}{V(\Delta)}. \tag{10.8}$$

The $V_{\theta_i}\left(E_{\boldsymbol{\theta}_{\sim i}}(\Delta|\theta_i)\right)$ term is the partial variance or the main effect caused by θ_i, and the inner expectation is the mean of the output calculated using all the changing values of parameter vector, $\boldsymbol{\theta}_{\sim i}$, i.e., all parameters except fixed θ_i. The second term at the right-hand side of Eq. (10.7) is more useful for understanding the definition of Eq. (10.8). If the value of parameter θ_i is known and it is fixed at the known value, the variance of Δ becomes from $V(\Delta)$ to $V_{\boldsymbol{\theta}_{\sim i}}(\Delta|\theta_i)$. Since the parameter value is unknown, the only realistic way of estimating the variance of Δ is to evaluate $E_{\theta_i}\left(V_{\boldsymbol{\theta}_{\sim i}}(\Delta|\theta_i)\right)$, the average variance with respect to parameter θ_i. The most important parameter is the one corresponding to the smallest $E_{\theta_i}\left(V_{\boldsymbol{\theta}_{\sim i}}(\Delta|\theta_i)\right)$, i.e., the largest $V_{\theta_i}\left(E_{\boldsymbol{\theta}_{\sim i}}(\Delta|\theta_i)\right)$ by virtue of Eq. (10.7). This is the reason that the first-order sensitivity index defined in Eq. (10.8) can be used to measure sensitivity of parameter θ_i.

Homma and Saltelli (1996) introduced the total-effective sensitivity index as

$$S_{T_i} = \frac{v_i + v_{ij} + \cdots v_{i,j,\ldots,k}}{v(\Delta)}. \tag{10.9}$$

It considers both the first-order effect of θ_i and the interactions of θ_i with other parameters. In order to calculate the total sensitivity index, the total variance is decomposed as

$$V(\Delta) = V_{\boldsymbol{\theta}_{\sim i}}\left(E_{\theta_i}(\Delta|\boldsymbol{\theta}_{\sim i})\right) + E_{\boldsymbol{\theta}_{\sim i}}\left(V_{\theta_i}(\Delta|\boldsymbol{\theta}_{\sim i})\right), \tag{10.10}$$

where the first term at the right-hand side measures the variance induced by $\boldsymbol{\theta}_{\sim i}$. With Eq. (10.10), S_{T_i} can be defined by ruling out the partial variance as

$$S_{T_i} = \frac{E_{\boldsymbol{\theta}_{\sim i}}\left(V_{\theta_i}(\Delta|\boldsymbol{\theta}_{\sim i})\right)}{V(\Delta)} = \frac{V(\Delta) - V_{\boldsymbol{\theta}_{\sim i}}\left(E_{\theta_i}(\Delta|\boldsymbol{\theta}_{\sim i})\right)}{V(\Delta)}. \tag{10.11}$$

The total-effect sensitivity index is also widely used in GSA. The common practice of using the Sobol' method is to evaluate both the first-order and total-effect sensitivity indices. If parameter interaction exists, the summation of the first-order index of all parameters is less than 100%, and the summation of the total-effect index is more than 100%.

3.2 MONTE CARLO IMPLEMENTATION AND TWO APPROXIMATION METHODS

Evaluating the aforementioned two indices requires calculating the mean and variance in the parameter space, and this is always done by using MC methods. The quasirandom sampling method is the most computationally efficient one among the existing MC methods. Following Saltelli et al. (2010), the mean and variance are evaluated via

$$V_{\theta_i}\left(E_{\boldsymbol{\theta}_{\sim i}}(\Delta|\theta_i)\right) = \frac{1}{n}\sum_{j=1}^{n} f(\mathbf{B}_j)\left(f\left(\mathbf{A}_{\mathbf{B},j}^i\right) - f(\mathbf{A}_j)\right), \tag{10.12}$$

and

$$E_{\boldsymbol{\theta}_{\sim i}}\left(V_{\theta_i}(\Delta|\boldsymbol{\theta}_{\sim i})\right) = \frac{1}{2n}\sum_{j=1}^{n} \left(f(\mathbf{A}_j) - f\left(\mathbf{A}_{\mathbf{B},j}^i\right)\right)^2, \tag{10.13}$$

where $\Delta = f(.)$ denotes a model execution for its parameters. The calculation requires two independent parameter sample matrices, $\mathbf{A}$ and $\mathbf{B}$, with the same dimension of $n \times d$, where n is the number of samples and d is the number of parameters. Matrix $\mathbf{A}_\mathbf{B}^i$ is the same as matrix $\mathbf{A}$ except that its ith column is from the ith column of matrix $\mathbf{B}$. Subscript j denotes the jth row of the corresponding matrix, i.e., the jth sample of the parameters. The MC implementation of Eqs. (10.12) and (10.13) requires $(2+d)n$ times of model executions, i.e., $2n$ executions corresponding to matrices $\mathbf{A}$ and $\mathbf{B}$ and dn executions using matrix $\mathbf{A}_\mathbf{B}^i$. The details of constructing the three matrices are present in Saltelli et al. (2004, 2010). Although the total NME only increases linearly with the number of model parameters, it is still computationally expensive, because the number of samples must be sufficiently large to attain convergence of the MC simulation.

Several methods have been developed to reduce the computational cost. An empirical method is to first conduct a Morris analysis to screen out unimportant parameters so that the Sobol' method is only applied to the important parameters selected by the Morris method (e.g., Chu-Agor et al., 2011; Zhao et al., 2011). This procedure assumes that the Morris method does not screen out important parameters, which, however, cannot be theoretically guaranteed. Another widely used approximation method is metamodeling to build cheap-to-compute surrogates or emulators of computationally expensive models so that performing a large NME is computationally affordable (O'Hagan, 2006). The methods of developing surrogates include Taylor series approximation (Hakami et al., 2003), response surface approximation (Helton and Davis, 2003), Fourier series (Saltelli et al., 1999), nonparametric regression (Helton, 1993; Storlie et al., 2009), kriging (Borgonovo et al., 2012; Lamoureux et al., 2014), Gauss process (Rasmussen and Williams, 2006), polynomial chaos expansion (Garcia-Cabrejo and Valocchi, 2014; Oladyshkin et al., 2012; Sudret, 2007), and sparse-grid collocation (SGC) (Buzzard, 2012; Buzzard and Xiu, 2011). However, the metamodeling methods may still need a relatively large NME to develop accurate surrogates, and the surrogate development is not always straightforward due to model nonlinearity (Razavi et al., 2012; Zhang et al., 2013; Zeng et al., 2016).

3.2.1 Sparse-Grid Collocation for Evaluating Mean and Variance

Dai and Ye (2015) developed another approach that uses the SGC method to evaluate the mean and variance terms directly without using MC approaches. The SGC techniques were originally developed for computing multidimensional integration (Smolyak, 1963), and have been shown to be efficient and effective tools to overcome the curse of dimensionality for high-dimensional numerical integration and interpolation (Barthelmann et al., 1999; Bungartz and Griebel, 2004; Gerstner and Griebel, 1998; Xiu and Hesthaven, 2005). The SGC techniques are particularly suitable for implementing the Sobol' method, because the mean and variance used in the sensitivity indices are multivariate integrals. Based on quadrature rules (e.g., Gerstner and Griebel, 1998), SGC evaluates the mean and variance of a quantity of interest at selected sparse grid (SG) points in the parameter space. Since the number of SG points (NSGP) is always significantly smaller than the MC points in parameter space, the corresponding NME is small. This use of SGC is similar to SGC applications to moment equations for estimating the mean and variance using quadrature rules directly without building surrogates (Lin and Tartakovsky, 2009, 2010; Lin et al., 2010; Shi et al., 2009).

Starting from an integral of a one-dimensional function $g(x)$, the general quadrature formula is (Gerstner and Griebel, 1998):

$$Qg := \sum_{i=1}^{N} \omega_i g(x_i) \qquad (10.14)$$

where x_i is the parameter values (i.e., quadrature points) chosen by following a quadrature rule, ω_i is corresponding weights, and N is number of quadrature points. For a d-dimensional function f, the tensor product of d quadrature formulas $\left(Q_{l_1} \otimes \cdots \otimes Q_{l_d}\right)$ is

$$\left(Q_{l_1} \otimes \cdots \otimes Q_{l_d}\right)f := \sum_{i_1=1}^{N_{l_1}} \cdots \sum_{i_d=1}^{N_{l_d}} \omega_{l_1 i_1} \cdots \omega_{l_d i_d} f\left(x_{l_1 i_1}, \cdots, x_{l_d i_d}\right) \tag{10.15}$$

where $x_{l_1} i_1, \cdots, x_{l_d} i_d$ are quadrature points chosen by a quadrature rule, $l_1, \ldots l_d$ are the precision levels for the quadrature rule, $\omega_{l_1 i_1}, \cdots, \omega_{l_d i_d}$ are the corresponding weights, and $N_{l_1}, \cdots, N_{l_d}$ are number of quadrature points for each dimension. The tensor product is still computationally expensive, because it suffers from the curse of dimensionality, i.e., the NME increases exponentially with the number of model parameters. This problem can be resolved by using the SGC method. Define a multiindex $k = (k_1, \cdots, k_d) \in \mathbb{N}^d$ with the norm of $\|k\| = k_1 + \cdots + k_d$. Instead of using the full tensor product quadrature rule, following the Smolyak or SG cubature rule, a d-dimensional function integral with precision level l can be approximated by (Gerstner and Griebel, 1998):

$$I(l, d)[f] = \sum_{l+1 \leq \|k\| \leq l+d} (-1)^{l+d-\|k\|} \binom{d-1}{\|k\|-l-1} \left(Q_{k_1} \otimes \cdots \otimes Q_{k_d}\right)f \tag{10.16}$$

The NME reduces from N^d for the full tensor product to $\frac{(2d)^l}{l!}$ for the SG tensor product (Novak and Ritter, 1999). The determination of precision level l has been discussed in Gerstner and Griebel (1998).

The SGC algorithm is flexible for GSA, because it allows using different quadrature rules for different types of parameter distributions. For example, if a parameter follows uniform distribution, the Clenshaw–Curtis rule can be used for evaluating the mean and variance. The quadrature points and weights of this rule for the integral $\int_{-1}^{1} g(x)dx$ are (Davis and Rabinowitz, 1975):

$$x_i = \cos\left(\frac{i\pi}{N}\right), \quad i = 0, 1, \ldots, N$$

$$\omega_0 = \omega_N = \frac{1}{N^2 - 1} \tag{10.17}$$

$$\omega_s = \omega_{N-s} = \frac{2}{N}\left(1 + \frac{1}{1-N^2}\cos\pi s + \sum_{j=1}^{N/2-1} \frac{2}{1-4j^2}\cos\frac{2j\pi s}{N}\right), \quad s = 1, \ldots, \frac{N}{2}.$$

where N is an even number. The Gauss–Hermite rule is chosen for the normal parameter distribution. For the integral $\int_{-\infty}^{\infty} e^{-x^2} g(x)dx$, the quadrature points are the solution of Hermite polynomial, $H_N(x)$, and weights are (Davis and Rabinowitz, 1967):

$$\omega_i = \frac{2^{N+1} N! \sqrt{\pi}}{\left[H_{N+1}(x_i)\right]^2} \tag{10.18}$$

In practical applications, if the integrals are different from the standard forms mentioned earlier for which the SGC quadrature rules were developed, change of variable is required to transform the integrals into the standard forms. Take as an example evaluation of the mean of function $h(y)$ with respect to random variable y that follows the normal distribution $N(\mu, \sigma)$,

$$E[h(y)] = \int_{-\infty}^{\infty} \frac{1}{\sigma\sqrt{2\pi}} \exp\left(-\frac{(y-\mu)^2}{2\sigma^2}\right) h(y) dy. \tag{10.19}$$

In order to use the Gauss–Hermite rule developed for the standard form of $\int_{-\infty}^{\infty} e^{-x^2} g(x) dx$, a change of variable is needed by defining

$$x = \frac{y-\mu}{\sqrt{2}\sigma} \text{ and } y = \sqrt{2}\sigma x + \mu, \tag{10.20}$$

with which Eq. (10.19) becomes

$$E[h(y)] = \int_{-\infty}^{\infty} \frac{1}{\sigma\sqrt{2\pi}} \exp(-x^2) h\left(\sqrt{2}\sigma x + \mu\right) dy = \frac{1}{\sqrt{\pi}} \int_{-\infty}^{\infty} \exp(-x^2) h\left(\sqrt{2}\sigma x + \mu\right) dx. \tag{10.21}$$

Applying the Gauss–Hermite rule to Eq. (10.21) leads to

$$E[h(y)] = \frac{1}{\sqrt{\pi}} \sum_{i=1}^{n} \omega_i h\left(\sqrt{2}\sigma x_i + \mu\right). \tag{10.22}$$

The quadrature points and the corresponding weights of other parameter distributions can be found in Davis and Rabinowitz (1967, 1975).

The SGC method is used for evaluating the mean and variance terms used in the definitions of the sensitivity index of the Sobol' method. To evaluate $V_{\theta_i}\left(E_{\theta_{\sim i}}(\Delta|\theta_i)\right)$ used in Eq. (10.8) for the first-order sensitivity index, the variance term is first expanded as

$$V_{\theta_i}\left(E_{\theta_{\sim i}}(\Delta|\theta_i)\right) = E_{\theta_i}\left(\left(E_{\theta_{\sim i}}(\Delta|\theta_i)\right)^2\right) - \left(E_{\theta_i}\left(E_{\theta_{\sim i}}(\Delta|\theta_i)\right)\right)^2, \tag{10.23}$$

and three SG integrations are used for evaluating the three mean terms. The first one approximates $E_{\theta_{\sim i}}(\Delta|\theta_i)$ as

$$E_{\theta_{\sim i}}(\Delta|\theta_i) \approx I_1(l_1, \ d-1)[f(\boldsymbol{\theta}_{\sim i})], \tag{10.24}$$

where l_1 is the level of precision of I_1. Note that the dimension is $d-1$ for $\boldsymbol{\theta}_{\sim i}$ (d being the dimension of $\boldsymbol{\theta}$), because θ_i is fixed. The second SG approximates $E_{\theta_i}\left(\left(E_{\theta_{\sim i}}(\Delta|\theta_i)\right)^2\right)$ as

$$E_{\theta_i}\left(\left(E_{\theta_{\sim i}}(\Delta|\theta_i)\right)^2\right) \approx I_2(l_2, 1)\left[(I_1(l_1, d-1)[f(\boldsymbol{\theta}_{\sim i})])^2\right], \tag{10.25}$$

where l_2 is the level of precision of I_2. Since the integral has only one random variable, θ_i, the dimension of I_2 is 1. $E_{\theta_i}\left(E_{\theta_{\sim i}}(\Delta|\theta_i)\right)$ is approximated as

$$E_{\theta_i}\left(E_{\theta_{\sim i}}(\Delta|\theta_i)\right) \approx I_3(l_3, 1)[I_1(l_1, d-1)[f(\boldsymbol{\theta}_{\sim i})]]. \tag{10.26}$$

where I_3 is another SG with precision level l_3.

To evaluate $E_{\theta_{\sim i}}\left(V_{\theta_i}(\Delta|\boldsymbol{\theta}_{\sim i})\right)$ for the total sensitivity index defined in Eq. (10.11), SG I_4 is developed for the mean term as:

$$E_{\theta_{\sim i}}\left(V_{\theta_i}(\Delta|\boldsymbol{\theta}_{\sim i})\right) \approx I_4(l_4, d-1)\left[V_{\theta_i}(\Delta|\boldsymbol{\theta}_{\sim i})\right] \tag{10.27}$$

where l_4 is the level of precision. By developing another SG for the variance term, $V_{\theta_i}(\Delta|\mathbf{\theta}_{\sim i}) = E_{\theta_i}(\Delta^2|\mathbf{\theta}_{\sim i}) - (E_{\theta_i}(\Delta|\mathbf{\theta}_{\sim i}))^2$, Eq. (10.27) becomes

$$E_{\mathbf{\theta}_{\sim i}}(V_{\theta_i}(\Delta|\mathbf{\theta}_{\sim i})) \approx I_4(l_4, d-1)\left[I_5(l_5, 1)\left[f^2(\theta_i)\right] - (I_6(l_6, 1)[f(\theta_i)])^2\right] \tag{10.28}$$

where I_5 and I_6 are two new SGs with precision levels of l_5 and l_6, respectively. To save computational cost, the total variance of output Δ is not evaluated by building new SGs but by using existing SGs as follows:

$$\begin{aligned} V(\Delta) &= E(\Delta^2) - (E(\Delta))^2 = E_{\mathbf{\theta}_{\sim i}}(E_{\theta_i}(\Delta^2|\mathbf{\theta}_{\sim i})) - (E_{\theta_i}(E_{\mathbf{\theta}_{\sim i}}(\Delta|\mathbf{\theta}_i)))^2 \\ &\approx I_4(l_4, d-1)\left[I_5(l_5, 1)\left[f^2(\theta_i)\right]\right] - (I_3(l_3, 1)[I_1(l_1, d-1)[f(\mathbf{\theta}_{\sim i})]])^2 \end{aligned} \tag{10.29}$$

Therefore, the total NME needed for calculating the global sensitivity indices of a single parameter is $m_1m_2 + m_1m_3 + m_4m_5 + m_4m_6$, where m_1 and m_4 are the NSGPs chosen for $\mathbf{\theta}_{\sim i}$, corresponding to levels l_1 and l_4, respectively, and m_2, m_3, m_5, and m_6 are the NSGPs chosen for θ_i, corresponding to levels l_2, l_3, l_5, and l_6, respectively. Since the values of m_1-m_6 are in general small, the total NMEs is still significantly smaller than $(2+d)n$ used in the quasirandom sampling method of MC simulations. The SGC algorithm is general in that it allows developing different SGs for different mean and variance terms involved in the Sobol' method.

For the numerical example discussed in Section 5, Table 10.2 lists the values of total-effect sensitivity index for precipitation (S_{T_P}) and hydraulic conductivity (S_{T_K}) evaluated for hydraulic

Table 10.2 NMCS and NME of quasirandom sampling method used to calculate total global sensitivity index for hydraulic head and ethene concentration of model r_1Z_1 under scenario 2 of the numerical example discussed in Section 5

NMCS	NME	S_{T_K}	S_{T_P}	NSGP	Level	NME	S_{T_K}	S_{T_P}
For Hydraulic Head at $x = 6000$ m								
50	200	0.160	0.640	9	1	18	0.325	0.791
100	400	0.240	0.878	25	2	50	0.268	0.735
1000	4000	0.265	0.767	49	3	98	0.268	0.735
10,000	40,000	0.251	0.732					
100,000	400,000	0.269	0.735					
1,000,000	4,000,000	0.268	0.735					
For Ethene Concentration at $x = 6000$ m and on 1000 days								
50	300	0.700	0.052	21	1	84	0.900	−0.047
100	600	1.105	0.043	145	2	590	0.947	0.035
1000	6000	0.849	0.039	651	3	2898	0.951	0.040
10,000	60,000	0.959	0.043	2277	4	11,880	0.952	0.042
100,000	600,000	0.960	0.043					
1,000,000	6,000,000	0.952	0.042					

NMCS, *number of Monte Carlo samples;* NME, *number of model executions;* NSGP, *number of sparse grid points.*
The NSGP for different levels of precision and the corresponding NME are also listed to show that NME is dramatically reduced in the sparse grid collocation method.

head and ethene (ETH) concentration of Model R1Z1 under Scenario 2 (the meaning of the model and scenario indices can be found in Section 5). The S_T values are evaluated for different numbers of MC samples and different NSGPs. In addition, the NMEs are also listed. For the quasirandom sampling method, as discussed previously, NME $= (2 + d)n$, where d is the number of parameters and n is the number of samples. For the SGC method, NSGP and NME depend on precision levels (l) and number of parameters (d). For example, to evaluate NSGP for evaluating the S_{T_K} of ETH concentration, if the precision levels are set as $l_4 = 1$ and $l_5 = 1$, NSGP is 21 and NME is 84, because the SG calculation needs to be repeated for each of the four parameters. NSGP is significantly larger for transport modeling than for flow modeling, because transport modeling involves four parameters but flow modeling only involves two parameters. Table 10.2 shows that the SGC algorithm can save dramatic computational cost when compared with the quasirandom sampling method. For flow modeling, only 50 model executions (with 25 SG points) are needed to achieve the same results obtained from 4 million model executions (with 1 million samples). For transport modeling, the SGC computational cost increases, and 11,880 model executions (with 2277 SG points) are needed, which is still significantly more efficient than the quasirandom sampling method. While 1 million samples may not be absolutely needed for the quasirandom sampling method, at least 100,000 samples are needed due to the well-known slow convergence of MC simulation. The dramatic saving of computational time may be related to the relative low nonlinearity of the model. Less computational saving is expected when model nonlinearity increases, and more research on this aspect is guaranteed in future studies.

3.2.2 Distributed Evaluation of Local Sensitivity Analysis

Rakovec et al. (2014) developed a hybrid local–global SA, and this method is referred to as the Distributed Evaluation of Local Sensitivity Analysis (DELSA). This method can be thought of as bridging the gap between Sobol' GSA and LSA, because it uses the results of LSA to evaluate the variance terms used in the first-order sensitivity index of the Sobol' method. In DELSA, the results of LSA are obtained at a single point in parameter space; the evaluation of the first-order sensitivity index is conducted at multiple points in the parameter space, and the average of the point evaluation is expected to be comparable to the Sobol' first-order sensitivity index. If DELSA is used to obtain a single global measure of sensitivity, the results are similar to the Morris method. Alternatively and as suggested by Rakovec et al. (2014), the entire distribution of local values can be considered to give considerably more insight about system dynamics. Rakovec et al. (2014) showed that DELSA has convergence properties that make it less computationally demanding than the Sobol' method. In practice, DELSA is used to produce parameter-space average values and is thus used primarily as a more computationally competitive alternative to the Sobol' method.

The DELSA method is described briefly here, using the presentation from Rakovec et al. (2014). For a first-order parameter SA, DELSA uses the first-order sensitivity index, which is calculated as

$$S_i = \frac{\left|\frac{\partial \Delta}{\partial \theta_i}\right|^2 s_i^2}{V(\Delta)}, \tag{10.30}$$

where $\partial \Delta / \partial \theta_i$ is an output of LSA and s_i^2 is the variance of the prior probability distribution on parameter θ_i. Using a uniform prior parameter distribution makes Eq. (10.30) equivalent to an equation presented by Sobol' and Kucherenko (2009), and makes the DELSA results comparable to Sobol' first-order parameter sensitivity indices (Rakovec et al., 2014). The $V(\Delta)$ term of Eq. (10.30) is the predictive variance of Δ, and it can be evaluated using the theory of first-order-second-moment (Hill and Tiedeman, 2007) as follows:

$$V(\Delta) = \left(\frac{\partial \Delta}{\partial \boldsymbol{\theta}}\right)^T (\mathbf{X}^T \boldsymbol{\omega} \mathbf{X})^{-1} \left(\frac{\partial \Delta}{\partial \boldsymbol{\theta}}\right)^T, \tag{10.31}$$

where $\mathbf{X}$ is the Jacobian matrix produced for LSA and $\boldsymbol{\omega}$ is a weight matrix. For the problem defined here, no observations are defined for which to calculate terms in the Jacobian—only the prior terms are involved (see Hill and Tiedeman, 2007 Appendix C). Thus, each diagonal term of the matrix $(\mathbf{X}^T \boldsymbol{\omega} \mathbf{X})$ corresponds to a parameter, and, for this problem, equals the reciprocal of the variance of the uniform distribution discussed earlier. Eq. (10.31) is a linear propagation (accomplished using the derivatives, $\partial \Delta/\partial \boldsymbol{\theta}$) of the parameter uncertainty expressed by $(\mathbf{X}^T \boldsymbol{\omega} \mathbf{X})^{-1}$ to obtain the variance of Δ. If the LSA is conducted at N points in parameter space using the forward difference method, the NME needed for evaluating Eq. (10.30) is $N(k+1)$, where k is the number of model parameters.

In addition to the statistics that are comparable to the Sobol' first-order sensitivity index, DELSA also provides sensitivity results distributed over parameter space. For example, Fig. 10.2A and B show how the single value results from Sobol' compare with a cumulative distribution of DELSA values for a rainfall-runoff model. The mean of the DELSA values provides a single value comparable to the Sobol' statistics, and, like the Sobol' statistics, is an average value over parameter space. Such averages provide very stable measures of parameter sensitivity. Another way to use the individual DELSA

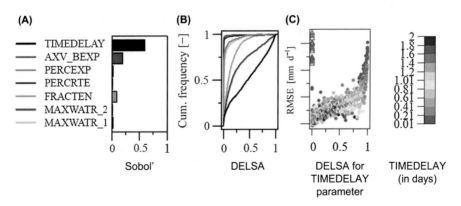

FIGURE 10.2

Important parameters identified in two ways. (A) Sobol' provides values that are averages for the defined parameter value ranges. (B) DELSA provides a distribution of values over the defined parameter value ranges. (C) DELSA results are investigated further to show that the most important parameter TIMEDELAY based on Sobol' is important for parameter values that produce a poor fit to the observations. This draws into question the Sobol' results. (C) Furthermore, the poorest fitting models are associated with values of the TIMEDELAY parameter that are substantially smaller than the 1-day time step used in the model. This suggests that the defined parameter range needs to be reconsidered.

values is to investigate how parameter sensitivity varies over parameter space. For example, Fig. 10.2C shows how the sensitivity values for the most sensitive parameter, TIMEDELAY, are associated with model fit. The results indicate that the importance shown by an averaged measure, such as that produced by the Sobol' method, can obscure important aspects of model sensitivity. Here, it becomes clear that the TIMEDELAY parameter is important mostly for models that do not fit the data very well. In addition, those models use a value of TIMEDELAY that is very small relative to the 1-day time step of the model. Indeed, the small value of TIMEDELAY relative to the time step of the model means that those model runs have unrealistic response functions and the results are not valid. Subsequent analysis of this model will require a modified range for the TIMEDELAY parameter that omits TIMEDELAY values less than about 0.5. This example suggests that DELSA shows considerable promise as a tool in SA.

4. SOBOL' METHOD FOR MULTIPLE MODELS AND SCENARIOS

As discussed in the introduction, it is necessary to extend the Sobol' method for multiple models and multiple scenarios. A recent development on this aspect is discussed in this section, using the presentation from Dai and Ye (2015). This section starts with a hierarchical framework for quantifying uncertainty in model parameters, structures, and scenarios, followed by the integration of the Sobol' method into the framework to conduct GSA for multiple models and scenarios.

4.1 HIERARCHICAL FRAMEWORK FOR UNCERTAINTY QUANTIFICATION

Fig. 10.3 illustrates the hierarchical framework for quantifying uncertainty in model scenarios, structures, and parameters. The framework includes two major components: uncertainty characterization and uncertainty quantification. The component of uncertainty characterization starts with addressing scenario uncertainty by considering multiple scenarios; Fig. 10.3 shows two scenarios (S_1 and S_2) as an example. Subsequently, model uncertainty is addressed by considering multiple models (e.g., M_1 and M_2 in Fig. 10.3). Although it happens often that the same models are used for different scenarios, different models may be needed for different scenarios, when scenario uncertainty affects model formulation. This is the first level of hierarchy in the framework. The conditioning of models on scenarios may affect model plausibility, as plausibility of a model may vary under different scenarios. The lowest level of uncertainty characterization is to address parametric uncertainty, and the parameters are specific to models. In other words, different models may have different parameters, which forms another level of hierarchy in the framework. After uncertainty characterization, the uncertainties are quantified in the opposite order, i.e., first quantifying parametric uncertainty, then parametric and model uncertainties together, and ultimately parametric, model, and scenario uncertainties.

Based on the hierarchical framework of model scenarios, structure, and parameters, the total variance of model output can be expressed as (Dai and Ye, 2015)

$$V(\Delta) = E_{\mathbf{S}}E_{\mathbf{M}|\mathbf{S}}E_{\boldsymbol{\theta}|\mathbf{M},\mathbf{S}}V(\Delta|\boldsymbol{\theta},\mathbf{M},\mathbf{S}) + E_{\mathbf{S}}E_{\mathbf{M}|\mathbf{S}}V_{\boldsymbol{\theta}|\mathbf{M},\mathbf{S}}E(\Delta|\boldsymbol{\theta},\mathbf{M},\mathbf{S})$$
$$+ E_{\mathbf{S}}V_{\mathbf{M}|\mathbf{S}}E_{\boldsymbol{\theta}|\mathbf{M},\mathbf{S}}E(\Delta|\boldsymbol{\theta},\mathbf{M},\mathbf{S}) + V_{\mathbf{S}}E_{\mathbf{M}|\mathbf{S}}E_{\boldsymbol{\theta}|\mathbf{M},\mathbf{S}}E(\Delta|\boldsymbol{\theta},\mathbf{M},\mathbf{S}),$$

(10.32)

where $\mathbf{M}$ and $\mathbf{S}$ are the sets of individual models, M, and individual scenarios, S, respectively, and $\mathbf{M}|\mathbf{S}$ and $\boldsymbol{\theta}|\mathbf{M},\mathbf{S}$ indicate the hierarchical relations that models are conditioned on scenarios and parameters

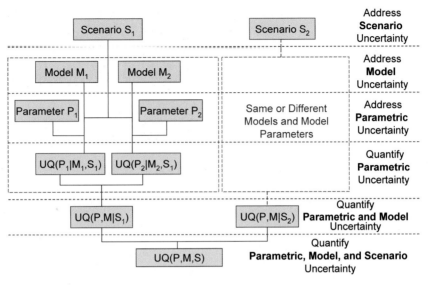

FIGURE 10.3

Hierarchical structure of characterization and quantification of scenario, model, and parametric uncertainties.

are conditioned on models and scenarios. Eq. (10.32) does not require that different scenarios have the same models. Instead, $\mathbf{M}$ is a union of all the models specific to the scenarios. For example, considering that models M_1 and M_2 are associated with scenario S_1 and models M_1 and M_3 with scenario S_2, $\mathbf{M}$ is a union of all the models, i.e., $\mathbf{M} = \{M_1, M_2, M_3\}$. Similarly, different models can have different parameters, and $\mathbf{\theta} = \mathbf{\theta}^{(1)} \cup \mathbf{\theta}^{(2)} \cdots \cup \mathbf{\theta}^{(K)}$, where $\mathbf{\theta}^{(k)}$ is the parameters of model M_k. These notations are used in the following discussion for evaluating the sensitivity indices. The total variance is decomposed into data variance, $E_{\mathbf{S}}E_{\mathbf{M}|\mathbf{S}}E_{\mathbf{\theta}|\mathbf{M},\mathbf{S}}V(\Delta|\mathbf{\theta}, \mathbf{M}, \mathbf{S})$; parametric variance, $E_{\mathbf{S}}E_{\mathbf{M}|\mathbf{S}}V_{\mathbf{\theta}|\mathbf{M},\mathbf{S}}E(\Delta|\mathbf{\theta}, \mathbf{M}, \mathbf{S})$; model variance, $E_{\mathbf{S}}V_{\mathbf{M}|\mathbf{S}}E_{\mathbf{\theta}|\mathbf{M},\mathbf{S}}E(\Delta|\mathbf{\theta}, \mathbf{M}, \mathbf{S})$; and scenario variance, $V_{\mathbf{S}}E_{\mathbf{M}|\mathbf{S}}E_{\mathbf{\theta}|\mathbf{M},\mathbf{S}}E(\Delta|\mathbf{\theta}, \mathbf{M}, \mathbf{S})$. The data variance here is treated as the variance of measurement error. Since it is independent of model parameters, structures, and scenarios (i.e., $E_{\mathbf{S}}E_{\mathbf{M}|\mathbf{S}}E_{\mathbf{\theta}|\mathbf{M},\mathbf{S}}V(\Delta|\mathbf{\theta}, \mathbf{M}, \mathbf{S}) = V(\Delta|\mathbf{\theta}, \mathbf{M}, \mathbf{S}) = C$), the data variance is a constant and not involved in the SA. The $E(\Delta|\mathbf{\theta}, \mathbf{M}, \mathbf{S})$ term is the model output of Δ given by a model under its associated parameters and scenario. Although the meaning of $E(\Delta|\mathbf{\theta}, \mathbf{M}, \mathbf{S})$ and $V(\Delta|\mathbf{\theta}, \mathbf{M}, \mathbf{S})$ may change for a stochastic model (e.g., kriging in Neuman et al., 2012; Lu et al., 2011, 2012), discussing it is beyond the scope of this study.

The GSA for identifying important parameters is focused only on $E_{\mathbf{S}}E_{\mathbf{M}|\mathbf{S}}V_{\mathbf{\theta}|\mathbf{M},\mathbf{S}}E(\Delta|\mathbf{\theta}, \mathbf{M}, \mathbf{S})$, because the variance of the other three terms is not caused by parametric uncertainty. With the introduced hierarchical structure and notations of models and scenarios, the first-order sensitivity index and total sensitivity index are redefined below for the three cases: (1) a single model and a single scenario, (2) multiple models but a single scenario, and (3) multiple models and multiple scenarios.

4.2 GLOBAL SENSITIVITY INDICES FOR SINGLE MODEL AND SINGLE SCENARIO

When considering a single model, M, and a single scenario, S, $E_S E_{\mathbf{M}|S} V_{\boldsymbol{\theta}|\mathbf{M},S} E(\Delta|\boldsymbol{\theta}, \mathbf{M}, \mathbf{S})$ becomes $V_{\boldsymbol{\theta}|M,S} E(\Delta|\boldsymbol{\theta}, \mathbf{M}, \mathbf{S})$. Similar to Eq. (10.8), the variance due to parametric uncertainty is decomposed as

$$V_{\boldsymbol{\theta}|M,S} E(\Delta|\boldsymbol{\theta}, M, S) = V_{\theta_i|M,S} E_{\boldsymbol{\theta}_{\sim i}|M,S}(E(\Delta|\boldsymbol{\theta}, M, S)|\theta_i) + E_{\theta_i|M,S} V_{\boldsymbol{\theta}_{\sim i}|M,S}(E(\Delta|\boldsymbol{\theta}, M, S)|\theta_i), \quad (10.33)$$

and the first-order sensitivity index is defined as

$$S_i = \frac{V_{\theta_i|M,S} E_{\boldsymbol{\theta}_{\sim i}|M,S}(E(\Delta|\boldsymbol{\theta}, M, S)|\theta_i)}{V_{\boldsymbol{\theta}|M,S} E(\Delta|\boldsymbol{\theta}, M, S)}. \quad (10.34)$$

Similarly, Eq. (10.10) can be rewritten as

$$\begin{aligned} V_{\boldsymbol{\theta}|M,S} E(\Delta|\boldsymbol{\theta}, M, S) = {} & E_{\boldsymbol{\theta}_{\sim i}|M,S} V_{\theta_i|M,S}(E(\Delta|\boldsymbol{\theta}, M, S)|\boldsymbol{\theta}_{\sim i}) \\ & + V_{\boldsymbol{\theta}_{\sim i}|M,S} E_{\theta_i|M,S}(E(\Delta|\theta, M, S)|\boldsymbol{\theta}_{\sim i}), \end{aligned} \quad (10.35)$$

and the total sensitivity index is defined as

$$S_{T_i} = \frac{E_{\boldsymbol{\theta}_{\sim i}|M,S} V_{\theta_i|M,S}(E(\Delta|\boldsymbol{\theta}, M, S)|\boldsymbol{\theta}_{\sim i})}{V_{\boldsymbol{\theta}|M,S} E(\Delta|\boldsymbol{\theta}, M, S)} = 1 - \frac{V_{\boldsymbol{\theta}_{\sim i}|M,S} E_{\theta_i|M,S}(E(\Delta|\boldsymbol{\theta}, M, S)|\boldsymbol{\theta}_{\sim i})}{V_{\boldsymbol{\theta}|M,S} E(\Delta|\boldsymbol{\theta}, M, S)}. \quad (10.36)$$

4.3 GLOBAL SENSITIVITY INDICES FOR MULTIPLE MODELS BUT SINGLE SCENARIO

When considering multiple models, $\mathbf{M}$, but a single scenario, S, $E_S E_{\mathbf{M}|S} V_{\boldsymbol{\theta}|\mathbf{M},S} E(\Delta|\boldsymbol{\theta}, \mathbf{M}, \mathbf{S})$ becomes $E_{\mathbf{M}|S} V_{\boldsymbol{\theta}|\mathbf{M},S} E(\Delta|\boldsymbol{\theta}, \mathbf{M}, S)$. It can be decomposed as

$$\begin{aligned} E_{\mathbf{M}|S} V_{\boldsymbol{\theta}|\mathbf{M},S} E(\Delta|\boldsymbol{\theta}, \mathbf{M}, S) = {} & E_{\mathbf{M}|S} E_{\theta_i|\mathbf{M},S} V_{\boldsymbol{\theta}_{\sim i}|\mathbf{M},S}(E(\Delta|\boldsymbol{\theta}, \mathbf{M}, S)|\theta_i) \\ & + E_{\mathbf{M}|S} V_{\theta_i|\mathbf{M},S} E_{\boldsymbol{\theta}_{\sim i}|\mathbf{M},S}(E(\Delta|\boldsymbol{\theta}, \mathbf{M}, S)|\theta_i) \end{aligned} \quad (10.37)$$

or

$$\begin{aligned} E_{\mathbf{M}|S} V_{\boldsymbol{\theta}|\mathbf{M},S} E(\Delta|\boldsymbol{\theta}, \mathbf{M}, S) = {} & E_{\mathbf{M}|S} E_{\boldsymbol{\theta}_{\sim i}|\mathbf{M},S} V_{\theta_i|\mathbf{M},S}(E(\Delta|\boldsymbol{\theta}, \mathbf{M}, S)|\boldsymbol{\theta}_{\sim i}) \\ & + E_{\mathbf{M}|S} V_{\boldsymbol{\theta}_{\sim i}|\mathbf{M},S} E_{\theta_i|\mathbf{M},S}(E(\Delta|\boldsymbol{\theta}, \mathbf{M}, S)|\boldsymbol{\theta}_{\sim i}) \end{aligned} \quad (10.38)$$

Accordingly, the first-order and total sensitivity indices for multiple models are defined as

$$S_i^{\mathbf{M}} = \frac{E_{\mathbf{M}|S} V_{\theta_i|\mathbf{M},S} E_{\boldsymbol{\theta}_{\sim i}|\mathbf{M},S}(E(\Delta|\boldsymbol{\theta}, \mathbf{M}, S)|\theta_i)}{E_{\mathbf{M}|S} V_{\boldsymbol{\theta}|\mathbf{M},S} E(\Delta|\boldsymbol{\theta}, \mathbf{M}, S)} \quad (10.39)$$

and

$$S_{T_i}^{\mathbf{M}} = \frac{E_{\mathbf{M}|S} E_{\boldsymbol{\theta}_{\sim i}|\mathbf{M},S} V_{\theta_i|\mathbf{M},S}(E(\Delta|\boldsymbol{\theta}, \mathbf{M}, S)|\boldsymbol{\theta}_{\sim i})}{E_{\mathbf{M}|S} V_{\boldsymbol{\theta}|\mathbf{M},S} E(\Delta|\boldsymbol{\theta}, \mathbf{M}, S)} \quad (10.40)$$

Eqs. (10.39) and (10.40) take into account the parameter influence under the individual models and provide a quantitative assessment of GSA with combined effects of uncertain parameters and models.

The mean of multiple models, $E_{\mathbf{M}|S}(\cdot)$, in Eqs. (10.39) and (10.40), is evaluated by the weighted average, i.e., $E_{\mathbf{M}|S}(\cdot) = \sum_k P(M_k|S)(\cdot)$, where $P(M_k|S)$ is the probability of model M_k under scenario S and satisfies $\sum_k P(M_k|S) = 1$. Based on the model averaging concept, Eq. (10.39) is evaluated as

$$S_i^{\mathbf{M}} = \frac{\sum_k P(M_k|S) V_{\theta_i|M_k,S} E_{\boldsymbol{\theta}_{\sim i}|M_k,S}(E(\Delta|\boldsymbol{\theta}, M_k, S)|\theta_i)}{\sum_k P(M_k|S) V_{\boldsymbol{\theta}|M_k,S} E(\Delta|\boldsymbol{\theta}, M_k, S)} \tag{10.41}$$

If parameter θ_i does not belong to model M_k, the $V_{\theta_i|M_k,S} E_{\boldsymbol{\theta}_{\sim i}|M_k,S}(E(\Delta|\boldsymbol{\theta}, M_k, S)|\theta_i)$ term is 0 and $S_i^{M_k} = 0$, because uncertainty in θ_i does not contribute to the predictive uncertainty of model M_k. Similarly, the total sensitivity index defined in Eq. (10.40) is evaluated as

$$S_{T_i}^{\mathbf{M}} = \frac{\sum_k P(M_k|S) E_{\boldsymbol{\theta}_{\sim i}|M_k,S} V_{\theta_i|M_k,S}(E(\Delta|\boldsymbol{\theta}, M_k, S)|\boldsymbol{\theta}_{\sim i})}{\sum_k P(M_k|S) V_{\boldsymbol{\theta}|M_k,S} E(\Delta|\boldsymbol{\theta}, M_k, S)} \tag{10.42}$$

If parameter θ_i does not belong to model M_k, the $V_{\theta_i|M_k,S}(E(\Delta|\boldsymbol{\theta}, M_k, S)|\boldsymbol{\theta}_{\sim i})$ term is zero and $S_{T_i}^{M_k} = 0$.

Comparing Eqs. (10.39) and (10.40) with Eqs. (10.34) and (10.36), respectively, shows that $S_i^{\mathbf{M}}$ and $S_{T_i}^{\mathbf{M}}$ of multiple models is not a weighted average of S_i and S_{T_i} of individual models, because model averaging is conducted for the variance of individual models and not for the index of individual models. Taking $S_i^{\mathbf{M}}$ as an example, its denominator is the variance of Δ averaged over all the models, and its nominator is the variance caused by parameter θ_i averaged over all the models. The basic idea behind the new sensitivity indices is that they evaluate parameter importance not for individual models but for all the alternative models. This avoids the problem of comparing the sensitivity indices of a parameter for different models. Therefore, the three questions raised in the introduction based on Table 10.1 are avoided by evaluating the variance terms over the multiple models. In the numerical example shown in Section 5, it shows that the new sensitivity indices avoid biased selection of important model parameters.

4.4 GLOBAL SENSITIVITY INDICES FOR MULTIPLE MODELS AND MULTIPLE SCENARIOS

When considering multiple models, $\mathbf{M}$, and multiple scenarios, $\mathbf{S}$, $E_{\mathbf{S}} E_{\mathbf{M}|\mathbf{S}} V_{\boldsymbol{\theta}|\mathbf{M},\mathbf{S}} E(\Delta|\boldsymbol{\theta}, \mathbf{M}, \mathbf{S})$ is decomposed as

$$
\begin{aligned}
E_{\mathbf{S}} E_{\mathbf{M}|\mathbf{S}} V_{\boldsymbol{\theta}|\mathbf{M},\mathbf{S}} E(\Delta|\theta, M, S) = {} & E_{\mathbf{S}} E_{\mathbf{M}|\mathbf{S}} E_{\theta_i|\mathbf{M},\mathbf{S}} V_{\boldsymbol{\theta}_{\sim i}|\mathbf{M},\mathbf{S}}(E(\Delta|\boldsymbol{\theta}, \mathbf{M}, \mathbf{S})|\theta_i) \\
& + E_{\mathbf{S}} E_{\mathbf{M}|\mathbf{S}} V_{\theta_i|\mathbf{M},\mathbf{S}} E_{\boldsymbol{\theta}_{\sim i}|\mathbf{M},\mathbf{S}}(E(\Delta|\boldsymbol{\theta}, \mathbf{M}, \mathbf{S})|\theta_i)
\end{aligned} \tag{10.43}
$$

and

$$E_S E_{\mathbf{M}|S} V_{\theta|\mathbf{M},\mathbf{S}} E(\Delta|\theta, \mathbf{M}, \mathbf{S}) = E_S E_{\mathbf{M}|S} E_{\theta_{\sim i}|\mathbf{M},\mathbf{S}} V_{\theta_i|\mathbf{M},\mathbf{S}} (E(\Delta|\theta, \mathbf{M}, \mathbf{S})|\theta_{\sim i})$$
$$+ E_S E_{\mathbf{M}|S} V_{\theta_{\sim i}|\mathbf{M},\mathbf{S}} E_{\theta_i|\mathbf{M},\mathbf{S}} (E(\Delta|\theta, \mathbf{M}, \mathbf{S})|\theta_{\sim i})\Big) \tag{10.44}$$

Accordingly, the first-order and total sensitivity indices for multiple models and multiple scenarios are defined as

$$
\begin{aligned}
S_i^{MS} &= \frac{E_S E_{\mathbf{M}|S} V_{\theta_i|\mathbf{M},\mathbf{S}} E_{\theta_{\sim i}|\mathbf{M},\mathbf{S}}(E(\Delta|\theta, \mathbf{M}, \mathbf{S})|\theta_i)}{E_S E_{\mathbf{M}|S} V_{\theta|\mathbf{M},\mathbf{S}} E(\Delta|\theta, \mathbf{M}, \mathbf{S})} \\[2mm]
&= \frac{\sum_S \sum_M P(S)P(M|S) V_{\theta_i|M,S} E_{\theta_{\sim i}|M,S}(E(\Delta|\theta, M, S)|\theta_i)}{\sum_S \sum_M P(S)P(M|S) V_{\theta|M,S} E(\Delta|\theta, M, S)}
\end{aligned}
\tag{10.45}
$$

and

$$
\begin{aligned}
S_{T_i}^{MS} &= \frac{E_S E_{\mathbf{M}|S} E_{\theta_{\sim i}|\mathbf{M},\mathbf{S}} E_{\theta_i|\mathbf{M},\mathbf{S}}(E(\Delta|\theta, \mathbf{M}, \mathbf{S})|\theta_{\sim i})}{E_S E_{\mathbf{M}|S} V_{\theta|\mathbf{M},\mathbf{S}} E(\Delta|\theta, \mathbf{M}, \mathbf{S})} \\[2mm]
&= \frac{\sum_S \sum_M P(S)P(M|S) E_{\theta_{\sim i}|M,S} V_{\theta_i|M,S}(E(\Delta|\theta, M, S)|\theta_{\sim i})}{\sum_S \sum_M P(S)P(M|S) V_{\theta|M,S} E(\Delta|\theta, M, S)}
\end{aligned}
\tag{10.46}
$$

Eqs. (10.45) and (10.46) take into account the parameter influence under the individual models and individual scenarios, and provide a quantitative assessment of GSA with combined effects of uncertain parameters, models, and scenarios. Again, S_T^{MS} of multiple scenarios is not a weighted average of S_T^M of individual scenarios.

Similar to the concept of model averaging, scenario averaging is applied here by replacing the mean for multiple scenarios, $E_S(\cdot)$, by the weighted average, i.e., $E_S(\cdot) = \sum_l P(S_l)(\cdot)$, where $P(S)$ is scenario probability and satisfies $\sum_l P(S_l) = 1$. Using the concepts of model averaging and scenario averaging, Eq. (10.45) can be evaluated as

$$
S_i^{MS} = \frac{\sum_l \sum_k P(S_l)P(M_k|S_l) V_{\theta_i|M_k,S_l} E_{\theta_{\sim i}|M_k,S_l}(E(\Delta|\theta, M_k, S_l)|\theta_i)}{\sum_l \sum_k P(S_l)P(M_k|S_l) V_{\theta|M_k,S_l} E(\Delta|\theta, M_k, S_l)}
\tag{10.47}
$$

If parameter θ_i does not belong to model M_k and scenario S_l, the term $V_{\theta_i|M_k,S_l} E_{\theta_{\sim i}|M_k,S_l}$ $(E(\Delta|\theta, M_k, S_l)|\theta_i)$ is 0, and $S_i^{M_k S_l} = 0$. Similarly, the total sensitivity index defined in Eq. (10.46) is evaluated as

$$
S_{T_i}^{MS} = \frac{\sum_l \sum_k P(S_l)P(M_k|S_l) E_{\theta_{\sim i}|M_k,S_l} V_{\theta_i|M_k,S_l}(E(\Delta|\theta, M_k, S_l)|\theta_{\sim i})}{\sum_l \sum_k P(S_l)P(M_k|S_l) V_{\theta|M_k,S_l} E(\Delta|\theta, M_k, S_l)}
\tag{10.48}
$$

If parameter θ_i does not belong to model M_k and scenario S_l, the term $V_{\theta_i|M_k,S}(E(\Delta|\theta,M_k,S)|\theta_{\sim i})$ is 0, and $S_{T_i}^{M_k} = 0$.

Eqs. (10.45) and (10.46) define the sensitivity indices not only for multiple models but also for multiple scenarios. Therefore, similar to the sensitivity indices defined in Eqs. (10.39) and (10.40) that avoid comparing parameter importance between different models, the sensitivity indices defined in Eqs. (10.45) and (10.46) avoid comparing parameter importance between different models and different scenarios. The new sensitivity indices thus provide an overall assessment of parameter importance when we do not know which model and modeling scenario should be used for simulating the system of interest.

5. SYNTHETIC STUDY WITH MULTIPLE SCENARIOS AND MODELS

Dai and Ye (2015) evaluated the global sensitivity indices given in Section 4 for a synthetic case of groundwater reactive transport modeling with three scenarios, four models, and six random parameters. The description of the numerical example and its results are based on the presentation of Dai and Ye (2015).

5.1 SYNTHETIC CASE OF GROUNDWATER REACTIVE TRANSPORT MODELING

In the synthetic domain of groundwater flow shown in Fig. 10.4, the unconfined groundwater aquifer of length L is under a steady-state condition and has a uniform precipitation, P, over the entire domain. Specification of the precipitation and the groundwater flow models are given below. A continuous contaminant source is placed at the center of the domain ($x = 5000$ m). Similar to Chen et al. (2013), a total of five chemical species are involved in the reactive transport modeling, and they are perchloroethene (PCE), trichloroethene (TCE), dichloroethene (DCE), vinyl chloride (VC), and ETH. These contaminants are subject to a single chain reaction (i.e., PCE → TCE → DCE → VC → ETH)

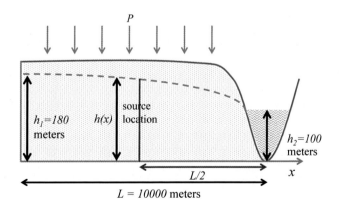

FIGURE 10.4

Diagram of the modeling domain of the synthetic study. $L = 1000$ m is the domain length. Precipitation is uniform for the entire domain. Constant head (h) boundary conditions are $h_1 = 180$ m and $h_2 = 100$ m. The continuous contaminant source is located in the middle of the domain.

described in the user's manual of **BIOCHLOR** (Aziz et al., 2000). The governing equations of the reactive transport modeling are as follows:

$$\frac{\partial c_1}{\partial t} = D\frac{\partial^2 c_1}{\partial x^2} - v\frac{\partial c_1}{\partial x} - k_1 c_1$$

$$\frac{\partial c_2}{\partial t} = D\frac{\partial^2 c_2}{\partial x^2} - v\frac{\partial c_2}{\partial x} + y_1 k_1 c_1 - k_2 c_2$$

$$\frac{\partial c_3}{\partial t} = D\frac{\partial^2 c_3}{\partial x^2} - v\frac{\partial c_3}{\partial x} + y_2 k_2 c_2 - k_3 c_3 \tag{10.49}$$

$$\frac{\partial c_4}{\partial t} = D\frac{\partial^2 c_4}{\partial x^2} - v\frac{\partial c_4}{\partial x} + y_3 k_3 c_3 - k_4 c_4$$

$$\frac{\partial c_5}{\partial t} = D\frac{\partial^2 c_5}{\partial x^2} - v\frac{\partial c_5}{\partial x} + y_4 k_4 c_4 - k_5 c_5$$

where $c_1 - c_5$ [M/L^3] are concentrations of PCE, TCE, DCE, VC, and ETH, respectively; D [L^2/T] is the hydrodynamic dispersion coefficient in the x-direction; v [L/T] is the seepage velocity; $k_1 - k_5$ [1/L] are the first-order degradation coefficients of the five species; and $y_1 - y_4$ are the yield coefficients of PCE, TCE, DCE, and VC, respectively. Since the velocity varies in space but constant velocity is needed to derive analytical solutions of Eq. (10.49), the averaged velocity is estimated as $v = \frac{\bar{q}}{\phi}$, where ϕ [−] is the porosity and $\bar{q}$ [L/T] is the harmonic mean of the specific discharge in the right half part of the domain (the half domain with pollution). The analytical solutions and their numerical implementation developed by Sun et al. (1999) are used in this study.

5.2 UNCERTAIN SCENARIOS, MODELS, AND PARAMETERS

The synthetic case considers three alternative scenarios of future precipitation. The three scenarios are wet, baseline, and dry scenarios, and they are referred to as Scenario 1, Scenario 2, and Scenario 3, respectively, for the convenience of following discussion. Precipitation of the baseline scenario (Scenario 2) is assumed to follow a normal distribution with the mean value of 1524 mm/year and standard deviation of 254 mm/year. In the wet and dry scenarios (Scenarios 1 and 3, respectively), the amount of precipitation is assigned to be 180% and 80% of the precipitation of the baseline scenario, respectively. The three distributions are truncated at 0 to ensure positive precipitation. The probability density functions (PDFs) of precipitation in the three scenarios are shown in Fig. 10.5. The figure shows that both the mean and variance increase from the dry to the wet scenario.

Two alternative groundwater flow models were built for the synthetic domain. The first model (denoted as one-zone model or Z_1) is based on the assumption that the domain is homogeneous with a constant hydraulic conductivity value. The corresponding mathematical model is

$$\frac{d^2\left(h(x)^2\right)}{dx^2} = -\frac{2w}{K}$$

$$h(x = 0) = h_1 \tag{10.50}$$

$$h(x = L) = h_2$$

where h [L] is the hydraulic head, x [L] is the distance from the left end of the domain, K [L/T] is the hydraulic conductivity constant over the domain, w [L/T] is the groundwater recharge rate, and h_1 [L]

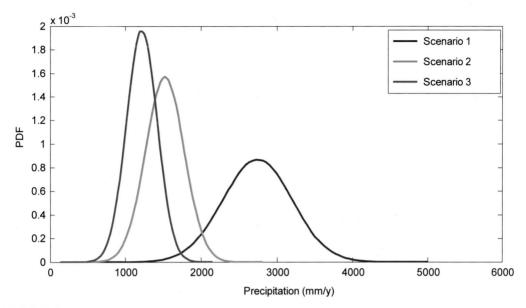

FIGURE 10.5

Probability density functions (PDFs) of precipitation under three scenarios: Scenario 1 (wet), Scenario 2 (baseline), and Scenario 3 (dry).

and h_2 [L] are the hydraulic heads at the left and right sides of the domain, respectively. The analytical solution of Eq. (10.50) is (Fetter, 2001)

$$h(x) = \sqrt{h_1^2 - \frac{(h_1^2 - h_2^2)x}{L} + \frac{w}{K}(L - x)x}$$

(10.51)

Using Darcy's law, the analytical solution of specific discharge, q [L/T], is

$$q(x) = \frac{1}{h}\left(\frac{K(h_1^2 - h_2^2)}{2L} - w\left(\frac{L}{2} - x\right)\right).$$

(10.52)

The second groundwater flow model (denoted as two-zone model or Z_2) was built with the assumption that the modeling domain consists of two vertically separated zones with the divide at the location $x = 7000$ m. The corresponding mathematical model is

$$\frac{d}{dx}\left(Kh(x)\frac{dh(x)}{dx}\right) = -w$$
$$h(x = 0) = h_1$$
$$h(x = L) = h_2$$

(10.53)

where $K = K_1$ [L/T] for $x \in [0, 7000]$ and $K = K_2$ [L/T] for $x \in [7000, L]$ are the hydraulic conductivities of the two zones. This model was solved numerically using the finite difference method.

Two alternative groundwater recharge models were considered in this study to convert precipitation to groundwater recharge. Considering recharge model uncertainty is necessary in groundwater modeling, because various techniques of recharge estimation have been developed and their recharge estimates can be substantially different (Scanlon et al., 2002). The first recharge model, denoted as R_1, is

$$M_1: w = \begin{cases} 16.88(P - 355.6)^{0.50} & P \geq 355.6 \\ 0 & \text{Otherwise} \end{cases}. \quad (10.54)$$

This model is based on the mass balance study of Thomas et al. (2009), which showed an exponential relation between precipitation and recharge. The other recharge model, denoted as R_2, is adapted from Krishna (1987),

$$M_2: w = \begin{cases} 0.15(P - 399.80) & P \geq 399.80 \\ 0 & \text{Otherwise} \end{cases}, \quad (10.55)$$

which relies on a linear relationship between precipitation and recharge. The coefficients of the two models are chosen to give sufficiently different recharge estimates by the two models. Fig. 10.6 plots the PDFs of the recharge estimates of the two models (R_1 is denoted as model 1, R_2 is denoted as model 2) under the three precipitation scenarios. The figure shows that the recharge estimate of recharge model R_1 is larger than that of recharge model R_2. For recharge model R_2 with linear relation between precipitation and recharge, the uncertainty in the recharge estimate is proportional to the uncertainty in precipitation.

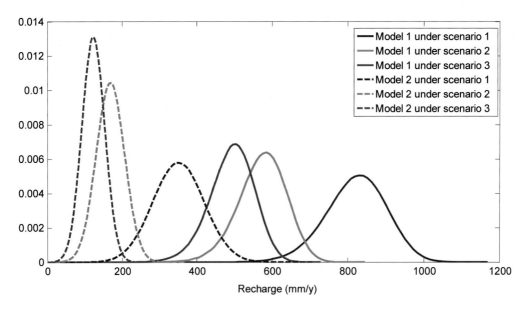

FIGURE 10.6

Probability density functions (PDFs) of groundwater recharge given by two models (models R1 and R2 are denoted as Model 1 and Model 2, respectively) under three precipitation scenarios.

Six parameters of the groundwater flow and transport models are considered to be random (model input factors are also treated as parameters in this study without loss of generality). The first parameter is precipitation, and its distributions under the three scenarios are discussed earlier. The next three are hydraulic conductivity, K, K_1, and K_2 (m/d), which are assumed to follow the normal distributions $N(15, 1)$, $N(20, 1)$, and $N(10, 1)$, respectively. These three normal distributions are truncated at the lower bound of 0. Although using lognormal distributions for hydraulic conductivity is more realistic, use of the normal distributions here is simply for method demonstration. Despite the fact that the results of SA may change significantly, we do not expect that the change would invalidate our new sensitivity indices. The fifth random parameter is hydrodynamic dispersion coefficient (m^2/d), and the uniform distribution $U(10, 10.1)$ is assumed. The last parameter is first-order degradation coefficient for PCE (1/d), and the normal distribution $N(0.05, 0.01)$ truncated at the lower bound of 0 is assumed.

Combining the two recharge models and the two groundwater models gives a total of four alternative models listed in Table 10.3: R_1Z_1, R_2Z_1, R_2Z_2, and R_2Z_2; each model has the same transport component given in Eq. (10.47). The manner of formulating the four models is similar to that of Ye et al. (2010) for quantifying model uncertainty in conceptualizing recharge component and hydrostratigraphic framework of the Death Valley Regional Flow system. For the four models, parametric uncertainty is conditioned on model uncertainty, from the hierarchical viewpoint discussed before. For example, uncertainty of hydraulic conductivity K is for models R_1Z_1 and R_2Z_1, and uncertainty of K_1 and K_2 are conditioned on models R_1Z_2 and R_2Z_2. Our new sensitivity indices are defined general, and different models can have different parameters.

5.3 TOTAL SENSITIVITY INDEX FOR HEAD UNDER INDIVIDUAL MODELS AND SCENARIOS

Table 10.3 lists the values of total sensitivity index (S_T) of precipitation (S_{T_P}) and hydraulic conductivity (S_{T_K}, $S_{T_{K_1}}$, and $S_{T_{K_2}}$) calculated for hydraulic head at the location $x = 6000$ m for the four models under the three scenarios. Because parameter K does not belong to models R_1Z_2 and R_2Z_2 and parameters K_1 and K_2 do not belong to models R_1Z_1 and R_2Z_1, the corresponding index values are 0, as discussed in Section 4. The table shows that, for each model and scenario, the sum of the total sensitivity indices for all parameters is slightly larger than 1, indicating that the interaction between the two parameters has a negligible contribution to head uncertainty. As a result, the first-order sensitivity index is almost identical to the total sensitivity index. Therefore, only the results of total sensitivity index are presented and discussed.

Table 10.3 shows that the index values may vary substantially for different models and scenarios. For example, S_{T_P} ranges from 5.96% for Model R_2Z_2 under Scenario 3 to 92.98% for Mode R_2Z_1 under Scenario 3, indicating that the importance of precipitation changes dramatically from unimportant to the most important. If the identification of important parameters is based on a single model and a single scenario (without considering model and scenario uncertainties), biased identification may occur. The variation of the total sensitivity index between different models and scenarios is physically meaningful. For the homogeneous models (R_1Z_1 and R_2Z_1), the increase of the index (S_{T_P}) of precipitation from Scenario 1 (wet scenario) to Scenario 3 (dry scenario) is attributed to the decrease of precipitation and thus the groundwater recharge from Scenario 1 to Scenario 3 (Fig. 10.6). In other words, impacts of precipitation on hydraulic head are larger, when the aquifer is relatively dry with a smaller amount

Table 10.3 Total sensitivity index of hydraulic conductivity (S_{T_K}, $S_{T_{K_1}}$, and $S_{T_{K_2}}$) and precipitation (S_{T_P}) calculated for hydraulic head at x = 6000 m under four models and three scenarios

	Scenario 1				Scenario 2				Scenario 3			
	R_1Z_1	R_2Z_1	R_1Z_2	R_2Z_2	R_1Z_1	R_2Z_1	R_1Z_2	R_2Z_2	R_1Z_1	R_2Z_1	R_1Z_2	R_2Z_2
S_{T_P} (%)	67.7	89.3	23.8	22.3	73.6	92.0	18.6	8.90	76.6	93.0	17.0	5.96
S_{T_K} (%)	32.4	11.0	0	0	26.8	8.64	0	0	23.6	7.27	0	0
$S_{T_{K_1}}$ (%)	0	0	3.15	8.58	0	0	5.79	13.7	0	0	6.99	15.3
$S_{T_{K_2}}$ (%)	0	0	73.1	69.2	0	0	75.7	77.5	0	0	76.5	79.2

After Model Averaging: Total Sensitivity Index for Multiple Models but Individual Scenarios

	Scenario 1	Scenario 2	Scenario 3
$S_{T_P}^{M}$ (%)	35.8	24.3	21.4
$S_{T_K}^{M}$ (%)	5.82	3.59	2.82
$S_{T_{K_1}}^{M}$ (%)	4.28	7.79	9.20
$S_{T_{K_2}}^{M}$ (%)	54.2	64.3	67.0

After Scenario Averaging: Total Sensitivity Index for Multiple Models and Multiple Scenarios

$S_{T_P}^{MS}$ (%)	28.1
$S_{T_K}^{MS}$ (%)	4.27
$S_{T_{K_1}}^{MS}$ (%)	6.77
$S_{T_{K_2}}^{MS}$ (%)	61.0

The indices for multiple models but individual scenarios are denoted as S_T^M; the indices for multiple models and multiple scenarios are denoted as S_T^{MS}.

of precipitation (or equivalently recharge). The variation pattern of the index (S_{T_K}) of hydraulic conductivity K is opposite. Also, for the homogeneous models (R_1Z_1 and R_2Z_1), S_{T_K} decreases from Scenario 1 (wet scenario) to Scenario 3 (dry scenario), because hydraulic conductivity becomes more important when recharge increases so that excessive water from precipitation can be discharged to the constant head boundary (Fig. 10.1).

For the heterogeneous models (R_1Z_2 and R_2Z_2), Table 10.3 reveals different variation patterns in the total sensitivity values from Scenario 1 to Scenario 3. It is observed that the importance of precipitation decreases in that S_{T_P} decreases from the range of 67.74%–92.98% for the homogeneous models to the range of 5.96%–23.75% for the heterogeneous models. Parameter K_2 becomes the most important parameter, which is reasonable because this parameter controls the amount of groundwater flow from the entire modeling domain to the river (Fig. 10.4). Different from the variation pattern for the homogeneous models, S_{T_P} decreases from Scenario 1 to Scenario 3, because the small parameter value of K_2 blocks the discharge to the river and the precipitation variation determines the precipitation influence on the hydraulic head at the location $x = 6000$ m, recalling that the precipitation variation is larger in Scenario 1 than in Scenario 3 (Fig. 10.5). This demonstrates again that parameter sensitivity indices change between the individual models and under different scenarios.

5.4 TOTAL SENSITIVITY INDEX FOR HEAD UNDER MULTIPLE MODELS BUT INDIVIDUAL SCENARIOS

To evaluate the total sensitivity index ($S_T^{\mathbf{M}}$ in Eq. (10.38)) listed in Table 10.3 for the four models but three individual scenarios, equal model probability (25%) is assigned to the four models for model averaging. It should be noted that $S_T^{\mathbf{M}}$ of the multiple models is not a weighted average of S_T of the individual models, because model averaging is conducted for the variance, not for the index, of the individual models. Table 10.3 demonstrates that the new sensitivity indices can be evaluated when different models have different number of parameters. For example, $S_{T_{K2}}^{\mathbf{M}}$ is evaluated for parameter K_2, which only belongs to models R_1Z_2 and R_2Z_2, and $S_{T_P}^{\mathbf{M}}$ is estimated for parameter P, which exists for all the four models. The model and scenario averaging techniques make it possible to compare the importance of different parameters that do not coexist in the same models.

The total sensitivity indices, $S_T^{\mathbf{M}}$, obtained after model averaging indicate that precipitation is the second most influential parameter, although it is the most influential parameter for models R_1Z_1 and R_2Z_1. This demonstrates that considering multiple models in the calculation of the sensitivity indices helps to prevent biased identification of important parameters based only on a single model. In other words, an important parameter of one model may not be important for multiple models, depending on the uncertainty magnitude of the parameter for the individual models.

The most influential parameter is parameter K_2, and it can be explained by Fig. 10.7A which plots the PDFs of the simulated hydraulic head by the four models under the three scenarios. The figure shows that, for a given scenario, the head uncertainty of the two-zone models is larger than that of the one-zone models. For parameters K_1 and K_2 of the two-zone models, K_2 has more influence on head simulation than K_1, because K_2 controls the groundwater flow to the river located at the right-hand side of the domain. The high sensitivity values of K_2 and P are useful for evaluating the relative plausibility of the models. For example, if head observations are available, they can be used to discriminate between the two recharge models and between the two groundwater flow models.

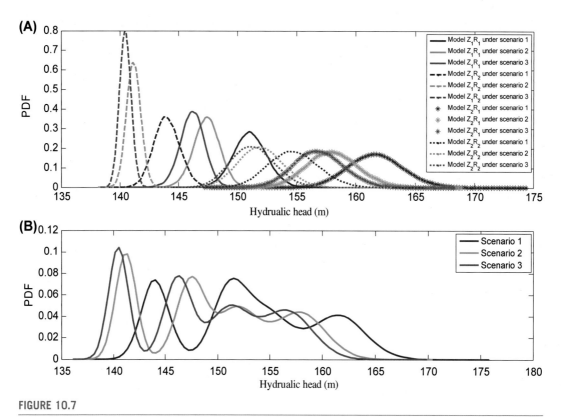

FIGURE 10.7

Probability density functions (PDFs) of hydraulic head at $x = 6000$ m simulated by (A) the two models under the three scenarios and (B) modeling averaging under the three scenarios.

5.5 TOTAL SENSITIVITY INDEX FOR HEAD UNDER MULTIPLE MODELS AND MULTIPLE SCENARIOS

For conducting scenario averaging, equal scenario probability (1/3) is assigned to the three scenarios. Table 10.3 lists the total sensitivity index, S_T^{MS}, of precipitation and hydraulic conductivity for multiple models and multiple scenarios. The sensitivity indices are the final measure of global parameter sensitivity with consideration of the combining effects of parameter, model, and scenario uncertainties. The S_T^{MS} values are similar to the S_T^M values of the individual scenarios, but different from the S_T values of individual models and scenarios. In particular, the S_T values of Model R_1Z_1 and R_2Z_1 under the three scenarios suggest that the precipitation is the most important parameter, whereas the S_T^{MS} (and S_T^M) values indicate that hydraulic conductivity K_2 is more important. It suggests that, without considering model and scenario uncertainties, it is likely that a biased identification of important parameter may occur.

The reason that the S_T^{MS} values are similar to the S_T^M values is that the head variance after model averaging is similar under the three scenarios, as shown in Fig. 10.7B. It suggests that scenario uncertainty plays a less important role than model uncertainty in terms of identifying important parameters.

This, however, is not a general case but specific to this synthetic example. The general uncertainty framework (Eq. (10.30)) and the new sensitivity indices are useful to evaluate relative contributions from different uncertainty sources to predictive uncertainty and to identify important parameters with consideration of the uncertainty sources.

5.6 TOTAL SENSITIVITY INDEX FOR ETHENE CONCENTRATION

The three sets (S_T, $S_T^{\mathbf{M}}$, and $S_T^{\mathbf{MS}}$) of total global sensitivity index are evaluated for ethene concentration at the location $x = 6000$ m after 1000 days, and their values are listed in Table 10.4. The table shows again that the sum of the four indices (i.e., S_{T_P} for precipitation, S_{T_K}, $S_{T_{K_1}}$, $S_{T_{K_2}}$ for hydraulic conductivity, S_{T_D} for dispersion coefficient, and $S_{T_{kl}}$ for first-order degradation coefficient of PCE) is slightly larger than 1, suggesting that the first-order and total sensitivity indices have similar values. Therefore, only the values of total sensitivity index are shown and discussed.

For the individual models and individual scenarios, the most important parameter is hydraulic conductivity (K for the one-zone models and K_2 for the two-zone models), indicating that hydraulic conductivity is the parameter that most influences the concentration simulation in this synthetic case. The S_T values of precipitation are small, ranging from 0.02% to 5.19%, which is reasonable because the variation of hydraulic gradient with precipitation is relatively small. The S_T values of dispersion coefficient are 0, which is attributed to the narrow range ($10.0-10.1$ m^2/d) of the parameter and the one-dimensional flow. The S_T values of degradation coefficient are also negligible, because, for most of the realizations, the plume of ethene (the last product of the chain reaction) has not reached the location $x = 6000$ m within 1000 days. If S_T is evaluated at another location $x = 5200$ m where ethene reaches this location after 1000 days for most of realizations, the S_T value of degradation coefficient can be significantly larger, e.g., 19.83% for model R_1Z_1 under Scenario 2. These results of SA suggest that, in the synthetic case, ethene transport at $x = 6000$ m and within 1000 days is largely affected by the flow parameters. This is reasonable because dispersive transport is negligible and advective transport is controlled by hydraulic conductivity and precipitation.

The $S_T^{\mathbf{M}}$ and $S_T^{\mathbf{MS}}$ values listed in Table 10.4 indicate that hydraulic conductivity (K for the one-zone models and K_2 for the two-zone models) is still the most important parameter after model and scenario averaging. However, the $S_{T_K}^{\mathbf{M}}$ values become dramatically smaller than the S_{T_K} values of the individual models, whereas the $S_{T_{K_2}}^{\mathbf{M}}$ values are similar to the $S_{T_{K_2}}$ values of the individual models. This leads to a conclusion that K_2 is significantly more important than K, which is consistent with the results listed in Table 10.3 for the sensitivity of hydraulic head to hydraulic conductivity. The values of $S_T^{\mathbf{MS}}$ are similar to the values of $S_T^{\mathbf{M}}$ under the individual scenarios, suggesting again that scenario uncertainty is less important than model uncertainty to ethene concentration at $x = 6000$ m after 1000 days. In this sense, identification of important parameters based on a single scenario is not biased. This, however, is unknown until the S_T, $S_T^{\mathbf{M}}$, and $S_T^{\mathbf{MS}}$ values are evaluated.

6. USING GLOBAL SENSITIVITY ANALYSIS FOR SATELLITE DATA AND MODELS

GSA has been widely used in numerical modeling with satellite data. For example, in the context of assimilating models with satellite, atmospheric, and surface observation data for ecology and paleoecology studies, Peng et al. (2011) used GSA to examine parameters sensitive to the data for the

Table 10.4 Total sensitivity index of model parameters calculated for ethene concentration at x = 6000 m under four models and three scenarios

	Scenario 1				Scenario 2				Scenario 3			
	R_1Z_1	R_2Z_1	R_1Z_2	R_2Z_2	R_1Z_1	R_2Z_1	R_1Z_2	R_2Z_2	R_1Z_1	R_2Z_1	R_1Z_2	R_2Z_2
S_{T_P} (%)	5.19	5.73	0.02	0.06	4.18	2.25	0.03	0.03	4.01	1.54	0.03	0.02
S_{T_K} (%)	94.7	94.1	0	0	95.2	97.2	0	0	95.5	97.9	0	0
$S_{T_{K_1}}$ (%)	0	0	1.97	1.38	0	0	1.81	0.97	0	0	1.30	0.71
$S_{T_{K_2}}$ (%)	0	0	98.0	98.7	0	0	98.4	99.0	0	0	99.1	99.6
S_{T_D} (%)	0.00	0.00	0.00	0.00	0.00	0.00	0.00	0.00	0.00	0.00	0.00	0.00
$S_{T_{k1}}$ (%)	0.90	0.85	0.13	0.16	0.90	0.74	0.15	0.17	0.90	0.71	0.15	0.17

After Model Averaging: S_T^M for Multiple Models but Individual Scenarios

	Scenario 1	Scenario 2	Scenario 3
$S_{T_P}^M$ (%)	0.37	0.28	0.26
$S_{T_K}^M$ (%)	5.72	8.49	9.48
$S_{T_{K_1}}^M$ (%)	1.37	1.02	0.92
$S_{T_{K_2}}^M$ (%)	92.43	90.36	89.61
$S_{T_D}^M$ (%)	0.00	0.00	0.00
$S_{T_{k1}}^M$ (%)	0.19	0.22	0.23

After Scenario Averaging: S_T^{MS} for Multiple Models and Multiple Scenarios

$S_{T_P}^{MS}$ (%)	0.30
$S_{T_K}^{MS}$ (%)	7.85
$S_{T_{K_1}}^{MS}$ (%)	1.11
$S_{T_{K_2}}^{MS}$ (%)	90.63
$S_{T_D}^{MS}$ (%)	0.00
$S_{T_{k1}}^{MS}$ (%)	0.21

purpose of parameter estimation. The applications of satellite SA are similar to those in other research fields. For example, to reduce the computational cost of variance-based GSA, the Morris method is first used to screen out nonimportant parameters so that the variance-based methods can be used only to important parameters. An example application of the two-step procedure is given in Lu et al. (2013) for the Australian community land surface model, which is a special case of the global land surface model that uses satellite data such as NASA's Shuttle Radar Terrain Mapping (Wood et al., 2011). In fact, the methods of GSA discussed in this chapter are mathematically general and can be applied to numerical models using satellite data in the manner described in Section 5 for the example application.

GSA for satellite data has its specific challenges that have not received sufficient attention. One of the challenges is the impacts of accuracy of satellite data on modeling results. For example, it is well known that MODIS data are subject to cloud coverage error, and Zhang et al. (2009) showed the impacts of missing MODIS data on phenological estimates. However, statistical characterization of the errors has not been well studied, and an inaccurate characterization of the data error may affect the results of SA. Taking LSA as an example, the variance of data error is always used to obtain a dimensionless sensitivity measure and can significantly affect the magnitude of the measure. The impacts of data accuracy may become more severe for GSA, because the data error may be amplified when it is propagated through the models for calculating variance of model predictions. Although the problem of data error is not specific to satellite data, it becomes more severe for satellite data because their accuracy cannot be easily characterized and validated. Solving this problem may require developing new statistical methods that are suitable to large data sets such as satellite data in future studies.

7. CONCLUSIONS AND PERSPECTIVES

This chapter discusses the use of GSA for identifying model parameters important to model outputs, which is necessary for model development, calibration, and improvement. The conventional methods of GSA address only parametric uncertainty to answer the following question: *How can we identify the important parameters, when we do not know the parameter values but do know the model structure and model scenario?* This chapter presents two widely used methods of GSA developed to answer this question. One is the Morris method that extends LSA from a single point (nominal parameter values) to multiple points in the parameter space. This method is conceptually straightforward, easy to implement, and computationally efficient. However, it does not provide a measure for quantitatively ranking the relative importance of model parameters. Therefore, the Morris method is always used as a screening method to screen important parameters out of unimportant parameters, although the screening can be somehow subjective if the mean and standard deviation of elementary effects of important parameters are not distinctively different from those of unimportant parameters. The other method of GSA discussed in this chapter is the Sobol' method, which is based on the variance decomposition method to identify the contribution to output uncertainty from individual model parameters and interactions between the parameters. The Sobol' method provides the first-order and total-effect sensitivity index, which can be used to quantitatively rank relative importance of model parameters. Therefore, the Sobol' method is theoretically superior to the Morris method. However, implementing the Sobol' method is computationally demanding than implementing the Morris method, because the two indices of the Sobol' method are always evaluated using MC approaches.

This chapter discusses two computationally efficient methods recently developed to implement the Sobol' method without using MC approaches. The two methods are the SGC methods developed by Dai and Ye (2015) and the DELSA method developed by Rakovec et al. (2014). The two methods are promising for breaking the computational barrier for applying GSA to real-world problems that are computationally expensive.

In addition to the two conventional methods of GSA, this chapter discusses a method developed by Dai and Ye (2015) to conduct GSA under uncertain model parameters, structures, and scenarios. This new method answers the following questions: *How can we identify the important parameters, if we do not know (1) the model scenarios under which model predictions will be made, (2) the model structures that will be used to make predictions, and (3) the model parameter values that will be used to make predictions?* The new method of Dai and Ye (2015) integrates the Sobol' method into the hierarchical framework for quantifying uncertainty in model scenarios, structures, and parameters. The framework is hierarchical because model structures are conditioned on model scenarios and model parameters are conditioned on model structures and scenarios. New definitions of the first-order and total-effect sensitivity indices are given for the following three situations: (1) single model and single scenario, (2) multiple models but single scenario, and (3) multiple models and multiple scenarios. Although not presented in this chapter, it is straightforward to define the sensitivity indices for the situation of single model but multiple scenarios. The numerical example of groundwater reactive transport modeling presented by Dai and Ye (2015) is included in this chapter to illustrate the use of the new sensitivity indices. The numerical example shows that, without considering model and scenario uncertainties, GSA focusing only on uncertain model parameters may results in biased selection of important parameters in the sense that the selected parameters are actually unimportant to the system of interest. The biased selection can be avoided by explicitly considering multiple models and model scenarios when evaluating the sensitivity indices.

GSA is an intriguing research topic, and there are still a large number of open questions to be answered. We present several viewpoints on development trends, research issues, and hotspots related to GSA:

1. Computationally efficient methods are needed for performing GSA, especially for real-world problems that are nonlinear and computationally demanding. Since most methods of GSA are implemented using MC approaches, the meta modeling approaches to build cheap but accurate surrogates is promising. After the surrogates are developed, the computational cost of executing the surrogates for MC simulations is negligible. However, the challenge to metamodeling approaches is that a large NME may still be needed to build accurate surrogates. This problem may be resolved by using parallel computing if the model executions needed for building the surrogates can be parallelized.

2. Parameter ranges used for GSA should be carefully selected. The current practice of choosing wide parameter ranges based on the literature data may not be appropriate, because certain parameter values may lead to physically unreasonable model outputs, although the parameter values themselves are physically reasonable. Physical constraints on model outputs should be examined for each specific problem and enforced during SA. Model outputs that violate the constraints should not be used for evaluating the statistics for SA.

3. More attention should be paid to develop methods of GSA for multiple model outputs, including different variables and/or the same variable at different locations and/or time. This is necessary

for multiobjective model calibration and experiment design, especially when the multiple model outputs are highly correlated and/or competitive. The current practice is to build a utility function for multiple model outputs and then to apply GSA for the utility function. However, the impacts of the utility function on the results of SA have not be investigated in depth.

4. LSA should be conducted before implementing time-consuming GSA. Despite the limitations of LSA discussed in the introduction, LSA can still provide a good amount of information on parameter importance. Hill et al. (2016) presented a numerical example in which local and global sensitivity analyses identified the same most and least important parameters. Given that LSA always takes a significantly small amount of computational time when compared with GSA, it is recommended to conduct LSA before conducting GSA.

ACKNOWLEDGMENTS

This work was supported in part by National Science Foundation NSF-EAR grant 1552329 and Department of Energy Early Career Award DE−SC0008272.

REFERENCES

Aziz, C.E., Newell, C.J., Gonzales, J.R., Haas, P.E., Clement, T.P., Sun, Y., 2000. BIOCHLOR Natural Attenuation Decision Support System: User's Manual, Version 1.0. U.S. EPA Office of Research and Development, Washington, DC.

Barthelmann, V., Novak, E., Ritter, K., 1999. High dimensional polynomial interpolation on sparse grid. Advances in Computational Mathematics 12, 273−288.

Beven, K., 2002. Towards a coherent philosophy for modeling the environment. Proceedings of the Royal Society of London A: Mathematical, Physical and Engineering Science 458 (2026), 2465−2484.

Beven, K., 2006. A manifesto for the equifinality thesis. Journal of Hydrology 320, 18−36.

Borgonovo, E., Castaings, W., Tarantola, S., 2012. Model emulation and moment-independent sensitivity analysis: an application to environmental modelling. Environmental Modelling & Software 34, 105−115.

Bredehoeft, J.D., 2003. From models to performance assessment: the conceptualization problem. Ground Water 41 (5), 571−577.

Bredehoeft, J.D., 2005. The conceptualization model problem-surprise. Hydrogeology Journal 13, 37−46.

Bungartz, H.J., Griebel, M., 2004. Sparse grids. Acta Numerica 13, 147−269.

Buzzard, G.T., 2012. Global sensitivity analysis using sparse grid interpolation and polynomial chaos. Reliability Engineering and System Safety 107, 82−89.

Buzzard, G.T., Xiu, D., 2011. Variance-based global sensitivity analysis via sparse grid interpolation and cubature. Communications in Computational Physics 9, 542−567.

Campolongo, F., Cariboni, J., Saltelli, A., 2007. An effective screening design for sensitivity analysis of large models. Environmental Modelling & Software 22, 1509−1518. http://dx.doi.org/10.1016/j.envsoft.2006.10.004.

Chen, X., Ng, B.M., Sun, Y., Tong, C.H., 2013. A computational method for simulating subsurface flow and reactive transport in heterogeneous porous media embedded with flexible uncertainty quantification. Water Resources Research 49, 5740−5755. http://dx.doi.org/10.1002/wrcr.20454.

Chu-Agor, M.L., Muñoz-Carpena, R., Kiker, G., Emanuelsson, A., Linkov, I., 2011. Exploring sea level rise vulnerability of coastal habitats through global sensitivity and uncertainty analysis. Environmental Modelling & Software 26 (5), 593−604. http://dx.doi.org/10.1016/j.envsoft.2010.12.003.

Dai, H., Ye, M., 2015. Variance-based global sensitivity analysis for multiple scenarios and models with implementation using sparse grid collocation. Journal of Hydrology 528, 286–300. http://dx.doi.org/10.1016/j.jhydrol.2015.06034.

Davis, P.J., Rabinowitz, P., 1967. Numerical integration. Blaisdell Pub. Co., Waltham, Massachusetts, p. 96.

Davis, P.J., Rabinowitz, P., 1975. Methods of Numerical Integration. Academic Press, New York, pp. 67–68.

Fetter, C.W., 2001. Applied Hydrogeology, fourth ed. Prentice-Hall, New Jersey, p. 143.

Garcia-Cabrejo, O., Valocchi, A., 2014. Global sensitivity analysis for multivariate output using polynomial chaos expansion. Reliability Engineering and System Safety 126, 25–36.

Gerstner, T., Griebel, M., 1998. Numerical integration using sparse grid. Numerical Algorithms 18, 209–232.

Hakami, A., Odman, M.T., Russell, A.G., 2003. High-order direct sensitivity analysis of multidimensional air quality models. Environmental Science & Technology 37, 2442–2452.

Helton, J.C., 1993. Uncertainty and sensitivity analysis techniques for use in performance assessment for radioactive waste disposal. Reliability Engineering and System Safety 42, 327–367.

Helton, J.C., Davis, F.J., 2003. Latin hypercube sampling and the propagation of uncertainty in analyses of complex system. Reliability Engineering and System Safety 81, 23–69.

Hill, M.C., Kavetski, D., Clark, M., Ye, M., Arabi, M., Lu, D., Foglia, L., Mehl, S., 2016. Practical use of computationally frugal model analysis methods. Ground Water 54 (2), 159–170. http://dx.doi.org/10.1111/gwat.12330.

Hill, M.C., Tiedeman, C.R., 2007. Effective Calibration of Ground Water Models, with Analysis of Data, Sensitivities, Predictions, and Uncertainty. John Wiley and Sons, New York, p. 480.

Homma, T., Saltelli, A., 1996. Importance measures in global sensitivity analysis of model output. Reliability Engineering and System Safety 52 (1), 1–17.

IPCC, 2000. In: Nakicenovic, N., Swart, R. (Eds.), IPCC Special Report: Emissions Scenarios. Cambridge University Press, UK, p. 570.

Krishna, R.K., 1987. Groundwater Assessment, Development and Management. Tata McGraw-Hill Publishing Co. Ltd, New Delhi, pp. 576–657.

Lamoureux, B., Mechbal, N., Masse, J.-R., 2014. A combined sensitivity analysis and kriging surrogate modeling for early validation of health indicators. Reliability Engineering and System Safety 130, 12–26.

Lin, G., Tartakovsky, A.M., 2009. An efficient, high-order probabilistic collocation method on sparse grids for three-dimensional flow and solute transport in randomly heterogeneous porous media. Advances in Water Resources 32 (5), 712–722, 1009 10.1016/j.advwatres.2008.09.003.

Lin, G., Tartakovsky, A.M., 2010. Numerical studies of three-dimensional stochastic Darcy's equation and stochastic advection-diffusion-dispersion equation. Journal of Scientific Computing 43, 92–117. http://dx.doi.org/10.1007/s10915-010-9346-5.

Lin, G., Tartakovsky, A.M., Tartakovsky, D.M., 2010. Uncertainty quantification via random domain decomposition and probabilistic collocation on sparse grids. Journal of Computational Physics 229 (19), 6995–7012. http://dx.doi.org/10.1016/j.jcp.2010.05.036.

Lu, D., Ye, M., Neuman, S.P., 2011. Dependence of Bayesian model selection criteria and Fisher information matrix on sample size. Mathematical Geosciences 43 (8), 971–993. http://dx.doi.org/10.1007/s11004-011-9359-0.

Lu, D., Ye, M., Neuman, S.P., Xue, L., 2012. Multimodel Bayesian analysis of data-worth applied to unsaturated fractured tuffs. Advances in Water Resources 35, 69–82. http://dx.doi.org/10.1016/j.advwatres.2011.10.007.

Lu, X., Wang, Y.-P., Ziehn, T., Dai, Y., 2013. An efficient method for global parameter sensitivity analysis and its applications to the Australian community land surface model (CABLE). Agricultural and Forest Meteorology 182–183, 292–303.

Meyer, P.D., Ye, M., Nicholson, T., Neuman, S.P., Rockhold, M., 2014. Incorporating scenario uncertainty within a hydrogeologic uncertainty assessment methodology. In: Mosleh, A., Wood, J. (Eds.), Proceedings of the International Workshop on Model Uncertainty: Conceptual and Practical Issues in the Context of Risk-informed Decision Making, International Workshop Series on Advanced Topics in Reliability and Risk Analysis. Center for Risk and Reliability, University of Maryland, College Park, MD, USA, pp. 99–119. ISSN:1084-5658.

Morgan, M.G., Henrion, M., 1990. Uncertainty: A Guide to Dealing with Uncertainty in Quantitative Risk and Policy Analysis. Cambridge University Press, New York.

Morris, M.D., 1991. Factorial sampling plans for preliminary computational experiments. Technometrics 33 (2), 161–174. http://www.jstor.org/stable/1269043.

Neuman, S.P., 2003. Maximum likelihood Bayesian averaging of alternative conceptual-mathematical models. Stochastic Environmental Research and Risk Assessment 17 (5), 291–305. http://dx.doi.org/10.1007/s00477-003-0151-7.

Neuman, S.P., Xue, L., Ye, M., Lu, D., 2012. Bayesian analysis of data-worth considering model and parameter uncertainties. Advances in Water Resources 36, 75–85. http://dx.doi.org/10.1016/j.advwatres.2011.02.007.

Novak, E., Ritter, K., 1999. Simple cubature formulas with high polynomial exactness. Constructive Approximation 15, 499–522.

O'Hagan, A., 2006. Bayesian analysis of computer code outputs: a tutorial. Reliability Engineering and System Safety 91, 1290–1300.

Oladyshkin, S., de Barros, F.P.J., Nowak, W., 2012. Global sensitivity analysis: a flexible and efficient framework with an example from stochastic hydrogeology. Advances in Water Resources 37, 10–22.

Peng, C.H., Guiot, J., Wu, H.B., Jiang, H., Luo, Y.Q., 2011. Integrating models with data in ecology and palaeoecology: advances towards a model-data fusion approach. Ecology Letters 14, 522–536.

Pohlmann, K., Ye, M., Reeves, D., Zavarin, M., Decker, D., Chapman, J., 2007. Modeling of Groundwater Flow and Radionuclide Transport at the Climax Mine Sub-CAU, Nevada Test Site. Desert Research Institute, Division of Hydrologic Sciences. Publication No. 45226, OE/NV/26383–06.

Pohlmann, K., Ye, M., 2012. Numerical Simulation of Inter-basin Groundwater Flow into Northern Yucca Flat, Nevada National Security Site, Using the Death Valley Regional Flow System Model. DOE/NV/26383–18. Nevada Site Office, National Nuclear Security Administration, U.S. Department of Energy,, Las Vegas, NV.

Rakovec, O., Hill, M.C., Clark, M.P., Weerts, A.H., Teuling, A.J., Uijlenhoet, R., 2014. Distributed evaluation of local sensitivity analysis (DELSA), with application to hydrologic models. Water Resources Research 50, 409–426. http://dx.doi.org/10.1002/2013WR014063.

Rasmussen, C.E., Williams, C.K.I., 2006. Gaussian Processes for Machine Learning. MIT Press, Cambridge, MA.

Razavi, S., Tolson, B.A., Burn, D.H., 2012. Review of surrogate modeling in water resources. Water Resources Research 48, W07401. http://dx.doi.org/10.1029/2011WR011527.

Razavi, S., Gupta, H.V., 2015. What do we mean by sensitivity analysis? The need for comprehensive characterization of "global" sensitivity in Earth and Environmental systems models. Water Resources Research 51, 3070–3092. http://dx.doi.org/10.1002/2014WR016527.

Running, S.W., Nemani, R.R., Heinsch, F.A., Zhao, M., Reeves, M., Hashimoto, H., 2004. A continuous satellite-derived measure of global terrestrial primary production. BioScience 54, 547–560.

Saltelli, A., Tarantola, S., Chan, K.P.-S., 1999. A quantitative model independent method for global sensitivity analysis of model output. Technometrics 41, 39–56.

Saltelli, A., Tarantola, S., Campolongo, F., Ratto, M., 2004. Sensitivity Analysis in Practice: A Guide to Assessing Scientific Models. John Wiley & Sons.

Saltelli, A., Annoni, P., Azzini, I., Campolongo, F., Ratto, M., Tarantola, S., 2010. Variance based sensitivity analysis of model output. Design and estimator for the total sensitivity index. Computer Physics Communications 181 (2), 259–270.

Scanlon, B.R., Healy, R.W., Cook, P.G., 2002. Choosing appropriate techniques for quantifying groundwater recharge. Hydrogeology Journal 10, 18–39.

Shi, L., Yang, J., Zhang, D., Li, H., 2009. Probabilistic collocation method for unconfined flow in heterogeneous media. Journal of Hydrology 365 (1–2), 4–10, 1086 10.1016/j.jhydrol.2008.11.012.

Smolyak, S.A., 1963. Quadrature and interpolation formulas for tensor products of certain classes of functions. Doklady Akademii Nauk SSSR 4, 240–243.

Sobol', I.M., 1993. Sensitivity analysis for nonlinear mathematical models. Mathematical Models and Computer Experiment 1 (4), 407–414.

Sobol', I.M., Kucherenko, S., 2009. Derivative based global sensitivity measures and their link with global sensitivity indices. Mathematics and Computers in Simulation 79 (10), 3009–3017. http://dx.doi.org/10.1016/j.matcom.2009.01.023.

Song, X., Zhang, J., Zhan, C., Xuan, Y., Ye, M., Xu, C., 2015. Global sensitivity analysis in hydrological modeling: review of concepts, methods, theoretical framework, and applications. Journal of Hydrology 523 (4), 739–757. http://dx.doi.org/10.1016/j.jhydrol.2015.02.013.

Storlie, C.B., Swiler, L.P., Helton, J.C., Sallaberry, C.J., 2009. Implementation and evaluation of nonparametric regression procedures for sensitivity analysis of computationally demanding models. Reliability Engineering and System Safety 94, 1735–1763.

Sudret, B., 2007. Global sensitivity analysis using polynomial chaos expansion. Reliability Engineering and System Safety 93, 964–979.

Sun, Y., Petersen, J.N., Clement, T.P., 1999. Analytical solutions for multiple species reactive transport in multiple dimensions. Journal of Contaminant Hydrology 25, 429–440.

Tang, J., Zhuang, Q., 2009. A global sensitivity analysis and Bayesian inference framework for improving the parameter estimation and prediction of a process-based Terrestrial Ecosystem Model. Journal of Geophysical Research 114, D15303. http://dx.doi.org/10.1029/2009JD011724.

Thomas, T.R., Jaiswal, K., Galkate, R., Singh, S., 2009. Development of a rainfall–recharge relationship for a fractured basaltic aquifer in Central India. Water Resources Management 23 (15), 3101–3119.

White, M.A., Thornton, P.E., Running, S.W., Nemani, R.R., 2000. Parameterization and sensitivity analysis of the BIOME-BGC terrestrial ecosystem model: net primary production controls. Earth Interactions 4 (3), 1–85.

Wood, E.F., et al., 2011. Hyperresolution global land surface modeling: meeting a grand challenge for monitoring Earth's terrestrial water. Water Resources Research 47, W05301. http://dx.doi.org/10.1029/2010WR010090.

Xiu, D., Hesthaven, J., 2005. High-order collocation methods for differential equations with random inputs. SIAM Journal on Scientific Computing 27 (3), 1118–1139.

Ye, M., Pohlmann, K.F., Chapman, J.B., Pohll, G.M., Reeves, D.M., 2010. A model-averaging method for assessing groundwater conceptual model uncertainty. Ground Water. http://dx.doi.org/10.1111/j.1745-6584.2009.00633.x.

Ye, M., Wang, L., Pohlmann, K.F., Chapman, J.B., 2016. Estimate groundwater interbasin flow using multiple models and multiple types of calibration data. Ground Water. http://dx.doi.org/10.1111/gwat.12422.

Zeng, X., Ye, M., Burkardt, J., Wu, J., Wang, D., 2016. Evaluating two sparse grid surrogates and two adaptation criteria for groundwater Bayesian uncertainty analysis. Journal of Hydrology 535, 120–134. http://dx.doi.org/10.1016/j.jhydrol.2016.01.058.

Zhang, G., Lu, D., Ye, M., Gunzburger, M., Webster, C., 2013. An adaptive sparse-grid high-order stochastic collocation method for Bayesian inference in groundwater reactive transport modeling. Water Resources Research 49, 1–22. http://dx.doi.org/10.1002/wrcr.20467.

Zhang, X., Friedl, M.A., Schaaf, C.B., 2009. Sensitivity of vegetation phenology detection to the temporal resolution of satellite data. International Journal of Remote Sensing 30 (8), 2061–2074.

Zhao, J., Scheibe, T.D., Mahadevan, R., 2011. Model-based analysis of the role of biological, hydrological and geochemical factors affecting uranium bioremediation. Biotechnology and Bioengineering 108 (7), 1537–1548.

OTHER SA METHODS: CASE STUDIES

SENSITIVITY AND UNCERTAINTY ANALYSES FOR STOCHASTIC FLOOD HAZARD SIMULATION

## 11

Z. Micovic[1], M.G. Schaefer[2], B.L. Barker[2]

BC Hydro, Burnaby, BC, Canada[1]; MGS Engineering Consultants, Olympia, WA, United States[2]

CHAPTER OUTLINE

Sensitivity Analysis in Earth Observation Modelling. http://dx.doi.org/10.1016/B978-0-12-803011-0.00011-2

1. INTRODUCTION

The traditional deterministic Inflow Design Flood (IDF) concept has been and is still being used to size dam spillways and appurtenant works to pass either a flood of predetermined probability of exceedance or the Probable Maximum Flood (PMF). The PMF is defined as the flood that may be expected from the most severe combination of critical meteorologic and hydrologic conditions that is reasonably possible in the drainage basin under study (FEMA, 2013). Note that the PMF is a deterministic concept, and its probability of occurrence is not explicitly defined. Theoretically, it represents the upper physical flood limit for a given watershed at a given season. In reality, PMF estimates are typically lower than the theoretical upper limit by some variable amount that depends on the available data, the chosen methodology, and the analyst's approach to deriving the estimate (Micovic et al., 2015).

The deterministic IDF standard is directly linked to the dam hazard classification so that low-hazard dams are designed using smaller IDF than high-hazard dams (i.e., multiple loss of life following a dam failure is certain). Consequently, the IDF used in the design of high-hazard dams is either the PMF or a very rare flood with return period ranging from 1000 to 10,000 years. For low-hazard dams, the IDF selection criteria vary, and typically include either a percentage of the PMF or return periods shorter than 1000 years (ICOLD, 2003).

Recently, an increasing number of dam owners started to apply various forms of risk-informed decision-making process in their dam safety assessments regarding flood hazard. The PMF concept cannot be used in risk analyses for dam safety because it is inherently deterministic. Risk-informed decision making requires development of full flood frequency curves (hydrologic hazard curves) for various flood characteristics up to and including extreme events, where scenarios of different hazards (not just overtopping) with different exceedance probabilities can be evaluated. Furthermore, the deterministic IDF standard itself is inadequate for flood hazard risk analyses with regard to dams or dam systems with seasonally fluctuating reservoir levels and active discharge control systems such as gated spillways. The IDF, by characterizing inflow to the reservoir, does not provide the necessary information (i.e., magnitude and exceedance probability) of the flood hazard in terms of hydraulic loadings acting on the dam itself (peak reservoir level).

Risk-informed decision making for dam safety requires more analysis effort than the IDF concept. To estimate the probability of dam overtopping due to floods, it is necessary to focus on estimating the probabilities of peak reservoir level. Note that the peak reservoir level, unlike the reservoir inflow, is not a natural and random phenomenon and its probability distribution cannot be computed analytically (e.g., by using statistical frequency analysis methods). The probability of the peak reservoir level is the combination of probabilities of all factors that influence it, including reservoir inflows, initial reservoir level, reservoir operating rules, system components failure, human error, measurement error, and unforeseen circumstances. Thus, the only solution for estimating the full probability distribution of the peak reservoir level is some kind of stochastic simulation that includes as many of these factors as possible. The main goal of stochastic flood hazard simulation is to carry out probabilistic analysis of various flood characteristics (inflow, outflow, peak reservoir level) resulting from floods on a dam/reservoir system and derive the continuous probability distributions, which could then be used to evaluate exceedance probabilities of various reservoir levels including the level corresponding to the dam crest (dam overtopping level) as well as the level resulting from the PMF. This way, different design criteria could be considered and evaluated at various flood frequency levels, departing from the widely used strict "pass/fail" deterministic design criteria.

2. BASIC PRINCIPLES OF STOCHASTIC APPROACH TO FLOOD HAZARD

In deterministic approaches, a particular flood characteristic (e.g., inflow, outflow, routed reservoir level) is the result of a fixed combination of meteorological, hydrological, and reservoir routing–related inputs. For instance, the peak reservoir level resulting from the PMF is derived using a fixed combination of the following inputs:

- Rainfall magnitude and its spatial and temporal distributions over the watershed (typically provided in form of probable maximum precipitation)
- Initial snowpack accumulation within the watershed
- Air temperature sequence for the duration of PMF event
- Initial soil moisture content of the watershed
- Initial reservoir level
- Availability and operating sequence of discharge facilities

On the other hand, stochastic approaches (e.g., Schaefer and Barker, 2002; Nathan et al., 2002; Paquet et al., 2013) treat those inputs as variables instead of fixed values considering the fact that any flood characteristic (e.g., the peak reservoir level) could be caused by an infinite number of different combinations of inputs. The variation of the flood-producing input parameters is achieved by stochastic sampling either from empirical distributions or from theoretical probability distributions fitted to observed data.

The one common aspect of all stochastic approaches to flood hazard for dam safety is the use of a deterministic watershed model (rainfall-runoff model) to convert rainfall and snowmelt/glacier melt into runoff from the watershed, which ultimately becomes the reservoir inflow. The watershed model is typically calibrated to satisfactorily represent hydrological behavior of the watershed over a long (25 years or more) continuous period for which historical record of climate input data is available. This creates a continuous database of watershed initial conditions, which can be stochastically sampled at any time/season of the year and combined with a rainfall event of a certain duration and probability, sampled from rainfall magnitude–frequency curve. The end result is thousands or millions of flood hydrographs ranging in magnitudes from common to extreme. The simulation for each year contains a set of climatic and storm parameters that were sampled through Monte Carlo procedures based on the historical record and collectively preserved dependencies among different hydrometeorological inputs. Simulated flood characteristics such as peak inflow, maximum reservoir release, inflow volume, and maximum reservoir level are the parameters of interest.

It is important to point out that the accuracy of flood modeling depends greatly on the accuracy of hydrometeorological input data used in the flood simulation process. Direct ground observations of hydrometeorological inputs used in flood modeling (e.g., rainfall, temperature, snowpack depth) are typically available as point measurements, often sparsely distributed within a watershed. To spatially distribute these point measurements, some kind of interpolation is required thereby reducing the accuracy and increasing the uncertainty associated with spatially estimated values, especially in areas with complicated topography and sparse point measurement stations. However, these uncertainties could be reduced by utilizing available remote sensing data from radar, radiosonde, and satellite observations. An example of a great data analysis tool used in flood modeling is the PRISM (Parameter-elevation Relationships on Independent Slopes Model) system (Daly et al., 1994), which takes point measurements of precipitation (and other parameters) and creates GIS-compatible data layers, which realistically represent the orographic processes so important in mountainous regions like the western United States and Canada. Also, use of radar data has greatly increased the resolution and accuracy of the spatial distribution of precipitation for historical storms. These data have improved the simulation of the spatial and temporal patterns of storms for use in stochastic modeling of floods. Similarly, remote sensing data from satellites provide spatial and temporal measurements of precipitation, air temperature, and snowpack areal extent. These data are utilized by atmospheric models for construction of time series of climatic variables through numerical reanalysis of historical conditions. Satellite data are also used in hydrologic modeling for identifying land use, forested areas, glaciers, and soil types, as well as for assessing seasonal watershed conditions for soil moisture and snowpack areal extent, which are critical for conducting realistic flood simulations. Radiosonde measurements of air temperature and dew points for selected pressure heights (e.g., 1000, 700, 500 mb) are used for estimating precipitable water. The main downside of radar and satellite observations is that they are not direct measurements and rely on complex algorithms converting radar reflectivities or satellite-measured radiances into rainfall rates at the ground. For example, some factors affecting accuracy of the radar rainfall measurements include the increase of the beam width and minimum detection height with distance from the radar and blockage/ground clutter from mountains and other obstacles, all leading to rainfall underestimation (Sene, 2013). Satellite rainfall measurements are unaffected by mountains and other obstacles and do not have spatial inconsistencies that affect radar, such as calibration differences and changes in radar beam height. However, the algorithms converting satellite signal to rainfall are typically less robust than those used for radar rainfall measurements (Scofield and Kuligowski, 2003). There is a plethora of research dealing with the assessment of remote sensing estimates of rainfall by comparing them with ground observations at different spatial and temporal scales. Examples include Wood et al. (2000) for radar-estimated rainfall, Stampoulis et al. (2013) for satellite-estimated rainfall, and Prat and Nelson (2015) for rainfall estimated by both radars and multisensor satellites. Furthermore, Sene (2013) provides a good summary of multisensor precipitation estimation techniques where the best features from each approach are combined to derive a single estimate; one example is the use of both geostationary and polar-orbiting satellite observations combined with available real-time rain gages and weather radar observations, and making use of the outputs from numerical weather prediction models. Based on the state of art in earth observations and their use as hydrometeorological inputs to flood modeling studies, it appears that the most sensible approach is to combine and complement existing ground point measurements with available spatially distributed remote sensing observations thereby reducing potential bias and overall uncertainty.

Note that it is extremely difficult if not impossible to accurately cover all aspects of stochastic flood simulation due to the enormous complexities of both the flood phenomenon and the dam/ reservoir system and all possible interactions among their respective inputs and components. That is why in practical applications not all flood-producing factors are modeled to the same extent—some are treated as stochastic variables, whereas some are fixed or not modeled at all. Brief descriptions of main aspects of stochastic simulation framework are provided in the following sections using the example of the BC Hydro's Bridge River hydroelectric system located approximately 200 km northeast of Vancouver, BC, Canada. The system consists of three dams and three reservoirs. The upstream portion of the Bridge River watershed is impounded by La Joie Dam and forms Downton Lake reservoir. All releases from the La Joie Dam, generation or spill, discharge into the Carpenter Lake reservoir formed by Terzaghi Dam. Releases from Terzaghi Dam can be either diverted into

Seton Lake reservoir formed by Seton Dam or discharged directly into Lower Bridge River. The total watershed area upstream of La Joie Dam is 1000 km^2, and the area for the local watershed between the Terzaghi and La Joie dams is 2720 km^2. The drainage area above Seton Dam is 1010 km^2. The Bridge River basin has rugged mountainous terrain with elevations ranging from 240 to 2900 m. The stochastic flood hazard simulation for the Bridge System was carried out using the latest version of the Stochastic Event Flood Model (SEFM) (Schaefer and Barker, 2015). The UBC Watershed Model (UBCWM) (Quick, 1995; Micovic and Quick, 2009) was used for the precipitation to runoff conversion.

2.1 STOCHASTIC SIMULATION OF RESERVOIR INFLOWS

Natural dependencies are prevalent throughout the collection of hydrometeorological variables. The natural dependencies/correlations are preserved in the sampling procedures with a particular emphasis on seasonal dependencies. For instance, the sampling of freezing levels is conditioned on both the month of occurrence and the 24-h precipitation magnitude. The watershed conditions for soil moisture, snowpack, and initial reservoir level are all interrelated and inherently correlated with the magnitude and sequencing of daily, weekly, and monthly precipitation. These interrelationships are established through calibration to historical streamflow records in long-term continuous watershed modeling and the state variables are stored for each day of the calibration period (typically 25 years or more). The hydrometeorological inputs to the Bridge River System stochastic flood model and the dependencies that exist in the stochastic simulation of a particular input are briefly described in the following sections. An in-depth discussion could be found in Micovic et al. (2016).

2.1.1 Storm Seasonality
Dependencies: Independent.
Probability model: 3-parameter Gamma distribution.

The seasonality of storm occurrence is defined by the monthly distribution of the historical occurrences of storms with widespread areal coverage that have occurred over the studied area. This information is used to select the date of occurrence of the storm for a given stochastic simulation. The basic concept is that the seasonality characteristics of extraordinary storms used in stochastic flood simulations should be the same as the seasonality of all significant storms in the historical record. The term "significant" is somewhat subjective, but it usually refers to storm events where precipitation maxima for a given storm duration exceeds a 10-year return period at three or more precipitation gages within the studied area. This criterion assures that only storms with both unusual precipitation amounts and broad areal coverage would be considered in the analysis. This procedure resulted in identification of 65 storm events in the period from 1919 to 2012. A probability plot was developed using numeric storm dates, and it was determined that the seasonality data could be well-described by a 3-parameter Gamma distribution with a skew coefficient of −0.50. A frequency histogram was then constructed based on the fitted Gamma distribution to depict the monthly distribution of the dates of extreme storms for input into SEFM.

2.1.2 Precipitation Magnitude—Frequency Relationship
Dependencies: Independent.
Probability model: 4-parameter Kappa distribution.

The magnitude of precipitation relevant to dam safety analyses is typically several orders of magnitude more extreme than has been observed in the historic record. As such, the estimation of this range of rainfall presents special difficulties and requires the extrapolation of relatively short historical data records. This extrapolation is rather challenging, especially considering that rainfall input is generally the most significant contributor to resulting stochastically derived flood hydrographs. Precipitation annual maxima series data were assembled for the critical storm duration (48 h in the case of Bridge River basin) from all stations within and near the watershed. This totaled 178 stations and 7589 station-years of record for stations with 15 or more years of record. The precipitation−frequency relationships for the Bridge System watersheds were developed through regional L-moment analyses (Hosking and Wallis, 1997) of point precipitation and spatial analyses of historical storms to develop point−area relationships and determine basin-average precipitation for the watershed using the 4-parameter Kappa distribution, which provided the best fit to the observed data sample.

2.1.3 Temporal and Spatial Distribution of Storms

Dependencies: Independent.
Probability model: Uniform distribution (each storm assumed equally likely).

Scalable spatial and temporal storm templates are needed for stochastic generation of storms. Stochastic storm generation is accomplished by linear scaling of the spatial and temporal storm patterns for a selected prototype storm. Specifically, the spatial and temporal storm templates are scaled by the proportion of the desired 48-h basin-average precipitation relative to the 48-h basin-average precipitation observed in a selected prototype storm. The historical storm record for the Bridge River basin for the 1985−2011 period was reviewed and 22 storms were identified for use in creating 22 spatial and temporal storm templates, which was sufficient to capture storm diversity over the basin. The 22 storms were sampled using a uniform distribution, with each storm having an equal probability of being selected.

2.1.4 Air Temperature and Freezing Level Temporal Patterns

Dependencies: Temperature and freezing level patterns are matched one to one with each prototype storm.
Probability model: Uniform, associated one to one with each storm temporal pattern.

Temporal patterns for air temperature and freezing level are used in computing snowmelt runoff. These temporal patterns were developed so that they could be rescaled by stochastically drawn values of air temperature and freezing level. The time series of air temperature at 1000 mb (i.e., at or near the sea level) is typically combined with temperature lapse rates to generate air temperature values within the full elevation range of a given watershed.

2.1.5 The 1000-mb Air Temperature Simulation

Dependencies: Storm magnitude.
Probability model: Physically based stochastic model.

Larger storm precipitation amounts are generally associated with higher 1000-mb dew point temperatures. This occurs because high levels of atmospheric moisture are needed to support large precipitation amounts and high levels of atmospheric moisture require higher air temperatures to sustain these moisture levels. Within the stochastic flood simulation framework, 1000-mb air temperatures during extreme storms could be simulated using a variety of approaches. The physically based probability model for 1000-mb dew point temperatures derived from monthly maximum dew point data (Hansen et al., 1994) was used for Bridge System.

2.1.6 Air Temperature Lapse Rates

Dependencies: Independent.

Probability model: Normal distribution.

Air temperature lapse rates are needed for the stochastic simulation of freezing levels and computation of air temperatures at various elevations within a given watershed. Analyses of upper air sounding data from Northwestern Washington and Central California stations revealed that air temperature lapse rates on the day of maximum 24-h precipitation for noteworthy storms were well described by the normal distribution. The mean value was found to be 5.1°C/1000 m, which is near the saturated pseudoadiabatic lapse rate. Similar results were found if examining the data from Washington or California separately, and the data from the two regions were combined to provide a larger sample for computing the distribution parameters.

2.1.7 Freezing Level

Dependencies: 1000-mb air temperature, air temperature lapse rate, and storm magnitude.

Probability model: Physically based stochastic model.

Freezing level on the day of maximum 24-h precipitation is used for scaling the indexed freezing level temporal pattern. Simulations are conducted by stochastically generating a 1000-mb air temperature and an air temperature lapse rate as described in previous sections, and computing the resulting freezing level. The computed freezing level is then used to scale the freezing level temporal pattern by adding the value of the computed freezing level to the indexed temporal pattern.

2.1.8 Watershed Model Antecedent Conditions Sampling

Dependencies: Storm seasonality.

Probability model: Resampling of historical conditions (October 1984 − present).

The UBCWM was calibrated to realistically simulate hydrological behavior of the Bridge River System watersheds over a continuous period (October 1, 1984, through September 30, 2012). The model state variables (snowpack accumulation, soil moisture conditions, base flow, etc.) were computed continuously by the UBCWM thereby creating a continuous database of watershed initial conditions that could be stochastically sampled at any time/season of the year and assumed to occur at the onset of a stochastically generated storm event. This approach ensured that the full range of synthetic storm events is simulated with the full range of watershed historically observed internal states (e.g., extreme rainfall on a dry watershed, rain-on-snow events, average rainfall on a saturated watershed).

2.1.9 Initial Reservoir Level

Dependencies: Storm seasonality and UBCWM antecedent conditions.

Probability model: Resampling of historical conditions with the current reservoir system operating rules (January 1992−present).

A resampling approach was used to set the reservoir elevation at the beginning of the simulation for the La Joie and Terzaghi dams. The reservoir level for Seton Dam varied little throughout the year and was set to a constant value of 236.20 m at the beginning of each simulation, which represents the mean elevation for the period of record. Reservoir system operating rules represent a compromise among the needs of various stakeholders. As a result, hydropower operation, flood protection, environmental constraints and recreational demands limit the reservoir operating range. It is therefore important to resample reservoir level data from the period in historic record that reflects current system/reservoir operating rules. In this case, this was the period after January 1992.

2.2 SIMULATION OF RESERVOIR OPERATION—FLOOD ROUTING

After thousands of inflow hydrographs have been stochastically derived and coupled with stochastically selected initial reservoir levels, they were routed through each of the three reservoirs in the Bridge River System to obtain peak reservoir levels and outflow hydrographs that flow into downstream reservoir(s). Flood routing simulations are carried out with the aim of realistically capturing the way the Bridge River hydroelectric system is operated during flood events ranging in annual exceedance probabilities (AEPs) from 1/2 to beyond 1/10,000. In reality this is a rather complex decision-making process involving various factors such as the inflow forecast, flood magnitude, downstream environmental conditions existing at the time of the inflow, human operator preferences, corporate pressures, and other site-specific and season-specific conditions.

2.3 SIMULATION PROCEDURE

The procedure for Bridge River System stochastic flood simulation involved five steps:

1. Select date of storm occurrence based on historical seasonality of storm occurrences
2. Select all parameters associated with the occurrence of the storm event
   a. Select the magnitude of the 48-h precipitation for the watersheds based on the 48-h basin-average precipitation—frequency relationship
   b. Select one of 22 prototype storms for describing the temporal and spatial distribution of the storm, and scale the prototype storm templates to have the selected 48-h basin-average precipitation amount
   c. Select 1000-mb air temperature from physically based probability temperature model for day of maximum 24-h precipitation in selected prototype storm
   d. Select air temperature lapse rate and compute reference freezing level for day of maximum 24-h precipitation for selected prototype storm based on 1000-mb air temperature and air temperature lapse rate
   e. Compute temperature temporal patterns using scaled 1000-mb air temperature and freezing level temporal patterns for selected prototype storm and compute hourly temperature time series for all elevation zones in each watershed
3. Establish antecedent watershed and reservoir conditions at the onset of storm event
   a. Select UBCWM antecedent condition file for the date that was selected for occurrence of extreme storm. This is selected from the database of antecedent condition files for the October 1, 1984, to September 30, 2015, period. This sets the antecedent snowpack, soil moisture, and other model state variables.
   b. Select initial reservoir level for La Joie and Terzaghi reservoirs for the date that was selected for occurrence of extreme storm. This is sampled from recorded reservoir level data for the period January 1, 1992, to December 31, 2012. The sampled year has similar antecedent precipitation as the year sampled for the UBCWM antecedent conditions.
4. Conduct watershed modeling by running the UBCWM and create reservoir inflows
5. Conduct reservoir routing of synthetic inflow floods
   a. Execute reservoir routing that implements current reservoir operational procedures. Routing starts with the most upstream dam (La Joie) and proceeds with each subsequent dam downstream. Reservoir inflow for downstream reservoirs is derived by combining flood releases from the upstream dam(s) and local reservoir inflows.

3. UNCERTAINTY ASSOCIATED WITH STOCHASTICALLY DERIVED FLOOD QUANTILES

In general, a calibrated watershed model and hydrometeorological inputs developed from historical data represent one plausible description of the "true state of nature" for the behavior of a watershed and flood generating mechanism. It must be recognized that uncertainties exist in the estimation of hydrometeorological inputs and watershed model parameters due to both aleatoric and epistemic reasons. Consequently, there are many alternative combinations of probabilistic and deterministic models and model parameters that could plausibly describe the "true state of nature." For instance, the extrapolation of plausible weather patterns is based on historical precedent. Therefore, the potential for future climate change is one of the uncertainties in the presented approach that needs to be handled via adoption of an appropriate level of conservatism in selection of design parameters, combined with a thorough uncertainty analysis. The goal of the uncertainty analysis is to derive a mean–frequency curve and uncertainty bounds for the various flood characteristics in a manner that reasonably captures the current understanding of the hydrologic behavior of the watershed as well as the effect of uncertainties in estimating the flood–frequency characteristics.

Stochastic flood modeling is technically sound in its principle since it attempts to derive probabilities of extreme floods from the physically plausible modeling framework (considering the available historical data and state-of-the-art hydrometeorological modeling tools). However, the physical complexity and number of parameters involved in the process of flood simulation and routing through a system of dams and reservoirs warrants caution when interpreting and communicating the results. Due to lack of data and incomplete knowledge of the processes involved, it is currently impossible to estimate flood frequency distributions with certainty, especially its upper tail, which is needed for dam safety risk assessments (e.g., floods with AEP of 10^{-4} and less). It is therefore prudent to use a parsimonious approach in the selection of hydrometeorological inputs and watershed model parameters for inclusion in the uncertainty analysis. For the Bridge River System stochastic flood simulation, this involved a two-step approach:

- Step 1 involved sensitivity analyses (and some engineering judgment) to determine which hydrometeorological inputs and model parameters have the greatest effect on the magnitude of the flood outputs of interest for the Bridge River System of dams and reservoirs
- In Step 2, the uncertainty analysis was performed on the inputs/parameters identified in Step 1

This two-step approach is outlined in the following sections, where five stochastic modeling components were identified through the global sensitivity analysis and subsequently included in the uncertainty analysis.

3.1 SENSITIVITY ANALYSIS

Global sensitivity analysis (Saltelli et al., 2001) was used to assess the sensitivity of the various flood characteristics to the various hydrometeorological inputs and watershed model parameters. This kind of sensitivity analysis is capable of examining sensitivity with regard to the full range of parameter distribution. As such, global sensitivity analysis can measure the effect of interactions between parameters and handle nonlinear behavior. In contrast, Local Sensitivity Analysis (e.g., "one-at-a-time" sampling) examines sensitivity only with regard to point estimates of parameter values, which results in the sensitivity measure being affected by the choice of parameter values.

Fig. 11.1 depicts examples of the type of scatterplots produced from the Monte Carlo simulations that were used to assess the sensitivity of the peak hourly reservoir inflow to hydrometeorological inputs such as freezing level and temporal distribution of storm precipitation. Similar scatterplots are produced for all relevant hydrometeorological inputs and rainfall-runoff model parameters with regard to three main flood characteristics, namely, reservoir inflow, dam outflow, and reservoir level.

Table 11.1 presents a qualitative listing of the sensitivity of the peak reservoir level to the various hydrometeorological inputs and watershed model components/parameters. Table 11.1 also contains a qualitative assessment of the relative magnitude of uncertainties for those inputs/parameters.

The relative rankings in Table 11.1 were reviewed, and candidates for inclusion in the uncertainty analysis were identified as those inputs/parameters in which there was both moderate to high sensitivity and a higher level of uncertainty in estimation of the input/parameter. This assessment resulted in identification of five components of the stochastic modeling framework to be included in the uncertainty analysis (Table 11.2).

3.2 UNCERTAINTY ANALYSIS

The first task in the process of uncertainty analysis is to identify sources of uncertainty and determine how they can be characterized in the analysis—by conventional probability distributions, empirical likelihood shape functions, or alternative choices of models for specific hydrological processes. It is important to distinguish between two main categories of uncertainty:

- Aleatoric uncertainty is irreducible and associated with natural variability of all flood-producing factors including both atmospheric processes and watershed hydrological response. It represents the uncertainty of the value of a variable or parameter due to chance. For example, flood magnitude at a given watershed at a specific time of year will vary not only due to atmospheric inputs (rainfall and snowmelt) but also due to the chance occurrence of prior climatic conditions and soil properties that led to soil moisture saturation level being what it was at the time of flood. Values of flood-producing factors are subject to chance, and the primary purpose (and greatest value) of stochastic flood modeling is to address the aleatoric uncertainties associated with the hydrometeorological inputs by treating those inputs as stochastic variables instead of fixed values.

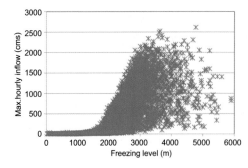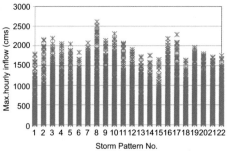

FIGURE 11.1

Scatterplots showing moderate to high sensitivity of peak hourly reservoir inflow to freezing level (left) and storm temporal distribution (right) for the La Joie Dam watershed.

Table 11.1 Qualitative sensitivity of maximum reservoir level at the three Bridge River System reservoirs to various hydrometeorological inputs and watershed model parameters

Model Component	Sensitivity of Flood Output to Model Component	Magnitude of Uncertainty	Comments
Storm Seasonality	Moderate	Moderate	Large sample set of storms
			Greater uncertainty for early-season storms
48-h Basin-Average Precipitation—Frequency Relationship for Watershed	High	Moderate to high	Large sample of annual maxima
			Uncertainty highest for extreme events
Temporal and Spatial Distribution of Storms	High	Low to moderate	Diverse sample of spatial and temporal patterns for 22 prototype storms
Antecedent Precipitation	Moderate	Low	Adequate sample of historical antecedent precipitation
Antecedent Soil Moisture Conditions	Moderate	Low	Adequate sample of historical antecedent soil moisture conditions
Baseflow	Low	Low	Adequate sample of historical baseflow conditions
1000-mb Air Temperature	Moderate	Low to moderate	Utilized in computing freezing level
Freezing Level	Moderate to high	Moderate	Runoff volume is sensitive to freezing level in winter months
Rainfall-Runoff Modeling	Moderate	Low to moderate	Long record of historical flows for watershed model calibration
Snowmelt Runoff Modeling	Moderate	Low to moderate	Long record of historical flows for watershed model calibration
Watershed Response to Fast Runoff	Moderate	Moderate to high	Long record of historical flows for watershed model calibration
			Uncertainty in response timing
Historical Flood—Frequency Relationship for Calibration of SEFM	Low to moderate	Low to moderate	Primarily used for validation of calibrated watershed model
Computation of Peak Reservoir Level via Spillway Stage—Discharge Curve	Low to moderate	Low to moderate	Stage—discharge curves based on field testing, and hydraulic model results available through standard industry handbooks

SEFM, *stochastic event flood model.*

- Epistemic uncertainty is associated with our lack of knowledge about a particular variable or process, and it may be reduced by a combination of research and additional data acquisition. Some typical sources of epistemic uncertainty in flood modeling include: rainfall-runoff model parameter uncertainty due to incomplete understanding of the underlying physics of watershed hydrological response; measurement errors in hydrometeorological inputs and representation of

Table 11.2 Model components selected for inclusion in the uncertainty analysis for derivation of flood frequency curves and uncertainty bounds

Model Component	Anticipated Relative Contribution to Total Uncertainty	Comments
48-h Basin-Average Precipitation—Frequency Relationship for Watershed	High	Uncertainty highest for extreme events
Watershed Response to Fast Runoff (UBCWM Timing Parameter)	Moderate to high	Affects magnitude and timing of reservoir inflow flood peak
1000-mb Air Temperature	Moderate	Components for computing air temperature and freezing level hourly time series during storms. They affect snowmelt runoff particularly at high elevations
Freezing Level	Moderate	
Storm Seasonality	Moderate	Uncertainty for months in which precipitation magnitudes are unrestricted

UBCWM, *UBC watershed model.*

watershed physical features; selection of inappropriate theoretical probability distribution for describing meteorological inputs (e.g., rainfall); and uncertainties in reservoir storage—elevation curve or spillway discharge rating curve used in flood routing.

The aim of the uncertainty analysis employing the Monte Carlo procedures is to derive a sample set of flood—frequency relationships for flood characteristics of interest by considering a sample set of "plausible model configurations." In this context, the term "plausible model configurations" represents alternative combinations of hydrometeorological inputs and alternative watershed model parameters that could reasonably describe the "true state of nature" and are selected from the Global Sensitivity Analysis. All the alternative combinations of hydrometeorological inputs and model parameters are "plausible" within the limits of sampling variability of historical data, state of knowledge of the hydrologic/hydraulic processes, and flood modeling experience/judgment of the analysts.

The flowchart in Fig. 11.2 describes the process of conducting an uncertainty analysis using a Monte Carlo framework and is based on the concept introduced by Nathan and Weinmann (2004). The inner loop is used to derive a flood—frequency relationship for flood characteristics for a given set of models/submodels and model parameters and explicitly incorporates aleatoric uncertainty. The outer loop represents alternative combinations of models/submodels and model parameters and represents epistemic uncertainty in the development of alternative plausible flood—frequency relationships for flood characteristics. However, dealing with reality is not quite as clean as indicated in the flowchart. There are aleatoric uncertainties, particularly associated with estimation of model parameters, that arise in the outer loop in assembling alternative configurations of the watershed model. Nonetheless, the flowchart provides a concise overview of the mechanics of conducting an uncertainty analysis.

A number of alternative combinations of models/submodels and model parameters could be assembled using a sampling methodology (e.g., Latin hypercube) to create a sample set of "plausible

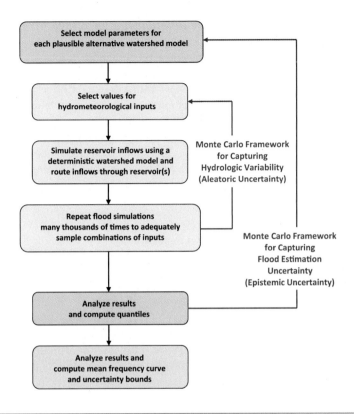

FIGURE 11.2

Flowchart for Monte Carlo frameworks for stochastic flood analysis and uncertainty analysis.

model configurations" representing reality. This sets the number of repetitions for the outer loop in Fig. 11.2. Typically 10—20 alternative model configurations are adequate to reasonably determine the mean flood—frequency relationship for flood characteristics and to characterize the magnitude of the uncertainty bounds. In this study, we used a practical option of assembling 11 alternative model configurations and computed the mean frequency curve and uncertainty bounds in a nonparametric manner by simple ranking flood outputs for specific AEP from the group of alternative model configurations. Using Cunnane (1978) nonparametric plotting-position formula resulted in 95th and 5th percentile flood outputs (e.g., peak reservoir levels) being the highest and lowest reservoir levels generated from the 11 model configurations, respectively. Similarly, the median value was the 6th largest value, and the mean value was computed from the 11 reservoir levels generated for a specific AEP.

3.3 CHARACTERIZATION OF UNCERTAINTIES FOR SELECTED MODEL COMPONENTS

Characterizations of uncertainties for each of the five selected model components included in the uncertainty analysis (Table 11.2) are described in the following sections.

3.3.1 The 48-hours Precipitation–Frequency Relationship for Bridge River System Watersheds

As mentioned in Section 2.1.2, the 48-h basin-average precipitation–frequency relationship was developed through regional analyses of point precipitation combined with spatial analyses of historical storms. Two stations (Downton Lake and Bralorne Upper) from within the Bridge basin were chosen as explanatory stations for the 48-h basin-average precipitation–frequency relationship. A multiple-regression relationship was developed between the 48-h precipitation maxima observed at explanatory meteorological stations and maximum 48-h basin-average precipitation for the Bridge River watershed observed in the 22 historical storms (Fig. 11.3). Monte Carlo methods were used to generate 48-h precipitation–frequency relationships for the watershed accounting for sampling variability and uncertainties in the estimation of the various parameters employed in the computation (Fig. 11.4). This approach addressed the following sources of uncertainty associated with the development of the watershed precipitation–frequency relationship:

- Estimate of mean for point precipitation used in regression
- Regional L-Cv
- Regional L-skewness
- Regional probability distribution
- Point to area regression

3.3.2 Watershed Response to Fast Runoff

Empirical likelihood functions were developed for the Fast Runoff Timing Constant (FRTK) parameter within the UBCWM to characterize uncertainties in the estimation of parameter values for

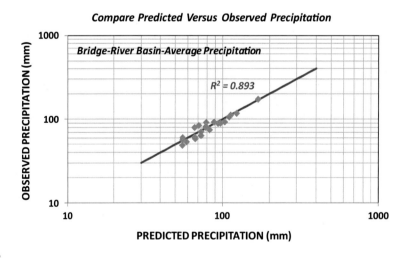

FIGURE 11.3

Comparison of predicted and measured 48-h basin-average precipitation for Bridge River watershed based on multiple-regression prediction equation.

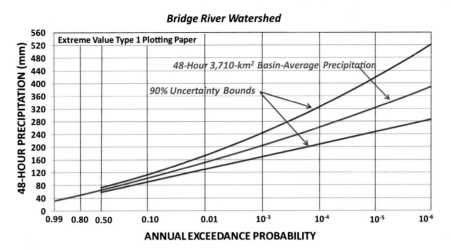

FIGURE 11.4

Computed 48-h precipitation–frequency relationship and 90% uncertainty bounds for the Bridge River watershed.

modeled Bridge River System watersheds. The FRTK parameter controls the timing of the watershed response to fast runoff generation and affects the magnitude and timing of the flood hydrograph peak. The best-estimate values for FRTK were determined through calibration of the watershed model to long-term streamflow time series and to historical floods. The shapes of the likelihood functions were based on experience gained from modeling and calibration of the UBCWM at basins throughout the world that were hydrologically similar to the Bridge River System. Fig. 11.5 shows the developed likelihood functions for both the Bridge and Seton watersheds. The watershed response for Bridge subbasins is faster reflecting their steeper slopes compared with the Seton watershed. The summary statistics for these likelihood functions (in days) were:

- Bridge subbasins (La Joie and Terzaghi): 0.5 for the mode/best estimate; 0.736 for the mean; and 0.357 for the standard deviation
- Seton: 1.0 for the mode/best estimate; 1.161 for the mean; and 0.430 for the standard deviation

From these summary statistics, 11 values were selected for each watershed in the Bridge River System using Latin hypercube sampling methods.

3.3.3 The 1000-mb Air Temperature and Freezing Level

Uncertainties in the freezing level for a given stochastic storm simulation were modeled through adjustment of the indexing value of the 1000-mb air temperature. This approach results in adjustment of the indexing value for the freezing level and for setting the maximum freezing level for a given month. The value of the 1000-mb air temperature adjustment was found to be 1.3°C through calibration to the historical flood–frequency relationship at La Joie watershed, which is the highest-elevation watershed in the Bridge System with the highest snowmelt runoff contribution. Uncertainty in the 1000 mb temperature adjustment was characterized as being equally likely over a range

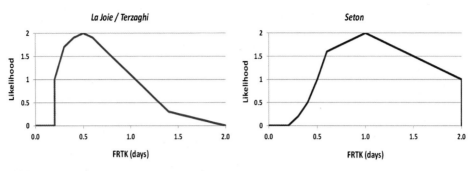

FIGURE 11.5

Likelihood functions for fast runoff timing constant (FRTK) for Bridge River System watersheds.

of $\pm$ 2.0°C centered on the calibration value. Table 11.3 lists the 11 pairings of 1000-mb air temperature adjustment and maximum freezing level for the cool season (Nov.–Apr.).

3.3.4 Storm Seasonality Where Precipitation Magnitudes Are Unrestricted

Precipitation magnitudes in SEFM simulations do not have an upper limit. However, restrictions are placed within SEFM on the months where precipitation magnitudes may exceed the Probable Maximum Precipitation (PMP) estimate. In general, precipitation magnitudes are allowed to exceed PMP estimates only in those months occupying the central body of the storm seasonality. This corresponds to the months of October through February for the Bridge River region. In those months which are external to the central body of the seasonality data, restrictions are placed on precipitation magnitudes and the upper limit to precipitation is set at the PMP estimate (Micovic et al., 2015) for a particular month.

There is uncertainty in the seasonality of long-duration storms with regard to the months in which precipitation magnitudes should be unrestricted. This uncertainty was modeled by considering precipitation magnitudes to be unrestricted for the central body of the seasonality distribution and then extending outward to consider additional months as unrestricted. Eleven sample sets for storm seasonality were created (Table 11.4) wherein several monthly groupings were applicable to several sample sets. In particular, the monthly groupings for sample sets 4, 5, 6, and 7 and the percent PMP restrictions generally correspond to the region's climatology.

4. RESULTS

Monte Carlo computer simulations were used to develop magnitude–frequency relationships for maxima of reservoir inflow, reservoir elevation, and dam outflow for the La Joie, Terzaghi, and Seton dams within the Bridge River hydroelectric system. These relationships were based on 10,000 computer simulations for each of the 11 "plausible model configurations." This approach is based on the total probability computation procedure developed by Nathan and Weinmann (2001), which greatly reduces the number of simulations that would otherwise have been required to develop the flood frequency relationships. In particular, it should be noted that each of the 11 "plausible model configurations" was composed of a random combination of the parameter values for the 11 sample sets, per standard Latin hypercube methodology.

Table 11.3 Uncertainty characteristics for sample set of 11 freezing level parameters for Bridge River System watersheds

Sample Set	1000 mb Temperature Adjustment (°C)	Freezing Level Adjustment (m)	Maximum Freezing Level (m)					
			November	December	January	February	March	April
1	−0.7	0	2700	2300	2100	2100	2200	2600
2	−0.3	0	2800	2400	2200	2200	2300	2800
3	0.1	0	3000	2500	2300	2300	2400	2900
4	0.5	0	3100	2600	2500	2400	2500	3000
5	0.9	0	3200	2800	2600	2500	2600	3200
6	1.3	0	3400	2900	2700	2700	2800	3300
7	1.7	0	3500	3000	2800	2800	2900	3500
8	2.1	0	3700	3200	3000	2900	3000	3600
9	2.5	0	3800	3300	3100	3100	3200	3800
10	2.9	0	4000	3500	3300	3200	3300	3900
11	3.3	0	4200	3600	3400	3400	3500	4100

Table 11.4 Restrictions on 48-h basin-average precipitation expressed as percentage of PMP

| Sample Set | Monthly Values of Maximum 48-h Precipitation Expressed as Percentage of PMP | | | | | | | | | |
	July	August	September	October	November	December	January	February	March	April
1	U	U	U	U	U	U	U	U	U	U
2, 10, 11	75%	U	U	U	U	U	U	U	U	U
3, 8, 9	52%	75%	U	U	U	U	U	U	U	75%
4, 5, 6, 7	52%	53%	77%	U	U	U	U	U	77%	54%

PMP, *probable maximum precipitation.*
"U" corresponds to precipitation magnitudes being unrestricted.

Flood outputs of interest from the stochastic flood model were presented as probability plots developed using a nonparametric plotting position. This approach avoids the problems often encountered in selecting and fitting a probability distribution, particularly for flood outputs such as reservoir levels that have been greatly affected by anthropogenic factors such as imposed reservoir operating procedures.

Each of the 11 plausible model configurations produces one flood—frequency relationship for a flood characteristic of interest. For example, Fig. 11.6 depicts the 11 flood—frequency relationships for the peak reservoir inflow for the 1000 km^2 La Joie Dam watershed computed using the SEFM methodology. The mean flood—frequency curves and uncertainty bounds for the peak reservoir inflow for La Joie Dam and peak reservoir elevation for Terzaghi Dam are shown in Figs. 11.7 and 11.8, respectively. Note that Fig. 11.8 is particularly important in terms of dam safety considerations and decisions because it provides information on the probability of dam overtopping.

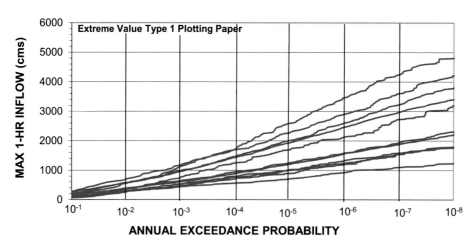

FIGURE 11.6

Example of simulated flood—frequency relationships for peak hourly reservoir inflow at La Joie Dam for 11 plausible model configurations.

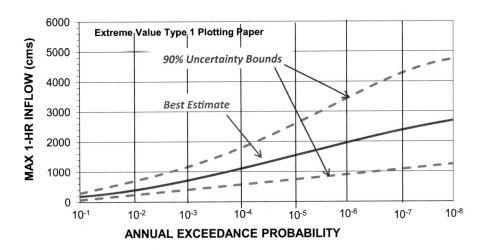

FIGURE 11.7

La Joie Dam, frequency curve for peak hourly reservoir inflow simulated by the SEFM.

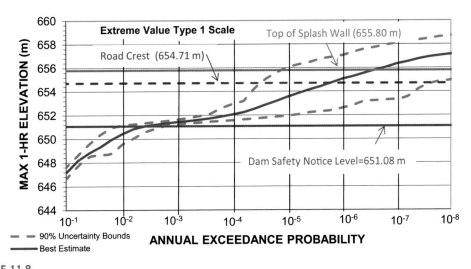

FIGURE 11.8

Terzaghi Dam, frequency curve for peak hourly reservoir level simulated by the SEFM.

5. EFFECT OF EARTH OBSERVATIONS ON UNCERTAINTY IN PROBABILISTIC FLOOD ESTIMATES

As mentioned earlier, aleatoric uncertainty is irreducible and is due to natural variability of all flood-producing factors, but the epistemic part of the total uncertainty is due to our lack of knowledge about a particular variable or process and could be reduced by a combination of research and additional data acquisition. Thus, earth observation methods and technologies play a crucial role in reducing uncertainty associated with results derived from flood simulation studies.

The presented study analyzed sensitivity of various flood characteristics to hydrometeorological inputs and watershed model parameters shown in Table 11.1, all of which were derived through some kind of earth observations. To get spatial representation of various hydrometeorological inputs over the Bridge System watershed area, it was necessary to combine ground point measurements with available spatially distributed remote sensing observations. For instance, since the Bridge River System is located in highly mountainous and remote terrain with relatively sparse climate station network, we applied the PRISM system, which combined point measurements of precipitation with gridded digital elevation data obtained by remote sensing, and derived faithful representation of the orographic enhancement of precipitation in a spatially distributed way. Where available, radar rainfall estimates were consulted in deriving the spatial and temporal distribution templates of 22 historical storms over the watershed. Upper air sounding data were used to complement and verify surface dew point temperature measurements needed for moisture maximization and freezing level assessments. Also, the UBCWM that was used to transform hydrometeorological inputs to watershed runoff entering the Bridge System reservoirs had to be calibrated so that the flood-generating process is simulated realistically. This involved using remote sensing data to establish physical description of the watershed (e.g., properly identifying topography, land use, forest cover, and glaciated areas) and verify model-calculated seasonal snowpack accumulation and subsequent depletion by satellite imagery. Finally, remote sensing data were used to produce reservoir bathymetry and calculating reservoir volumes, which was essential in routing stochastically derived floods through the reservoirs.

The process of combining and complementing surface point measurements with spatially distributed remote sensing observations effectively reduced the epistemic uncertainty component of total uncertainty associated with stochastic flood simulation for the Bridge River System. The effect of each particular observation/measurement on total uncertainty was not explicitly quantified; however, it was implicitly accounted through the selected plausible model configurations included in the presented two-step approach consisting of global sensitivity and uncertainty analyses.

6. CONCLUDING REMARKS

Risk-informed decision framework for dam safety requires the development of full flood frequency curves for various flood characteristics, where scenarios of different hazards (not just overtopping) with different exceedance probabilities can be combined and assessed. The classic frequency analysis (i.e., fitting a theoretical probability distribution to a sample set of observed data) could only be applied to the reservoir inflow since it is the only flood characteristic that could be assumed to be an independent and identically distributed random variable. In addition, this assumption only applies to natural reservoir inflows, i.e., unaffected by upstream regulation. Other flood characteristics such as dam outflow and reservoir elevation are affected by reservoir operating rules and procedures and cannot be analyzed by classic frequency analysis. The problem becomes even more complex in a system of multiple dams and reservoirs where the inflow to a downstream reservoir is greatly affected by operating procedures of the upstream projects.

Therefore, the only way to derive the full probability distribution of nonrandom flood characteristics (e.g., peak reservoir level, maximum dam outflow discharge, and peak reservoir inflow to a downstream reservoir in a system of reservoirs) is through stochastic simulation. This study shows the results of stochastic flood simulation carried out on a complex system of three dams and three reservoirs with seasonally fluctuating elevations and active discharge control systems (e.g., gated spillways). Due to system complexities, lack of observed data, and incomplete knowledge of the processes involved, a parsimonious approach was utilized for the selection of hydrometeorological inputs and various model parameters for inclusion in the uncertainty analysis. The approach consisted of two main steps: Step 1 involved global sensitivity analyses to determine which inputs and model parameters have the greatest effect on the magnitude of the Bridge River System flood characteristics, and Step 2 involved performing the uncertainty analysis on the inputs/parameters identified in Step 1.

It should be recognized that the physical complexity and large number of parameters involved in the process of flood simulation and routing through a system of dams and reservoirs prevents complete accuracy in determining flood quantiles, especially at the upper tail of probability distributions. Additionally, the flood exceedance probabilities of interest for dam safety risk assessments are very small, typically 1/1000 or less and the estimate uncertainties for these floods are the largest. That is why it is very important to estimate associated uncertainties when communicating flood frequency results and dam safety risk assessments to various stakeholders (e.g., downstream communities that could be seriously endangered by a dam failure and regulatory agencies).

Finally, there are additional uncertainties not included in this analysis that could dramatically affect the flood frequency results. These uncertainties include spillway gates operations failure, human error in reservoir operations, telemetry errors, landslides into reservoir, debris jams, as well as additional project-specific adverse circumstances. Accounting for those factors would further increase both the complexity of simulation and the total estimated uncertainty. The industry is slowly moving toward complete system simulation, where every single component of the hydroelectric system will be included in the simulation along with its interaction with other components, thereby providing more accurate values of probabilities of failure of each system component as well as the entire system. However, with the current state of the art, that appears to be a daunting endeavor and will remain impracticable in the near future. Meanwhile, the presented approach is useful and a necessary contribution to dam safety risk assessments, where the continuous probability distribution is derived for each flood characteristic of interest along with estimates of its uncertainty bounds. That way, different design criteria and failure scenarios with associated risks and uncertainties can be considered and evaluated at various flood frequency levels, departing from the widely used strict "pass/fail" criteria such as the IDF.

REFERENCES

Cunnane, C., 1978. Unbiased plotting positions — a review. Journal of Hydrology 37, 205—222.

Daly, C., Neilson, R.P., Phillips, D.L., 1994. A statistical-topographic model for mapping climatological precipitation over mountainous terrain. Journal of Applied Meteorology 33, 140—158.

FEMA, 2013. Selecting and Accommodating Inflow Design Floods for Dams. US Federal Emergency Management Agency. Report P-94.

Hansen, E.M., Fenn, D.D., Corrigan, P., Vogel, J.L., Schreiner, L.C., Stodt, R.W., 1994. Probable Maximum Precipitation for Pacific Northwest States — Columbia River (Including Portions of Canada), Snake River and Pacific Coastal Drainages. Hydrometeorological Report No. 57. US National Weather Service, Silver Spring, MD, US.

Hosking, J.R.M., Wallis, J.R., 1997. Regional Frequency Analysis – An Approach Based on L-Moments. Cambridge University Press, Cambridge, UK.

ICOLD, 2003. Dams and Floods: Guidelines and Case Histories, Bulletin 125. International Committee on Large Dams, Paris, France.

Micovic, Z., Hartford, D.N.D., Schaefer, M.G., Barker, B.L., 2016. A non-traditional approach to the analysis of flood hazard for dams. Stochastic Environmental Research and Risk Assessment 30, 559–581.

Micovic, Z., Quick, M.C., 2009. Investigation of the model complexity required in runoff simulation at different time scales. Hydrological Sciences Journal 54 (5), 872–885.

Micovic, Z., Schaefer, M.G., Taylor, G.H., 2015. Uncertainty analysis for probable maximum precipitation estimates. Journal of Hydrology 521, 360–373.

Nathan, R.J., Weinmann, P.E., 2001. Estimation of large to extreme floods, book VI. In: Australian Rainfall and Runoff – A Guide to Flood Estimation. The Institution of Engineers, Canberra, Australia.

Nathan, R.J., Weinmann, P.E., 2004. An improved framework for the characterisation of extreme floods and for the assessment of dam safety. In: Hydrology: Science & Practice for the 21st Century. Proceedings of British Hydrological Society, London, UK, vol. 1, pp. 186–193.

Nathan, R.J., Weinmann, P.E., Hill, P.I., 2002. Use of a Monte Carlo framework to characterise hydrological risk. ANCOLD Bulletin (122), 55–64.

Paquet, E., Garavaglia, F., Gailhard, J., Garçon, R., 2013. The SCHADEX method: a semi-continuous rainfall-runoff simulation for extreme flood estimation. Journal of Hydrology 495, 23–37.

Prat, O.P., Nelson, B.R., 2015. Evaluation of precipitation estimates over CONUS derived from satellite, radar, and rain gauge data sets at daily to annual scales (2002–2012). Hydrology and Earth System Sciences 19, 2037–2056.

Quick, M.C., 1995. The UBC watershed model. In: Singh, V.J. (Ed.), Computer Models of Watershed Hydrology. Water Resources Publications, Highlands Ranch, CO, US, pp. 233–280.

Saltelli, A., Chan, K., Scott, E.M., 2001. In: Sensitivity Analysis, first ed. John Wiley and Sons, New York, NY, US.

Schaefer, M.G., Barker, B.L., 2002. Stochastic event flood model (SEFM). In: Singh, V.J., Frevert, D.K. (Eds.), Mathematical Models of Small Watershed Hydrology and Applications. Water Resources Publications, Highlands Ranch, CO, US, pp. 707–748.

Schaefer, M.G., Barker, B.L., 2015. Stochastic Event Flood Model – User's Manual. MGS Engineering Consultants, Inc., Olympia, WA, US.

Scofield, R.A., Kuligowski, R.J., 2003. Status and outlook of operational satellite precipitation algorithms for extreme precipitation events. Weather Forecast 18, 1037–1051.

Sene, K., 2013. Flash Floods: Forecasting and Warning. Springer, Dordrecht, Netherlands.

Stampoulis, D., Anagnostou, E.N., Nikolopoulos, E.I., 2013. Assessment of high-resolution satellite-based rainfall estimates over the mediterranean during heavy precipitation events. Journal of Hydrometeorology 14, 1500–1514.

Wood, S.J., Jones, D.A., Moore, R.J., 2000. Accuracy of rainfall measurement for scales of hydrological interest. Hydrology and Earth System Sciences 4, 531–543.

SENSITIVITY OF WELLS IN A LARGE GROUNDWATER MONITORING NETWORK AND ITS EVALUATION USING GRACE SATELLITE DERIVED INFORMATION

12

V. Uddameri[1], A. Karim[1], E.A. Hernandez[1], P.K. Srivastava[2,3]

Texas Tech University, Lubbock, TX, United States[1];
NASA Goddard Space Flight Center, Greenbelt, MD, United States[2];
Banaras Hindu University, Varanasi, Uttar Pradesh, India[3]

CHAPTER OUTLINE

Sensitivity Analysis in Earth Observation Modelling. http://dx.doi.org/10.1016/B978-0-12-803011-0.00012-4

1. INTRODUCTION

Groundwater is the most extracted raw material with global extraction rates fast approaching 1000 km^3/year (NGWA, 2016). Groundwater is often the only reliable source of water supply in rural arid and semiarid regions, and over 60% of the total groundwater production in the world is used to sustain irrigated agriculture (Zektser and Everett, 2004). Groundwater provides ready access to water without the need for expensive infrastructure; it is reliable and serves as a buffer against droughts (Calow et al., 2010). However, given the decentralized nature of the resource, it is not as regulated as other sources of water (Lund et al., 2014). Consequently, groundwater has been overexploited in many parts of the world (Moench, 1992; Fornés et al., 2005; Scanlon et al., 2012). Intensive groundwater development has played a major role in improving global food security and has contributed to the economic development and well-being of rural economies in both developing and developed nations alike (Kong et al., 2015; Marston et al., 2015). However, most aquifers undergo minimal recharge and groundwater supplies cannot be easily augmented (Steward et al., 2013). As such, current use of groundwater threatens the availability of this resource for future generations (Uddameri, 2005).

There is growing recognition that groundwater resources must be actively managed to ensure their long-term viability and to assure that the benefits of the resources are equitably available to all water users in any given region (Gleeson et al., 2012; Singh, 2014; Uddameri et al., 2014). Given the diffuse nature of groundwater resources and its importance to a large number of users, decentralized, participatory groundwater management schemes are viewed as having the greatest opportunities for success (Figureau et al., 2015). Groundwater management must be strongly rooted in science, and decisions must be based on observed responses of the aquifer to natural and anthropogenic stresses. A well-designed groundwater monitoring network is critical to understanding the behavior of the aquifer. Over the past two decades, there has been a concerted effort worldwide to install new networks (Hsu, 1998; Thomas and Arnold, 2015) as well as to upgrade and modernize existing ones (Lee et al., 2007; Zhou et al., 2013). Procedures and protocols to design monitoring networks using statistical power analysis, geostatistics, and multicriteria decision-making approaches have also been proposed in recent times to guide groundwater monitoring network development efforts (Uddameri and Andruss, 2014a,b).

Groundwater level measurements collected from monitoring wells are used to generate maps to identify areas where water levels are falling rapidly and also to estimate the amount of water that is left in storage (McGuire, 2014). Institutions that are responsible for the management of groundwater use water level measurements to ascertain whether their regulatory practices are helpful in meeting the long-term management goals for the aquifer. For example, in the state of Texas, groundwater management is carried out by locally elected political subdivisions called groundwater conservation districts (GCDs). The GCDs within a region are also grouped into planning bodies called groundwater management areas (GMAs). GMAs are required by statute to establish the "desired future conditions (DFC)" of all aquifers under their jurisdiction. The GMAs overlying the High Plains aquifer (also referred to as the Ogallala aquifer) in Texas have generally adopted either a 50/50 or a 60/50 rule, which requires that at least 50% (or 60%) of the saturated thickness is available in 2060 (50 years from 2010). In other words, groundwater use in the area, over the 50-year period, is restricted to 50% (or 40%) of the total saturated thickness that was available in the year 2010. Groundwater-related statutes in Texas also require GMAs and GCDs to monitor the aquifer to ensure that the adopted DFCs are being met.

Groundwater monitoring is expensive, and as private wells are used to obtain water level data (Batt et al., 2006), the sampling is also affected by site accessibility constraints. Although the tendency is to obtain as many measurements as possible within the constraints of time and resources, it is critical that monitoring redundancy is minimized. Collection of superfluous data adds little information while increasing costs and does not add significantly to the final products derived from groundwater level measurements (e.g., contour plots) that aid in groundwater management. Understanding the worth of data collected at each well within a monitoring network helps prioritize which wells should absolutely be monitored and also provides an indication of the minimum number of wells required within a monitoring network. The identification of critical wells within a network can be viewed as a sensitivity analysis problem. If the final derived product (e.g., water availability) is very sensitive to measurements at a well, then sampling at that well becomes important to reliably test management goals. A randomization-based sensitivity analysis procedure is developed here to assess the sensitivity of wells in a large monitoring network. The methodology also provides uncertainty bounds for estimated water levels at unmonitored locations and as such can be used to refine groundwater monitoring networks.

The refinement of an existing groundwater monitoring network requires decision making under uncertainty. Improper removal of wells may prove costly if critical wells that should have otherwise be retained are excluded from the framework (Type 1 error) and/or when a large number of insensitive wells are retained within the network (Type 2 error). Clearly, Type 1 error increases the risk of making a bad decision and Type 2 error adds unnecessary redundancy and monitoring costs (Uddameri and Andruss, 2014b). However, verification of an existing network is difficult a priori (i.e., before the decisions to eliminate or add monitoring wells are made), and therefore monitoring network refinements are often carried out in an ad hoc fashion with limited corroboration from independent sources.

Information from Earth observation satellites often provide a convenient way to corroborate and extend the utility of ground-based measurements. Although remote sensing has been used extensively in surface water hydrology studies, its benefits have not been fully realized in groundwater applications (Becker, 2006). However, recent advances, especially the Gravity Recovery and Climate Experiment (GRACE), has opened up new opportunities to use Earth observations to study groundwater flow behavior. GRACE-derived information has been used to study groundwater storage changes in several large basins such as the High Plains aquifer (Rodell and Famiglietti, 2002); Mississippi River basin (Rodell et al., 2007), and aquifers in India (Rodell et al., 2009), to name a few. Recently, Richey et al. (2015) used GRACE to quantify renewable groundwater stresses on the 37 largest aquifers in the world. In addition to estimating changes in hydrologic storage, GRACE-derived data sets have also been used to evaluate and improve hydrologic components of climate models (e.g., Niu and Yang, 2006; Swenson and Milly, 2006). GRACE data have also been used to study droughts (Long et al., 2013; Thomas et al., 2014). Assimilation or fusion of GRACE-based and model-derived data for enhancing hydrologic predictions has also been an active area of interest (AOI) (Schumacher et al., 2016; Kumar et al., 2016).

Building upon these lines of inquiry, GRACE-derived data are hypothesized here to be useful to evaluate the sensitivity of observation wells in a monitoring network. An innovative sensitivity analysis—based procedure has been developed here to prioritize monitoring wells. The study also seeks to establish whether significant spatial relationships exist between sensitivity measures developed to prioritize monitoring wells and terrestrial water storage (TWS) anomaly which is the basic hydrologic parameter derived from GRACE satellite remote sensing data. A strong correlation between the sensitivity measures and GRACE data would help validate the groundwater monitoring well prioritization protocol developed here and also indicate the utility of GRACE data to design groundwater monitoring networks. The Southern High Plains aquifer that lies within the state of Texas is used to illustrate the concepts and methods developed as part of this study.

2. METHODOLOGY

2.1 SATURATED THICKNESS: A FUNDAMENTAL MEASURE OF GROUNDWATER AVAILABILITY

As well installation is expensive, regional monitoring networks often make use of existing wells on public and private lands. Therefore, a large regional-scale monitoring network can be envisioned to contain a finite number of wells, w, that are randomly scattered over an AOI. Water levels (WL) below ground surface (BGS) are measured from a prescribed datum (e.g., mean sea level), at these irregularly placed wells, and used to calculate parameters such as aquifer volume (water availability) that are of interest to regulators and policy makers. Groundwater levels are often measured with respect to a local land surface datum (LSD). The LSD in turn is referenced to a global datum such as mean sea level (MSL), which allows hydraulic heads (H) at different wells to be measured from the same datum. The bottom elevation (BT) of the aquifer (measured with respect to the global datum) can be ascertained from geological surveys and can be used to obtain the saturated thickness (ST) at any given point in time t, in an unconfined aquifer.

$$H_{t,(MSL)} = LSD_{(MSL)} - WL_{(t,BGS)} \tag{12.1}$$

$$ST_t = H_t - BT_{(MSL)} \tag{12.2}$$

The recoverable amount of water at any time, t, can be obtained from the knowledge of specific yield (S_y) as:

$$V = S_y A_c ST_t \tag{12.3}$$

where, A_c is the cross-sectional area over which the groundwater extraction is taking place.

As the saturated thickness and volume extracted from storage are to be calculated over the entire AOI, the hydraulic heads or saturated thickness measured at wells are interpolated onto a regular (often equally spaced) grid. The recoverable groundwater (a measure of water availability) can be obtained by summing the saturated thickness over all cells in the grid.

2.2 INVERSE DISTANCE WEIGHTING APPROACH FOR OBTAINING REGIONAL GROUNDWATER AVAILABILITY FROM WELL MEASUREMENTS

A variety of deterministic and stochastic interpolation methods are available for projecting values measured from a set of irregular points onto a regularly spaced grid. These methods including Voronoi plots, linear and spline interpolation, inverse distance weighting (IDW), and kriging (Isaaks and Srivastava, 2001). Kriging and IDW are perhaps the two most commonly used methods for interpolation (Li and Heap, 2011). Simple and ordinary kriging are stochastic methods that essentially use the regression framework to fit a surface through discrete point measurements. Kriging methodologies provide not only an estimate at an unmeasured point but also the standard error of that estimate. Therefore, the estimate and the uncertainty associated with it can both be ascertained. However, the approach makes several assumptions about the underlying data and its spatial structure, which are not always valid in real-world applications. Furthermore, the theoretical variogram does not fully capture all the observed variability in the empirical variogram, and several subjective decisions are necessary during the variogram modeling phase.

The IDW on the other hand is a deterministic technique that is based on the first law of geography, which states that all things are related to each other and that things closer are related to an even greater extent. As the name suggests, IDW computes the estimate at an unmeasured point as a weighted average of other surrounding points where the variable of interest has been measured. The reciprocal of the distance between an observation and the unmeasured point is used as the weighting factor, which results in the nearest points having a greater influence on the estimation process (Eq. 12.5).

$$H_{t,k} = \frac{\sum_{i=1}^{n} w_{ik} H_{t,i}}{\sum_{i=1}^{n} w_{ik}} \quad \forall i = 1, ..., n \tag{12.4}$$

where:

$$w_{ik} = \frac{1}{[d_{ik} + \varepsilon]^m} \quad \text{and} \quad d_{ik} = \sqrt{(x_i - x_k)^2 + (y_i - y_k)^2} \tag{12.5}$$

where H_t is the hydraulic head, w is the weight, and n is the number of nearest neighbors used in the estimation process. The measured points (wells) are indexed as i, and the unmeasured point of interest is indexed as k; m is the power and ε is a small tolerance added to avoid division by zero. The power, m, is often taken to be 2 (Shepard, 1968). Optimal values for both m and n can be obtained by splitting the data into training and testing data sets and calculating the combination that yields the lowest mean square error on the testing data set or using a cross-validation technique. The kd-tree algorithm provides a fast search technique to find the nearest neighbors (Bentley, 1975) and is often used with the IDW method (Qian et al., 2012) and adopted here as well.

2.3 ASSESSMENT OF INDIVIDUAL WELL SENSITIVITY USING THE JACKKNIFE APPROACH

The Jackknife approach can be used to identify the sensitivity of individual wells within the groundwater monitoring network. In the Jackknife approach, the hydraulic head (saturated thickness) estimates are made at all (unmonitored) locations on a regular grid using the complete network of, w, wells. Next a well, −i, is removed from the network and hydraulic head estimates are computed on the regular grid using this (w−1) well network. The process is sequentially repeated by taking one well offline at a time and making necessary head estimates on the regular grid. Thus, an empirical distribution for the hydraulic head can be developed at each point on the regular grid. This empirical distribution can also be used to obtain summary statistics and moments (Efron and Gong, 1983). The Jackknife approach is more convenient than the random bootstrap method because it reasonably preserves the spatial correlation at each realization as only one well is removed at a time from the original data set.

The empirical distribution constructed using Jackknife resampling can be used to evaluate the adequacy of the monitoring network. For example, the standard deviations can be mapped to identify areas where estimation errors are high. High estimation errors arise due to an inadequate number of monitoring wells in close proximity to the grid point of interest. In addition, it is proposed here that the sensitivity of the individual well on the overall head estimation be obtained by summing the absolute deviations at each location normalized with respect to the Jackknife estimator of the hydraulic head at that location. Therefore, the mean absolute deviation (MAD)—based sensitivity coefficient of the ith well is obtained as:

$$S_{mad,i} = \frac{1}{J} \sum_{j=1}^{J} \left(\frac{|H_{t,j}^{all} - H_{t,j}^{-i}|}{H_{i,j}} \right) \quad (12.6)$$

where, S_{mad} is the MAD—based sensitivity coefficient. It is evident from Eq. (12.6) that a well is clearly influential if its exclusion causes significant deviations in the estimated heads. The proposed sensitivity coefficient is also dimensionless and as such can be used to rank wells according to their influence. The mean squared deviation sensitivity coefficient, S_{msd}, can also be defined in a similar manner as follows:

$$S_{msd,i} = \frac{1}{J} \sum_{j=1}^{J} \left(\frac{\sqrt{(H_{t,j}^{all} - H_{t,j}^{-i})^2}}{H_{i,j}} \right) \quad (12.7)$$

Again, the sensitivity coefficient based on mean square deviation S_{mad} is also dimensionless and can be used to rank wells within a network with regard to their relative influence. Both these metrics essentially measure the distance between the estimate obtained with all the wells and the one obtained by excluding a specific well. Although these measures are novel, the developed metrics are similar in spirit to distance—based measures used to identify influencing points in linear regression (Cook, 1979). The utility of these metrics to evaluate a groundwater monitoring network is illustrated next using a case study from the Southern High Plains of Texas (SHP).

2.4 GRACE SATELLITE AND GLOBAL HYDROLOGIC DATA

GRACE consists of a pair of two satellites that were launched in March 2002. These satellites act in unison and orbit one behind another at an approximate distance of 220 km. Small changes in the distances between the two satellites caused due to gravitational anomalies are accurately measured and used to construct monthly maps of the Earth's average gravitational field (http://earthobservatory.nasa.gov/Features/GRACE/). Short-term changes in gravity fields are largely influenced by the movement of water mass that forms a relatively thin fluid envelope of the Earth. The TWS is a vertically integrated measure of water storage in the soil moisture, groundwater, biomass, snow, and rivers. Algorithms have therefore been developed to estimate the monthly TWS variations using information provided by the GRACE satellites (Swenson and Wahr, 2006). Postprocessed TWS anomalies (i.e., variations from a baseline time period January 2004—December 2009) aggregated on a $1° × °1$ grid are publicly available from: http://grace.jpl.nasa.gov/data/get-data/monthly-mass-grids-land/(Swenson, 2012).

3. STUDY AREA

The SHP is a semiarid region with extremely limited surface water resources (Fig. 12.1). Groundwater from the Ogallala aquifer is used extensively for irrigated agriculture, and the region produces nearly 25% of the cotton, 15% of the wheat and 10% of the peanuts produced in the United States. Although

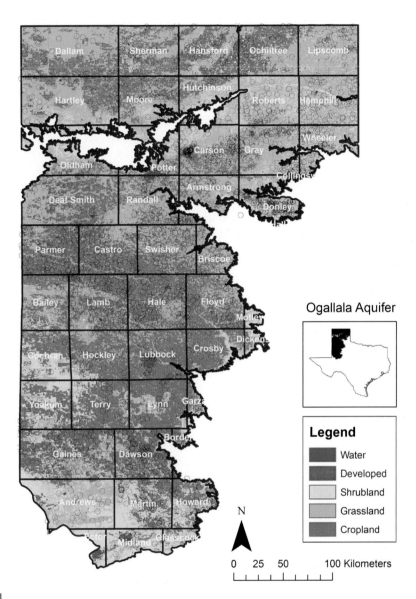

FIGURE 12.1

Ogallala aquifer extent in the Texas Panhandle (light gray circles indicate wells).

only 20% of the entire Ogallala aquifer lies within Texas, over 5 million acre-feet of water is extracted from the aquifer in Texas, annually. Groundwater levels in the Ogallala aquifer have undergone significant depletion at several locations, particularly in the southern end of the aquifer. Enhancing the long-term viability of the Ogallala aquifer is viewed as perhaps the largest single water management concern of the United States (Upendram and Peterson, 2007) and has resulted in a call for a concerted effort in conserving and extending the useful life of the largest aquifer in North America (Sophocleous, 2012).

As stated previously, current groundwater management policies in Texas require that total production in the aquifer over the next 50 years be restricted to 40–50% of the total aquifer volume that was available in 2010. The Texas Water Development Board (TWDB) and its cooperators conduct annual water level measurements in the area. The water level measurements for the year 2014 and other well information were extracted from the TWDB groundwater database (TWDB, 2015), and the 1917 wells that were monitored in that year are used for analysis here. As can be seen from Fig. 12.2, the monitoring density can vary widely across the 48 counties that the aquifer underlies.

Being a semiarid region that receives minimal snowfall and with extremely scant surface water resources, the TWS variations in the study area are largely a function of changes in soil moisture and groundwater resources. Separation of soil moisture and groundwater storage components requires either the availability of comprehensive soil moisture data sets (Swenson et al., 2008) or advanced modeling techniques (Breña-Naranjo et al., 2014) and is outside the scope of this study. Clearly, soil moisture exhibits seasonal behavior due to climatic factors (rainfall, evapotranspiration) as well as irrigation practices. As such, monthly TWS values for the year 2014 were averaged to reduce the effects of soil moisture variability on the observed variability in the TWS. A 1600 m × 1600 m grid was overlaid on the aquifer, and the centroid corresponding to each cell was taken as a set of regularly placed points upon which to interpolate groundwater levels. The 1600 m × 1600 m grid (approximately 640 acres) corresponds well with the dimensions of the agricultural sections common in the Western United States and resulted in 46,000 cells over the study area. Furthermore, the adopted grid is consistent with the discretization of the regional groundwater flow models developed for the region (Blandford et al., 2003; Dutton et al., 2004). The specific yield and the aquifer bottom thickness

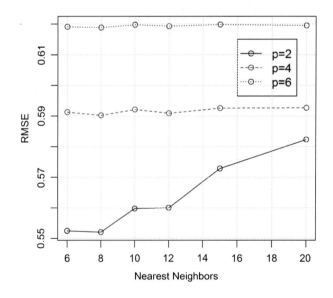

FIGURE 12.2

Root mean square error (RMSE) for each power for various numbers of nearest neighbors.

information were extracted from these models and used to estimate the water available in storage and extractions over time.

The area intersected by the Ogallala aquifer in each cell was computed using ArcGIS 10.2 (ESRI Inc., Redlands, CA, US). Custom scripts were developed in R programming environment (Team, 2015) to perform IDW and other required calculations. nearest neighbors and their distances were identified using the kd-tree algorithm (Arya et al., 1998) as it is computationally efficient and performs the required calculations in O(w log(w)) time.

4. RESULTS AND DISCUSSION
4.1 MONITORING NETWORK AND ESTIMATION GRID CONFIGURATION ANALYSIS

A total of 1917 wells were monitored in the year 2014, and there were on an average 40 wells per county. However, as seen from Fig. 12.1, their distribution is not uniform. The well densities range from 1 well per 20 km^2 in Dawson County to 1 well per nearly 384 km^2 in Potter County. Five counties (Collingsworth, Ector, Glasscock, Hall, and Motley) had no monitoring wells. These counties encompass approximately 1.16% of the total area of the Ogallala aquifer in Texas. On average, there is another monitoring well within 3.6 km of any given well but the closest well could be as far as 18 km.

The monitoring wells were randomly divided into training (70%) and testing (30%) data sets. The training data set was used to perform IDW with varying levels of nearest neighbors and predict values of the testing data set. The root mean square error between the observed values and the predicted values for the testing data sets was the lowest when the power was equal to 2 (Euclidian distance) and the number of nearest neighbors was equal to 8 (Fig. 12.2); therefore, these values were used in all subsequent analysis.

The cumulative distribution function for the closest well to the 1600 m $\times$ 1600 m grid where water availability calculations are performed is depicted in Fig. 12.3. The nearest monitoring well to the grid centroid varies from 20 m to as much as 30 km. This wide range in the nearest distances implies that the estimates associated with different grid cells are going to have variable levels of uncertainty.

4.2 PREDICTIONS OF SATURATED THICKNESS AND AVAILABLE GROUNDWATER RESOURCES

The inverse distance approach was used to estimate the saturated thickness profile of the aquifer (shown in Fig. 12.4) for the year 2014. Based on these estimates and the spatially varying specific yields, the total amount of recoverable water is estimated to be about 49.632 million hectare-meter (MHM). This prediction compares well with the availability estimate presented in McGuire et al. (2003) and McGuire (2014). The predicted estimate is higher in magnitude than that projected by Steward and Allen (2015) based on their logistic regression analysis. However, as can be seen from Fig. 12.4, the water availability is not uniform in the study area. Areas in Terry, Yoakum, and Swisher counties have saturated thicknesses less than 10 m, which is often considered the practical lower limit for sustainably producing water for irrigation purposes (Hristovska et al., 2010). It can also be seen that significant portions of the aquifer have saturated thicknesses in the range of 10–25 m, which under the current groundwater management goals would be unsuitable for irrigated agriculture over the next

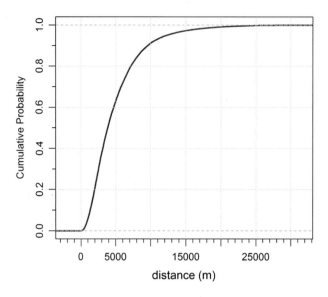

FIGURE 12.3

Cumulative distribution of wells based on distance to nearest neighbor.

50-year period. A clear water availability gradient can be inferred from Fig. 12.4 in which northern counties have greater amounts of water than counties in the southern parts of the state.

4.3 UNCERTAINTY IN ESTIMATED SATURATED THICKNESS AND WATER AVAILABILITY

Fig. 12.5 depicts the standard deviation of the saturated thickness estimates obtained using the Jackknife resampling procedure. For the most part, the standard deviation is less than 1 m, which indicates that the present network provides reasonably accurate estimates of saturated thicknesses over a large portion of the aquifer. Fig. 12.5 also suggests that generally speaking, larger variances (>3 mm) correspond to areas where there are fewer wells. The standard deviation well density plot (Fig. 12.6) also corroborates that standard deviations are lower when the well density is greater than 0.04 wells/km². However, the relationship is nonlinear indicating that variability is not a function of saturated thickness alone. In particular, larger variances are noted in portions of Hale, Castro, and Lamb counties despite relatively higher well densities. A comparison of Figs. 12.1 and 12.5 suggests that the variance in estimates is generally higher in areas with intense agricultural activities, which is to be expected because the aquifer is stressed during the growing season and may not have fully recovered when measurements are made in the late December—early March time frame.

The baseline water availability estimated using the complete network of 1917 wells was equal to 49.63 MHM. The estimated recoverable water estimates were higher than the baseline values for all Jackknife realizations with an upper limit of 50.39 MHM or 1.53% increase over the baseline (see Fig. 12.7). This result suggests that removal of any well from the network will lead to overestimation

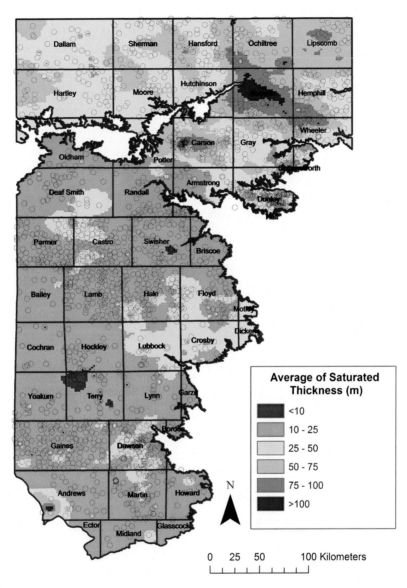

FIGURE 12.4

Saturated thickness of Ogallala aquifer in Texas.

of the available groundwater, which can also be interpreted to mean that addition of wells could potentially lower the current estimate of groundwater availability. The standard deviation plot (Fig. 12.5) identifies several candidate locations to augment the existing groundwater monitoring network.

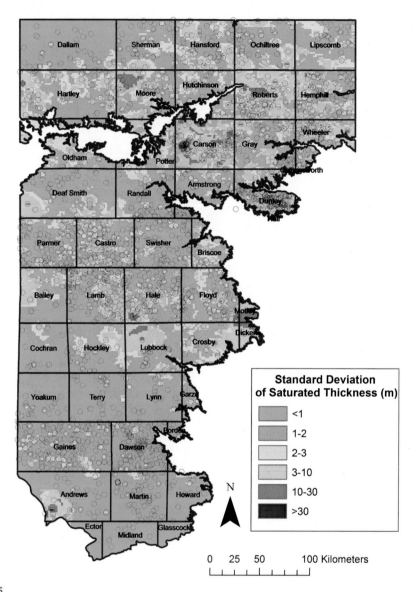

FIGURE 12.5

Standard deviation of the saturated thickness estimates obtained using Jackknife resampling.

Comparison of Figs. 12.4 and 12.5 suggest that the prediction variability is generally higher in areas with higher estimated saturated thickness. This result indicates that regions with higher saturated thicknesses have a greater leverage on the predicted variance than those with relatively lower water

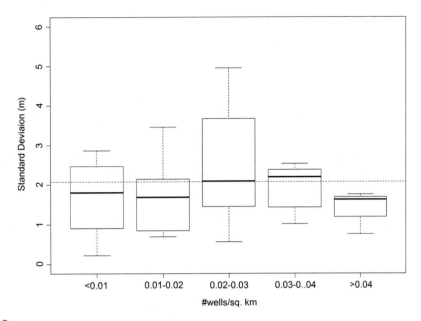

FIGURE 12.6

Standard deviation of saturated thickness estimates among various well densities. Based on wells monitored in the year 2014.

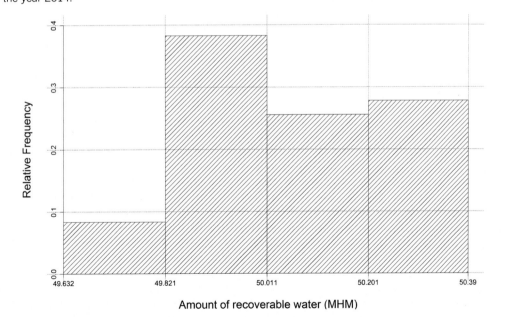

FIGURE 12.7

Frequency with which recoverable water was estimated as more than the baseline amount of 49.63 MHM.

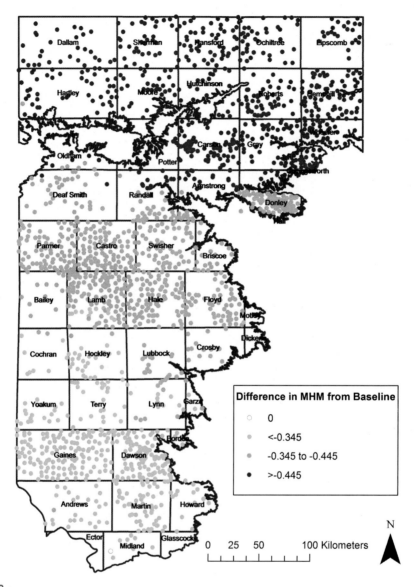

FIGURE 12.8

Relative importance of wells based on change in MHM from the baseline.

availability. From a geographic point of view, higher saturated thicknesses are found in the northern sections of the study area, and as such monitoring in the northern portions of the aquifer are relatively more important than monitoring in the southern portions. The influence of individual wells on the

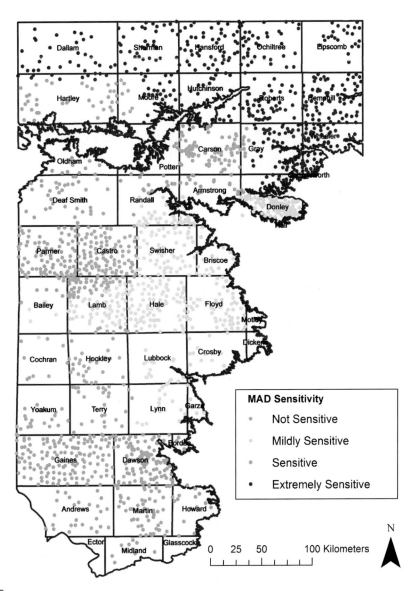

FIGURE 12.9

The mean absolute deviation (MAD) sensitivity coefficient.

estimated water availability (recoverable water) is schematically depicted in Fig. 12.8, which depicts the relative importance of wells in the northern counties in accurately estimating the amount of available water in the aquifer.

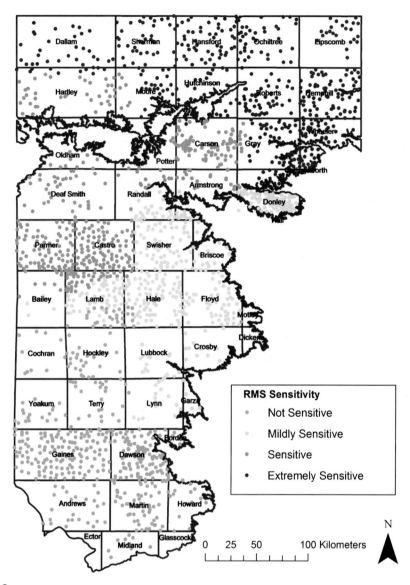

FIGURE 12.10

The root mean square (RMS) sensitivity measure.

4.4 SENSITIVITY OF INDIVIDUAL WELLS

The MAD sensitivity measure (Eq. 12.6) and the root mean square (RMS) sensitivity measure (Eq. 12.7) for different wells in the network are depicted in Figs. 12.9 and 12.10, respectively. The sensitivity

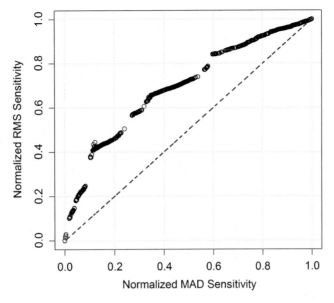

FIGURE 12.11

Normalized mean absolute deviation (MAD) sensitivity versus normalized RMS sensitivity.

measures were categorized into five classes by equally subsetting the output space of these metrics. The absolute sensitivity measures ranged from 0.04–17.65 for MAD sensitivity and 0.002–0.127 for RMS sensitivity. Both the sensitivity measures were normalized to a 0–one scale (0 = minimum value and 1 = maximum value) and are plotted in Fig. 12.11 to highlight the monotonic relationship between the two sensitivity measures. The RMS sensitivity measure magnifies the larger errors more so than the MAD-based sensitivity measure and as such causes the observed behavior in Fig. 12.11.

The normalized sensitivity drops much more rapidly across the wells using the MAD measure than the RMS measure. Therefore, if these measures are employed to reduce the size of the network, the MAD sensitivity measure would result in a larger number of deletions for a given cutoff. For example, if wells with normalized sensitivity less than or equal to 0.25 are to be eliminated then the MAD measure would cause the removal of 460 wells (23.9%), whereas the RMS measure would lead to a removal of 159 wells (8.4%) from the existing network.

4.5 EVALUATION OF SENSITIVITY MEASURES AGAINST GRACE-DERIVED DATA

The average TWS anomaly for the year 2014 is depicted in Fig. 12.12. A north–south spatial gradient in the GRACE observed water storage anomalies is evident in the figure. The anomalies correspond to deviations from the 2004–2009 averages and as such are representative of the depletions occurring in the Ogallala aquifer within Texas. Comparison of TWS anomalies in Fig. 12.12 to the computed sensitivity coefficient in Figs 12.9 and 12.10 indicate a strong correlation between the computed sensitivities and GRACE observed anomalies. The Spearman rank correlation coefficient between the TWS anomaly and the MAD sensitivity measure was fairly high (R = 0.75), as was the case between

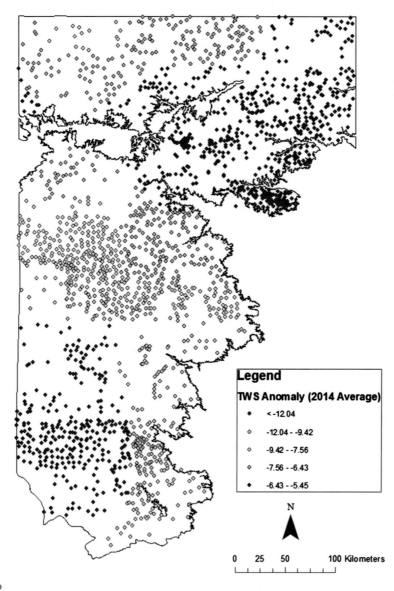

FIGURE 12.12

GRACE—derived terrestrial water storage (TWS) anomalies at monitoring wells.

the TWS anomaly and the RMS sensitivity measure (R = 0.75). It is likely that this correlation can be further improved if the TWS anomaly was further decomposed into groundwater and soil moisture components. Nonetheless, the strong correlation indicates that even raw (undecomposed) TWS anomaly can be useful to identify critical monitoring locations and help with the refinement of groundwater monitoring networks.

5. SUMMARY AND CONCLUSIONS

IDW is a commonly used technique for interpolation in environmental and water resources investigations. However, the approach is deterministic and as such does not provide an estimate of variability of the estimated values. The goal of this study was to integrate Jackknife randomization method with the IDW procedure to estimate uncertainty associated with estimation. As the Jackknife procedure only excludes one well at a time, the spatial correlation structure is largely preserved during each realization. The simple bootstrap approach does not preserve the autocorrelation structure, and algorithms that seek to generate realizations with spatial autocorrelation are cumbersome to apply. However, the independence among realizations is a key assumption of the Jackknife procedure. The approach does not make any distributional assumptions and helps generate empirical distribution functions of estimates. As such, the proposed approach is useful when the observations are skewed and do not meet the assumptions of normality required for parametric analysis. The empirical distribution functions generated using the Jackknife approach were used in this study to develop an innovative sensitivity analysis approach to prioritize monitoring wells within a network. Two sensitivity based influence measures have been proposed in this study to identify the importance of monitoring wells in a network. The sensitivity measure based on MAD was noted to have a steeper response than the RMS-based sensitivity measure and would generally eliminate a greater number of wells for a specified sensitivity cutoff. The sensitivity measures developed here are analogous to the distance-based measures (e.g., Cook's distance) that are commonly used to identify influencing variables in a linear regression. The developed sensitivity analysis−based prioritization approaches were used to understand the adequacy of current monitoring efforts, identify critical areas for monitoring, and prioritize wells based on their importance in estimating saturated thicknesses and the recoverable amount of groundwater in the Ogallala aquifer within Texas. The framework is, however, general and can be applied to other monitoring networks as well.

Independent verification of the developed sensitivity measures were carried out using TWS data obtained from GRACE satellite systems. A strong correlation between the sensitivity measures and TWS anomaly provided additional confirmation that the critical wells identified for estimating groundwater availability were also the ones showing greatest responses to changes in TWS. The complete network provided the lowest estimate of water availability and removal of wells led to a higher (less conservative) estimate. Therefore, additional monitoring, especially in areas with higher estimated variability, is recommended. Reducing monitoring efforts in the southern portions and shifting the resources to northern areas would lead to a more robust estimation of water availability in the aquifer.

The results of the study indicate a spatial gradient in the importance of wells, with wells in the northern region being more important to the network than those in the south. The importance of a well generally correlates well with the saturated thickness indicating that regions with greater saturated thickness have a greater leverage on the network. The estimated variability in saturated thickness is also a function of the estimated saturated thickness. While increasing the well density generally decreased the estimated variability, wells in areas with intensive agriculture also exhibited greater variability despite significant local redundancy in the network. This result may arise from highly nonlinear water table dynamics induced by agricultural activities.

REFERENCES

Arya, S., Mount, D.M., Netanyahu, N.S., Silverman, R., Wu, A.Y., 1998. An optimal algorithm for approximate nearest neighbor searching fixed dimensions. Journal of the ACM (JACM) 45 (6), 891−923.

Batt, A.L., Snow, D.D., Aga, D.S., 2006. Occurrence of sulfonamide antimicrobials in private water wells in Washington County, Idaho, USA. Chemosphere 64 (11), 1963−1971.

Becker, M.W., 2006. Potential for satellite remote sensing of ground water. Ground Water 44 (2), 306−318.

Bentley, J.L., 1975. Multidimensional binary search trees used for associative searching. Communications of the ACM 18 (9), 509−517.

Blandford, T., Blazer, D., Calhoun, K., Dutton, A., Naing, T., Reedy, R., Scanlon, B., 2003. Groundwater Availability of the Southern Ogallala Aquifer in Texas and New Mexico—Numerical Simulations through 2050: Final Report Prepared for the Texas Water Development Board by Daniel B. Stephens & Associates: Inc.

Breña-Naranjo, J.A., Kendall, A.D., Hyndman, D.W., 2014. Improved methods for satellite-based groundwater storage estimates: a decade of monitoring the high plains aquifer from space and ground observations. Geophysical Research Letters 41 (17), 6167−6173.

Calow, R.C., MacDonald, A.M., Nicol, A.L., Robins, N.S., 2010. Ground water security and drought in Africa: linking availability, access, and demand. Ground Water 48 (2), 246−256.

Cook, R.D., 1979. Influential observations in linear regression. Journal of the American Statistical Association 74 (365), 169−174.

Dutton, A., Tinker, S.W., John, A., Katherine, G., 2004. Adjustment of Parameters to Improve the Calibration of the Ogallala-Northern Portion Model of the Ogallala Aquifer, Panhandle Water Planning Area. Texas Water Development Board.

Efron, B., Gong, G., 1983. A leisurely look at the bootstrap, the jackknife, and cross-validation. The American Statistician 37 (1), 36−48.

Figureau, A.-G., Montginoul, M., Rinaudo, J.-D., 2015. Policy instruments for decentralized management of agricultural groundwater abstraction: a participatory evaluation. Ecological Economics 119, 147−157.

Fornés, J.M., la Hera, Á., Llamas, M.R., 2005. The silent revolution in groundwater intensive use and its influence in Spain. Water Policy 7 (3), 253−268.

Gleeson, T., Alley, W.M., Allen, D.M., Sophocleous, M.A., Zhou, Y., Taniguchi, M., VanderSteen, J., 2012. Towards sustainable groundwater use: setting long-term goals, backcasting, and managing adaptively. Ground Water 50 (1), 19−26.

Hristovska, T., Watkins, K., Anders, M., Karov, V., 2010. The impact of saturated thickness and water decline rate on reservoir size and profit. Rice Research Studies. Available online: http://arkansasagnews.uark.edu/591-46.pdf.

Hsu, S.-K., 1998. Plan for a groundwater monitoring network in Taiwan. Hydrogeology Journal 6 (3), 405−415.

Isaaks, E.H., Srivastava, R.M., 2001. An Introduction to Applied Geostatistics. 1989. Oxford University Press, New York, US.

Kong, X., Zhang, X., Lal, R., Zhang, F., Chen, X., Niu, Z., Song, W., 2015. Groundwater depletion by agricultural intensification in China's HHH plains, since 1980s. Advances in Agronomy 135, 59−106.

Kumar, S.V., Zaitchik, B.F., Peters-Lidard, C.D., Rodell, M., Reichle, R., Li, B., Jasinski, M., Mocko, D., Getirana, A., De Lannoy, G., 2016. Assimilation of gridded grace terrestrial water storage estimates in the north American land data assimilation system. Journal of Hydrometeorology. http://dx.doi.org/10.1175/JHM-D-15-0157.1.

Lee, J.Y., Yi, M.J., Yoo, Y.K., Ahn, K.H., Kim, G.B., Won, J.H., 2007. A review of the national groundwater monitoring network in Korea. Hydrological Processes 21 (7), 907−919.

Li, J., Heap, A.D., 2011. A review of comparative studies of spatial interpolation methods in environmental sciences: performance and impact factors. Ecological Informatics 6 (3), 228−241.

Long, D., Scanlon, B.R., Longuevergne, L., Sun, A.Y., Fernando, D.N., Save, H., 2013. GRACE satellite monitoring of large depletion in water storage in response to the 2011 drought in Texas. Geophysical Research Letters 40 (13), 3395−3401.

Lund, J., Medellín-Azuara, J., Harter, T., 2014. Why California's Agriculture Needs Groundwater Management. CaliforniaWaterBlog. com.

Marston, L., Konar, M., Cai, X., Troy, T.J., 2015. Virtual groundwater transfers from overexploited aquifers in the United States. Proceedings of the National Academy of Sciences of the United States of America 112 (28), 8561−8566.

McGuire, V.L., Johnson, M.R., Schieffer, R.L., Stanton, J.S., Sebree, S.K., Verstraeten, I.M., 2003. Water in Storage and Approaches to Ground-Water Management, High Plains Aquifer, 2000, vol. 1243. US Geological Survey.

McGuire, V.L., 2014. Water-Level Changes and Change in Water in Storage in the High Plains Aquifer, Predevelopment to 2013 and 2011−13 (No. 2014-5218). US Geological Survey.

Moench, M.H., 1992. Chasing the watertable: equity and sustainability in groundwater management. Economic and Political Weekly A171−A177.

NGWA, 2016. Facts About Global Groundwater Usage. Retrieved from Westerville, OH 43081−48978, US.

Niu, G.-Y., Yang, Z.-L., 2006. Assessing a land surface model's improvements with GRACE estimates. Geophysical Research Letters 33 (7).

Qian, Z., Cao, L., Su, W., Wang, T., Yang, H., 2012. Recent Advances in Computer Science and Information Engineering, vol. 1. Springer Science & Business Media.

Richey, A.S., Thomas, B.F., Lo, M.H., Reager, J.T., Famiglietti, J.S., Voss, K., Swenson, S., Rodell, M., 2015. Quantifying renewable groundwater stress with GRACE. Water Resources Research 51 (7), 5217−5238.

Rodell, M., Famiglietti, J., 2002. The potential for satellite-based monitoring of groundwater storage changes using GRACE: the High Plains aquifer, Central US. Journal of Hydrology 263 (1), 245−256.

Rodell, M., Velicogna, I., Famiglietti, J.S., 2009. Satellite-based estimates of groundwater depletion in India. Nature 460 (7258), 999−1002.

Rodell, M., Chen, J., Kato, H., Famiglietti, J.S., Nigro, J., Wilson, C.R., 2007. Estimating groundwater storage changes in the Mississippi River basin (US) using GRACE. Hydrogeology Journal 15 (1), 159−166.

Scanlon, B.R., Faunt, C.C., Longuevergne, L., Reedy, R.C., Alley, W.M., McGuire, V.L., McMahon, P.B., 2012. Groundwater depletion and sustainability of irrigation in the US High Plains and Central Valley. Proceedings of the National Academy of Sciences of the United States of America 109 (24), 9320−9325.

Schumacher, M., Kusche, J., Döll, P., 2016. A systematic impact assessment of GRACE error correlation on data assimilation in hydrological models. Journal of Geodesy 1−23.

Shepard, D., 1968. A two-dimensional interpolation function for irregularly-spaced data. In: Paper Presented at the Proceedings of the 1968 23rd ACM National Conference, pp. 517−524.

Singh, A., 2014. Simulation and optimization modeling for the management of groundwater resources. II: combined applications. Journal of Irrigation and Drainage Engineering. http://dx.doi.org/10.1061/(ASCE) IR.1943-4774.0000689.

Sophocleous, M., 2012. Retracted: conserving and extending the useful life of the largest aquifer in North America: the future of the High Plains/Ogallala aquifer. Ground Water 50 (6), 831−839.

Steward, D.R., Bruss, P.J., Yang, X., Staggenborg, S.A., Welch, S.M., Apley, M.D., 2013. Tapping unsustainable groundwater stores for agricultural production in the High Plains Aquifer of Kansas, projections to 2110. Proceedings of the National Academy of Sciences of the United States of America 110 (37), E3477−E3486.

Steward, D.R., Allen, A.J., 2015. Peak groundwater depletion in the High Plains Aquifer, projections from 1930 to 2110. Agricultural Water Management. http://dx.doi.org/10.1016/j.agwat.2015.10.003.

Swenson, S.C., 2012. GRACE Monthly Land Water Mass Grids NETCDF RELEASE 5.0. Ver. 5.0. PO.DAAC, CA, USA. Available at: 10.5067/TELND-NC005.

Swenson, S.C., Milly, P., 2006. Climate model biases in seasonality of continental water storage revealed by satellite gravimetry. Water Resources Research 42 (3). http://dx.doi.org/10.1029/2005WR004628.

Swenson, S., Wahr, J., 2006. Post-processing removal of correlated errors in GRACE data. Geophysical Research Letters 33 (8). http://dx.doi.org/10.1029/2005GL025285.

Swenson, S., Famiglietti, J., Basara, J., Wahr, J., 2008. Estimating profile soil moisture and groundwater variations using GRACE and Oklahoma Mesonet soil moisture data. Water Resources Research 44 (1). http://dx.doi.org/10.1029/2007WR006057.

Team, R.C., 2015. R: A Language and Environment for Statistical Computing [Internet], 2013. R Foundation for Statistical Computing, Vienna, Austria. Document freely available on the internet at: http://www.r-project.org.

Thomas, A.C., Reager, J.T., Famiglietti, J.S., Rodell, M., 2014. A GRACE-based water storage deficit approach for hydrological drought characterization. Geophysical Research Letters 41 (5), 1537—1545.

Thomas, J.C., Arnold, L.R.R., 2015. Installation of a Groundwater Monitoring-Well Network on the East Side of the Uncompahgre River in the Lower Gunnison River Basin, Colorado, 2012. US Geological Survey Data Series 923 (2015).

TWDB, 2015. Groundwater Database Reports. Retrieved from: https://www.twdb.texas.gov/groundwater/data/gwdbrpt.asp.

Uddameri, V., 2005. Sustainability and groundwater management. Clean Technologies and Environmental Policy 7 (4), 231—232.

Uddameri, V., Andruss, T., 2014a. A GIS-based multi-criteria decision-making approach for establishing a regional-scale groundwater monitoring. Environmental Earth Sciences 71 (6), 2617—2628.

Uddameri, V., Andruss, T., 2014b. A statistical power analysis approach to estimate groundwater-monitoring network size in Victoria County Groundwater Conservation District, Texas. Environmental Earth Sciences 71 (6), 2605—2615.

Uddameri, V., Hernandez, E.A., Estrada, F., 2014. A fuzzy simulation—optimization approach for optimal estimation of groundwater availability under decision maker uncertainty. Environmental Earth Sciences 71 (6), 2559—2572.

Upendram, S., Peterson, J.M., 2007. Irrigation technology and water conservation in the high plains aquifer region. Journal of Contemporary Water Research & Education 137 (1), 40—46.

Zektser, I.S., Everett, L. G., 2004. Groundwater Resources of the World and Their Use. UNESCO IHP-VI Series on Groundwater No 6, UNESCO, Paris, France. http://unesdoc.unesco.org/images/0013/001344/134433e.pdf.

Zhou, Y., Dong, D., Liu, J., Li, W., 2013. Upgrading a regional groundwater level monitoring network for Beijing Plain, China. Geoscience Frontiers 4 (1), 127—138.

MAKING THE MOST OF THE EARTH OBSERVATION DATA USING EFFECTIVE SAMPLING TECHNIQUES

J. Indu[1], D. Nagesh Kumar[2]

Indian Institute of Technology Bombay, Mumbai, India[1]; Indian Institute of Science, Bangalore, India[2]

CHAPTER OUTLINE

1. INTRODUCTION: LOOKING FROM ABOVE

Earth observation (EO) refers to the process of collecting, processing, measuring, modeling, and disseminating information by studying Earth's air, water, and land surfaces using data collected from satellites, buoys, weather stations, and other devices. Hydrological modeling has been sought after as the most viable alternative to address the absence of basin-wide hydrological data. The spatiotemporal variability of hydrological variables such as runoff, infiltration rate, evapotranspiration, and stream-flow is being increasingly studied using meteorological forcing data (e.g., precipitation, wind speed, temperature). For measuring hydrological data, the historical practice has been to use in situ monitoring stations and ground instrumentation networks. With the declining density of ground-based monitoring stations, collecting and monitoring hydrological variables like precipitation and

Sensitivity Analysis in Earth Observation Modelling. http://dx.doi.org/10.1016/B978-0-12-803011-0.00013-6

streamflow has become increasingly difficult. The gaps within in situ measurements are being bridged using widely available satellite-based forcing products (Gebregiorgis et al., 2012; Gebregiorgis and Hossain, 2011, 2013; Hong et al., 2004; Nijssen and Lettenmaier, 2004; Kamal-Heikman et al., 2007). These products include the Climate Prediction Center morphing technique CMORPH (Joyce et al., 2004; Joyce and Xie, 2011), Precipitation Estimation from Remotely Sensed Imagery Using Artificial Neural Networks (PERSIANN; Hsu et al., 1997; Hong et al., 2004), and Tropical Rainfall Measuring Mission (TRMM)−based 3B42RT (Huffman et al., 2010). In addition, new missions are also being proposed solely dedicated toward measurement of hydrological variables like precipitation and streamflow [Global Precipitation Measurement (GPM) mission] (Smith et al., 2007), streamflow (Surface Water and Ocean Topography mission) (Alsdorf et al. 2007), and soil moisture (Soil Moisture Active and Passive mission) (Entekhabi et al. 2010).

Extended knowledge of the hydrologic processes and state variables have dramatically stemmed from remote sensing (RS) data with fine spectral and spatial resolution, which has enabled to assess crucial hydrologic variables such as surface soil moisture, surface temperature, reflectivity, snow cover, and vegetation (Karaska et al., 2004; Dozier et al., 2009; Lindsay and Dhun, 2015). With space-borne observations rapidly becoming a valuable tool for global climate change studies, the number of satellites being launched have tremendously increased as a function of launch rates and mission longevity. The Earth Observation Satellites (EOS) that are currently in orbit carry sensors that measure different regions of the electromagnetic spectrum like visible, infrared, and microwave. The passive microwave radiometers measure the outgoing longwave radiation emanating from the Earth's surface that pass through the atmosphere. Some of the newly launched satellites also employ active sensors that emit energy and record the reflected/backscattered response from the Earth's surface.

The National Aeronautics and Space Administration (NASA) and the Master Catalog of National Space Science Data Centre have mentioned around 7075 spacecrafts (launched between October 4, 1957, and December 31, 2013). Of these nearly 12% were found to be directly contributing toward Earth science (NSSDC, 2014). The Television Infrared Observation (TIROS) satellite was the first mission dedicated to meteorology that produced the first satellite image of the Earth on April 1, 1960. Shortly afterward, the NIMBUS satellite was launched in 1964, which spawned a series of satellites capable of monitoring atmospheric composition and oceanic biological processes using a synergistic combination of visible, infrared, and microwave instruments. Following the TIROS mission, many meteorological satellites were launched into space between 1969 and 1996 by the Soviet Union. The limitations of these Meteor series of satellites were in terms of their inability to differentiate between land cover types. Unlike the television camera−based systems flown on Landsat and the red and near-infrared cameras on India's Bhaskara satellites, the later generations of meteorological satellites have focused on carrying multispectral instruments thereby offering the potential for global land cover mapping. Significant contributions in this regard were extended by the Advanced Very High Resolution Radiometer (AVHRR) and the Landsat series of satellites.

The key aim is to create near-real-time images of the Earth's surfaces that are easily accessible. For more than five decades, the National Oceanic and Atmospheric Administration (NOAA) has been dedicated toward collecting, processing, and distributing data for hazard assessment, weather forecasting, and events of significant environmental relevance like El Nino, La Nina, and ocean acidification. The endless streams of information relayed back from these satellites form the backbone for weather forecasting and climate research. Information about the oceans, land surfaces, and atmosphere like wind patterns and clouds are utilized to produce accurate forecasts. The weather alerts relayed via communication media like television, radio, and web pages all rely on NOAA satellite data. Till date, most of the satellites that have been solely launched for monitoring the Earth's atmosphere are either designed with the purpose of weather forecasting or toward research applications. With the launch of more and more satellites, more and more data are being generated. Hence, developing a data management technology plays a crucial role not only for the acquisition and storage/accessing of data but also for their intelligent usage in forecast studies.

One such process that involves ingesting millions of near-real-time observations to generate initial states of weather and climate forecasts is through data assimilation techniques. An intelligent accommodation of all the observation data depends crucially on the data selection techniques. Sensitivity looks at the changes in a forecast owing to changes in the initial conditions. The adjoint approach of data assimilation efficiently estimates the sensitivities of a forecast (Baker and Daley, 2000; Fourrie et al., 2002). This approach involves computing weights for the observations after removing background residuals. In weather modeling and forecast studies, a crucial role is played by the initial conditions that affect the forecast accuracy. When data assimilation is aimed at generating weather forecasts, sensitivity analysis (SA) helps understand the forecast errors. Literature presents different approaches for SA pertaining to data assimilation. For example, Torn and Hakim (2008) utilize sample statistics to estimate the relationships between forecast metrics and initial conditions; Ancell and Hakim (2007) have compared ensemble sensitivity with adjoint sensitivity.

2. DATA ASSIMILATION

Forecasting the state of complex systems like that of the atmospheric variables either requires extrapolating past values of the variables assuming that its historical curve is smooth or using a priori knowledge about the dynamics of the system by simulating the evolution of the concerned variable. Before the era of numerical computing, the former approach was predominantly followed. The present approach involves using high computational power. The fundamental limitation underlying both the approaches remains the poor predictability of the weather system. An optimal solution for this problem would be data assimilation i.e., combining the most exhaustive theoretical and observational knowledge and using all the available past and present observations. Data assimilation refers to a set of statistical techniques that are aimed at improving knowledge regarding the past, present, and future states of a system using experimental data and the theoretical a priori knowledge of the system. Due to the complexity of the underlying process, the use of an elementary technique such as regression might not be suitable.

Ascertaining the physical, chemical, and biological processes of the Earth system requires availability of high-level data products. Tackling the effects of environmental threats and climate change impacts requires skillful reliable predictions so that timely guidance can be provided to policy makers and society against the impacts of calamities. Mathematical models encapsulate the understanding of the evolution of various components of the Earth systems and of the interactions between them. For prediction, the most important prerequisite is an input for the best estimate representing the initial state of the system. The sensitivity of initial conditions can make the forecast to depend on how closely the grid-based initial state is estimated from observation data. For satellite data, the issue is compounded as the observed quantity is mainly spectral radiance/brightness temperature, which varies with respect to frequency and hence can only be indirectly related to the model parameters.

Then how do we arrive at the best estimate of the Earth system state using the available observational data? How do we ensure that the time series of the estimated states obey the governing equations responsible for evolution of the system? The answer to these questions lies in data assimilation. The key idea is not to accumulate and absorb all the available data and pass them through a computer program. The technique involves a careful procedure that evolves from the existing knowledge of the measurement process, uncertainty in measurement, governing equations of system, and their expected errors as approximated on a computer.

Assimilation of observations distributed in time can be performed using many techniques. One of the widely used methods for data assimilation in numerical weather prediction is the ensemble Kalman filter. In ensemble Kalman filter, the effect of approximate value of error covariances is estimated prior to the analysis step. If E indicates the expectation, ϕ refers to the model state at a particular time, and the notations f, t, a denote the forecasted, true, and analyzed states, Eqs. (13.1) and (13.2) represent the prior and posterior error covariances as:

$$P^F = E\left[\left(\phi^f - \phi^t\right)\left(\phi^f - \phi^t\right)^T\right] \tag{13.1}$$

$$P^a = E\left[\left(\phi^a - \phi^t\right)\left(\phi^a - \phi^t\right)^T\right] \tag{13.2}$$

As the true state is unknown, in ensemble Kalman filters, the ensemble covariance is used so that the mean state is taken as an estimate for the truth.

The need is to combine observational data with data from forecast using a numerical model so as to produce an optimal estimate of the evolution of the system. The model should be capable of interpolating data for the data-devoid regions in space and time. When the word "optima" is used, it indicates the statistical basis of the most advanced implementations of the method. When the word "optimal" is used, it largely refers to estimating the system state using a weighted combination of observations and numerical forecasts. For example, consider the task of useful representation to predict the Earth system, which involves atmosphere, land, and ocean systems. This mammoth task might need a billion numbers and an accurate error characterization of the essential variables. Therefore, their evolution is an extremely challenging task requiring cutting-edge scientific computing. This is an essential big data challenge requiring its direct assimilation. The topic of data assimilation has received much interest being a catalyst technology. The EO data were underexploited for decades due to the nonexistence of data assimilation techniques, which are available at present, to make the best use of the available data. Proper data assimilation results in the following far-fetched benefits.

2.1 ERROR TRACKING

Regular comparison of numerical forecasts with observational data leads to examining of extremely valuable error statistics. This enormously contributes toward improving the weather forecasting models.

2.2 BUILDING A DATABASE

A synergistic use of different versions of observing systems will have different virtues and deficiencies. They can be judiciously utilized to get an optimized value of the resulting data set.

2.3 FROM DATA-RICH REGIONS TO DATA-POOR REGIONS

Data assimilation provides a means to propagate information from data-rich regions in space and time to data-poor regions. This is vital for fully exploiting satellite information, which suffers from sampling issues. Estimation of unobserved parameters is rendered possible by relying on the relationships expressed in the governing equations of the model.

2.4 DECIDING FACTOR IN DESIGNING FUTURE SYSTEMS

On launching a new satellite, scientists and researchers attempt at the incremental value of the new instrument. Data assimilation provides an essential means to answer such questions and target future observations toward meaningful features of societal concern like an extreme calamity.

The applications of data assimilation in Earth sciences have received tremendous impetus owing to their effect on societal applications, which include weather forecasting, seasonal climate prediction, and land surface modeling. A revolutionary technology in meteorology and oceanography that enables long-range predictions of climate draws heavily on the information from various Earth observing satellites. These in turn need to be synthesized using coupled atmosphere—ocean models. An example of this prediction is the 1997 El Nino event, which was detected by various satellites including the ERS-2. The detected anomaly was warming of the ocean surfaces in the equatorial Eastern Pacific. This anomaly when fed into a coupled ocean model successfully predicted the onset of the event. Another interesting application of data assimilation is in the field of inverse modeling to infer quantities that are not well measured. Satellite measurements of atmospheric aerosols can help quantify the sources and sinks of the important greenhouse gases especially in regions devoid of ground-based measurements. The resulting application extends well beyond atmospheric or oceanographic sciences. Another interesting application is the knowledge regarding near-surface soil moisture, which is not only important for hydrological cycle but also essential for agriculture applications and climate predictions. Data assimilation is at the forefront of scientific computing, enabling tackling of enormous intellectual and technological challenges to better the design and deploy future operations satellites.

Information regarding the Earth's system and related processes can be acquired by observations, which in turn help understand the spatiotemporal relationships between variables typically expressed using model equations. The information procured by in situ observations using either ground-based stations, aircraft, and satellites tend to complement each other. This is essentially because in situ observations have high spatial and temporal resolution but lack in terms of area covered. However, satellite-based measurements have high spatial coverage but limitations in terms of spatiotemporal resolution. Often the ground-based in situ data are used to calibrate satellite data. These in situ observations also tend to be characterized by errors and gaps in acquisition on a spatial and temporal scale. To get a more complete picture of a physical process, the gaps in observation data need to be filled, for which a model is needed (Lahoz et al., 2010a). The model can be either one that utilizes the spatial and temporal autocorrelation of observation data in a geostatistical framework or a general circulation model or a land surface model that considers energy transport between land surface and the atmosphere. Before adopting any modeling approach, SA needs to be executed to evaluate the suitability of the model to reproduce the physical process/variable with the desired accuracy. SA is essentially an integral check of a simulation model. The ability to estimate the impact of observations on a forecast metric is beneficial for thinning large observation database to a few highly sensitive ones during the process of data assimilation. The following section addresses some of the sampling techniques generally employed for filling gaps in data acquired from either in situ observation sites or satellite-borne sensors.

3. SAMPLING SCHEMES

Satellite observations have become an indispensable source for observing properties of Earth's land surface and atmosphere for more than two decades. In satellite-based data like soil moisture, temperature, aerosols, and more recently greenhouse gases, the existence of data gaps is a source of concern in many products from EOS and represents a large source of uncertainty in analyzing the spatiotemporal variability of large data sets. These data gaps in satellite-borne information are mainly caused due to intrinsic effects owing to their orbital geometry. Other reasons may be attributed to cloud contamination/instrument failure etc. The growing volume and diversity of satellite data sets require a means of efficiently propagating the available information from data-rich regions in space and time to data-poor regions. This is vital for fully exploiting satellite information, which suffer from sampling issues. Some of the effective sampling techniques popularly adopted pertaining to data assimilation are discussed in the following paragraphs.

Data assimilation and sensitivity go hand in hand. In hydrology, SA is invaluable for identifying the salient model parameters and improving the model structure. When the modeler has a clear understanding about all the parameters used as input, a focus on sensitive parameters can result in reduced uncertainty (Lenhart et al., 2002). Therefore, SA is useful not only for model development but also for model validation and reduction of uncertainty (Hamby, 1994). Various spatial sampling schemes enable to examine the variations in environmental properties of interest (Minasny and McBratney, 2007; Lin et al., 2009). These schemes utilize different sampling techniques, which may be random, systematic, stratified, or nested sampling (Thompson, 1992; Fortin and Edwards, 2001), which typically transform the data sets using a series of steps to get a complete description of the phenomena of interest (Fortin and Edwards, 2001). Usually, mathematical and computational problems, revolving around issues of practical importance involve interpolation of data for nonmeasured regions and generation of realistic scenarios that obey the data (Cressie, 1993; Lin et al., 2009; Hristopulos and Elogne, 2009). Difficulty arises when the objective is accurate representations of small-scale cloud microphysical processes such as drizzle formation using numerical models of the atmosphere. Using a grid box averaging mechanisms as input to a microphysical parameterization process cannot aid in providing accurate fields to drive the microphysics (Sommeria and Deardorff, 1977; Rotstayn, 2000; Pincus and Klein, 2000; Larson et al., 2001). This is because the grid boxes themselves are too small to fully resolve the cloud microphysical processes and computational constraints offered by the nonlinear microphysical formulas add to the existing complexities. The subgrid variability must be taken into account and parameterizations need to be used, which range from simplified analytic formulas to complex numerical subroutines that represent a myriad of atmospheric processes such as conversion of cloud droplets to drizzle drops and activation of cloud condensation nuclei.

4. BOOTSTRAP SAMPLING

Introduced first by B. Efron (1979, 1981, 1982) and further developed by Efron and Tibshirani (1994), the bootstrap approach has been extensively used in statistical sciences within a few decades. In general, two approaches are being described to model problems using bootstrap by Efron and Tibshirani (1994). A statistician would mainly be more concerned about summarizing the sample-based study and try to generalize the findings in a scientific manner using sample statistics. The magnitude of fluctuations of the sample statistic from one sample to another can very well be used for assessing the margin of its errors. A picture of all possible variations of the sample statistic presented in the form of a probability distribution results in a sampling distribution. The bootstrap approach is a very general and popular resampling procedure for arriving at the distribution of sample statistics based on identically and independent distributed samples.

For example, let $a_1, a_2, a_3, \ldots \ldots a_n$ represent independent random variables having a distribution function say F. Let μ be the location estimate, which in turn is the function of the random variables $A_1, A_2, A_3, \ldots \ldots A_n$ and has a probability distribution, which is determined by n and F. Bootstrap resampling technique can either be parametric or nonparametric in nature. The core idea in a parametric bootstrap approach is to simulate the data from $F(a/\eta^*)$, where η stands for an unknown parameter and η^* represents a good estimate of η. This represents the case wherein the distribution F is known. Suppose the structure of F is completely unknown, then the idea of nonparametric bootstrap needs to be adopted so as to simulate data from the empirical cumulative distribution function (CDF) of F_n. For detailed discussion on the same, Efron and Tibshirani (1993) may be referred. With respect to EOS data, the bootstrap technique is mostly used to produce a very large number of surrogate replicates of a sample statistic, which are in turn computed by resampling with replacement from the available sample data. The sample statistic computed from these large number of bootstrap samples/phantom samples can be used to declare the confidence limits of the unknown population. The primary application of bootstrap technique involves approximating the standard error of a sample estimate, bias correction, creation of confidence intervals, percentiles, etc. For example, the mean of sampling distribution of $\alpha^\wedge$ often tends to vary from α, which can be written as:

$$Bias(\hat{\alpha}) = E(\hat{\alpha}) - \alpha \approx O(1/n) \tag{13.3}$$

This can be written in Eq. (13.4) as:

$$\frac{1}{N} \sum_{i=1}^{N} \alpha_i^* - \hat{\alpha} = Bias(\hat{\alpha}) \tag{13.4}$$

Here, α_i^* are the bootstrap copies of $\overset{\wedge}{\alpha}$ wherein the basis is on the standard bootstrap approach of replacing the population by the empirical population of the sample.

A second approach of bootstrapping involves resampling the residuals and is characterized as bootstrapping residuals. A similar approach can be employed to draw a random sample of residuals with replacement so that the same be added back to the model predictions to form a bootstrap sample. The sample may be drawn either from an empirical distribution of residuals or a parametric model of the distribution. The underlying core assumption is that the residuals are independent and identically distributed. McRoberts (2009) explains methods of studying residuals in the case of heteroscedasticity.

5. LATIN HYPERCUBE SAMPLING

Natural phenomena will usually be spatially heterogeneous, and their associated study is important for geographical data analysis and environmental sciences (Csillag et al., 1996; Goodchild and Haining, 2004; Green and Plotkin, 2007). The elements of geographical variation involve global variance (namely, the population variance) and the spatial distribution of global variation (i.e., the population autocorrelation) (Wang et al., 2010). Among the popular methods to deal with spatial heterogeneity are spatially stratified sampling in design-based sampling wherein the heterogeneous area is divided into subareas that are homogeneous rather than the area as a whole, thereby reducing the total variance.

Latin hypercube sampling (LHS), also known as stratified sampling technique, is used to generate multivariate samples of statistical distributions (McKay et al., 1979; Minasny and McBratney, 2006; Ireland et al., 2015; Petropoulos et al., 2009c). For example, consider a variable Y, which is a function of multiple variables like z_1, z_2, z_3, ...z_n. Often, it might be of interest to know the variability of Y when z's vary according to some assumed joint probability distribution. LHS sampling selects k different values from each of the n variables z_1, z_2, ...z_n using the following approach. Every time, the range of each variable will be divided into k nonoverlapping intervals based on equal probability. From each interval, one value is selected at random with respect to the probability density in the interval. The k such values obtained for z_1 are randomly paired with the k values of z_2. These in turn are paired in a random manner with the k values of z_3 to form k triplets, and so on, until the kn tuplets are formed. These tuplets represent the kn-dimensional input vectors. This sample can be thought of as forming a kxn matrix of input where the ith row contains values of each of the n input variables to be used in the ith run. A more extensive discussion on LHS sampling can be found in Iman et al. (1981a,b). This approach is comparable to uniform sampling from the quantiles of the distribution followed by inverting the cdf to get the actual distribution values that the quantiles represent. Stratifying the cdf into equally probable regions ensures that the random model inputs will essentially include values having a low probability of occurrence, which at the same time have high consequence to the model outputs. An example would be soil types that possess high moisture-holding capacity but with low saturated hydraulic conductivity, which although are encountered infrequently, will be of high consequence in estimating the runoff.

A persistent problem will be that at each time step, the process of sampling will tend to induce statistical noise into the model. This can be tackled by means of a variance reduction method (Raisanen and Barker, 2004). The LHS technique reduces variance by preventing the sample points from clumping together in sample space. For example, if the LHS scheme is being utilized for an application related to atmospheric sciences like modeling of cloud microphysics, the dependable variables will be total water mixing ratio, liquid water potential temperature, vertical velocity, cloud droplet number mixing ratio, and drizzle mixing ratio. The goal of the LHS technique will be to essentially spread out the sample points in space so that the high as well as the low values of each variate be rightly contained in the sample. In the context of ensemble soil moisture data assimilation, the computation burden increases due to the large state—space dimensionality, the complexity and nonlinearity of models, requirements of multiple model runs, etc. The literature presents several studies that stress on the dependability of LHS as a useful tool to represent hydrologic model parameter uncertainty and in geostatistics for simulation of Gaussian random fields (Beven and Freer, 2001; Christiaens and Feyen, 2002; van Griensven et al., 2006; Abbaspour et al. 2007; Minasny and McBratney, 2002; Pebesma and Heuvelink, 1999; Zhang and Pinder, 2004).

6. CASE STUDY USING BOOTSTRAP SAMPLING

Rainfall products derived from microwave RS are known to suffer from various sources of uncertainties, a major portion of which can be attributed to sampling uncertainty. Infrequent satellite visits cause difficulty in measuring the spatiotemporal variability of rainfall. In order to provide quantitative confidence on the rainfall estimates from satellites, estimating sampling errors is crucial. Sampling errors not only depend on the type of satellites (e.g., geostationary or low Earth orbiting) but also on precipitation type, season, etc. TRMM is a low Earth orbiting satellite that carries an active sensor called precipitation radar (PR) and a passive sensor called TRMM Microwave Imager (TMI). To date, studies related to TRMM sampling uncertainty have focused on comparing rainfall rates from a surface-based dense network of rain gauges with those from TRMM. This technique suffers from a disadvantage for regions lacking in in situ validation data. To circumvent this issue, attempts have been made to estimate sampling errors on a global scale using available satellite rainfall data products themselves. Recently, Iida et al., (2010) have developed a technique using bootstrap approach to evaluate the relative sampling errors of PR-observed trimonthly rainfall. This technique has successfully evaluated relative sampling errors globally for $5^0 \times 5^0$ grid sizes without relying on rainfall from a dense network of ground instruments. Indu and Nagesh Kumar (2014) have investigated the potential of this approach in estimating sampling uncertainty for Mahanadi basin in India. Case study results are being presented for rainfall analyses for a 6-year data period from TRMM. Seasonal rainfall estimates are being analyzed here from June 2002 to September 2006. Three data products are being examined, namely, the TMI-derived 2A12 data, the PR-derived 2A25 data, and the TMI-PR combined data of 2B31. Post 2001 data products are considered for the analysis owing to the TRMM orbital boost from 350 to 402.5 km in August 2001, which altered the data quality significantly.

6.1 METHODOLOGY AND RESULTS

In the present study, the individual overlapping coincident snapshots of rain events in various orbital passes are examined during 6 years of daily seasonal rainfall over Mahanadi basin for a grid size of $1^0 \times 1^0$. If $i = 1, 2, 3........n$ denotes the number of times the satellite visits an area ($1^0 \times 1^0$ grid), $R(1), R(2), R(3)...........R(n)$ be the corresponding area averaged rain rates in millimeters per hour from all footprints that are in the box, and $N(1), N(2), N(3)..............N(n)$ be the number of PR footprints associated with each visit, then the average seasonal rainfall (mm) observed within each grid (of area A) can be expressed as:

$$R_s = \frac{\sum_{i=1}^{n} N(i) * R(i)}{\sum_{i=1}^{n} N(i)} * 24 * 122 \qquad (13.5)$$

The value of relative sampling error σ can be calculated using say 1000 bootstrap samples of R_S. In the bootstrap technique by Iida et al. (2010), standard deviation is regarded as a measure of sampling error and the value of relative standard deviation is considered as a measure of the relative sampling error.

The relative sampling errors (%) were estimated for 1^0 grid boxes over Mahanadi basin using bootstrap technique for seasonal rainfall during the data period of 6 years [2002 to 2006]. Comparative evaluation of results (Fig. 13.1) shows that the sampling uncertainty from the TMI-derived 2A12 data product is less compared with that from the 2A25 and 2B31 data products.

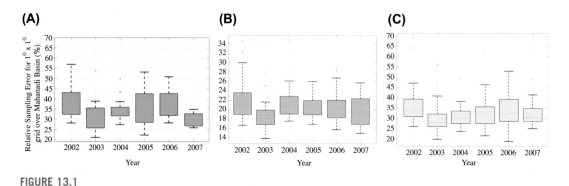

FIGURE 13.1

Time series for Mahanadi basin for $1^0 \times 1^0$ grid size for the 11-year data period showing Relative sampling errors (%) for (A) 2A25, (B) 2A12, and (C) 2B31.

This can be attributed to difference in sampling swath between the instruments of TMI and PR. The TMI data swath of 780 km enables a superior overview of the synoptic rainfall events than PR (having data swath of only 215 km). The wider TMI swath enables a greater number of observations to be available for each spatial domain (grid box). This implies that despite the limitation of TMI radiometer–retrieved rainfall from the 2A12 product, their land rainfall will possess comparatively low bias due to sampling uncertainty when compared with the 2A25 and 2B31 products. In order to study the effect of sampling type on relative sampling errors, the approach was conducted using LHS.

Fig. 13.2 shows the relative sampling error for $2^0 \times 2^0$ grid boxes over India for the 2002 monsoon season [June, July, August, September (JJAS)]. Theoretical dependence of relative sampling error is on rainfall variability and the space time correlation length of rainfall. In addition, it also depends on the mean rainfall amounts and the sampling frequency. From Fig. 13.2 it can be observed that the relative sampling error distribution over the case study region of Mahanadi basin falls within the range of 16–27%. Indu and Nagesh Kumar (2014) have studied the dependency of sampling uncertainty on different rainfall regimes and grid sizes.

The suitability of various sampling techniques requires examining various sampling techniques that have potentially greater flexibility in simulating the relative sampling errors. Results for 2A12 data product are show in Fig. 13.3. It can be observed that relative sampling errors estimated using the LHS technique were similar to those from bootstrap technique. The LHS technique tends to slightly overestimate the relative sampling errors. This slight overestimation pertaining to orbital data products needs to be further studied before any clear conclusion can be drawn of its significance. Pertaining to the case study region, whether the results obtained arise owing to the limitation of the LHS technique, due to the type of data samples, or due to the low number of simulations needs to be further studied before making conclusions.

The present study examined the efficiency of bootstrap sampling in calculating the relative sampling errors over the catchment of Mahanadi basin, India, for three TRMM orbital data products based on radar (PR), radiometer (TMI), and the combined product of both (PR-TMI). Results indicate that the radiometer-derived rainfall estimates from 2A12 incurred the least range of sampling uncertainty when compared with 2A25 and 2B31 rainfall estimates. Also, retrieval of sampling uncertainty using

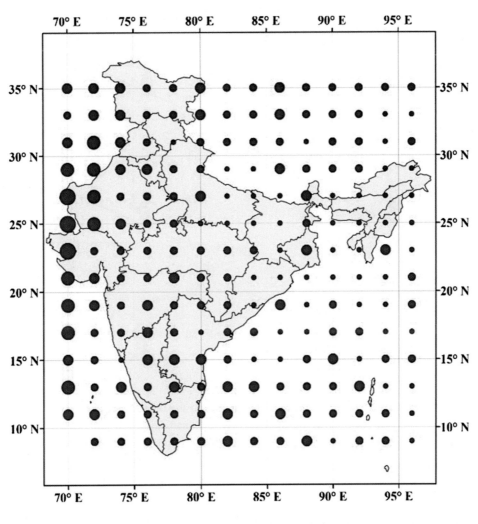

Sampling Error (%) for 2^0 x 2^0 grid size

16-27   27-36   37-51   53-73   85-113

FIGURE 13.2

Sampling error using PR 2A25 data over India for the JJAS 2002 data period (JJAS 2002 refers to the time period of June, July, August, September (Indian Summer Monsoonal Months) of the year 2002.).

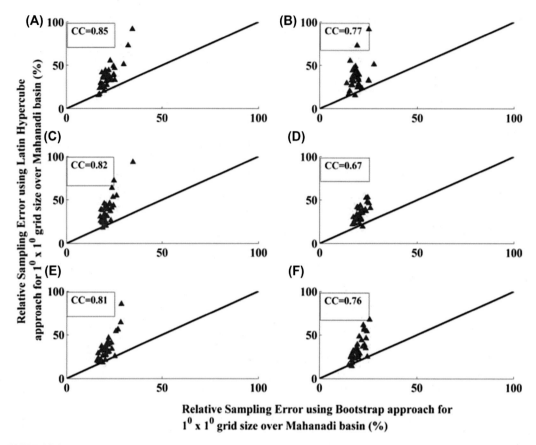

FIGURE 13.3

Scatter plots showing relative sampling errors estimated using the bootstrap approach versus those estimated using the LHS technique for 2A12 product during the years (A) 2002, (B) 2003, (C) 2004, (D) 2005, (E) 2006, and (F) 2007. *CC*, correlation coefficient; continuous line denotes one-to-one line.

LHS was shown to have implications for wide-scale assessment of satellite rainfall retrievals pertaining to hydrological applications. The present methodology can also be implemented for future missions of GPM.

7. CONCLUSIONS

Instrumental data sets obtained in real time are likely to suffer from sampling issues because of the uneven spatiotemporal coverage. The RS data depending on the type of instrument employed will largely be influenced by atmospheric conditions, clouds, aerosols, and heavy precipitation. Instrument malfunction during extreme weather scenario such as hurricanes and tornadoes also hamper the temporal coverage of the data. Usually, the greatest proliferation of uncertainty in data volume occurs owing to the intermittent nature of satellite data.

The unique perspective from space is that the launch of each satellite will be accompanied by massive data with a unique sampling pattern, which pose problems to the traditional data assimilation techniques. With the increasing volumes of data made available from the space in real time, ignoring the imprecise information can no longer be afforded. This is where data assimilation derives its strength from. The present chapter has emphasized on the need of data assimilation techniques for satellite data. The importance of sampling techniques to propagate data from data-rich regions to data-poor regions has also been discussed. The present chapter briefly summarizes the bootstrap and LHS techniques. Case study results highlighting the importance of bootstrap sampling technique for uncertainty estimate of satellite orbital data have been presented. With the ever-increasing need to process real-time data for societal applications, the importance of parametric and nonparametric approaches for gap filling cannot be underestimated.

REFERENCES

Abbaspour, K.C., Yang, J., Maximov, I., Siber, R., Bogner, K., Mieleitner, J., Zobrist, J., Srinivasan, R., 2007. Modeling hydrology and water quality in the pre-alpine/alpine Thur watershed using SWAT. Journal of Hydrology 333, 413–430.

Alsdorf, D.E., Rodriguez, E., Lettenmaier, D.P., 2007. Measuring surface water from space. Reviews of Geophysics 45, RG2002. http://dx.doi.org/10.1029/2006RG000197.

Ancell, B., Hakim, G.J., 2007. Comparing adjoint and ensemble sensitivity analysis with applications to observations targeting. Monthly Weather Review 135, 4117–4134.

Baker, N.L., Daley, R., 2000. Observation and background adjoint sensitivity in the adaptive observation-targeting problem. Quarterly Journal of the Royal Meteorological Society 126, 1431–1454.

Beven, K.J., Freer, J., 2001. Equifinality, data assimilation, and uncertainty estimate in mechanistic modelling of complex environmental systems. Journal of Hydrology 249, 11–29.

Christiaens, K., Feyen, J., 2002. Constraining soil hydraulic parameter and output uncertainty of the distributed hydrological MIKE SHE model using the GLUE framework. Hydrological Processes 16 (373), 391.

Cressie, N.A., 1993. Statistics for Spatial Data, revised ed. Wiley, NY.

Csillag, F., Kertesz, M., Kummert, A., 1996. Sampling and mapping of heterogeneous surfaces: multi-resolution tiling adjusted to spatial variability. International Journal of Geographical Information Systems 10, 851–875.

Dozier, J., Green, R.O., Nolin, A.W., Painter, T.H., 2009. Interpretation of snow properties from imaging spectrometry. Remote Sensing of Environment 113, S25–S37. http://dx.doi.org/10.1016/j.rse.2007.07.029.

Efron, B., 1979. Bootstrap methods: another look at the jackknife. The Annals of Statistics 7, 1–26.

Efron, B., 1981. Nonparametric estimates of standard error: the jackknife, the bootstrap and other methods. Biometrika 63, 589–599.

Efron, B., 1982. The jackknife, the bootstrap, and other resampling plans. Society of Industrial and Applied Mathematics CBMS-NSF Monographs 38.

Efron, B., Tibshirani, R.J., 1994. An Introduction to the Bootstrap. CRC Press, Boca Raton, FL.

Efron, B., Tibshirani, R.J., 1993. An Introduction to the Bootstrap. Chapman & Hall, New York.

Entekhabi, D., Njoku, E.G., Neill, P.E.O., Kellogg, K.H., Crow, W.T., Edelstein, W.N., Entin, J.K., Goodman, S.D., Jackson, T.J., Johnson, J., Kimball, J., Piepmeier, J.R., Koster, R.D., Martin, N., McDonald, K.C., Moghaddam, M., Moran, S., Reichle, R., Shi, J.-C., Spencer, M.W., Thurman, S.W., Tsang, L., Van Zyl, J., 2010. The soil moisture active passive (SMAP) mission. Proceedings of the IEEE 98, 704–716. http://dx.doi.org/10.1109/Jproc.2010.2043918.

Fortin, M.J., Edwards, G., 2001. In: Hunsaker, C.T., Goodchild, M.F., Friedl, M.A., Case, T.J. (Eds.), A Cognitive View of Spatial Uncertainty. Springer, New York, pp. 133–157.

Fourrie, N., Doerenbecher, A., Bergot, T., Joly, A., 2002. Adjoint sensitivity of the forecast to TOVS observations. Quaterly Jornal of the Royal Meteorological Socity 128, 2759–2777.

Gebregiorgi, A.S., Peters-Lidard, G., Tian, Y., Hossain, F., 2012. Tracing hydrologic model simulation error as a function of satellite rainfall estimation bias components and land use and land cover conditions. Water Resources Research. http://dx.doi.org/10.1029/2011WR011643.

Gebregiorgis, A.S., Hossain, F., 2013. Estimation of satellite rainfall error variance using readily available geophysical features. IEEE Geoscience and Remote Sensing. http://dx.doi.org/10.1109/TGRS.2013.2238636.

Gebregiorgis, A.S., Hossain, F., 2011. How much can a priori hydrologic model predictability help in optimal merging of satellite precipitation products? Journal of Hydrometeorology. http://dx.doi.org/10.1175/JHM-D-10-05023.1.

Goodchild, M., Haining, R.P., 2004. GIS and Spatial Data Analysis: Converging Perspectives. Papers in Regional Science 83, 363–385.

Green, J.L., Plotkin, J.B., 2007. A statistical theory for sampling species abundances. Ecology Letters 10, 1037–1045.

Van Griensven, A., Meixner, T., Grunwald, S., Bishop, T., Diluzio, M., Srinivasan, R., 2006. A global sensitivity analysis tool for the parameters of multi-variable catchment models. Journal of Hydrology 324 (1–4), 10–23.

Hamby, D.M., 1994. A review of techniques for parameter sensitivity analysis. Environmental Monitoring and Assessment 32 (2), 135–154.

Hong, Y., Hsu, K., Gao, X., Sorooshian, S., 2004. Precipitation estimation from remotely sensed imagery using artificial neural network – cloud classification system. Journal of Applied Meteorology 43 (12), 1834–1853.

Hristopulos, D.T., Elogne, S.N., 2009. Computationally efficient spatial interpolators based on Spartan spatial random fields. IEEE Transactions on Signal Processing 57, 3475–3487.

Hsu, K., Gao, X., Sorooshian, S., Gupta, H.V., 1997. Precipitation estimation from remotely sensed information using artificial neural networks. Journal of Applied Meteorology 36 (No. 9), 1176–1190.

Huffman, G.J., Adler, R.F., Bolvin, D.T., Nelkin, E.J., 2010. The TRMM multi-satellite precipitation analysis (TMPA). In: Hossain, F., Gebremichael, M. (Eds.), Satellite Rainfall Applications for Surface Hydrology. Springer Verlag (Chapter 1), ISBN: 978-90-481-2914-0, 3–22.

Iida, Y., Kubota, T., Iguchi, T., Oki, R., 2010. Evaluating sampling error in TRMM/PR rainfall products by the bootstrap method: estimation of the sampling error and its application to a trend analysis. Journal of Geophysical Research 115. http://dx.doi.org/10.1029/2010JD014257. D22119.

Iman, R.L., Helton, J.C., Campbell, J.E., 1981a. An approach to sensitivity analysis of computer models: Part I – 24. Introduction, input variable selection and preliminary variable assessment. Journal of Quality Technology 13, 174–183.

Iman, R.L., Helton, J.C., Campbell, J.E., 1981b. An approach to sensitivity analysis of computer models: Part II – ranking of input variables, response surface validation, distribution effect and technique synopsis. Journal of Quality Technology 13, 232–240.

Indu, J., Nagesh Kumar, D., 2014. Evaluation of TRMM PR sampling error over a subtropical basin using bootstrap technique, IEEE Transactions on Geoscience and Remote Sensing. IEEE 52 (11), 6870–6881. http://dx.doi.org/10.1109/TGRS.2014.2304466.

Ireland, G., Petropoulos, G.P., Carlson, T.N., Purdy, S., 2015. Addressing the ability of a land biosphere model to predict key biophysical vegetation characterisation parameters with Global Sensitivity Analysis Environmental Modelling & Software, 65, 94–107. http://dx.doi.org/10.1016/j.envsoft.2014.11.010.

Joyce, R.J., Xie, P., 2011. Kalman filter based CMORPH. Journal of Hydrometeorology 12, 1547–1563.

Joyce, R.J., Janowiak, J.E., Arkin, P.A., Xie, P., 2004. CMORPH: a method that produces global precipitation estimates from passive microwave and infrared data at high spatial and temporal resolution. Journal of Hydrometeorology 5, 487–503.

Kamal-Heikman, S., Derry, L.A., Stedinger, J.R., Duncann, C.C., 2007. A simple predictive tool for lower Brahmaputra River basin monsoon flooding. Earth Interactions 11.

Karaska, M.A., Huguenin, R.L., Becham, J.L., Wang, M., Jensen, J.R., Kaufmann, R.S., 2004. AVIRIS measurements of chlorophyll, suspended minerals, dissolved organic carbon, and turbidity in the Neuse River, North Carolina. Photogrammetric Engineering and Remote Sensing 70, 125−133.

Lin, G.F., Chen, G.R., Wu, M.C., Chou, Y.C., 2009. Effective forecasting of hourly typhoon rainfall using support vector machines. Water Resources Research 45, W08440. http://dx.doi.org/10.1029/2009WR007911.

Lahoz, W.A., Walker, S.−E., Dammann, D., 2010. The NILU SURFEX-enkf Land Data Assimilation System. NILU Technical report TR 2/2010, Jan 2010. Available from: http://www.nilu.no.

Larson, K.J., Basagaoglu, H., Marino, M.A., 2001. Prediction of optimal safe groundwater yield and land subsidence in the Los Banos-Kettleman City area, California, using a calibrated numerical simulation model. Journal of Hydrology 242 (1−2), 79−102.

Lenhart, T., Eckhardt, K., Fohrer, N., Frede, H.-G., 2002. Comparison of two different approaches of sensitivity analysis. Physics and Chemistry of the Earth 27, 645−654.

Lindsay, J.B., Dhun, K., 2015. Modelling surface drainage patterns in altered landscapes using LiDAR. International Journal of Geographical Information Science 29, 397−411. http://dx.doi.org/10.1080/13658816.2014.975715.

McKay, M., Beckman, R., Conover, W., 1979. A comparison of three methods for selecting values of output variables in the analysis of output from a computer code. Technometrics 21 (2), 239−245.

McRoberts, R.E., 2009. Diagnostic tools for nearest neighbors techniques when used with satellite imagery. Remote Sensing of Environment 113, 489−499.

Minasny, B., McBratney, A.B., 2006. A conditional Latin hypercube method for sampling in the presence of ancillary information. Computer & Geosciences 32, 1378−1388.

Minasny, B., McBratney, A.B., 2007. Chapter 12. Latin hypercube sampling as a tool for digital soil mapping. In: Lagacherie, P., McBratney, A.B., Voltz, M. (Eds.), Digital Soil Mapping, an Introductory Perspective, Developments in Soil Science, vol. 31. Elsevier, Amsterdam, pp. 153−166.

Minasny, B., McBratney, A.B., 2002. Uncertainty analysis for pedotransfer functions. European Journal of Soil Science 53, 417−430.

Nijssen, B., Lettenmaier, D.P., 2004. Effect of precipitation sampling error on simulated hydrological fluxes and states: anticipating the Global Precipitation Measurement satellites. Journal of Geophysical Research 109. http://dx.doi.org/10.1029/2003JD003497. D02103.

NSSDC, 2014. The NASA Master Directory Held at the NASA Space Science Data Center. http://nssdc.gsfc.nasa.gov/nmc/SpacecraftQuery.jsp.

Pebesma, E., Heuvelink, B.M., 1999. Latin hypercube sampling of Gaussian random fields. Technometrics 41, 303−312.

Petropoulos,, G., Wooster, M.J., Kennedy, M., Carlson, T.N., Scholze, M., 2009c. A global sensitivity analysis study of the 1d SimSphere SVAT model using the GEM SA software. Ecological Modelling, 220 (19), 2427−2440. http://dx.doi.org/10.1016/j.ecolmodel.2009.06.006.

Pincus, R., Klein, S.A., 2000. Unresolved spatial variability and microphysical process rates in large-scale models. Journal of Geophysical Research 105, 27,059−27,065.

Raisanen, P., Barker, H.W., 2004. Evaluation and optimization of sampling errors for the Monte Carlo independent column approximation. Quarterly Journal of the Royal Meteorological Society 130, 2069−2085.

Rotstayn, L.D., 2000. On the "tuning" of autoconversion parameterizations in climate models. Journal of Geophysical Reserach 105, 15495−15508.

Smith, E.A., et al., 2007. International global precipitation measurement (GPM) program and mission: an overview. In: Measuring Precipitation from Space: EURAINSAT and the Future.

Sommeria, G., Deardorff, J.W., 1977. Subgrid-scale condensation in models of nonprecipitating clouds. Journal of the Atmospheric Sciences 34, 344−355.

Thompson, S.K., 1992. Sampling. John Wiley & Sons, Inc, New York, p. 334.

Torn, R.D., Hakim, G.J., 2008. Ensemble-based sensitivity analysis. Monthly Weather Review 136, 663—677.

Wang, J.F., Haining, R.P., Cao, Z.D., 2010. Sampling surveying to estimate the mean of a heterogeneous surface: reducing the error variance through zoning. International Journal of Geographical Information Science 24, 523—543.

Zhang, Y., Pinder, G.F., 2004. Latin-hypercube sample-selection strategies for correlated random hydraulic-conductivity fields. Water Resources Research 39 (Art. No. 1226).

ENSEMBLE-BASED MULTIVARIATE SENSITIVITY ANALYSIS OF SATELLITE RAINFALL ESTIMATES USING COPULA MODEL

S. Moazami[1], S. Golian[2]

Islamshahr Branch, Islamic Azad University, Islamshahr, Tehran, Iran[1]; *Shahrood University of Technology, Shahrood, Iran*[2]

CHAPTER OUTLINE

1. INTRODUCTION

In recent years, the rainfall rate retrieved from remotely sensed fields has drawn a great deal of interest. Satellite-based algorithms that incorporate passive microwave (MW) and infrared (IR) data provide more accurate estimates of precipitation throughout the globe. However, due to the indirect nature of satellite rainfall estimates (SREs), their use is subject to error and uncertainty. Therefore, in this study a method based on copula functions has been developed to adjust the bias of SREs. Indeed, copula is a statistical measure that represents the joint distribution of random variables, each variable being marginally uniformly distributed and examining the association or dependence between many variables. However, the focus of sensitivity analysis in this study is on the prediction uncertainty of the simulated input variables. The consideration of simulation models and the explicit description of the uncertainty around all input variables characterize global sensitivity analysis. Sensitivity analysis was originally created by Tomovic and Vukobratovic (1972) to deal simply with uncertainties in the input variables and model parameters (Saltelli, 1999). Nevertheless, sensitivity analysis can be distinguished from uncertainty analysis. Sensitivity analysis procedures explore and quantify the impact of possible errors in input data on predicted model outputs and system performance indices. Simple sensitivity analysis procedures can be used to illustrate either graphically or numerically the consequences of alternative assumptions about the future. Uncertainty analyses employing probabilistic descriptions of model inputs can be used to derive probability distributions of model outputs and system performance indices (Loucks and Van Beek, 2005).

The observed bias of satellite data is obtained for each event by comparing between rain gauge measurement as reference true data and satellite estimation at each selected pixel. Note that for the biases of different events observed at each pixel, different probability distribution functions (PDFs) may be fitted, and thus the biases of each pixel are considered as an input variable in a multivariate copula model. Copulas can model multivariate random variables with different marginal distributions while also describing the dependence structure of input variables. On the other hand, in this study, the authors used an ensemble-based model, which can provide more accurate delineation of precipitation uncertainty than a single realization (Dai et al., 2014, 2016). An ensemble of precipitation fields consists of a large number of realizations; each realization represents a possible rainfall event (Aghakouchak, 2010). Rainfall realizations are simulated in which satellite estimates are perturbed with a stochastically generated error component. Obviously, generating a more accurate set of random error fields may result in true possible realizations. Copulas that can preserve the spatial dependence of the simulated random variables are a robust tool to generate rainfall error over different pixels concurrently. However, the choice of copula itself plays an important role in this regard and highlights the sensitivity of the model output to changes in each copula parameter. For this purpose, we developed two elliptical copula models named Gaussian copula and t-copula. Therefore, it will be possible to analyze the sensitivity of model output to different types of copula and to also assess the robustness of each copula in adjusting the bias of SREs.

In the current research, since the biases of SREs are derived from different pixels and each pixel is termed as an input variable, the accurate estimate of correlations between variables is also important in the uncertainty analysis (Wang et al., 2010). The marginal effects of input variables on the predicted probabilities of uncertainty are the main disadvantage of classical multivariate distributions such as bivariate normal, log-normal, and gamma in which the modeling of the dependence structure between variables is not independent of the choice of the marginal distributions (Genest and Favre, 2007; Dupuis, 2007). However, the advent of a copula-based method allows to incorporate correlations while also isolating the marginal effects of each input variable.

The copula method for multivariate analysis is somewhat new in hydrology. Favre et al. (2004) mentioned the advantages of using copulas in modeling the dependence between random variables when compared with other traditional methods. De Michele and Salvadori (2003) utilized copulas to model the dependence between rainfall duration and intensity and declared that, in this way, both properties of marginal distributions (assumed to be of generalized Pareto type) and dependence between storm duration and intensity were preserved. Dupuis (2007) modeled the dependence structure of correlated hydrological variables using the copula method. Genest and Favre (2007) reviewed inference methods for copulas based on rank methods. Zhang and Singh (2007) utilized four Archimedean copulas to derive bivariate rainfall frequency distributions. They employed the bivariate distributions to determine bivariate joint and conditional return periods of rainfall data in the Amite River basin, Louisiana, United States. They found that Frank and Ali-Mikhail-Hag families fitted better on data compared with Gumbel—Hougaard and Cook—Johnson families. Shiau (2006) took advantage of two-dimensional copulas to model the joint drought duration and severity distributions at Wushantou station, Southern Taiwan. The author selected the Galambos copula family based on maximum log-likelihood criterion. The copula fitting for drought duration and severity were quite satisfactory. Golian et al. (2012) used the copula method to investigate the joint response of key hydrologic variables, including total precipitation depths and the corresponding simulated peak discharges of a medium-sized watershed for different antecedent soil moisture conditions. Moazami et al. (2014) develop a copula-based ensemble simulation method for analyzing the uncertainty and adjusting the bias of two high-resolution satellite precipitation products (PERSIANN and TMPA-3B42).

2. SATELLITE RAINFALL ESTIMATES

Lack of a reliable and extensive observing system is one of the most important challenges in rainfall analysis, hydrologic predictions, and water resources management in many parts of the world. In fact, the availability of high-quality ground rainfall data is very limited across the most regions of developing countries. As an alternative source, satellite precipitation products, which provide high spatial coverage of input data for various hydrologic models, can be useful for data-sparse and ungauged basins (Moazami et al., 2013).

The satellite rainfall products used in this study are based on precipitation estimation from Remote Sensing Information using Artificial Neural Network (PERSIANN) (Sorooshian et al., 2000) and Tropical Rainfall Measuring Mission (TRMM) Multi-satellite Precipitation Analysis (TMPA)—adjusted product (3B42) (Huffman et al., 2007). The PERSIANN system uses neural network function classification/approximation procedures to compute the estimate of rainfall rate at each 0.25 degree × 0.25 degree pixel of the IR brightness temperature image provided by geostationary satellites. An adaptive training feature facilitates updating of the network parameters whenever independent estimates of rainfall are available. The PERSIANN system was based on geostationary IR imagery and later extended to include the use of both IR and daytime visible imagery. Rainfall products are available from 50°S to 50°N globally (Moazami et al., 2014).

TMPA provides global precipitation estimates from a wide variety of meteorological satellites (Huffman et al., 2010). Indeed, the TMPA estimates are available in the form of two products, a near-real-time version (3B42RT) (about 6 h after real time) covering the global latitude belt from 60°N to 60°S and a gauge-adjusted post—real-time research version (3B42) (approximately 10—15 days after the end of each month) within the global latitude belt ranging between 50°N and 50°S. Both 3B42RT and 3B42 have 3-h temporal and 0.25 degree × 0.25 degree spatial resolution. The 3B42RT uses the TRMM Combined Instrument data set, which includes the TRMM precipitation radar and the TRMM Microwave Imager, to calibrate precipitation estimates derived from available Low Earth Orbit MW radiometers. The 3B42RT then merges all the estimates at 3-h intervals; and the gaps in the analyses are filled using Geostationary Earth Orbit IR data regionally calibrated to the merged MW product. The 3B42 adjusts the monthly accumulations of the 3-h fields from 3B42RT based on a monthly gauge analysis, including the Global Precipitation Climatology Project (Gebremichael et al., 2005) 1 degree × 1 degree monthly rain gauge analysis and the Climate Assessment and Monitoring System 5 degree × 5 degree monthly rain gauge analysis (Jiang et al., 2012). Note that the daily satellite rainfall data employed in the current work are computed by aggregating 3-h temporal resolution data over 24 hours for both PERSIANN and TMPA-3B42 products (Moazami et al., 2014).

3. METHODOLOGY OF ENSEMBLE-BASED MULTIVARIATE ANALYSIS

Due to the large space−time variability of precipitation, bias correction of the ensemble forecasts has been proved to be a challenging task. Hence, ensemble-based statistical approaches have been used in the correction of bias and generation of reliable forecast in hydrology. Fig. 14.1 displays a general diagram of the method proposed in this chapter, and the steps in brief are as follows.

3.1 CALCULATING BIASES

For each rainfall event, the difference between rain gauge measurements as reference surface rainfall data and SRE at each pixel is considered and termed as observed bias:

$$Bias_{i,j} = (R_{obs})_{i,j} - (R_{SRE})_{i,j} \tag{14.1}$$

where $Bias_{i,j}$ is the calculated bias at the ith pixel and rainstorm j and $(R_{obs})_{i,j}$ and $(R_{SRE})_{i,j}$ are the observed rainfall and rainfall from satellite, respectively, at the ith pixel and rainstorm j.

3.2 FITTING MARGINAL DISTRIBUTION FUNCTION

Considering the calculated biases at each pixel, a PDF was fitted based on the maximum likelihood (ML) method for parameter estimation. Best distribution function was selected based on Kolmogrov−Smirnov (K-S) goodness-of-fit test.

The K-S test is defined as a nonparametric test to measure the distance between the empirical and calculated CDF (cumulative distribution function) (Stephens, 1974):

$$M = M\{|E(x) - R(x)|\} \tag{14.2}$$

where $E(x)$ and $R(x)$ are the empirical and calculated CDFs, respectively. The null hypothesis for this test assumes that the variable is a part of the calculated distribution.

Other tests, e.g., Anderson−Darling (A-D) and Akaike Information criterion (AIC) can also be used for selecting the best distribution function.

3.3 JOINT DISTRIBUTION FUNCTION OF BIASES USING COPULA

3.3.1 Copulas

Copula is a very useful function to implement efficient algorithms for simulating joint distributions in a more realistic way. In fact, copulas are able to model the dependence structure independent of the

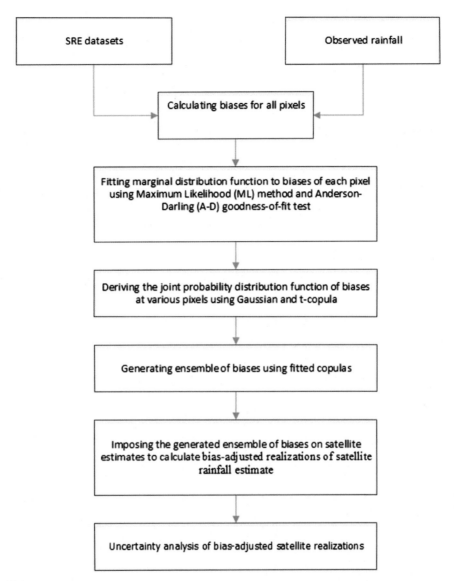

FIGURE 14.1

Schematic diagram for the methodology of multivariate bias analysis using the copula method.

marginal distributions. It is then possible to build multidimensional distributions with different margins, the structure of dependence being mathematically formalized through the copula (Favre et al., 2004).

In a case with two random variables, Sklar's theorem (Sklar, 1959) states that if $H(x, y)$ is a joint distribution function of random variables, namely, X and Y with marginal distributions $F_X(x)$ and $F_Y(y)$, respectively, then there exists a copula such that:

$$H(x, \ y) = C\{F_X(x), \ F_Y(y)\} \quad x, \ y \in R \tag{14.3}$$

where $C:[0, \ 1]^2 \to [0, \ 1]$ is a copula. If $F_X(x)$ and $F_Y(y)$ be continuous, then C is unique. Conversely, for any univariate distribution $F_X(x)$ and $F_Y(y)$ and any copula C, the function $H(x, \ y)$ defined the previous equation is a two-dimensional distribution function with $F_X(x)$ and $F_Y(y)$ as the marginal distributions. The detailed proof of Sklar's theorem can be found in Schweizer and Sklar (1983).

Under the assumption that the marginal distributions are continuous with probability density functions $f_X(x)$ and $f_Y(y)$, the joint probability density function then becomes:

$$h_{X,Y}(x, \ y) = c(f_X(x), \ f_Y(y)) \cdot f_X(x) f_Y(y) \tag{14.4}$$

where c is the density function of C, defined as

$$c(u, \ v) = \frac{\partial^2 C(u, \ v)}{\partial u \cdot \partial v}, \tag{14.5}$$

Also, for two random variables, namely, X and Y, the joint probability $\Pr(X \geq x, Y \geq y)$ is calculated as follows:

$$\Pr(X \geq x, \ Y \geq y) = 1 - F_X(x) - F_Y(y) + C(F_X(x), \ F_Y(y)) \tag{14.6}$$

The conditional distribution function for V given $U = u$ is denoted by $C_u(v)$ and is expressed by:

$$C_u(v) = P[V \leq v | U \leq u] = \frac{\partial C(u, \ v)}{\partial u} \tag{14.7}$$

Also, the empirical copula frequency function can be calculated mathematically by:

$$\widehat{C}_i = \frac{number \ of \ (x_j, \ y_j) \ such \ that \ x_j < x_i \ and \ y_j < y_i}{n - 1} \tag{14.8}$$

$$i, j = 1, 2, \ldots, n$$

where n is the number of observation.

Joe (1997), Nelsen (2006) and Cherubini et al. (2004) provided a number of one-parameter families of copulas.

There are several methods to estimate copula parameters including the ML approach, pseudo-ML method, and inference function of marginal distribution (IFM). To estimate copula parameter, θ, the IFM method proposed by Joe (1997) can be employed (Nelsen, 2006; Zhang and Singh, 2007). First, parameters of univariate marginal distributions are estimated using the ML method, then estimation of the copula dependence parameters is performed as follows:

$$\ln L\big[f_{X,Y}(x, \ y; \ \alpha, \ \beta, \ \theta)\big] = \ln \sum_{i=1}^{n} c_\theta(F_X(x_i; \ \alpha), F_Y(y_i; \ \beta); \ \theta) + \sum_{i=1}^{n} \big[\ln f_X(x_i; \ \alpha) + \ln f_Y(y_i; \ \beta)\big]$$

$$\tag{14.9}$$

where α and β are parameters of $f_X(x)$ and $f_Y(y)$ marginal distributions, θ is the $C_\theta(F_X(x), F_Y(y))$ copula parameter, and n is the number of observations.

For selection of the best copula in modeling-dependent structure, the AIC (Akaike, 1974) criterion was adopted (Shiau, 2006; Zhang and Singh, 2007).

3.3.2 Gaussian Copula

The multivariate Gaussian (normal) copula (MGC) method in which the dimensionless random numbers are generated simultaneously in all spatially dependent regions is expressed as the following equation:

$$C_R(\overrightarrow{u}) = \varphi_R\big(\varphi^{-1}(u_1),\ \varphi^{-1}(u_2), \ldots, \varphi^{-1}(u_n)\big) \tag{14.10}$$

where φ_R is the standardized multivariate normal distribution with correlation matrix R and $\varphi^{-1}(u_i)$ is the inverse of the standard univariate normal distribution function φ.

3.3.3 t-Copula

The t-copula, also known as Student copula, is an elliptical copula based on Student distribution. The t-copula can be thought of as representing the dependence structure implicit in a multivariate t-distribution. It is a model that has received much attention, particularly in the context of modeling multivariate hydrological data (for example, see Aghakouchak et al., 2010a,c). A d-dimensional t-copula is defined on a d-dimensional random vector $\overrightarrow{X} = (X_1,\ X_2, \ldots,\ X_d)'$ as follows:

$$C_{v,p}^t(u) = \int_{-\infty}^{t_v^{-1}(u_1)} \cdots \int_{-\infty}^{t_v^{-1}(u_d)} \frac{\Gamma\left(\frac{v+d}{2}\right)}{\Gamma\left(\frac{v}{2}\right)\sqrt{(\pi v)^d |P|}} \left(1 + \frac{x'P^{-1}x}{v}\right)^{-\frac{v+d}{2}} dx \tag{14.11}$$

where v is the degree of freedom, P is the correlation matrix implied by the dispersion matrix $\sum$, and t_d^{-1} is the inverse of a standard univariate t-student (t_v) distribution.

3.4 ENSEMBLE BIAS SIMULATION

Let C^n be the copula of a multivariate n-dimensional distribution $G \equiv (G_1, \ldots, G_n)$ where $G_1, \ldots, G_n$ are the marginal distributions. To obtain a simulated field of $x \equiv (x_1, \ldots, x_n)$ with marginals of $G_1, \ldots, G_n$, the following three steps are required:

1. Estimate the parameter of the copula C^n.
2. Simulate uniform random variables $y(y_1, \ldots, y_n)$ using the copula C^n.
3. Transform the univariate marginals to $G_1, \ldots, G_n$ using Sklar's theorem (Sklar, 1959):
   $x_i = G_i^{-1}(y_i)$.

It is worth pointing out that $G_1, \ldots, G_n$ do not necessarily need to have the same distribution family (Aghakouchak, 2010). As mentioned before, since copula is invariant to transformations (one of the main advantages of copula), the simulated random variables will have the same spatial dependence structure as those of the input data (Moazami et al., 2014).

3.4.1 Generating an Ensemble of Satellite Rainfall Realizations

In this research, the copula-based model utilized the observed biases as input data. These biases were extracted from the comparison between SRE and ground station of daily rainfall events over 20 study pixels. Since the bias at each pixel can be considered as a variable, a multivariate 20-dimensional Gaussian copula and a t-copula were implemented to generate multiple random bias fields. A parametric approximation that uses A-D goodness-of-fit test was applied to fit the best PDF to each variable as its marginal distribution. Additionally, the ML Estimation method was used to calculate the parameters of each distribution.

Using MGC and t-copulas, ensembles of bias fields were generated randomly for each pixel. Note that the computed linear correlation matrix of the transformed marginals can simulate dependent random variables. Then, the outlier data were removed from the generated biases using the Mahalanobis distance as a well-known criterion for the detection of multivariate outliers. [Regarding the Mahalanobis distance method, interested readers are referred to the original publications discussed by Werner (2003), Hodge and Austin (2004), Ben-Gal (2005), and Filzmoser et al. (2005)]. Thereafter, the generated biases were imposed on the original satellite rainfall estimate (OSRE) (before any adjustment of bias through the developed model in the current study) to obtain an ensemble of bias-adjusted realizations of SRE (BASRE) at each pixel.

The following equation describes the general formulation of the proposed method:

$$P_{BASRE_i} = P_{OSRE_i} + GBias_i \tag{14.12}$$

where P_{BASRE_i} is an ensemble of bias-adjusted satellite rainfall realizations at ith pixel, P_{OSRE_i} is the OSRE at the ith pixel, and $GBias_i$ is the randomly generated bias fields based on observed biases at the ith pixel. It should be noted that in this study, all the selected daily rainfall events show a positive value of observed rainfall; however, the satellite data can be either positive or zero. Also, if the rainfall value becomes negative after bias adjustment, it will be set to zero (Moazami et al., 2014).

3.5 UNCERTAINTY ANALYSIS PROCEDURE

The procedure implemented here to analyze the uncertainty associated with satellite estimates is based on ensemble simulation. The model output sensitivity is quantified by the 80% prediction uncertainty band (80PPU) calculated at the 10% lower and 90% upper limit levels of the simulated ensemble. The 80PPU band means that 80% of observed data fall within this band and 20% of the inappropriate simulations are not considered. The 80PPU band was selected after a trade-off between different percentages.

In stochastic simulations where predicted output is given by a prediction uncertainty band instead of a signal, two different indices, including P-factor and R-factor can be used to compare observations with simulations (Schuol et al., 2008). In the current work, P- and R-factors are obtained to assess the accuracy of the simulated ensembles with respect to the observed data. The P-factor is the percentage of pixels in which the observed values are bracketed by the 80PPU band. The maximum value of the P-factor is 100%, which ideally brackets all the observed data in the 80PPU band. The R-factor is the ratio of the average distance between the 10th and 90th percentiles and the standard deviation of the corresponding measured variable (Abbaspour et al., 2007). The desired value of R-factor in practical studies is 1, which implies that the simulations are in best agreement with the observations. The R-factor is calculated as:

$$R - \text{factor} = \frac{\frac{1}{k} \sum_{i=1}^{k} (Y_U - Y_L)_i}{\sigma_{obs}} \tag{14.13}$$

where Y_U and Y_L represent the upper (90%) and lower (10%) boundaries, respectively, of the 80PPU; k is the number of pixels (here 20); and σ_{obs} indicates the standard deviation of the observed rainfall value of a single daily event for all the 20 pixels (Abbaspour et al., 2007).

The goodness of calibration and prediction uncertainty is judged based on the closeness of the average value of P-factor to 100% and that of the R-factor to 1 (Schuol et al., 2008; Yang et al., 2008). The sensitivity of each P- and R-factor depends on the randomly generated biases via copula. Thus a reasonable set of random generation results in a better pair of P- and R-factors. As in this research there were several daily events of input data, different pairs of P- and R-factors have been created, each of them associated with a simulated ensemble of an individual event. Then the average value of each factor was obtained and considered as a tool to evaluate the robustness of the proposed model. It is noted that the aforementioned steps were implemented for several sets of randomly generated bias fields and then an appropriate set among them was selected based on the best values of both P- and R-factors.

4. APPLICATION (CASE STUDY) AND RESULTS

To elaborate the application of the copula method for uncertainty analysis of SRE products, we used daily rain gauge observations provided by Iran Water Resources Management Co. in Khuzestan Province in the southwest part of Iran. Fig. 14.2A and B shows the location of Khuzestan Province in Iran and the rain gauges distributed on it, respectively. This region was selected due to its different topography and climate conditions. Khuzestan has a total area of 63,238 km^2 and contains more than 30% of the total surface water resources of the country. Moreover, the geography of Khuzestan encompasses terrain from the plains in the southern to the mountains in the northern parts of the province. Over this area the average value of rainfall has been reported in the range between 250 and 400 mm.

Khuzestan consists of around 100 pixels (corresponding to the PERSIANN and TMPA pixels); also, there are 80 rain gauges across this region. To determine the appropriate pixels for this study, a set of 20 pixels with the largest number of reference ground data associated with the daily rainfall events during the study period (2003–06) was selected (Fig. 14.2C) (Moazami et al., 2014).

The main purpose of this research is to develop a model for uncertainty analysis of biases from two widely used SRE products taking into account the spatial dependency between variables of different pixels. In addition, the proposed model applies an ensemble-based approach, which would lead to a sensitivity analysis method based on the use of copulas. The evaluation is implemented during a period of 3 years (2003–06) including six rainy months (winter and spring seasons) in each year over the twenty 0.25 degree × 0.25 degree pixels.

4.1 EVALUATION OF BIAS-ADJUSTED SATELLITE RAINFALL ESTIMATES

To evaluate the accuracy of the simulated ensembles, three types of continuous statistical indices, including Bias, RMSE (root mean square error), and CC (correlation coefficient) were employed.

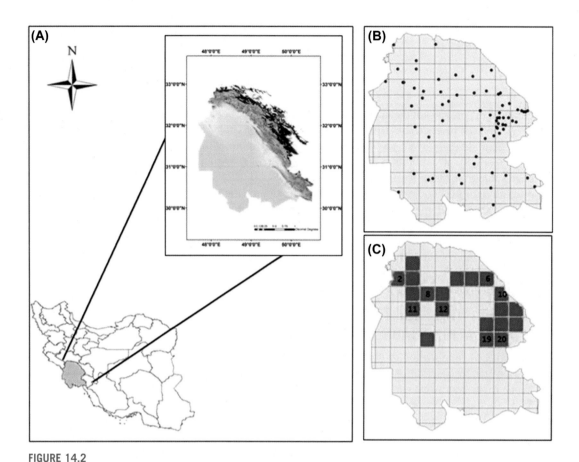

FIGURE 14.2

Study area: (A) Study area of Khuzestan province, (B) location of rain gauges and satellite pixels over the study area, and (C) selected satellite pixels for simulation (Moazami et al., 2014).

These indices were obtained for both original (OSRE) and bias-adjusted realizations (BASRE) of satellite precipitation products separately by comparing them with the rain gauge observations as true data. Note that the BASRE was derived based on both Gaussian and t-copulas. Figs. 14.3 and 14.4 provide the values of the three indices for the PERSIANN and TMPA-3B42 products, respectively. In these figures the values of three indices associated with the OSRE are the average values of input daily events over each pixel, and the values associated with the BASRE are the average values of 50% quantiles of bias-adjusted ensembles (each ensemble with around 1000 members has a value of 50% quantile) over each pixel. As shown, for all three indices the estimates of both PERSIANN and TMPA-3B42 are improved after bias adjustment. Also, it can be seen that t-copula performs slightly better compared with Gaussian copula. Table 14.1, in addition, indicates that the ensembles generated by t-copula improve the satellite estimates more than those generated by Gaussian copula for both PERSIANN and TMPA-3B42 products.

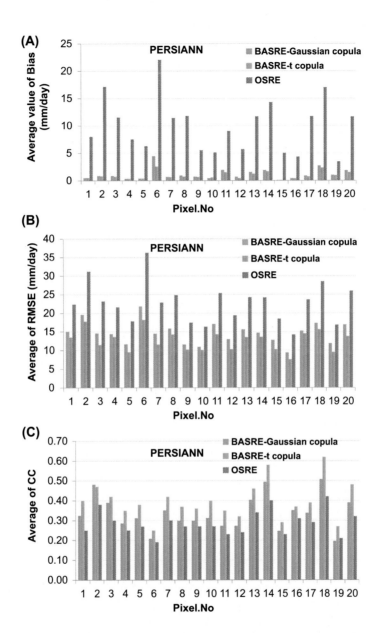

FIGURE 14.3

Comparison between OSRE and BASRE (the 50% quantile of simulated realizations) using Gaussian and t-copulas; the average values of (A) Bias, (B) RMSE, and (C) CC for input daily rainfall events over each pixel for PERSIANN product. *BASRE*, bias-adjusted realizations of satellite rainfall estimate; *CC*, correlation coefficient; *OSRE*, original satellite rainfall estimate; *RMSE*, root mean square error.

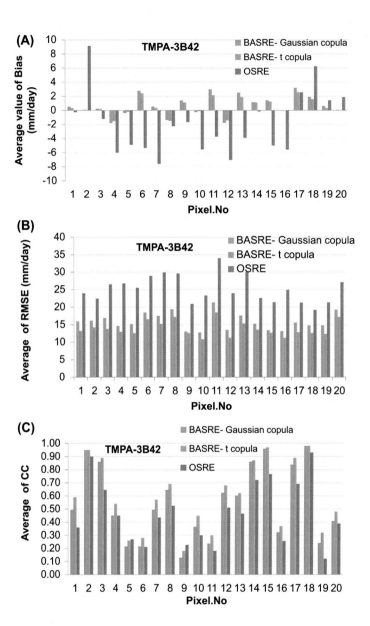

FIGURE 14.4

Comparison between OSRE and BASRE (the 50% quantile of simulated realizations) using Gaussian and t-copulas; the average values of (A) Bias, (B) RMSE, and (C) CC for input daily rainfall events over each pixel for TMPA-3B42 product. *BASRE*, bias-adjusted realizations of satellite rainfall estimate; *CC*, correlation coefficient; *OSRE*, original satellite rainfall estimate; *RMSE*, root mean square error.

Table 14.1 Comparison of the Average Values of Statistical Indices of Input Daily Rainfall Events Over 20 Pixels for the OSRE and BASRE (the 50% Quantile of Simulated Realizations) Using Gaussian and t-Copulas.

	Satellite Products	Bias (mm/day) (OSRE)	Bias (mm/day) (BASRE)	Improved Bias (%)	RMSE (mm/day) (OSRE)	RMSE (mm/day) (BASRE)	Improved RMSE (%)	CC (OSRE)	CC (BASRE)	Improved CC (%)
Gaussian copula	PERSIANN	10.01	1.16	88.41	22.73	14.68	35.42	0.29	0.34	17.24
	TMPA-3B42	−1.95	0.70	64.10	25.18	15.95	36.66	0.47	0.54	14.89
t-copula	PERSIANN	10.01	0.95	90.51	22.73	12.59	44.61	0.29	0.40	37.93
	TMPA-3B42	−1.95	0.54	72.31	25.18	13.84	45.04	0.47	0.59	25.53

BASRE, bias-adjusted realizations of satellite rainfall; CC, correlation coefficient; OSRE, original satellite rainfall estimate; RMSE, root mean square error.

4.2 COPULA-BASED BIAS SIMULATION

In this study, a 20-dimensional MGC and t-copula were utilized to simulate the ensembles of biases from SREs for 20 pixels. Copulas are invariant to monotonic transformations, and thus the simulated random biases will have the same spatial dependence structure as that of the observed biases. In order to display this feature of copula, scatterplots of the observed and copula-based randomly generated biases for two pixels with the highest and lowest CC are compared in Figs. 14.5 and 14.6. As was revealed in

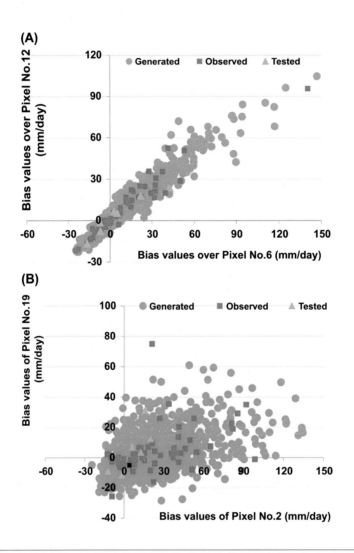

FIGURE 14.5

Scatterplots of bias pairs for the input observed events (*red symbols*), the six tested events (*green symbols*), and the generated samples by t-copula (*blue symbols*) associated with PERSIANN product over (A) two pixels with the highest value of the correlation coefficient (CC), and (B) two pixels with the lowest value of the CC.

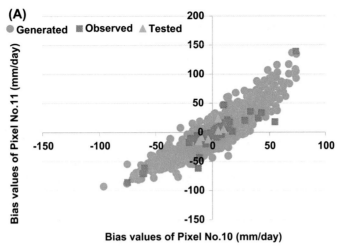

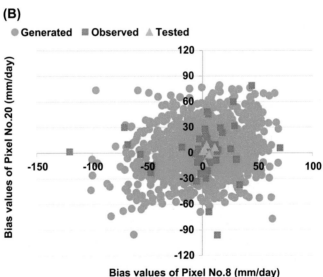

FIGURE 14.6

Scatterplot of bias pairs for the input observed events (*red symbols*), the six tested events (*green symbols*), and the generated samples by t-copula (*blue symbols*) associated with TMPA-3B42 product over (A) two pixels with the highest value of the correlation coefficient (CC), and (B) two pixels with the lowest value of the CC based on the observed events.

the previous section, t-copula performed better than Gaussian and hence its results are demonstrated. Scatterplots have been provided for two pixels with the highest and also two pixels with the lowest CCs among 20 study pixels. As shown in these figures, the correlations between the observed biases (marked with red symbols) are reasonably preserved in the generated biases (marked with blue symbols) for both highest and lowest values of CC between two pixels. Furthermore, the scatterplots of the bias values for the six tested events, which have not been included in the input data, are presented in Figs. 14.5 and 14.6 (marked with green symbols). The results presented here indicate that the t-copula can simulate reasonable ensembles over pixels with different dependencies as the correlation of generated data is similar to that of the observed one.

4.3 TESTING THE DEVELOPED MODEL AND UNCERTAINTY ANALYSIS

The robustness of copula-based model proposed in this study was tested for events that have not been considered as input data in ensemble simulation. For this purpose, an appropriate set of generated biases (based on its P- and R-factors) was imposed on the original form of satellite estimates to simulate an ensemble of bias-adjusted form. Table 14.2 displays the computed P- and R-factors of six tested events for PERSIANN and TMPA separately. In addition, Fig. 14.7 shows the result of the simulated rainfall ensembles using Gaussian and t-copulas for PERSIANN and TMPA-3B42 over events 3 and 5 in all 20 study pixels. In these figures, the solid red lines show the OSREs, the solid blue lines express the rainfall values derived from rain gauges, the gray areas represent the 80PPU bands of the simulated ensembles of BASREs, and the solid green lines indicate the 50% quintiles of simulated realizations. As shown, the BASREs using Gaussian and t-copulas (gray bands) reasonably encompass the ground reference measurements (solid blue lines). However, t-copula performs better estimation than Gaussian for both satellite products.

The results obtained here were based on an approach that was similar to that of Aghakouchak et al. (2010a—c) since it made use of the copula technique to generate an ensemble of rainfall realizations. However, Aghakouchak et al. (2010a—c) aimed to apply copulas in simulating the multivariate error fields of radar rainfall (RR) estimates, although the intention of this study was to correct the bias of satellite precipitation estimates probabilistically. In addition, Dai et al. (2014, 2016) analyzed the uncertainties associated with RR estimates based on the t-copula and proposed ensemble generator for RR using copula and autoregressive model. Also, some literature have dealt with the uncertainties of SRE algorithms. Hong et al. (2006b) developed a methodology based on Monte Carlo approach to quantify satellite-based precipitation estimation error characteristics and to assess the influence of the error propagation on hydrological simulation. Liu et al. (2015) proposed a parametric joint PDF model to analyze and study the 9-year satellite-derived precipitation data sets of CMORPH, PERSIANN, and the real-time TRMM product 3B42, version 7 (TRMM-3B42-RTV7). The aim of their study was to investigate how the error—intensity relationship varies with the season or region. Hossain and Anagnostou (2006) developed a two-dimensional satellite rainfall error model (SREM2D) for simulating ensembles of satellite rain fields. They generated random error fields of SREs by Monte Carlo simulation of given realizations. Teo and Grims (2007) described an approach for estimating the uncertainty of satellite-based rainfall values using ensemble generation of rainfall fields based on stochastic calibration. They obtained the correct spatial correlation structure within each ensemble member by the use of geostatistical sequential simulation. Taking into account the aforementioned studies, one can see that the application of copulas is very limited in characterizing the bias of SREs.

Table 14.2 Simulated Ensembles *P*- and *R*-factors of the Six Tested Events for Both PERSIANN and TMPA-3B42 Products Using Gaussian and t-Copulas

Satellite Products		PERSIANN						TMPA-3B42			
	Tested Events	Event 1	Event 2	Event 3	Event 4	Event 5	Event 6	Event 1	Event 2	Event 3	Event 4
Gaussian copula	P-factor	65%	65%	80%	75%	50%	65%	80%	70%	80%	70%
	R-factor	2.16	1.28	1.14	1.35	1.07	1.59	1.94	1.57	1.12	1.22
t-copula	P-factor	70%	75%	85%	80%	65%	75%	85%	80%	95%	80%
	R-factor	2.37	1.12	1.14	1.4	1.35	2.2	1.84	1.37	1.21	1.22

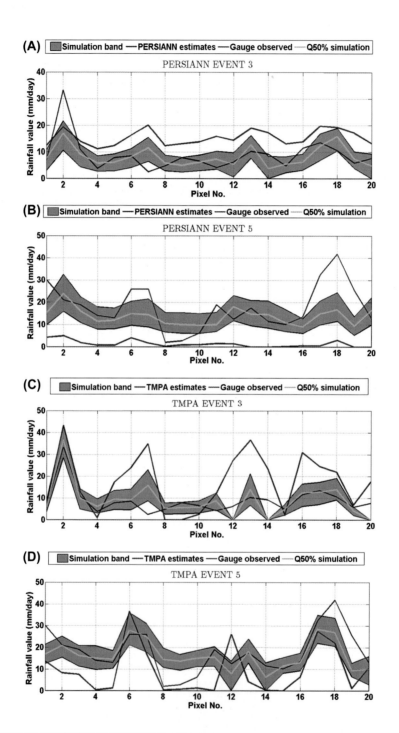

FIGURE 14.7

Comparison between satellite estimates (*red line*), rain gauge observations (*blue line*), and 80% confidence band associated with bias-adjusted rainfall estimates (*gray band*) of the third and fifth tested daily events over 20 study pixels. (A, B) Gaussian copula for PERSIANN, (C, D) Gaussian copula for TMPA-3B42, (E, F) t-copula for PERSIANN, and (G, H) t-copula for TMPA-3B42.

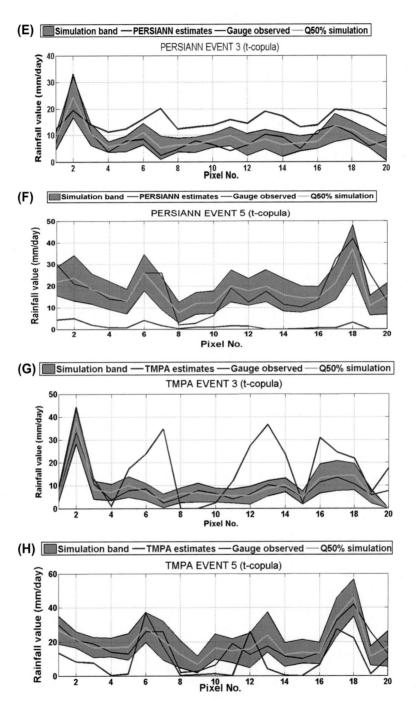

FIGURE 14.7 Cont'd

5. CONCLUSIONS AND FUTURE DIRECTIONS

Since uncertainties of precipitation as a major input data can propagate into hydrological and meteorological models, the efficiency of hydrological modeling and also successive planning of water resources definitely depend on the accuracy of rainfall estimates. Detecting the spatial and temporal variability of precipitation is crucial for meteorological prediction; however, the conventional rain gauges and ground-based radar due to sparse network are not able to provide uninterrupted data, particularly over remote regions and complex terrain. Therefore, satellite-based techniques that provide extended precipitation coverage can be appropriate complements to rain gauge and radar measurements and are increasingly applied to atmospheric and hydrological applications at different space—time scales (Hong et al., 2006a,b). However, the indirect nature of satellite estimates makes using them subject to evaluation and uncertainty analysis.

In this chapter, a probability method based on simulation ensembles was proposed, which aimed to adjust the bias and analyze the uncertainty of SREs using multivariate copulas. The main advantage of copulas is to preserve the spatial dependency among variables independent of their marginals, which is useful in simulation of multivariate random fields. It is to be noted that the copula model output is almost sensitive to the PDF of input variables, copula parameters, and also different kinds of copula family. Thus, the sensitivity analysis of each randomly simulated realization will lead to the selection of an appropriate ensemble that has the most similarity to the observed true data.

In this study, to measure quantitatively the uncertainty of the simulated realizations, two factors (P and R) were computed for each ensemble by comparing it with the observed data. Then, these factors were calibrated in a trade-off process to more accurately generate 80PPU band of realizations. With respect to the three statistical indices of Bias, RMSE, and CC to evaluate the performance of the realizations, the model (particularly t-copula) is robust in improving the satellite estimates. However, in future, it would be interesting to analyze the technique sensitivity to different input variables such as various types of precipitation (frontal, stratiform, orographic, and convective) that can occur in different climatic conditions during a special season. Also, assessing the sensitivity of the model to a number of rain gauges located at any pixel and alleviating the effect of gauge uncertainties for evaluating different satellite precipitation products would be of great interest.

Finally, with respect to the extreme events, i.e., floods and droughts, using ensemble-based models can result in extreme prediction uncertainty and its associated risks for a specified precipitation (Moazami et al., 2014).

REFERENCES

Abbaspour, K.C., Yang, J., Maximov, I., Siber, R., Bogner, K., Mieleitner, J., Zobrist, J., Srinivasan, R., 2007. Modelling hydrology and water quality in the pre-alpine/alpine Thur watershed using SWAT. Journal of Hydrology 333, 413—430.

AghaKouchak, A., 2010. Simulation of Remotely Sensed Rainfall Fields Using Copulas. Institut für Wasserbau der Universität Stuttgart.

Aghakouchak, A., Habib, E., Bárdossy, A., 2010a. A comparison of three remotely sensed rainfall ensemble generators. Atmospheric Research 98 (2—4), 387—399.

AghaKouchak, A., Bárdossy, A., Habib, E., 2010b. Conditional simulation of remotely sensed rainfall data using a non-Gaussian v-transformed copula. Advances in Water Resources 33, 624—634.

AghaKouchak, A., Bárdossy, A., Habib, E., 2010c. Copula-based uncertainty modelling: application to multi-sensor precipitation estimates. Hydrological Processes 24, 2111−2124. http://dx.doi.org/10.1002/hyp.7632.

Akaike, H., December 1974. A new look at the statistical model identification. IEEE Transactions on Automatic Control AC-19 (6).

Ben-Gal, I., 2005. Outlier detection. In: Maimon, O., Rockach, L. (Eds.), Data Mining and Knowledge Discovery Handbook: A Complete Guide for Practitioners and Researchers. Kluwer Academic Publishers, ISBN 0-387-24435-2.

Cherubini, U., Luciano, E., Vecchiato, W., 2004. Copula Methods in Finance. John Wiley & Sons Ltd, The Atrium, Southern Gate, Chichester, West Sussex PO19 8SQ, England, ISBN 0-470-86344-7.

Dai, Q., Han, D., Rico-Ramirez, M., Srivastava, P.K., 2014. Multivariate distributed ensemble generator: a new scheme for ensemble radar precipitation estimation over temperate maritime climate. Journal of Hydrology 511, 17−27.

Dai, Q., Han, D., Zhuo, L., Zhang, J., Islam, T., Srivastava, P.K., 2016. Seasonal ensemble generator for radar rainfall using copula and autoregressive model. Stochastic Environmental Research and Risk Assessment 30 (1), 27−38.

De Michele, C., Salvadori, G., 2003. A Generalized Pareto intensity-duration model of storm rainfall exploiting 2-Copulas. Journal of Geophysical Research 108 (D2), 4067−4077. http://dx.doi.org/10.1029/2002JD002534.

Dupuis, D.J., 2007. Using copulas in hydrology: benefits, cautions, and issues. Journal of Hydrologic Engineering 12, 381−393. http://dx.doi.org/10.1061/_ASCE_1084−0699_2007_12:4_381.

Favre, A.C., El Adlouni, S., Perreault, L., Thie'monge, N., Bobe'e, B., 2004. Multivariate hydrological frequency analysis using copulas. Water Resources Research W01101. http://dx.doi.org/10.1029/2003WR002456.

Filzmoser, P., 2005. A Multivariate Outlier Detection Method. Department of Statistics and Probability Theory, Vienna, Austria. Technical report.

Gebremichael, M., Krajewski, W.F., Morrissey, M.L., Huffman, G.J., Adler, R.F., 2005. A detailed evaluation of GPCP one-degree daily rainfall estimates over the Mississippi River Basin. Journal of Applied Meteorology and Climatology 44 (5), 665−681.

Genest, C., Favre, A.-C., 2007. Everything you always wanted to know about copula modeling but were afraid to ask. Journal of Hydrologic Engineering 12 (4), 347−368.

Golian, S., Saghafian, B., Farokhnia, A., 2012. Copula-based interpretation of continuous rainfall−runoff simulations of a watershed in northern Iran. Canadian Journal of Earth Sciences 49, 681−691.

Hodge, V.J., Austin, J., 2004. A Survey of Outlier Detection Methodologies. Kluwer Academic Publishers. Dept. of Computer Science, University of York, York, UK.

Hong, Y., Adler, R., Huffman, G., 2006a. Evaluation of the potential of NASA multi-satellite precipitation analysis in global landslide hazard assessment. Geophysical Research Letters 33, L22402. http://dx.doi.org/10.1029/GL028010.

Hong, Y., Hsu, K., Moradkhani, H., Sorooshian, S., 2006b. Uncertainty quantification of satellite precipitation estimation and Monte Carlo assessment of the error propagation into hydrologic response. Water Resources Research 42, W08421. http://dx.doi.org/10.1029/WR004398.

Hossain, F., Anagnostou, E.N., June 2006. A two-dimensional satellite rainfall error model. IEEE Transactions on Geoscience and Remote Sensing 44 (6), 1511.

Huffman, G., Adler, R., Bolvin, D., Gu, G., Nelkin, E., Bowman, K., Stocker, E., Wolff, D., 2007. The TRMM multi-satellite precipitation analysis: quasi global, multi-year, combined-sensor precipitation estimates at fine scale. Journal of Hydrometeorology 8, 38−55.

Huffman, G.J., Adler, R.F., Bolvin, D.T., Nelkin, E.J., 2010. The TRMM multisatellite precipitation analysis (TMPA). In: Hossain, F., Gebremichael, M. (Eds.), Chapter in Satellite Applications for Surface Hydrology. Springer.

Jiang, S., Ren, L., Hong, Y., Yong, B., Yang, X., Yuan, F., Ma, M., 2012. Comprehensive evaluation of multi-satellite precipitation products with a dense rain gauge network and optimally merging their simulated hydrological flows using the Bayesian model averaging method. Journal of Hydrology 452−453, 213−225.

Joe, H., 1997. Multivariate Models and Dependence Concepts. Chapman Hall, London.

Liu, Y., Jimenez, E., Hussaini, M.Y., Ökten, G., Goodrick, S., 2015. Parametric uncertainty quantification in the Rothermel model with randomized quasi-Monte Carlo methods. International Journal of Wildland Fire 24, 307–316.

Loucks, D.P., Van Beek, E., 2005. Water Resources Systems Planning and Management. © UNESCO, ISBN 92-3-103998-9.

Moazami, S., Golian, S., Kavianpour, M.R., Hong, Y., 2013. Comparison of PERSIANN and V7 TRMM multi-satellite precipitation analysis (TMPA) products with rain gauge data over Iran. International Journal of Remote Sensing 34 (22), 8156–8171. http://dx.doi.org/10.1080/01431161.2013.833360.

Moazami, S., Golian, S., Kavianpour, M.R., Hong, Y., 2014. Uncertainty analysis of bias from satellite rainfall estimates using copula method. Atmospheric Research 137, 145–166.

Nelsen, R., 2006. An Introduction to Copulas (Springer Series in Statistics). Springer Verlag, New York.

Saltelli, A., 1999. Evaluation of Sensitivity and Uncertainty Analysis Methods in a Quality Assessment Framework with Application to Environmental and Business Statistics. Lot 14 of EUROSTAT SUP COM 1997, JRC reference: Administrative Arrangement, 13592-1998-01 A1CA ISP LU, Final Report, Prepared by JRC-ISIS-SAIE-UASA, June, 17-23, 1999, European Commission, Joint Research Centre, Institute for Systems, Informatics and Safety, T.P. 361 – 21020 Ispra (VA) – Italy.

Schuol, J., Abbaspour, K.C., Yang, H., Srinivasan, R., Zehnder, A.G.B., 2008. Modeling blue and green water availability in Africa. Water Resources Research 44, W07406. http://dx.doi.org/10.1029/2007WR006609.

Schweizer, B., Sklar, A., 1983. Probabilistic Metric Spaces. Elsevier Science Publishing Co., Inc. Republished in 2005 by Dover Publications, Inc. with a new preface, errata, notes, and supplementary references.

Shiau, J.T., 2006. Fitting drought duration and severity with two dimensional copulas. Water Resources Management 20 (5), 795–815. http://dx.doi.org/10.1007/s11269-005-9008-9.

Sklar, A., 1959. Fonctions de R'epartition 'a n Dimensions et Leurs Marges, 8. Publications de l'Institut de Statistique de L'Universit'e de, Paris, pp. 229–231.

Sorooshian, S., Hsu, K., Gao, X., Gupta, H.V., Imam, B., Braithwaite, D., 2000. Evaluation of the PERSIANN system satellite-based estimates of tropical rainfall. Bulletin of the American Meteorological Society 81 (9), 2035–2046.

Stephens, M.A., 1974. EDF statistics for goodness of fit and some comparisons. Journal of the American Statistical Association 69, 730–737.

Teo, C.K., Grimes, D.I.F., 2007. Stochastic modelling of rainfall from satellite data. Journal of Hydrology 346 (1–2), 33–50.

Tomovic, R., Vukobratovic, M., 1972. General Sensitivity Theory (Modern Analytic and Computational Methods in Science and Mathematics). Published by Elsevier Science Ltd. ISBN 10: 0444001085 ISBN 13: 9780444001085.

Wang, X., Gebremichael, M., Yan, J., 2010. Weighted likelihood copula modeling of extreme rainfall events in Connecticut. Journal of Hydrology 390, 108–115.

Werner, M., 2003. identification of Multivariate Outliers in Large Data Sets (A thesis submitted to the University of Colorado at Denver in partial fulfillment of the requirements for the degree of Doctor of Philosophy, Applied Mathematics).

Yang, J., Reichert, P., Abbaspour, K.C., Xia, J., Yang, H., 2008. Comparing uncertainty analysis techniques for a SWAT application to the Chaohe Basin in China. Journal of Hydrology 358, 1–23.

Zhang, L., Singh, V.P., 2007. Bivariate rainfall frequency distributions using Archimedean copulas. Journal of Hydrology 332, 93–109.

SOFTWARE TOOLS IN SA FOR EO

EFFICIENT TOOLS FOR GLOBAL SENSITIVITY ANALYSIS BASED ON HIGH-DIMENSIONAL MODEL REPRESENTATION

15

T. Ziehn[1], A.S. Tomlin[2]

CSIRO, Aspendale, VIC, Australia[1]; University of Leeds, Leeds, United Kingdom[2]

1. INTRODUCTION

With more and more computer power becoming available there is a general trend to increase the complexity of models used in a range of scientific fields by adding more processes and interactions at higher spatial and temporal resolutions. However, this increase in complexity does not necessarily improve the overall performance of the model in terms of its ability to predict observational data, or even more importantly, under extrapolation to conditions where no observations exist. This stems from the fact that a large number of processes can usually not be resolved at a fundamental level and therefore have to be parameterized. In most cases we do not know the exact value of the assigned parameters and prior parameter values are therefore based on expert knowledge. In some cases these parameters can be determined via experiments that are designed to isolate them (such as in gas-phase chemical kinetics) and in other cases via theoretical modeling (e.g., via quantum methods), but in many cases parameters can be little more than an informed guess. The large uncertainties associated with prior parameter values also lead to large variations in the model output. Therefore to produce models that are robust enough to be predictive for conditions outside of which they were validated, it becomes more and more important to focus on the reduction of input parameter uncertainties.

Parameter optimization methods are very useful in improving model performance, because they provide an objective way of constraining models against observations and in this way are able to reduce the parameter uncertainties (Tarantola, 2005). In addition to in situ data, Earth observation (EO) data can be used as a further observational constraint on modeled processes. For example, simulations of carbon and water fluxes with global land surface models (LSMs) exhibit large uncertainties, which are partly due to uncertainties in parameter values of the processes formulated within these models. The assimilation of the fraction of absorbed photosynthetically active radiation provided by the European Space Agency's (ESA's) Medium Resolution Imaging Spectrometer can improve the match to independent observations (Kaminski et al., 2012). However, in most cases, optimization methods cannot cope with the large number of uncertain parameters because the problem is underdetermined, i.e., the total number of parameters is greater than the number of observations available, and ill-posed, i.e., several different solutions exist that are all equally consistent with the observations. This problem is, for example, commonly faced by atmospheric inversion studies (Kaminski et al., 1999; Ziehn et al., 2011a). Furthermore, not all parameters can be constrained by a certain set of observations (Ziehn et al., 2011b). It is therefore crucial to first identify the most important or most sensitive parameters so that only these parameters are included within the optimization process. This is usually achieved by applying sensitivity analysis (SA).

SA methods can be broadly divided into local and global methods. Local methods are mainly suitable for linear models since they only investigate the parameter sensitivity at a fixed point, usually at the best estimate for the parameter. Since they are gradient-based methods, they assume small parameter variations, and hence unless they are applied for a large number of nominal parameter values, they cannot explore the impact of large parameter uncertainties. The focus in this work is on a class of global SA methods, i.e., methods that cover the whole domain of parameter uncertainties. Global SA can be performed in many different ways, and there is no one method that is suitable for all models in all fields. The choice of method also depends on the question being asked and the amount of information required. For example, some methods such as screening methods (Morris, 1991; Campolongo et al., 2007), offer a ranking of the parameters, but without being able to quantify the relative importance of each parameter or parameter interactions. However, they can be less computationally demanding than other global methods, particularly for systems with a smaller number of parameters.

To provide a measure of a parameter's importance one needs to calculate a sensitivity measure. This is commonly expressed through sensitivity indices that describe the relative contribution of one or more model parameters to the overall model output uncertainty. Variance-based sensitivity indices such as the Sobol sensitivity indices (Sobol, 2001) can be derived through the analysis of variance decomposition approach, describing how much of the model output variance is caused by the variance of an input parameter or set of parameters. The calculation of the Sobol sensitivity indices can be computationally very expensive, especially when the model takes a significant amount of time for a single run (Chan et al., 2000). However, a number of metamodel approaches are available that allow the derivation of the same sensitivity indices in a much more efficient way. Such approaches include high-dimensional model representation (HDMR) methods, Gaussian Process Emulator (GPE) methods, and methods based on Polynomial Chaos Expansions (PCE).

GPE methods (Oakley & O'Hagan, 2002, 2004) are based on the assumption that given a set of target outputs $y = f(\mathbf{x})$, the value of y at an unknown value of $\mathbf{x}$ follows a multivariate Gaussian distribution. According to Saltelli et al. (2008), since Gaussian emulators attempt to interpolate the mapping from $\mathbf{x}$ to $f(\mathbf{x})$ by applying a Gaussian kernel of the same dimension as that of the input parameter space, the methods can suffer from overparametrization. In practice, these methods have mainly been used for systems with a low number of input parameters and are more suitable for systems with a small number of main effects and only weak parameter interactions (Saltelli et al., 2008; Petropoulos et al., 2014, 2015). In PCE methods, the uncertainty in $\mathbf{x}$ is expressed as a polynomial expansion of basis random variables (Reagan et al., 2005; Najm et al., 2009; Sheen et al., 2009). The overall output variance may then be represented as the sum over terms involving the coefficients of the equivalent expansion.

Here, we introduce the HDMR method (Rabitz et al., 1999). Similar to the GPE and PCE methods, the HDMR method leads to a metamodel that can be used not only to explore the input−output relationship of complex models but also to efficiently calculate sensitivity indices. In this study, we focus on the model inputs that are fixed in time (but may vary spatially) and therefore refer to them as model parameters or input parameters.

A general overview of HDMR is given in the following section. We then introduce the graphical user interface (GUI)-HDMR software package (Ziehn and Tomlin, 2009), which is based on random sampling (RS)-HDMR discussed in Section 3. The software is freely available and could potentially be applied to any model or data set. A selection of case studies from a wide range of applications is presented in Section 4 to underline the usefulness of the GUI-HDMR software and to highlight some common features that occur within global sensitivity studies.

2. HIGH-DIMENSIONAL MODEL REPRESENTATION

The HDMR method is a set of tools explored by Rabitz et al. (1999) to express the input parameter−output relationship of a complex model with a large number of parameters. The mapping between the model parameters $x_1,...,x_n$ and the output $f(\mathbf{x}) = f(x_1,...,x_n)$ in the domain R^n can be written in the following form:

$$f(\mathbf{x}) = f_0 + \sum_{i=1}^{n} f_i(x_i) + \sum_{1 \leq i < j \leq n} f_{ij}(x_i, x_j) + ... + f_{12...n}(x_1, x_2, ..., x_n) \tag{15.1}$$

Here f_0 denotes the mean effect (zeroth order), which is a constant. The function $f_i(x_i)$ is a first-order term giving the effect of parameter x_i acting independently (although generally nonlinearly) upon the output $f(\mathbf{x})$. The function $f_{ij}(x_i,x_j)$ is a second-order term describing the cooperative effects of the parameters x_i and x_j upon the output $f(\mathbf{x})$. The higher order terms reflect the cooperative effects of increasing numbers of model parameters acting together to influence the output $f(\mathbf{x})$. The HDMR expansion is computationally very efficient if higher order parameter interactions are weak and can therefore be neglected. For many systems, an HDMR expansion up to second order already provides satisfactory results and a good approximation of $f(\mathbf{x})$ (Li et al., 2001). Where it does not, appropriate transformations of the outputs can sometimes be used to help build a low-order HDMR model.

There are two commonly used HDMR expansions. Cut-HDMR depends on the value of $f(\mathbf{x})$ at a specific reference point $\bar{\mathbf{x}}$, and RS-HDMR depends on the average value of $\bar{\mathbf{x}}$ over the whole domain, where the average is usually obtained over a suitable random sample.

RS-HDMR is computationally more efficient for models with a large number (more than 10) of parameters, because the sample size N required for RS-HDMR does not directly depend on the dimension of the parameter space. Previous research has shown that even better convergence properties can be achieved by using a quasi-random sampling (QRS) method rather than a straightforward RS method (Sobol', 1967; Kucherenko, 2007).

2.1 RANDOM SAMPLING—HIGH-DIMENSIONAL MODEL REPRESENTATION

The component functions of the HDMR expansion in Eq. (15.1) are determined through an averaging process from the same RS sample with size N. First, all parameters x_i are rescaled such that $0 \leq x_i \leq 1$ for all i. The output function is then defined in the unit hypercube $K^n = \{(x_1, \dots, x_n), i = 1, \dots, n\}$. The zeroth-order term f_0 is approximated by the average value of $\mathbf{x}^{(s)} = \left(x_1^{(s)}, x_2^{(s)} \dots, x_n^{(s)}\right)$ for all $s=1,2,\dots,N$

$$f_0 \approx \frac{1}{N} \sum_{s=1}^{N} f\left(\mathbf{x}^{(s)}\right) \tag{15.2}$$

with N being the number of sampled model runs. The higher order component functions are approximated by orthonormal polynomials:

$$f_i(x_i) \approx \sum_{r=1}^{k} \alpha_r^i \varphi_r(x_i) \tag{15.3}$$

$$f_{ij}(x_i, x_j) \approx \sum_{p=1}^{l} \sum_{q=1}^{l'} \beta_{pq}^{ij} \varphi_p(x_i)\varphi_q(x_j) \tag{15.4}$$

$$\dots$$

where k,l,l' represents the order of the polynomial expansion, α_r^i and β_{pq}^{ij} are constant coefficients to be determined, and $\varphi_r(x_i)$, $\varphi_p(x_i)$, and $\varphi_q(x_j)$ are the basis functions (Li et al., 2002a). The approximation of the component functions reduces the sampling effort dramatically so that only one set of random or quasirandom samples N is necessary to determine all RS-HDMR component functions.

The expansion coefficients α_r^i and β_{pq}^{ij} can be determined by a minimization process and Monte Carlo (MC) integration (Li et al., 2002a), which leads to:

$$\alpha_r^i \approx \frac{1}{N} \sum_{s=1}^{N} f\left(\mathbf{x}^{(s)}\right) \varphi_r\left(x_i^{(s)}\right) \tag{15.5}$$

$$\beta_{pq}^{ij} \approx \frac{1}{N} \sum_{s=1}^{N} f\left(\mathbf{x}^{(s)}\right) \varphi_p\left(x_i^{(s)}\right) \varphi_q\left(x_j^{(s)}\right) \tag{15.6}$$

The HDMR expansion can be used for many purposes, for example, as a metamodel that approximates the model behavior or for calculating partial variances used for SA. The focus is here on the latter, and the calculation of the sensitivity indices is described in the following section.

2.2 GLOBAL SENSITIVITY ANALYSIS USING RANDOM SAMPLING—HIGH-DIMENSIONAL MODEL REPRESENTATION

Sobol's method (Sobol, 2001) is commonly used in global SA to calculate the partial variances and sensitivity indices. It is conceptually the same as the RS-HDMR approach. However, once the RS-HDMR expansion is calculated, the partial variances D_i, D_{ij}, ... for SA purposes are easily obtained from Li et al. (2002b):

$$D_i \approx \sum_{r=1}^{k_i} \left(\alpha_r^i\right)^2 \tag{15.7}$$

$$D_{ij} \approx \sum_{p=1}^{l_i} \sum_{q=1}^{l_j} \left(\beta_{pq}^{ij}\right)^2 \tag{15.8}$$

$$\cdots$$

Then by normalizing the partial variances by the overall variance D of the model output we finally get the sensitivity indices:

$$S_{i_1,\ldots,i_s} = \frac{D_{i_1,\ldots,i_s}}{D}, \quad 1 \leq i_1 < \ldots < i_s \leq n \tag{15.9}$$

The first-order sensitivity index S_i measures the main effect of the input parameter x_i on the output, or in other words the fractional contribution of x_i to the variance of $f(\mathbf{x})$. The second-order sensitivity index S_{ij} measures the interaction effect of x_i and x_j on the output, and so on. The sensitivity indices can be used for importance ranking for the input parameters for each of the model outputs and therefore as a focus for model improvement.

2.3 VARIANCE REDUCTION METHODS

The determination of the expansion coefficients (i.e., α_r^i and β_{pq}^{ij}) is based on MC integration. The error of the MC integration controls the accuracy of the RS-HDMR expansion. Variance reduction methods can be applied to improve the accuracy of the MC integration. Two methods have been successfully applied in connection with the RS-HDMR method: the correlation method as suggested by Li et al. (2003) and the ratio control variate method suggested by Li and Rabitz (2006). In both cases the determination of the expansion coefficients becomes an iterative process and requires an analytical reference function $h(x)$. This function has to be similar to $f(\mathbf{x})$, and as shown in Li et al. (2003) and Li and Rabitz (2006), a truncated RS-HDMR expansion can be used as a reference function whose expansion coefficients were calculated by direct MC integration.

The application of a variance reduction method allows improvements in the accuracy of the metamodel without increasing the sample size N. This is very beneficial for practical applications of RS-HDMR especially for high-dimensional systems.

3. GRAPHICAL USER INTERFACE-HIGH-DIMENSIONAL MODEL REPRESENTATION SOFTWARE

The GUI-HDMR software was developed to combine existing HDMR tools and new HDMR extensions in one package and to make them freely available to all users. The software is written in Matlab and requires the standard Matlab package. A more detailed description of the software including performance tests using analytical test functions can be found in Ziehn and Tomlin (2009).

3.1 OVERVIEW

A flow chart of the GUI-HDMR software is presented in Fig. 15.1. The software comes with a GUI, but it can also be used without the GUI as a script-based approach. In both cases the user has to supply two external files. The first one contains the rescaled parameter values ($0 \leq x_i \leq 1$), which were used to run the model, and the second one contains the corresponding output values. Both files have to be provided as ASCII text files within a matrix format. The rows represent the sample number ($1...N$), and the columns stand for the different parameters ($i = 1...n$) or the number of considered outputs. If only one output is considered, then the output file is simply a column vector.

The set of parameter values can be any uniformly distributed MC sample (if the parameters can be controlled) or measured values (if using experimental data). If the parameters can be controlled, then a QRS method such as the Sobol' sequence (Sobol', 1967) is preferable. This guarantees that the input space is covered more uniformly than by using random values, and it provides a better convergence rate.

Additionally, a second set of parameter and output values may be provided in the same format as the first set. To show the accuracy of the constructed metamodel, the relative error (RE) and the probability density function (PDF) are calculated for a number of samples. If a different set of

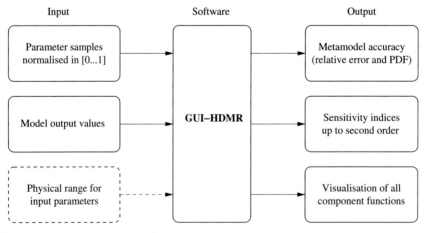

FIGURE 15.1

GUI-HDMR flow chart. *GUI*, graphical user interface; *HDMR*, high-dimensional model representation; *PDF*, probability density function.

parameter and output values is provided, then the metamodel is tested against values that were not used to construct the HDMR expansion. If no second set is provided, then the same set of parameter and output values is used for the accuracy test. Optionally one can also provide the original parameter ranges to produce the plots of the component functions with the correct ranges instead of the rescaled ones.

A fully functional metamodel can be constructed using only the one set of samples provided. Additionally, sensitivity indices of first- and second-orders can be calculated. It is also possible to plot the relationship of each parameter or the interaction of two parameters, which is embodied in the HDMR component functions [see Eqs. (15.3) and (15.4)], on the model output and to compare against conventional scatter plots. The advantage of plotting the HDMR component functions is that they are able to demonstrate the isolated effect of a parameter or parameter interaction on the model output, which is usually hidden in the conventional scatter plot, because it shows the influence of other parameters as well. The component functions also highlight any nonlinearities in the response of the output to changes in the input parameter across its defined range of uncertainty.

3.2 EXTENSIONS TO STANDARD HIGH-DIMENSIONAL MODEL REPRESENTATION

The set of HDMR tools as described is Section 2 uses the same polynomial order for all first- and second-order component functions, respectively. However, because the HDMR component functions are independent, the order of the polynomial approximation should be chosen separately for each component function. Consequently, an optimization method has been developed by Ziehn and Tomlin (2008a), which automatically chooses the best polynomial order for each of the component functions. The computational effort to calculate the optimal order for all polynomials is low, and therefore it is worthwhile to calculate it for improving the accuracy of the final metamodel.

Furthermore, a threshold was introduced by Ziehn and Tomlin (2008b) exclude unimportant component functions from the HDMR expansion. The idea of this approach is to reduce the number of component functions to be approximated by polynomials and therefore to reduce numerical errors. This is particularly useful if the model has a large number of parameters (greater than 20) and only small higher order effects.

3.3 SOFTWARE FEATURES

The GUI-HDMR software provides the main window from which the calculations can be started and progress shown. Three other windows can be accessed from the main window: a setup window, a window showing the optimal polynomial orders, and the accuracy of the metamodel and a window for SA.

Fig. 15.2 shows the setup window for the HDMR analysis. The sample input parameter and output files can be loaded into the workspace. If one wants to verify the constructed metamodel with a different sample set than the one used for constructing the metamodel, then a different set of sample input and output files can be provided for the accuracy test. Optionally the ranges for the parameters can be loaded to produce the plots with the correct ranges. In the settings section, one has to state the number of samples to be used to construct the HDMR metamodel and the number of samples to be used for the accuracy test. The maximum polynomial order for the approximation of the component

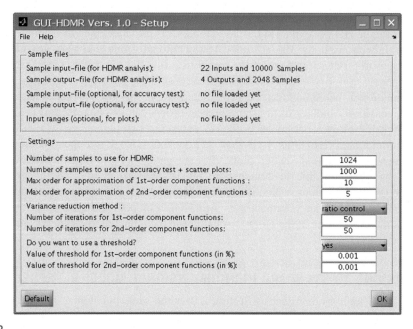

FIGURE 15.2

GUI-HDMR setup window: input parameter and output files can be loaded and settings can be applied for the HDMR analysis. *GUI*, graphical user interface; *HDMR*, high-dimensional model representation.

functions can be defined separately for first- and second-order. Orthonormal polynomials up to tenth order are supported (Note that the computational time increases if using higher order polynomials. This is especially true for the second-order component functions, because a large number of expansion coefficients have to be calculated). A variance reduction method can be chosen, and the number of iterations can be entered for the calculation of the α and β expansion coefficients. It is also possible to apply a threshold that assures that only the component functions having a contribution larger than the threshold are considered. The threshold is given in %, and its application is explained in detail in Ziehn and Tomlin (2008b).

After the calculations are finished and the metamodel has been successfully constructed, the results window (Fig. 15.3) gives an overview of how many component functions have been approximated and by what polynomial order. Furthermore, the RE and PDF are calculated to show the accuracy of the first- and second-order metamodels.

The SA window as shown in Fig. 15.4 presents the ranking of the five most important parameters based on the first-order sensitivity indices. The ranking of the five most important parameter interactions, based on the second-order sensitivity indices, can also be displayed. Additionally, all component functions can be plotted and compared with scatter plots. It is also possible to save all produced plots as Portable Network Graphics (png), Portable Document Format (pdf), and Encapsulated PostScript (eps). Finally, the whole Matlab workspace can be saved and loaded into the GUI-HDMR software again for further analysis and plots.

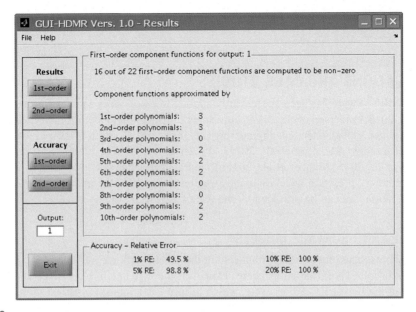

FIGURE 15.3

GUI-HDMR results window: the optimal order is shown for the component functions that were approximated by polynomials. To assess the accuracy of the HDMR metamodel, the relative error is also stated. *GUI*, graphical user interface; *HDMR*, high-dimensional model representation.

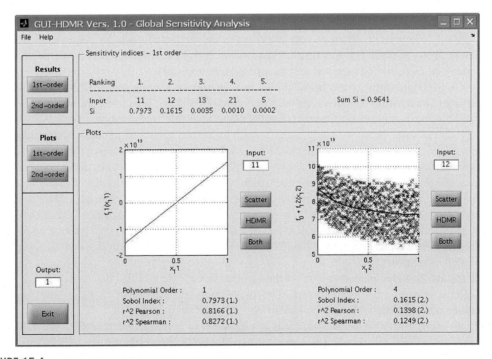

FIGURE 15.4

GUI-HDMR sensitivity analysis window: the ranking of the five most important parameters based on the first-order sensitivity indices is shown. Additionally, all component functions can be plotted. *GUI*, graphical user interface; *HDMR*, high-dimensional model representation.

4. APPLICATIONS AND CASE STUDIES

The GUI-HDMR software has been successfully applied in many areas such as Earth system science (i.e., Lu et al., 2013; Anderson et al., 2014), atmospheric dispersion modeling (i.e., Ziehn and Tomlin, 2008a; Ziehn et al., 2009a), chemical engineering (i.e., Ziehn and Tomlin, 2008b; Ziehn et al., 2009b), computational biology (i.e., Tam et al., 2015; Münzer et al., 2015a), biomedical engineering (i.e., Fuentes-Garí et al., 2015; Münzer et al., 2015b), and neuroscience (i.e., Mihalas et al., 2011; Chen et al., 2013). Here we present a selection of case studies to highlight the main features of the software and the information that can be provided by the use of global SA methods.

4.1 GLOBAL LAND SURFACE MODELS

The terrestrial biosphere plays an important role in the global carbon cycle and has a great impact on the accumulation of CO_2 in the atmosphere. However, there is much uncertainty about the size of natural sinks of the terrestrial carbon cycle, which in turn has a major impact on climate predictions (Zaehle et al., 2005). The large variations in the predictions of the future atmospheric CO_2 load result not only from differences between the models (Friedlingstein et al., 2006) but also from uncertainties of the process parameters of global LSMs (Knorr and Heimann, 2001).

Global LSMs traditionally provide the surface boundary conditions and surface fluxes of heat and water into the atmosphere. Over the past two decades, additional processes, such as biogeochemical cycles (Wang et al., 2010), vegetation dynamics (Krinner et al., 2005), and fires (Thonicke et al., 2001), have been implemented into global LSMs to address a wide range of issues in climate change and natural resource management. Nearly all global LSMs use about 5–30 different plant functional types (PFTs) to represent different vegetation types on land, and each PFT can have up to 60 parameters (Wang et al., 2011). As a result, most global LSMs have several hundred parameters, which are in many cases poorly constrained and therefore quite uncertain.

The GUI-HDMR software has been applied to a number of leading global LSMs to identify parameters that are sensitive to a given set of observations. The most sensitive parameters can then be constrained using parameter optimization methods and available measurements such as in situ and remotely sensed data. Here, we analyze the sensitivity of the simulated annual gross primary productivity (GPP) and latent heat flux (LE) to 22 biophysical parameters for each of the 10 PFTs used in the Australian community LSM CABLE (Wang et al., 2011; Kowalczyk et al., 2006).

Using the Sobol sequence we generated 1024 samples and ran CABLE for each sample independently. First- and second-order sensitivity indices were calculated for each parameter for annual GPP and LE at the PFT level and globally using a subset of the 1024 samples and all samples together. We found that the sensitivity indices calculated based on 256 samples were not significantly different from the sensitivity indices calculated based on all 1024 samples, which indicates that the sample size is more than adequate.

The results show that more than 80% of the variance in GPP for all PFTs can be explained by first-order effects of the same 3 parameters. The most important parameter is V_{cmax}, the maximum carboxylation rate of the leaves at the canopy top at a leaf temperature of 25°C, which on its own contributes more than 50% to the variance in annual GPP. In contrast to annual GPP, the ranking of important parameters for annual LE varies across the PFTs. More than 70% of the variance in simulated annual LE can be explained by first-order effects of three to five parameters. Second-order effects are small among all parameters (2−10%).

If the effect of a parameter on a model is linear, then no matter what part of the input domain the focus is on, the same width of input uncertainty would lead to the same variance in the output and hence the same parameter importance ranking. However, if the effect of a parameter on the model output is nonlinear, then the ranking based on the sensitivity indices will not only depend on the uncertainty range assigned for this parameter but also the location of this range, since the slope of the component function changes throughout the range. This is shown for V_{cmax} in Fig. 15.5 for global annual GPP. Using all 1024 samples the GUI-HDMR software approximates the response of global annual GPP to V_{cmax} by a fourth-order polynomial. Compared with a linear approximation, the slope of

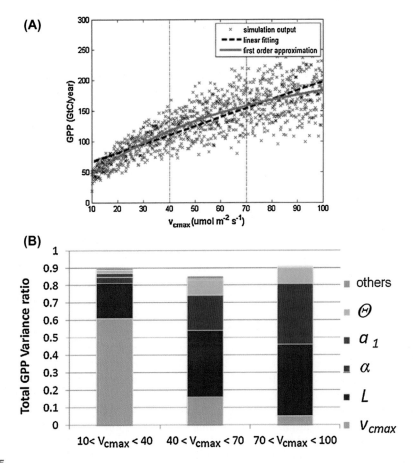

FIGURE 15.5

(A) Variation of global annual GPP with V_{cmax}. The gray solid line represents the sum of the zeroth- and first-order terms (fourth-order polynomial) for V_{cmax} in the HDMR expansion. The dashed black line represents a linear fit. (B) First-order sensitivity indices of the five most important parameters for each of the three ranges of V_{cmax}. *GPP*, gross primary productivity; *HDMR*, high-dimensional model representation.

the first-order effect is greater at lower V_{cmax} values and smaller at higher V_{cmax} values (Fig. 15.5A). To demonstrate the effect of varying the uncertainty range for V_{cmax} on its sensitivity index, we divide the 1024 simulations into three groups depending on the value of V_{cmax}: (1) $10 \leq V_{cmax} < 40$ μmol/m^2/s, (2) $40 \leq V_{cmax} < 70$ μmol/m^2/s, and (3) $70 \leq V_{cmax} < 100$ μmol/m^2/s. As shown in Fig. 15.5B, V_{cmax} is the most sensitive parameter in group (1) contributing by more than 60% to the overall variance in annual GPP, but is ranked only fifth in group (3) with a contribution of only about 4% to the overall variance.

The presence of nonlinear responses highlights the importance of assessing the sensitivity of model parameters over their whole uncertainty range by using global SA tools such as GUI-HDMR. The HDMR component functions are particularly useful in this case because they provide a visual picture of the response to isolated parameter changes across the whole uncertainty range. Further details of this study can be found in Lu et al. (2013).

4.2 ATMOSPHERIC DISPERSION MODELS

Atmospheric dispersion models are used in a wide variety of applications to help assess the links between ambient pollutant concentrations and emissions. Such models can then be employed to help evaluate pollution mitigation strategies such as emission reduction programs or traffic management strategies. In order for models to contribute usefully to such decision-making processes, the user would like to be confident that their sensitivity to possible changes in emissions is greater than their sensitivity to uncertain model parameters. This is certainly not a given, since such models usually represent a number of complex and coupled processes including physical flows, turbulent mixing, chemical reactions, and deposition. The description of all these processes requires parameterizations, and the quantification of each parameter will contain uncertainties. We therefore present a case study here as an example of turbulent, reactive dispersion models where the aim is to predict the mean concentration and level of concentration fluctuation for key pollutants NO_2 and O_3 depending on a range of NO_x ($NO + NO_2$) emission rates and background ozone concentrations.

Detailed, time-resolved models of reactive turbulent processes are computationally very expensive, and hence for pollutant dispersion applications, time-averaged models with parameterizations of turbulence are often used for predicting near-field concentrations. Lagrangian stochastic models, based on marked particle approaches (Thomson, 1987), are one such choice of time-averaged turbulent flow models and can provide predictions of both mean concentrations and concentration fluctuations in atmospheric plumes for both primary and secondary pollutant species (Sawford, 2004; Cassiani et al., 2005; Garmory et al., 2006; Middleton et al., 2008). Lagrangian particle models compute trajectories of a large number of discrete so-called particles (not real particles but each representing a unit of concentration) to describe the transport and turbulent diffusion of tracers in the atmosphere. Within such approaches, parameterizations of turbulent length and time scales are needed along with other parameters such as emission rates, background pollutant concentrations, and reaction rate constants. If models of this type are to be used effectively within the development of pollution control strategies, then even when taking into account uncertainties in model parameters, we would wish the variance in predicted concentrations to be as small as possible. SA can help us to determine which parameters contribute the most to this variance and therefore provides useful information to help us constrain the model better through, for example, additional field measurements, satellite or airborne measurements, or fundamental parameter studies.

A combined Lagrangian stochastic model with a micromixing submodel (Dixon and Tomlin, 2007; Ziehn et al., 2009a) is used to investigate a reactive plume of nitrogen oxides (NO_x) released into a turbulent grid flow doped with ozone (O_3). Sensitivities to the model input parameters are explored for several NO_x concentration scenarios based on wind tunnel experiments (Brown and Bilger, 1996). The experiments themselves did not include photolysis reactions due to the absence of ultraviolet light and hence the reactants only underwent the reaction $NO + O_3 = NO_2 + O_2$. However, we include photolysis processes here for O_3 and NO_2 within a chemical scheme consisting of eight reactions between species NO, O, O_3, and NO_2. The case study is intended to represent near-field mixing that may occur in the atmosphere close to, for example, a NO_x point source or a traffic source. However, the principles could be extended to larger scale models, such as regional air quality models, where satellite observations could be used to provide constraints on predicted concentrations. A Gaussian distribution is assumed for the source of NO_x, which is mainly NO but with varying amounts of NO_2 ranging from 0% to 20% of the total NO_x. Since the scenario is based on conditions within a wind tunnel, the wind direction is not varied within this study. Of course in the real world, wind direction may highly influence near-field pollutant concentrations.

A total of 22 parameters were assumed to be uncertain, among them turbulence parameters, temperature-dependent reaction rate parameters, photolysis rates, temperature, the fraction of NO in total NO_x at the source, and the background concentration of O_3 (see Ziehn et al. (2009a) for details). The ranges for the uncertain parameters were defined according to a minimum and maximum value, assuming equal probability throughout the ranges based on a comparison of information available from a wide range of literature studies. In reactive dispersion models, it is important to investigate the combined effects of important physical and chemical parameters because the simulated concentrations downwind from the source are influenced by both dispersion (atmospheric mean advection and turbulent mixing) and chemical reactions. In particular, the parameterization of the mixing model used within the Lagrangian stochastic approach can affect the overall reaction rates since it influences the concentrations of species involved in the reactions.

Two different NO_x scenarios were simulated representing: (1) a high NO_x, highly polluted situation where the source concentration was 1.26×10^{16} molecules/cm^3/500 ppmv and (2) a lower source concentration for NO_x of 1.26×10^{15} molecules/cm^3/50 ppmv. The target outputs chosen for the SA are the mean concentrations of the pollutants NO_2 and O_3 at the plume center line and the variance of fluctuations in NO_2 concentration represented by γ'_{NO_2} at different distances x away from the point source. We focus on a subset of the results here. A quasirandom sequence of up to $N = 1024$ was used for the RS-HDMR analysis of first- and second-order sensitivity indices. The maximum order for the approximation of the polynomials used was chosen to be 10 for first-order and 3 for second-order component functions with a threshold of 1% for the exclusion of unimportant component functions from the HDMR expansion.

The accuracy of the HDMR metamodel fit has been shown in previous work to depend on the complexity of the model response as well as the sample size used for the polynomial fits (Ziehn and Tomlin, 2008b, 2009). In general, where the model response involves higher order terms (i.e., parameter interactions) a larger number of samples are required to accurately fit the HDMR expansion. A common approach is to start with a small sample size and increase it until the sensitivity results converge. In this case due to the smaller contribution of higher order terms, a fairly small sample size ($N = 512$) was sufficient to achieve convergence except for the γ'_{NO_2} analysis where $N = 1024$ was required. However, the accuracy of the results for the SA was also seen to depend on the setup of the

Lagrangian stochastic model. Various numbers of particles (50,000 up to 1,000,000) within the marked particle model, and different simulation end times (200–1200 s) were tested, and the mean concentrations after each time step were recorded. It was found that a large number of particles (1,000,000) was necessary, in connection with a long simulation end time of $t = 1200$ s (12,000 time steps), to ensure that the predicted mean concentrations reached convergence. If convergence was not reached, then spurious sensitivity indices could be produced, particularly for the minor contributing parameters. This case study highlights the importance of reducing numerical noise within the sample, which can contribute to the output variance, but cannot be attributed by the HDMR method to any of the input parameters, leading to spurious results. In cases where a good metamodel fit cannot be achieved, even as the sample size is increased, then the user should check for numerical noise issues.

Table 15.1 shows the first-order sensitivity indices for the mean O_3 concentration $\overline{\Gamma}_{O_3}$ at the plume center for both source NO_x scenarios. For the high NO_x scenario we expect low predicted $\overline{\Gamma}_{O_3}$ within the plume due to a high reaction rate of O_3 with NO. In total, seven parameters are responsible for the variance of the output. The most important parameter is the activation energy term E/R for the Arrhenius expression for the reaction rate for $NO + O_3 = NO_2$, which contributes more than 40% to the overall variance and has a much stronger influence than the preexponential factor (A) for the reaction. This indicates that to reduce the variance in predicted $\overline{\Gamma}_{O_3}$, further fundamental kinetic studies are required for this reaction, particularly better quantification of its temperature dependence. The most important turbulence parameter from the Lagrangian stochastic model is α, which defines the relationship between turbulence time scales (total turbulent kinetic energy k and its dissipation rate ε) and the mixing time scale t_m at each point in the flow via the following equation (Dixon and Tomlin, 2007):

$$t_m = \alpha \frac{k}{\varepsilon} \tag{15.10}$$

Table 15.1 First-Order Sensitivity Indices for $\overline{\Gamma}_{O_3}$ at Plume Center for Both Source NO_x Scenario

Parameter	$S_i\ (x = 2.2\ \text{m})$ High NO_x	$S_i\ (x = 5.8\ \text{m})$ High NO_x	$S_i\ (x = 2.2\ \text{m})$ Lower NO_x	$S_i\ (x = 5.8\ \text{m})$ Lower NO_x
E/R for $NO + O_3 = NO_2$	0.41 (1)	0.44 (1)	0.14 (2)	0.19 (2)
Mixing time scale coefficient α	0.30 (2)	0.25 (2)	0.11 (3)	0.11 (4)
Background concentration for O_3	0.11 (3)	0.07 (4)	0.69 (1)	0.59 (1)
Fraction of NO in total NO_x	0.06 (4)	0.09 (3)	0.02 (5)	0.02 (5)
Temperature	0.03 (5)	0.04 (5)	0.01 (6)	0.02 (6)
Structure function coefficient c_0	0.03 (6)	0.04 (6)	0.02 (4)	0.02 (3)
A factor for $NO + O_3 = NO_2$	0.03 (7)	0.03 (7)	0.01 (7)	0.01 (7)
$\sum S_i$	0.97	0.96	0.97	0.97

The numbers in brackets indicate the ranking of the parameter. Sample size N = 512.

The Lagrangian structure function coefficient c_0, which determines the effective turbulent diffusion in velocity space, shows a very low sensitivity, indicating that the extent of mixing defined within the model has a much larger influence on the prediction of secondary species such as O_3. Parameter interactions were not found, as shown by the total first-order sensitivity coefficients adding to almost 1. The relative importance of the parameters, and even the ranking, changes with growing distance x from the source. For example, the fraction of NO in total source NO_x becomes more important at further distances x from the source.

Fig. 15.6 shows the component functions and scatter plots for these two most important parameters at 2.2 m from the source for the high NO_x scenario. The relationship between E/R for the reaction $NO + O_3 = NO_2$ and $\overline{\Gamma}_{O_3}$ is nearly linear (positive), whereas the relationship between α and $\overline{\Gamma}_{O_3}$ is fairly nonlinear (positive). One striking feature of these plots is the width of the distribution in O_3 concentrations. A factor of 8 difference is seen between the lowest and highest values, indicating a high degree of variance within the predicted outputs. This highlights the importance of providing better constraints on the key input parameters in order to improve model robustness. Different approaches can be applied for achieving better parameter constraints. One method would be to perform further fundamental parameter studies. Another would be to combine modeling with various observational data sets such as from ground-based, airborne, or satellite observations within optimization or data assimilation approaches.

For the lower initial source NO_x scenario a significant change in the ranking of the parameters is seen (Table 15.1). Plume center $\overline{\Gamma}_{O_3}$ is now most sensitive to the background O_3 concentrations, although further away from the NO_x source the activation energy for the reaction of NO with O_3 increases in relative importance. Overall, the model predictions for $\overline{\Gamma}_{O_3}$ are shown to be more sensitive to chemical and physical parameters rather than the turbulence parameterizations, which is a positive finding from the point of view of robustness of the Lagrangian model. However, the analysis

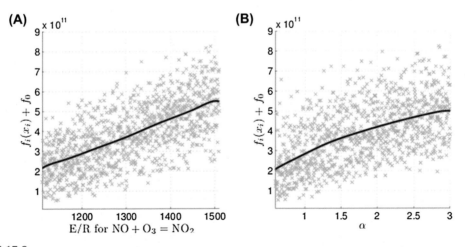

FIGURE 15.6

First-order component functions and scatter plots for (A) E/R for $NO + O_3 = NO_2$ and (B) the mixing time scale coefficient α with respect to $\overline{\Gamma}_{O_3}$ at the plume center for the higher NO_x scenario and $x = 2.2$ m. The mean f_0 is added to f_i for comparison with scatter plot.

demonstrates that accurate determination of field parameters (e.g., background pollutant concentrations), and one particular chemical reaction rate is required for modeling similar scenarios in the real world. The analysis helps to narrow down the group of parameters required for further study. Neither was temperature, for example, found to be important in this high-NO_x situation, nor were the rates of the other chemical reactions included in the scheme. Perhaps surprisingly, the relative fraction of NO_2 within the source NO_x was also of fairly low importance, reaching a maximum of 9% at the farthest distance from the source.

Table 15.2 shows the estimated sensitivity indices for $\overline{\Gamma}_{NO_2}$ at the plume center line for the two different source NO_x scenarios. For the distances studied (i.e., near-field dispersion), the relative fraction of $NO{:}NO_2$ at the source is the most influential parameter. Fig. 15.7 shows the HDMR component functions and scatter plots for the top two ranked parameters for the higher NO_x scenario. The responses are fairly linear although as the fraction of NO_2 in source NO_x increases (as NO reduces to 80%), the scatter in the data increases, indicating the increasing influence of other parameters. This feature, termed heteroscedasticity, occurs commonly within global sensitivity studies and is not a feature that could be explored through linear methods. In this case, the behavior suggests that for higher levels of primary NO_2 (i.e., a larger amount of NOx released as NO_2) the output variance of the modeled peak $\overline{\Gamma}_{NO_2}$ increases. Overall, the variability in predicted $\overline{\Gamma}_{NO_2}$ can span a factor of 10 and stresses the importance of determining the level of primary NO_2 at the source. This could be of particular significance for traffic-related sources where NO_2 emissions depend highly on factors such as fuel type (e.g., higher for diesel than for petrol) and pollution control strategies. For example, the fraction of primary NO_2 emissions may increase as the proportion of particle controlled diesel vehicles increases in the overall vehicle fleet (Carslaw, 2005).

Background [O_3] has a very low significance on the predicted $\overline{\Gamma}_{NO_2}$, although this increases for the lower NO_x scenario for larger distances from the source. Therefore for high-NO_x situations, near-field $\overline{\Gamma}_{NO_2}$ is more strongly influenced by emission profiles than by secondary formation through chemical

Table 15.2 First-Order and Most Significant Second-Order Sensitivity Indices for $\overline{\Gamma}_{NO_2}$ at Plume Center

Parameter	S_i ($x = 5.8$ m) High NO_x	S_i ($x = 2.2$ m) Lower NO_x	S_i ($x = 5.8$ m) Lower NO_x
Fraction of NO in total NO_x (x_{22})	0.88 (1)	0.86 (1)	0.64 (1)
c_0 (x_{11})	0.06 (2)	0.06 (2)	0.10 (3)
α (x_{12})	0.02 (3)	0.04 (3)	0.10 (2)
E/R for $NO + O_3 = NO_2$	–	0.01 (4)	0.08 (4)
Background concentration for O_3	–	–	0.02 (5)
$\sum S_i$	0.96	0.96	0.95
$S_{11,22}$	0.02	0.01	0.01
$S_{12,22}$	0.01	–	–
$\sum S_i + \sum S_{ij}$	0.98	0.98	0.97

The numbers in brackets indicate the ranking of the parameter. Sample size N = 512.

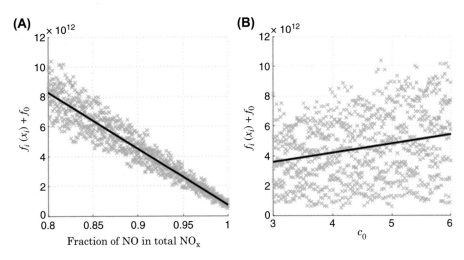

FIGURE 15.7

First-order component functions and scatter plots for (A) fraction of NO in total NO_x and (B) the structure function coefficient c_0 with respect to $\overline{\Gamma}_{NO_2}$ at the plume center for the higher NO_x scenario and $x = 4.8$ m. The mean f_0 is added to f_i for comparison with scatter plot.

processes. At larger distances from the source and therefore lower [NO_x], this mixing of background O_3 into the plume has an increasingly important effect on NO_2 formation. The influence of the activation energy for the formation of NO_2 from NO and O_3 also increases at larger distances from the source.

Finally, we assess the influences on predicted variance in the extent of NO_2 fluctuations γ'_{NO_2} along the plume center line for this case study. The component functions and scatter plots for the two main parameters are shown in Fig. 15.8. The feature of heteroscedasticity is strongly present for this output. The majority of the variance in the center line γ'_{NO_2} is seen for the higher fractions of NO_2 within the NO_x source and higher values of the mixing time scale coefficient, which would increase the mixing time scale t_m. Were the source to be 100% NO, then low fluctuations in NO_2 concentrations would be seen. However, as NO_2 increases to 20% of the source NO_x, a wide range of levels of NO_2 fluctuation are present, depending on the selection of the mixing time scale coefficient. The results suggest that within Lagrangian stochastic models, where turbulent mixing processes have to be parameterized, careful selection of the mixing time scale is required to accurately predict the fluctuations in secondary pollutant concentrations and due to turbulence. However, the sensitivity indices shown in Tables 15.1 and 15.2 suggest that this selection would not highly influence the prediction of mean concentrations of secondary species such as NO_2 and O_3.

4.3 APPLICATION FOR EARTH OBSERVATIONS

The application of the GUI-HDMR software package is not restricted to a certain type of model or data and could therefore potentially be used for a large variety of applications including the coupling of models with remotely sensed data. However, computer resources might be a limiting factor for models with long simulation times (i.e., hours or days), since the HDMR method requires the user to run the model for a

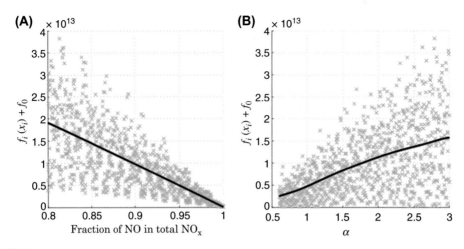

FIGURE 15.8

First-order component functions and scatter plots for (A) fraction of NO in total NO_x and (B) the mixing time scale coefficient α with respect to γ'_{NO_2} at the plume center for the higher NO_x scenario and $x = 4.8$ m. The mean f_0 is added to f_i for comparison with scatter plot.

large number of samples. Importantly, the required sample size does not directly depend on the parameter space dimension, as is the case with other global SA methods, and for a large number of case studies where the GUI-HDMR software has been applied, a sample size in the order of a couple of hundred was sufficient. All required model runs are independent and can therefore be performed in parallel.

Potential applications of the GUI-HDMR software include the assimilation of EO data into LSMs. As we have demonstrated in the first case study (Section 4.1), there is an ongoing need to improve the reliability of those models by providing better estimations of the process parameters involved. The GUI-HDMR software can be used to identify important parameters, which can then be constrained by EO data. The benefits and limitations of using EO data in LSMs have been addressed in a number of studies. For example, Land Cover and Land Use (LCLU) classification of multitemporal Landsat satellite data was used in Dieye et al. (2012) to assess the sensitivity of soil organic carbon modeled by the Global Ensemble Biogeochemical Modeling System (GEMS). The study focuses on the Senegal region and demonstrates a significant dependency not only on the LCLU classification errors but also on where the LCLU classes occur relative to the other GEMS model inputs. Barbu et al. (2011) investigate the joint assimilation of a soil wetness index (SWI) product together with leaf area index (LAI) in an LSM. It is demonstrated that a significant improvement of the root zone soil water content is obtained by assimilating dimensionless root zone SWI data, whereas a lower impact is observed when assimilating in situ data.

5. SUMMARY AND CONCLUSIONS

We have introduced the GUI-HDMR software package, which is based on a metamodeling approach and allows the calculation of global sensitivity indices up to the second-order. All sensitivity indices and parameter—output relationships can be estimated from only one set of (quasi) random samples. In many cases, a modest sample size is usually sufficient for the SA, which makes the GUI-HDMR software particularly suited for complex models with a large number of input parameters.

Here, we presented case studies from two different fields—atmospheric dispersion modeling and global land surface modeling. Key parameters and parameter interactions were identified by using only around 500 or less model simulations. This shows the effectiveness of the GUI-HDMR software, particularly for applications in which first-order effects are dominant. The exploration of higher order (i.e., second-order) effects may require a larger sample size, and here a combination of screening methods and the GUI-HDMR software might be beneficial as demonstrated in (Ziehn et al., 2009b). Although the GUI-HDMR software has not been applied in the field of EOs yet, it is well suited to handle high dimensions and nonlinearities, as are common in assimilating remotely sensed data.

REFERENCES

Anderson, B., Borgonovo, E., Galeotti, M., Roson, R., 2014. Uncertainty in climate change modeling: can global sensitivity analysis be of help? Risk Analysis 34 (2), 271–293. http://dx.doi.org/10.1111/risa.12117. URL: http://dx.doi.org/10.1111/risa.12117.

Brown, R., Bilger, R., 1996. An experimental study of a reactive plume in grid turbulence. Journal of Fluid Mechanics 312, 373–407.

Barbu, A.L., Calvet, J.-C., Mahfouf, J.-F., Albergel, C., Lafont, S., 2011. Assimilation of soil wetness index and leaf area index into the ISBA-A-gs land surface model: grassland case study. Biogeosciences 8 (7), 1971–1986. http://dx.doi.org/10.5194/bg-8-1971-2011. URL: http://www.biogeosciences.net/8/1971/2011/.

Campolongo, F., Cariboni, J., Saltelli, A., 2007. An effective screening design for sensitivity analysis of large models. Modelling, computer-assisted simulations, and mapping of dangerous phenomena for hazard assessment Environmental Modelling & Software 22 (10), 1509–1518. http://dx.doi.org/10.1016/j.envsoft.2006.10.004. URL: http://www.sciencedirect.com/science/article/pii/S1364815206002805.

Chan, K., Tarantola, S., Saltelli, A., Sobol, I.M., 2000. Variance based methods. In: Saltelli, A., Chan, K., Scott, E.M. (Eds.), Sensitivity Analysis. John Wiley & Sons.

Chen, X., Mihalas, S., Niebur, E., Stuphorn, V., 2013. Mechanisms underlying the influence of saliency on value-based decisions. Journal of Vision 13 (12), 18. http://dx.doi.org/10.1167/13.12.18. URL:http://dx.doi.org/10.1167/13.12.18.

Cassiani, M., Franzese, P., Giostra, U., 2005. A PDF micromixing model of dispersion for atmospheric flow. Part I: development of the model, application to homogeneous turbulence and to neutral boundary layer. Atmospheric Environment 39 (8), 1457–1469. http://dx.doi.org/10.1016/j.atmosenv.2004.11.020. URL: http://www.sciencedirect.com/science/article/pii/S1352231004010994.

Carslaw, D.C., 2005. Evidence of an increasing NO_2/NO_X emissions ratio from road traffic emissions. Atmospheric Environment 39 (26), 4793–4802. http://dx.doi.org/10.1016/j.atmosenv.2005.06.023. URL: http://www.sciencedirect.com/science/article/pii/S1352231005005443.

Dixon, N., Tomlin, A., 2007. A lagrangian stochastic model for predicting concentration fluctuations in urban areas. Atmospheric Environment 41 (37), 8114–8127. http://dx.doi.org/10.1016/j.atmosenv.2007.06.033. URL: http://www.sciencedirect.com/science/article/pii/S1352231007006000.

Dieye, A.M., Roy, D.P., Hanan, N.P., Liu, S., Hansen, M., Touré, A., 2012. Sensitivity analysis of the GEMS soil organic carbon model to land cover land use classification uncertainties under different climate scenarios in senegal. Biogeosciences 9 (2), 631–648. http://dx.doi.org/10.5194/bg-9-631-2012. URL: http://www.biogeosciences.net/9/631/2012/.

Fuentes-Garí, M., Misener, R., García-Munzer, D., Velliou, E., Georgiadis, M.C., Kostoglou, M., Pistikopoulos, E.N., Panoskaltsis, N., Mantalaris, A., 2015. A mathematical model of subpopulation kinetics for the deconvolution of leukaemia heterogeneity. Journal of The Royal Society Interface 12 (108). http://dx.doi.org/10.1098/rsif.2015.0276.

Friedlingstein, P., Cox, P., Betts, R., Bopp, L., von Bloch, W., Brovkin, V., Cadule, P., Doney, S., Eby, M., Fung, I., Bala, G., John, J., Jones, C., Joos, F., Kato, T., Kawamiya, M., Knorr, W., Lindsay, K., Matthews, H.D., Raddatz, T., Rayner, P., Reick, C., Roeckner, E., Schnitzler, K.-G., Schnur, R., Strassmann, K., Weaver, A.J., Yoshikawa, C., Zeng, N., 2006. Climate—carbon cycle feedback analysis: results from the C^4MIP model intercomparison. Journal of Climate 19, 3337—3353. http://dx.doi.org/10.1175/JCLI3800.1.

Garmory, A., Richardson, E., Mastorakos, E., 2006. Micromixing effects in a reacting plume by the stochastic fields method. Atmospheric Environment 40 (6), 1078—1091. http://dx.doi.org/10.1016/j.atmosenv.2005.11.002. URL: http://www.sciencedirect.com/science/article/pii/S1352231005010617.

Kaminski, T., Knorr, W., Scholze, M., Gobron, N., Pinty, B., Giering, R., Mathieu, P.-P., 2012. Consistent assimilation of MERIS FAPAR and atmospheric co_2 into a terrestrial vegetation model and interactive mission benefit analysis. Biogeosciences 9 (8), 3173—3184. http://dx.doi.org/10.5194/bg-9-3173-2012. URL: http://www.biogeosciences.net/9/3173/2012/.

Kaminski, T., Heimann, M., Giering, R., 1999. A coarse grid three-dimensional global inverse model of the atmospheric transport: 1. Adjoint model and Jacobian matrix. Journal of Geophysical Research: Atmospheres 104 (D15), 18535—18553. http://dx.doi.org/10.1029/1999JD900147. URL. http://dx.doi.org/10.1029/1999JD900147.

Kucherenko, S., 2007. Application of Global Sensitivity Indices for measuring the effectiveness of Quasi-Monte Carlo methods and parameter estimation. In: Proceedings of the Fifth International Conference on Sensitivity Analysis of Model Output, pp. 35—36.

Knorr, W., Heimann, M., 2001. Uncertainties in global terrestrial biosphere modeling: 1. A comprehensive sensitivity analysis with a new photosynthesis and energy balance scheme. Global Biogeochemical Cycles 15 (1), 207—225. http://dx.doi.org/10.1029/1998GB001059. URL: http://dx.doi.org/10.1029/1998GB001059.

Krinner, G., Viovy, N., de Noblet-Ducoudr, N., Oge, J., Polcher, J., Friedlingstein, P., Ciais, P., Sitch, S., Prentice, I.C., 2005. A dynamic global vegetation model for studies of the coupled atmosphere-biosphere system, gB1015, Global Biogeochemical Cycles 19 (1). http://dx.doi.org/10.1029/2003GB002199. URL: http://dx.doi.org/10.1029/2003GB002199.

Kowalczyk, E.A., Wang, Y.P., Law, R.M., Davies, H.L., McGregor, J.L., Abramowitz, G., 2006. The CSIRO atmosphere biosphere land exchange (CABLE) model for use in climate models and as an offline model. In: CSIRO Marine and Atmospheric Research Technical Paper 013.

Li, G., Rosenthal, C., Rabitz, H., 2001. High dimensional model representations. The Journal of Physical Chemistry A 105 (33), 7765—7777. http://dx.doi.org/10.1021/jp010450t. URL: http://dx.doi.org/10.1021/jp010450t.

Li, G., Wang, S.-W., Rabitz, H., 2002a. Practical approaches to construct RS-HDMR component functions. The Journal of Physical Chemistry A 106 (37), 8721—8733. http://dx.doi.org/10.1021/jp014567t. URL: http://dx.doi.org/10.1021/jp014567t.

Li, G., Wang, S.-W., Rabitz, H., Wang, S., Jaff, P., 2002b. Global uncertainty assessments by high dimensional model representations (HDMR). Chemical Engineering Science 57 (21), 4445—4460. http://dx.doi.org/10.1016/S0009-2509(02)00417-7. URL: http://www.sciencedirect.com/science/article/pii/S0009250902004177.

Li, G., Rabitz, H., Wang, S.-W., Georgopoulos, P.G., 2003. Correlation method for variance reduction of Monte Carlo integration in RS-HDMR. Journal of Computational Chemistry 24 (3), 277—283. http://dx.doi.org/10.1002/jcc.10172. URL: http://dx.doi.org/10.1002/jcc.10172.

Li, G., Rabitz, H., 2006. Ratio control variate method for efficiently determining high-dimensional model representations. Journal of Computational Chemistry 27 (10), 1112—1118. http://dx.doi.org/10.1002/jcc.20435. URL: http://dx.doi.org/10.1002/jcc.20435.

Lu, X., Wang, Y.-P., Ziehn, T., Dai, Y., 2013. An efficient method for global parameter sensitivity analysis and its applications to the Australian community land surface model (CABLE). Agricultural and Forest Meteorology 182—183 (0), 292—303. http://dx.doi.org/10.1016/j.agrformet.2013.04.003. URL: http://www.sciencedirect.com/science/article/pii/S0168192313000804.

Morris, M.D., 1991. Factorial sampling plans for preliminary computational experiments. Technometrics 33 (2), 161—174. URL: http://www.jstor.org/stable/1269043.

Münzer, D.G.G., Kostoglou, M., Georgiadis, M.C., Pistikopoulos, E.N., Mantalaris, A., 2015a. Cyclin and DNA distributed cell cycle model for GS-NS0 cells. PLoS Computational Biology 11 (2), e1004062. http://dx.doi.org/10.1371/journal.pcbi.1004062.

Münzer, D.G., Ivarsson, M., Usaku, C., Habicher, T., Soos, M., Morbidelli, M., Pistikopoulos, E., Mantalaris, A., 2015b. An unstructured model of metabolic and temperature dependent cell cycle arrest in hybridoma batch and fed-batch cultures. Biochemical Engineering Journal 93, 260−273. http://dx.doi.org/10.1016/j.bej.2014.10.013. URL: http://www.sciencedirect.com/science/article/pii/S1369703X14002976.

Mihalas, S., Dong, Y., von der Heydt, R., Niebur, E., 2011. Mechanisms of perceptual organization provide auto-zoom and auto-localization for attention to objects. Proceedings of the National Academy of Sciences of the United States of America 108 (18), 7583−7588. http://dx.doi.org/10.1073/pnas.1014655108. URL: http://www.pnas.org/content/108/18/7583.abstract.

Middleton, D., Jones, A., Redington, A., Thomson, D., Sokhi, R., Luhana, L., Fisher, B., 2008. Lagrangian modelling of plume chemistry for secondary pollutants in large industrial plumes. Atmospheric Environment 42 (3), 415−427. http://dx.doi.org/10.1016/j.atmosenv.2007.09.056. http://www.sciencedirect.com/science/article/pii/S1352231007008680.

Najm, H.N., Debusschere, B.J., Marzouk, Y.M., Widmer, S., Le Matre, O.P., 2009. Uncertainty quantification in chemical systems. International Journal for Numerical Methods in Engineering 80 (6−7), 789−814. http://dx.doi.org/10.1002/nme.2551. URL: http://dx.doi.org/10.1002/nme.2551.

Oakley, J., O'Hagan, A., 2002. Bayesian inference for the uncertainty distribution of computer model outputs. Biometrika 89 (4), 769−784. http://dx.doi.org/10.1093/biomet/89.4.769. URL: http://biomet.oxfordjournals.org/content/89/4/769.abstract.

Oakley, J.E., O'Hagan, A., 2004. Probabilistic sensitivity analysis of complex models: a Bayesian approach. Journal of the Royal Statistical Society: Series B (Statistical Methodology) 66 (3), 751−769. http://dx.doi.org/10.1111/j.1467-9868.2004.05304.x. URL: http://dx.doi.org/10.1111/j.1467-9868.2004.05304.x.

Petropoulos, G.P., Griffiths, H.M., Carlson, T.N., Ioannou-Katidis, P., Holt, T., 2014. SimSphere Model Sensitivity Analysis Towards Establishing its Use for Deriving Key Parameters Characterising Land Surface Interactions. Geoscientific Model Development 7, 1873−1887. http://dx.doi.org/10.5194/gmd-7-1873-2014.

Petropoulos, G.P., Ireland, G., Griffiths, H., Kennedy, M.C., Ioannou-Katidis, P., Kalivas, D.K.P., 2015. Extending the Global Sensitivity Analysis of the SimSphere model in the Context of its Future Exploitation by the Scientific Community. Water MDPI 7, 2101−2141. http://dx.doi.org/10.3390/w7052101.

Reagan, M.T., Najm, H.N., Pbay, P.P., Knio, O.M., Ghanem, R.G., 2005. Quantifying uncertainty in chemical systems modeling. International Journal of Chemical Kinetics 37 (6), 368−382. http://dx.doi.org/10.1002/kin.20081. URL: http://dx.doi.org/10.1002/kin.20081.

Rabitz, H., Aliş, O.F., Shorter, J., Shim, K., 1999. Efficient input-output model representations. Computer Physics Communications 117 (12), 11−20. http://dx.doi.org/10.1016/S0010-4655(98)00152-0. URL: http://www.sciencedirect.com/science/article/pii/S0010465598001520.

Sobol, I., 2001. Global sensitivity indices for nonlinear mathematical models and their Monte Carlo estimates. The Second IMACS Seminar on Monte Carlo Methods Mathematics and Computers in Simulation 55 (13), 271−280. http://dx.doi.org/10.1016/S0378-4754(00)00270-6. URL: http://www.sciencedirect.com/science/article/pii/S0378475400002706.

Saltelli, A., Ratto, M., Andres, T., Campolongo, F., Cariboni, J., Gatelli, D., Saisana, M., Tarantola, S., 2008. Global Sensitivity Analysis: The Primer. John Wiley & Sons, Ltd.

Sheen, D.A., You, X., Wang, H., Lovås, T., 2009. Spectral uncertainty quantification, propagation and optimization of a detailed kinetic model for ethylene combustion. Proceedings of the Combustion Institute 32 (1), 535−542. http://dx.doi.org/10.1016/j.proci.2008.05.042. URL: http://www.sciencedirect.com/science/article/pii/S1540748908001958.

Sobol', I.M., 1967. On the distribution of points in a cube and the approximate evaluation of integrals. USSR Computational Mathematics and Mathematical Physics 7 (4), 86−112. http://dx.doi.org/10.1016/0041-5553(67)90144-9. URL: http://www.sciencedirect.com/science/article/pii/0041555367901449.

Sawford, B., 2004. Micro-mixing modelling of scalar fluctuations for plumes in homogeneous turbulence. Flow, Turbulence and Combustion 72 (2–4), 133–160. http://dx.doi.org/10.1023/B: APPL.0000044409.74300.db. URL: http://dx.doi.org/10.1023/B%3AAPPL.0000044409.74300.db.

Tarantola, A., 2005. Inverse problem theory and methods for model parameter estimation. Society for Industrial and Applied Mathematics. http://dx.doi.org/10.1137/1.9780898717921. URL: http://epubs.siam.org/doi/abs/10.1137/1.9780898717921.

Tam, Z.Y., Gruber, J., Halliwell, B., Gunawan, R., 2015. Context-dependent role of mitochondrial fusion–fission in clonal expansion of mtDNA mutations. PLoS Computational Biology 11 (5), e1004183. http://dx.doi.org/10.1371/journal.pcbi.1004183.

Thonicke, K., Venevsky, S., Sitch, S., Cramer, W., 2001. The role of fire disturbance for global vegetation dynamics: coupling fire into a dynamic global vegetation model. Global Ecology and Biogeography 10 (6), 661–677. http://dx.doi.org/10.1046/j.1466-822X.2001.00175.x. URL: http://dx.doi.org/10.1046/j.1466-822X.2001.00175.x.

Thomson, D.J., 1987. Criteria for the selection of stochastic models of particle trajectories in turbulent flows. Journal of Fluid Mechanics 180, 529–556. http://dx.doi.org/10.1017/S0022112087001940. URL: http://journals.cambridge.org/article_S0022112087001940.

Wang, Y.P., Law, R.M., Pak, B., 2010. A global model of carbon, nitrogen and phosphorus cycles for the terrestrial biosphere. Biogeosciences 7 (7), 2261–2282. http://dx.doi.org/10.5194/bg-7-2261-2010. URL: http://www.biogeosciences.net/7/2261/2010/.

Wang, Y.P., Kowalczyk, E., Leuning, R., Abramowitz, G., Raupach, M.R., Pak, B., van Gorsel, E., Luhar, A., 2011. Diagnosing errors in a land surface model (CABLE) in the time and frequency domains. g01034, Journal of Geophysical Research: Biogeosciences 116 (G1). http://dx.doi.org/10.1029/2010JG001385. URL: http://dx.doi.org/10.1029/2010JG001385.

Ziehn, T., Knorr, W., Scholze, M., 2011a. Investigating spatial differentiation of model parameters in a carbon cycle data assimilation system, gB2021, Global Biogeochemical Cycles 25 (2). http://dx.doi.org/10.1029/2010GB003886. URL: http://dx.doi.org/10.1029/2010GB003886.

Ziehn, T., Scholze, M., Knorr, W., 2011b. Development of an ensemble-adjoint optimization approach to derive uncertainties in net carbon fluxes. Geoscientific Model Development 4 (4), 1011–1018. http://dx.doi.org/10.5194/gmd-4-1011-2011. URL: http://www.geosci-model-dev.net/4/1011/2011/.

Ziehn, T., Tomlin, A., 2009. GUI–HDMR– A software tool for global sensitivity analysis of complex models. Environmental Modelling & Software 24 (7), 775–785. http://dx.doi.org/10.1016/j.envsoft.2008.12.002. URL: http://www.sciencedirect.com/science/article/pii/S1364815208002168.

Ziehn, T., Tomlin, A., 2008a. Global sensitivity analysis of a 3D street canyon model—Part I. The development of high dimensional model representations. Atmospheric Environment 42 (8), 1857–1873. http://dx.doi.org/10.1016/j.atmosenv.2007.11.018. URL: http://www.sciencedirect.com/science/article/pii/S1352231007010643.

Ziehn, T., Tomlin, A.S., 2008b. A global sensitivity study of sulfur chemistry in a premixed methane flame model using HDMR. International Journal of Chemical Kinetics 40 (11), 742–753. http://dx.doi.org/10.1002/kin.20367. URL: http://dx.doi.org/10.1002/kin.20367.

Ziehn, T., Dixon, N.S., Tomlin, A.S., 2009a. The effects of parametric uncertainties in simulations of a reactive plume using a Lagrangian stochastic model. Atmospheric Environment 43 (37), 5978–5988. http://dx.doi.org/10.1016/j.atmosenv.2009.07.060. URL: http://www.sciencedirect.com/science/article/pii/S1352231009006827.

Ziehn, T., Hughes, K.J., Griffiths, J.F., Porter, R., Tomlin, A.S., 2009b. A global sensitivity study of cyclohexane oxidation under low temperature fuel-rich conditions using HDMR methods. Combustion Theory and Modelling 13 (4), 589–605. http://dx.doi.org/10.1080/13647830902878398. URL: http://www.tandfonline.com/doi/abs/10.1080/13647830902878398.

Zaehle, S., Sitch, S., Smith, B., Hatterman, F., 2005. Effects of parameter uncertainties on the modeling of terrestrial biosphere dynamics, gB3020, Global Biogeochemical Cycles 19 (3). http://dx.doi.org/10.1029/2004GB002395. URL: http://dx.doi.org/10.1029/2004GB002395.

A GLOBAL SENSITIVITY ANALYSIS TOOLBOX TO QUANTIFY DRIVERS OF VEGETATION RADIATIVE TRANSFER MODELS

16

J. Verrelst[1], J.P. Rivera[2]

Universitat de València, València, Spain[1]; Centro de Investigación Científica y de Educación Superior de Ensenada, Ensenada, Mexico[2]

CHAPTER OUTLINE

1. INTRODUCTION

Since the advent of optical remote sensing, physically based radiative transfer models (RTMs) have helped in understanding light interception by plant canopies and interpreting vegetation reflectance in terms of biophysical characteristics (Jacquemoud et al., 2009). RTMs attempt to describe absorption and scattering and are useful in a wide range of applications, including designing vegetation indices, performing sensitivity analyses, developing inversion procedures to accurately retrieve vegetation properties from remotely sensed data, and generating artificial scenes as observed by an optical sensor. In the development of new optical sensors, plant and atmospheric RTMs are currently used in an end-to-end

Sensitivity Analysis in Earth Observation Modelling. http://dx.doi.org/10.1016/B978-0-12-803011-0.00016-1

simulator that functions as a virtual laboratory in any aspect of sensor specifications and data processing. For instance, plant and atmosphere RTMs are heavily used in the preparation of European Space Agency's (ESA's) eighth Earth Explorer mission FLEX [Fluorescence Explorer] (Rivera et al., 2014c).

In all of these Earth observation (EO) studies, an important requisite is knowledge of the key input RTM variables driving the spectral output in a specific spectral region. As such, a sufficiently realistic simplified model that is driven by only the key variables makes exploration of a broad range of target and observation conditions easier and more effective (Mousivand et al., 2014). To achieve this, a sensitivity analysis (SA) needs to be applied.

A SA evaluates the relative importance of each input variable in a model and can be used to identify the most influential variables affecting model outputs (Saltelli et al., 1999; Wainwright et al., 2014; Ireland et al., 2015; Petropoulos et al., 2015). Hence, SA can be applied with RTMs to identify the key determinants of outputs such as reflectance and fluorescence. Less influential variables can also be identified and be safely set to default values under relatively wide ranges of conditions. In general, SA methods may be categorized as either *local* or *global*. Local sensitivity analysis (LSA) methods are often referred to as "one-factor-at-a-time", because they involve changing one input variable at a time while holding all others at their central values, then measuring variation in the outputs. A drawback of LSA methods is that they are informative only at the central point where the calculation is executed and do not encompass the entire input variable space. Thus, LSA methods are inadequate for analyzing complex models having many variables, and they may be highly dimensional and/or nonlinear (Saltelli and Annoni, 2010; Yang, 2011; Nossent et al., 2011).

Unlike LSA, global sensitivity analysis (GSA) explores the full input variable space. The contribution of each input variable to the variation in outputs is averaged over the variation of all input variables, i.e., all input variables are changed together (Saltelli et al., 1999). In general, variance-based SA methods aim to quantify the amount of variance that each input variable contributes to the unconditional variance of the model output (Nossent et al., 2011; Petropoulos et al., 2009c). GSA is thus required to identify the driving variables of an RTM. GSA can be applied to identify the key determinants of spectral outputs such as reflectance, fluorescence, and radiance. However, it can also enable the identification of noninfluential variables, which can then be safely set to arbitrary values under relatively wide ranges of conditions. Although generally regarded as a more comprehensive method, GSA is computationally intensive, and such methods may be perceived as overly complicated. To the best of our knowledge, currently no user-friendly toolbox exists that enables applying GSA methods to RTMs. Yet, the development of a GSA toolbox dedicated to the analysis of RTMs may greatly benefit the use of RTMs in all kinds of remote sensing applications.

Over the past few years, various leaf and canopy RTMs have been brought together and standardized within a single scientific graphical user interface (GUI) toolbox called ARTMO (Automated Radiative Transfer Models Operator) (Verrelst et al., 2012). In ARTMO, RTMs can be operated in a semiautomatic fashion for any kind of optical sensor operating in the visible, near-infrared (NIR) and shortwave infrared (SWIR) range (400−2500 nm). Therefore, having a diverse range of RTMs with varying complexity at hand, this platform can perfectly serve as a benchmark for the development and evaluation of a GSA toolbox. This chapter introduces a novel GSA toolbox developed within the ARTMO framework.

This chapter is organized as follows. Section 2 first briefly summarizes the theory of variance-based GSA. Section 3 then outlines the ARTMO software package, and Section 4 introduces ARTMO's GSA toolbox. Section 5 subsequently provides a few case studies of analyzed RTMs using the GSA toolbox. Finally, Section 6 provides a discussion on potential applications and Section 7 concludes the chapter.

2. VARIANCE-BASED GLOBAL SENSITIVITY ANALYSIS

Since the pioneering work of Sobol, the most popular variance-based methods include the Sobol' method (Sobol', 1990), the Fourier Amplitude Sensitivity Test (FAST) (Cukier et al., 1973), and a modified version of the Sobol' method proposed by Saltelli et al. (2010). This modification has been demonstrated to be effective in identifying the so-called Sobol's sensitivity indices. These indices quantify both the main sensitivity effects (first-order effects) (i.e., the contribution to the variance of the model output by each input variable), and total sensitivity effects (i.e., the first-order effect plus interactions with other input variables) of input variables (Song et al., 2012). This method has been implemented into the GSA toolbox. A description according to Song et al. (2012) is outlined.

Formally, given a model $Y = f(X)$, where Y is the model output, $X = (X_1, X_2,..., X_k)$ is the input parameter vector. A variance decomposition of f suggested by Sobol' (1990) is:

$$V(Y) = \sum_{i=1}^{k} V_i + \sum_{i=1}^{k} \sum_{j=i+1}^{k} V_{ij}... + V_{1,...,k} \tag{16.1}$$

where X is rescaled to a k-dimensional unit hypercube Ω^k, $\Omega^k = |X|0 \leq X_i \leq 1, i = 1,...,k\}$; $V(Y)$ is the total unconditional variance; V_i is the partial variance or "main effect" of X_i on Y and given by $V_i = V[E(Y|X_i)]$; and V_{ij} is the joint impact of X_i and X_j on the total variance minus their first-order effects.

Here, the first-order sensitivity index S_i and total effect sensitivity index S_{Ti} are given as (Saltelli et al., 2008):

$$S_i = \frac{V_i}{V(Y)} = \frac{V[E(Y|X_i)]}{V(Y)} \tag{16.2}$$

$$S_{Ti} = S_i + \sum_{j \neq i} S_{ij} + \cdots = \frac{E[V(Y|X_{\sim i})]}{V(Y)} \tag{16.3}$$

where $X_{\sim i}$ denotes variation in all input variables and X_i, S_{ij} is the contribution to the total variance by the interactions between variables.

Following Saltelli et al. (2010), to compute S_i and S_{Ti} two independent input variable sampling matrices P and Q with dimension (N, k) were created, where N is the sample size and k is the number of input variables. Each row in matrix P and Q represents a possible value of X. The variable ranges in the matrices are scaled between 0 and 1. The Monte Carlo approximations for $V(Y)$, S_i, and S_{Ti} are defined as follows (Nossent et al., 2011; Saltelli et al., 2010):

$$\widehat{f_0} = \frac{1}{N} \sum_{j=1}^{N} f(P)_j \tag{16.4}$$

$$\widehat{V}(Y) = \frac{1}{N} \sum_{j=1}^{N} \left(f(P)_j\right)^2 - \hat{f}_0^2 \tag{16.5}$$

$$\widehat{S_i} = \frac{1}{N} \sum_{j=1}^{N} \frac{f(Q)_j \left(f\left(P_Q^{(i)}\right)_j - f(P)_j \right)}{\widehat{V}(Y)} \tag{16.6}$$

$$\widehat{S_{Ti}} = \frac{1}{2N} \sum_{j=1}^{N} \frac{\left(f(P)_j - f\left(P_Q^{(i)}\right)_j \right)^2}{\widehat{V}(Y)} \tag{16.7}$$

where $\widehat{\cdots}$ is the estimate, $\widehat{f_0}$ is the estimated value of the model output, and $P_Q^{(i)}$ represents all columns from P except the ith column, which is from Q, using a radial sampling scheme (Saltelli and Annoni, 2010). To compute S_i and S_{Ti} simultaneously, a scheme suggested by Saltelli (2002) was used, which reduced the model runs to $N(k + 2)$.

To sample the P and Q matrices the Sobol' quasi-random sampling sequence (Sobol', 1967) was used. This sequence helps to distribute the sampling points as uniformly as possible in the variable space to avoid clustering and increases the convergence rate (Saltelli et al., 2008). Therefore, the use of these sequences enhances the convergence of the Monte Carlo integrals. The Monte Carlo integration, and thus the Sobol's SA, normally converges at a rate of $1/\sqrt{n}$, whereas Sobol' quasi-random sampling enhances this to almost $1/n$ (Nossent et al., 2011).

3. RADIATIVE TRANSFER MODELS AND ARTMO

This section introduces plant RTMs. Radiative transfer is the physical phenomenon of energy transfer in the form of electromagnetic radiation. The propagation of radiation through a vegetated medium is affected by absorption, emission, and scattering processes (e.g., Jacquemoud et al., 2009). Radiative transfer modeling of a vegetated medium is based on physical laws and takes into account physical processes describing the interaction of radiation with the diverse components of nature, e.g., at foliage, canopy, and atmosphere levels. Accordingly, RTMs can take place at the leaf, canopy or atmosphere scale. Leaf optical models simulate the radiative transfer interactions of light and leaf biochemical properties. Canopy RTMs describe the interaction between light and plant canopy and are of importance for extracting canopy biophysical parameters. These models provide the linkage with observations from remote optical sensors by simulating top-of-canopy (TOC) reflectance. When inverting canopy models against a remote sensing image, they allow retrieving structural characteristics (e.g., Rivera et al., 2013). When a canopy is coupled with a leaf optical model, leaf biochemical attributes can also be retrieved from a remote sensing image (e.g., Jacquemoud et al., 1995). More advanced RTMs, such as soil-vegetation-atmosphere-transfer (SVAT) models, typically combine leaf, canopy, and atmospheric characteristics. These models typically consist of several submodels and are able to calculate vegetation—light interaction processes at the ecosystem scale.

The ARTMO toolbox, freely downloadable at http://ipl.uv.es/artmo, was essentially developed to streamline the use of a variety of leaf, canopy, and SVAT RTMs in one scientific GUI toolbox (Verrelst et al., 2012). The toolbox was subsequently expanded with all types of postprocessing tools and toolboxes. The toolbox is designed modularly, i.e., RTMs or postprocessing tools can be added or removed by simply adding or removing their folders within ARTMO's structure. In short,

ARTMO permits the user (1) to choose between leaf and canopy RTMs of a low to high complexity (e.g., the leaf models PROSPECT-4, PROSPECT-5, DLM; the canopy models 4SAIL, FLIGHT, INFORM; and the SVAT model SCOPE); (2) to specify or select spectral band settings specifically for various existing air- and space-borne sensors or user-defined settings, typically for future sensor systems; (3) to simulate large data sets of TOC reflectance spectra for sensors sensitive in the optical range (400−2500 nm); (4) to generate look-up tables (LUTs), which are stored in a relational SQL database management system (MySQL, version 5.5 or higher; local installment required); and finally (5) to configure and run various retrieval scenarios using EO reflectance data sets for biophysical variable mapping applications (Verrelst et al., 2015b).

ARTMO has been developed in Matlab (2011 version or higher). Fig. 16.1 presents ARTMO's v. 3.14 main window and a systematic overview of the drop-down menu below. To start with, in the main window a new project can be initiated, a sensor chosen, and a comment added, whereas all processing modules are accessible through drop-down menus at the top bar.

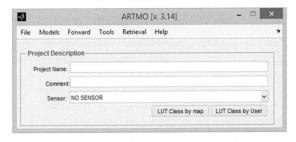

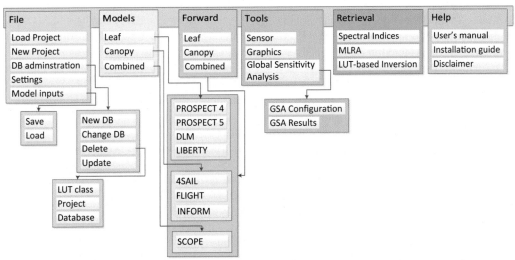

FIGURE 16.1

Screenshot of ARTMO's v3.14 main window (top) and schematic overview of its drop-down menu (bottom).

A first rudimentary version of ARTMO was developed with the purpose of automating mapping biophysical parameters through RTM inversion (Verrelst et al., 2012). The toolbox has been improved and expanded since then, such as the implementation of retrieval toolboxes. These toolboxes allow for the mapping of biophysical parameters by making use of RTMs in a semiautomatic fashion. They are based on parametric and nonparametric regression as well as physically based inversion using an LUT and led to the development of a: (1) "Spectral Indices assessment toolbox" (Rivera et al., 2014b); (2) "Machine Learning Regression Algorithm toolbox" (Rivera et al., 2014a); and, (3) "LUT-based inversion toolbox" (Rivera et al., 2013). Beyond retrieval, several RTM postprocessing modules have been developed. ARTMO v3.14, with the most important novel inclusion of the "Global Sensitivity Analysis (GSA) Toolbox," is formally presented in this chapter.

4. GLOBAL SENSITIVITY ANALYSIS TOOLBOX

The GSA toolbox is essentially built out of two modules (GUIs): (1) the GSA-configuring module and the (2) GSA results visualization module. Both modules are outlined.

4.1 GLOBAL SENSITIVITY ANALYSIS CONFIGURATION MODULE

The core component that enables selecting an RTM and configuring the GSA is displayed in Fig. 16.2. The GSA configuration module can be broken down into three parts: (1) main settings, (2) RTM input settings, and (3) RTM output settings.

In the first panel the main settings of configuring a GSA need to be given. These include: (1) to give a project name (e.g. 'GSA_PROSAIL'), (2) to choose a GSA method, (3) to choose the number of samples per variable, (4) to choose the RTM to be analyzed, and optionally, (5) to choose for resampling of the spectra to a sensor setting. These general settings are briefly outlined as follows:

1. *Name*: The GSA toolbox stores the Sobol's sensitivity indices (S_i and S_{Ti}) results in a MySQL database that is running underneath ARTMO. By entering a name, the general information of the GSA exercise is stored in a MySQL overview table. GSA sensitivity results are stored in an associated table.
2. *GSA method*: In this first version, only the GSA method of Saltelli et al. (2010) has been successfully implemented. This method is most commonly used in GSA studies and delivers robust results. Alternative GSA methods such as extended FAST and Sobol have been explored but led to unstable results and were therefore discarded.
3. *Number of samples*: Here the number of samples (N) per input variable (k) can be entered. Following Saltelli (2002), in total $N(k + 2)$ simulations are generated.
4. *RTM*: In this first version the following RTMs can be analyzed with the GSA toolbox:
   a. The leaf models: PROSPECT4, PROSPECT5, DLM, and LIBERTY.
   b. The leaf and canopy models: PROSPECT4-SAIL, PROSPECT5-SAIL, LIBERTY-SAIL, PROSPECT4-INFORM, PROSPECT5-INFORM, LIBERTY-INFORM, and the SVAT model SCOPE. Note that a canopy model is always coupled with a leaf model because a canopy model requires the input of leaf optical data.
5. *Sensor*: By selecting a sensor, GSA results will be generated only for the spectral bands according to the chosen optical sensor. This option allows identifying driving RTM variables for spectral

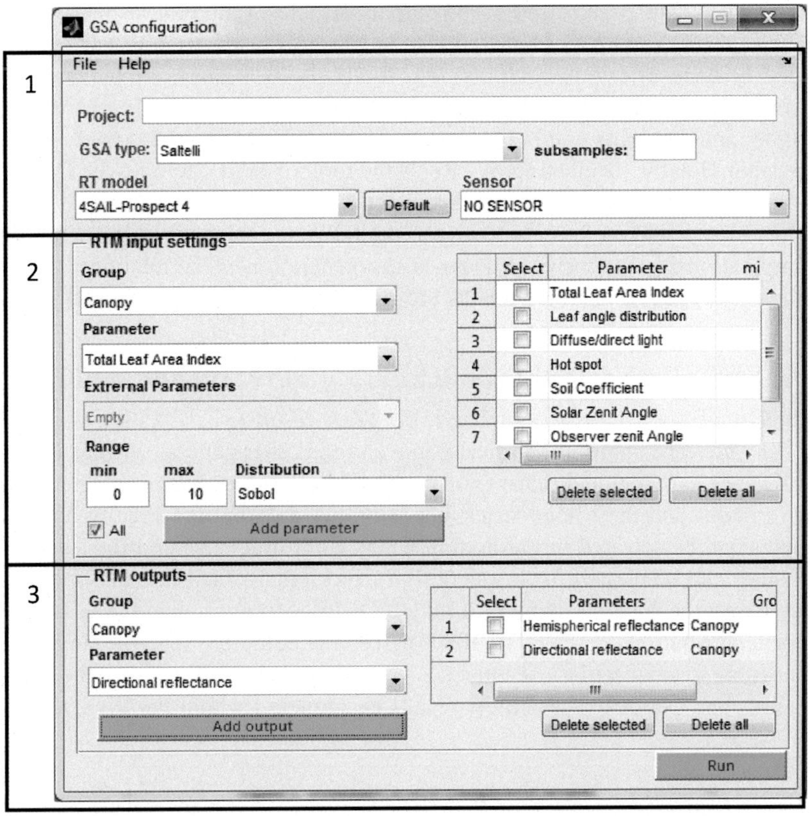

FIGURE 16.2

Screenshot of GSA configuration module.

data as observed by a sensor. In ARTMO's Sensor module any kind of sensor within the 400–2400 nm can be configured. By default the following operative and forthcoming satellite optical sensors are provided: LANDSAT-7, SPOT-4, CHRIS-M3, MODIS, MERIS, Sentinel-2, Sentinel-3 OLCI, and SLSTR.

In the second panel, the RTM input setting can be configured. Depending on the complexity of the selected RTM, inputs are organized per "Group." This is especially the case for the SVAT model SCOPE that is built of several submodels that are thematically grouped (e.g., soil, leaf, canopy, micrometeorology). Otherwise, only the group "Inputs" will appear. The RTM input variables are listed in the drop-down menu below. The input variables have to be selected one-by-one or all together. Thereby, for each variable the minimum and maximum boundaries and a data distribution have to be given. Default values are provided and by default the Sobol quasi-random sequence is provided, and a uniform, normal, and exponential distribution can also be selected. By clicking on "Add parameter" the variable with their boundaries is then entered into the right panel. The boundaries can be modified, and the inserted variables can be removed.

The bottom panel enables selection of the output variables. Often RTMs provide more than one output variable, usually spectral output (e.g., reflectance, transmittance), and in case of SCOPE outputs can also be fluxes (i.e., point outputs). The toolbox allows analyzing the relative contribution of each entered input variable to each given output product. Accordingly, the user can select to include multiple model outputs within the same analysis. Selected model outputs will appear in the right-bottom panel. Finally, by clicking on "Run," the toolbox starts running the GSA. Accordingly, first the simulations will be generated and subsequently the Sobol' indices will be calculated. This can take a while, depending on the complexity of the RTM and the entered sampling size. However, the most computationally expensive process is the generation of simulations; the GSA itself is executed rather fast. Finally, obtained results are stored in the MySQL database.

4.2 GLOBAL SENSITIVITY ANALYSIS RESULTS VISUALIZATION MODULE

Once the whole analysis has been completed, the GSA results visualization window will appear (Fig. 16.3). This interface shows the project name and the other general settings (GSA type, RTM, number samples per variable, and Sensor) of the analyzed GSA experiment. The panel below lists the inserted variables and their boundaries and sampling distribution. Finally, the analyzed RTM output attributes can be selected in the bottom panel. First-order (S_i) and total-order (S_{Ti}) Sobol's sensitivity indices can be plotted for inspection or exported. To facilitate the interpretation of obtained results across the spectral domain, it may be helpful to normalize the GSA results, i.e., the sum of relative contributions is scaled to 100% across the complete spectral range.

It is also possible to view earlier-executed GSA exercises. By scrolling down the drop-down menu, all GSA exercises and associated sensitivity results encountered within the MySQL database can be

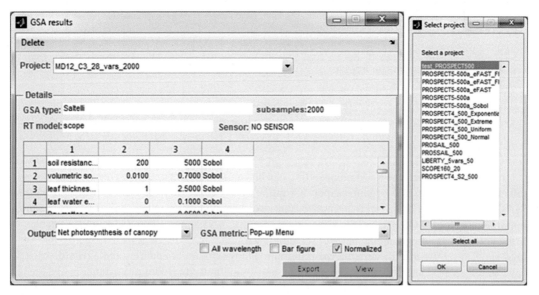

FIGURE 16.3

Screenshot of GSA results visualization module (left) and overview of projects to be deleted (right).

accessed. Also, when clicking on "Delete," a GUI appears that provides an overview of all earlier-executed GSA projects. Selected projects are then deleted.

5. CASE STUDIES

Having the GSA toolbox outlined, its functioning will be demonstrated through the analysis of some RTMs. Total sensitivity indices (S_{Ti}) will be calculated for the leaf model PROSPECT-4, the leaf and canopy model PROSAIL (i.e., PROSPECT coupled with the canopy model SAIL), and the SVAT model SCOPE. These RTMs are briefly described as follows:

1. The leaf optical PROSPECT models (Jacquemoud et al., 2009) are among the most popular leaf RTMs; likely due to their availability and low number of input variables. PROSPECT-4 is currently one of the most widely used leaf optical models and is based on earlier PROSPECT versions. The model calculates leaf reflectance and transmittance as a function of its biochemistry and anatomical structure. It consists of four parameters, namely, leaf structure, chlorophyll content (C_{ab}), equivalent water thickness (C_w), and dry matter content (Cd). PROSPECT-4 simulates directional hemispherical reflectance over the solar spectrum from 400 to 2500 nm at a fine spectral resolution of 1 nm. PROSPECT-5 is an extension of PROSPECT-4, and additionally includes the parameters total carotenoid content (Car) and brown pigments (Bp) (Feret et al., 2008).

2. At the canopy scale, SAIL is the most popular canopy RTM, also likely due to its availability and rather low number of input variables (Jacquemoud et al., 2009). SAIL is based on a four-stream approximation of the RT equation, in which case one distinguishes two direct fluxes (incident solar flux and radiance in the viewing direction) and two diffuse fluxes (upward and downward hemispherical flux) (Verhoef et al., 2007). SAIL inputs consist of leaf area index (LAI), leaf angle distribution (LAD), ratio of diffuse and direct radiation, soil coefficient, hot spot, and sun−target−sensor geometry. Spectral input consists of leaf reflectance and transmittance spectra and a soil reflectance spectrum. The leaf optical properties can come from a leaf RTM such as PROSPECT. By coupling PROSPECT with SAIL (PROSAIL), a leaf-canopy model is generated that allows analyzing the impact leaf biochemical and structural variables at the canopy scale.

3. SCOPE is a vertical integrated radiative transfer and energy balance SVAT model (Van Der Tol et al., 2009; Van Der Tol et al., 2014). It calculates radiation transfer in a multilayer canopy to obtain reflectance and fluorescence in the observation direction as a function of the solar zenith angle (SZA) and leaf inclination distribution. The distribution of absorbed radiation within the canopy is calculated with the SAIL model (Verhoef, 1984). The distribution of absorbed radiation is further used in a micrometeorological representation of the canopy for the calculation of photosynthesis, sun-induced chlorophyll fluorescence, and latent and sensible heat. The fluorescence and thermal radiation emitted by individual leaves is finally propagated though the canopy, again with the SAIL modeling concept (Van Der Tol et al., 2009). A leaf biochemical module enables to calculate the photosynthetic rate and the fraction of absorbed light returning as fluorescence, as a function of APAR, temperature, relative humidity, and the concentrations of CO_2 and O_2. A PROSPECT-like leaf optical model enables the propagation of leaf physiological mechanisms such as fluorescence through the leaf and eventually through the canopy. See Van Der Tol et al. (2014) for more details.

5.1 EXPERIMENTAL SETUP

Prior to interpreting the generated Sobol's sensitivity indices, the impact of different sampling distributions on the GSA outputs was tested. Its impact appeared to be marginal, therefore only the default Sobol' quasi-random sampling sequence was used in subsequent exercises. In each GSA exercise, each variable was sampled 5000 times.

RTM input ranges for PROSPECT-4 and PROSPECT-5 (2 additional variables) and SAIL are provided in Table 16.1.

5.2 RESULTS

5.2.1 PROSPECT S_{Ti} Results

Given that the Sobol' total-order effect (S_{Ti}) of a variable includes the first-order plus all the interaction effects, this sensitivity index is of most interest to disentangle the functioning of an RTM. For the PROSPECT leaf optical models, normalized S_{Ti} results for leaf reflectance and transmittance along the 400–2500 nm region are displayed in Fig. 16.4. Because the relative contributions sum up to 100%, the role of each input variable along the spectral range can be intuitively inspected. For instance, PROSPECT-4 S_{Ti} results clarify that the relative role of C_{ab} is only exhibited in the visible part, and the importance of the structural variables N and Cd is exhibited across the whole spectral range. Leaf water content (C_w) only governs reflectance and transmittance from around 1200 nm onward. Note that reflectance and transmittance results hardly differ. Yet water content plays a somewhat more dominant role in leaf transmittance. PROSPECT-5 consists of two extra variables, Bp and Car. Particularly, Bp play a dominant role in the visible part of the spectrum

Table 16.1 PROSPECT-4, PROSPECT-5, and SAIL input variables boundaries					
Model		**Variable**	**Unit**	**Min.**	**Max.**
Leaf: PROSPECT					
−4	N	Leaf structural variable	[−]	1	4
	C_{ab}	Chlorophyll a + b content	µg/cm^2	0	100
	C_w	Equivalent water thickness	g/cm^2 or cm	0.0001	0.05
	Cd	Dry matter content	g/cm^2	0.0001	0.05
−5	Car	Carotenoids	µg/cm^2	0	10
	Bp	Brown pigments	g/cm^2	0	10
Canopy: SAIL					
	LAI	Total leaf area index	m^2 m^2	0	10
	LAD	Leaf angle distribution	degree	0	90
		Hot spot	[−]	0	1
		Diffuse/direct light	[−]	0	100
		Soil coefficient	[−]	0	1
	SZA	Solar zenith angle	Degree	0	90
	RAA	Relative azimuth angle	Degree	0	180

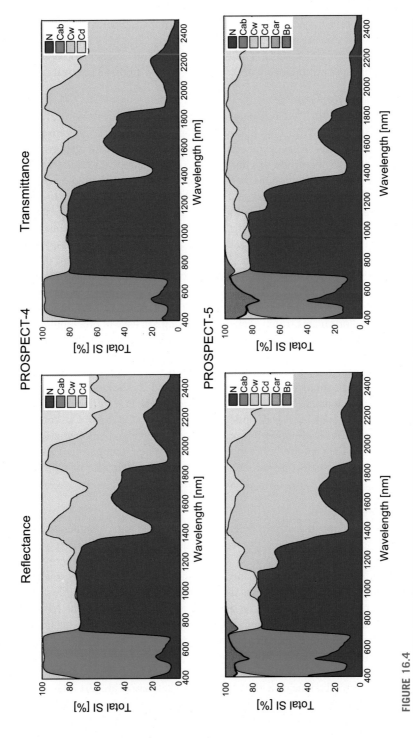

FIGURE 16.4

S_{Ti} results for PROSPECT-4 (top) and PROSPECT-5 (bottom) reflectance (left) and transmittance (right) outputs.

(400–700 nm), both in reflectance and transmittance. Compared with PROSPECT-4, it can be observed that Bp suppress somewhat the dominant role of C_{ab}. Car play a much smaller role, hardly 1%, and only in the 400–550 nm spectral window. Overall, results suggest the utility of GSA, e.g., when aiming to develop spectral indices with sensitivity to leaf pigments or water content.

5.2.2 PROSAIL S_{Ti} Results

When subsequently upscaling to TOC reflectance by coupling the leaf PROSPECT models with the canopy model SAIL, i.e., PROSAIL, both leaf and canopy input variables govern the variability of directional and hemispherical TOC reflectance. Normalized S_{Ti} results are displayed in Fig. 16.5. It can be noted that in all cases the prime driving variable is the canopy structural variable LAI. LAI quantifies leaf density and is a key variable in many surface and climate studies (e.g., Jacquemoud et al., 2009). The PROSAIL S_{Ti} results indicate that LAI alone can explain up to 50% of the total variability (i.e., with interactions), especially in the SWIR (1400–3000 nm). Its dominance can be explained by two factors. First, LAI controls the amount of leaf elements and consequently controls the amount of spectrally distinct soil reflectance propagating through the canopy (in case of low LAI). Second, the introduction of leaf elements enables the interactions with leaf optical properties, i.e., a higher LAI will imply more light absorption due to the absorption properties of leaf elements and scattering effects. Another driving variable, especially for directional TOC reflectance, is LAD, which controls the orientation of the leaves (Verhoef, 1984). LAD, like LAI, regulates to an extent whether TOC reflectance originates from vegetation only (e.g., in case of erectophile LAD) or is also mixed with soil reflectance (e.g., in case of planophile LAD). Other drivers are soil coefficient (regulates the contribution of wet and dry soil portions) and SZA. Also, the PROSPECT leaf optical properties play a prominent role. The patterns of leaf biochemical variables are similar as those for PROSPECT alone, but their relative importance is suppressed by the dominance of the aforementioned driving canopy variables, especially in case of directional TOC reflectance. Remarkably, for hemispherical TOC reflectance, the leaf optical properties N, C_{ab}, and C_w continue to play a dominant role. Another difference between directional reflectance and hemispherical reflectance is that for the latter SZA appears to be more important. When comparing PROSPECT-4 with PROSPECT-5 the dominance of Bp prevails. It suppresses the dominance of C_{ab}, apparently even more than when comparing both leaf optical models alone.

5.2.3 SCOPE S_{Ti} Results

The following case study moves away from reflectance and instead aims to understand the underlying mechanisms of simulated solar-induced chlorophyll fluorescence (SIF). SIF represents absorbed radiant energy by chlorophyll pigments that is not converted into chemical energy (or heat) but is reemitted as radiation. Although the emitted canopy-leaving SIF flux is relatively small compared with reflected sunlight (about 1–5% in the NIR), it is a broadband spectrum within the 650–800 nm spectral window with two emission maxima: in the red around 690 nm and in the NIR around 740 nm (Papageorgiou and Govindjee, 2004; Baker, 2008). Thanks to progress in technology and data processing, only more recently developed imaging spectrometers are able to capture this subtle signal. The challenge now lies in correctly interpreting the SIF signal, especially when aiming to exploit SIF and its relationships with canopy photosynthesis, for instance, by linking SIF with gross primary production (GPP) as is currently done on a global scale (Frankenberg et al., 2011; Guanter et al., 2012).

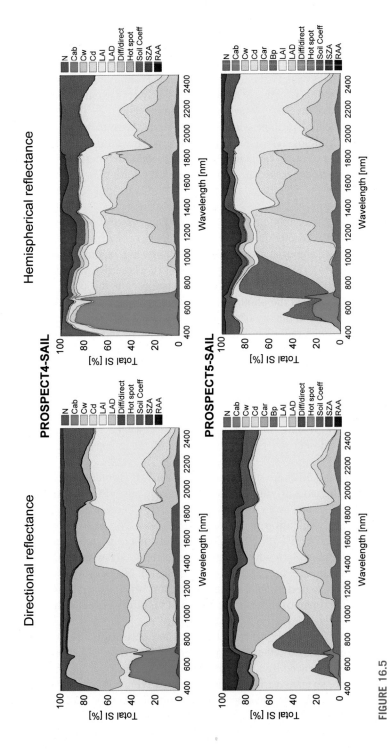

FIGURE 16.5

S_{Ti} results for PROSPECT-4 coupled with SAIL (top) and PROSPECT-5 coupled with SAIL (bottom) directional reflectance (left) and hemispherical reflectance (right) outputs.

The use of an advanced RTM such as SCOPE can be very helpful in resolving open questions about how to interpret the SIF signal. SCOPE incorporates leaf biochemical submodels that regulate photosynthesis and fluorescence emission. By applying a GSA exercise on SCOPE, S_{Ti} results can reveal driving variables of SIF as well as of photosynthesis of the canopy, and so assist in disentangling SIF—photosynthesis relationships. Because SCOPE consists of over 30 input variables, many of them may not be needed for the study of SIF—photosynthesis relationships. The GSA exercise will enable the identification of noninfluential variables. In this exercise, the large majority of SCOPE variables were included, i.e., leaf physiological, leaf optical, canopy structural, and micrometeorological variables. To keep the amount of input variables manageable, the following variables have been kept fixed: geometry, soil, and aerodynamic and canopy geometry variables. See Table 16.2 for the variables' boundaries. Again, the Saltelli et al. (2010) method was applied, and each variable was sampled 2000 times.

Fig. 16.6 shows the S_{Ti} SIF results along the 650—800 nm spectral window. It is immediately evident from visual analysis that despite the many input variables, only a few variables drive simulated SIF. These are (sorted by most important variables first): shortwave incoming radiation (Rin), LAI, atmospheric vapor pressure (ea), C_{ab}, and air temperature (Ta). These five variables alone contribute about 94.5% of the normalized S_{Ti} along the spectral window. On the other hand, the key leaf physiological variable that regulates the photosynthetic capacity, V_{cmo}, only accounts for about 1%. Hence, S_{Ti} results underline the complexity of the SIF signal, especially when aiming to relate SIF to photosynthesis. Surface incoming radiation can be readily obtained, and thus in principle there is no need to include Rin in the analysis. Apart from that, it is very likely that relationships are confounded by variations in vegetation, such as leaf C_{ab} and canopy structure, here expressed as LAI but also LAD (not included in this analysis).

For a more in-depth disentanglement of the complex relationships between SIF and photosynthesis, a final GSA exercise was employed. Apart from spectral outputs, SCOPE also calculates fluxes over the whole simulated canopy system, such as the total fluorescence (F_{total}), i.e., integral of the fluorescence broadband signal, and the net photosynthesis of the canopy (NPC), which is closely related to GPP.

The S_{Ti} results for F_{total} and NPC are displayed in Fig. 16.7. It is interesting to observe that SIF and NPC identified mostly the same driving variables. Both F_{total} and NPC are primarily governed by micrometeorological variables such as Ta, ea, Rin, atmospheric CO_2 concentration, wind speed (u), and the biochemical variable regulating photosynthetic capacity V_{cmo}, and LAI. These variables count up to 87.1% for F_{total} and 67.7% for NPC. Consequently, the joint importance of these driving variables may to a great extent explain why it is possible to relate SIF with NPC (e.g., as shown in Frankenberg et al., 2011; Guanter et al., 2012 on a global scale). Noteworthy, V_{cmo} appears to be a driving variable of NPC, as expected, but considerably less than F_{total}. That suggests that the relationships between SIF and photosynthesis are not so straightforward, and thus care is required when interpreting SIF.

Moreover, some clear differences between both fluxes are observable. Most obviously, F_{total} is much more driven by atmospheric variables such as incoming shortwave radiation (Rin) and Ta than NPC. In turn, NPC is more influenced by a diversity of leaf biochemistry (V_{cmo}, m, kV) leaf optical, and structural variables. Since so many variables play a role in driving NPC, it suggests that it is a more complex flux than is F_{total}. Hence, to develop unbiased relationships toward NPC based on SIF measurements remains a challenging task.

Table 16.2 SCOPE v. 153 used input variables boundaries; the remaining SCOPE variables have been kept fixed

	Variable	Unit	Min	Max
Leaf Physiology				
M	Ball–Berry stomatal conductance parameter	[−]	2	20
kV	Extinction coefficient for a vertical profile of V_{cmo} (maximum value of V_{cmo} occurs at the top of the canopy)	[−]	0	0.8
Res	Parameter for dark respiration $(Rd = Rd_{param}*V_{cmo})$	[−]	0.001	0.03
V_{cmo}	Maximum carboxylation capacity (at optimum temperature)	μmol/ms	0	200
Leaf Optical				
N	Mesophyll structural parameter in PROSPECT	[−]	1	2.5
C_w	Water content in PROSPECT	g/cm^2	0	0.1
Cd	Dry matter content in PROSPECT	g/cm^2	0	0.05
Cs	Senescence factor PROSPECT	[−]	0	0.9
C_{ab}	Chlorophyll content in PROSPECT	μg/cm^2	0	80
Canopy				
Lw	Leaf width	m	0.01	0.1
LAI	Leaf area index	m^{2}m^2	0	7
Hc	Canopy height	m	0.1	2
Micrometeorologic				
P	Air pressure	hPa	300	1090
U	Wind speed	m/s	0	50
Oa	O_2 concentration in the air	ppm	0	220
ea	Atmospheric vapor pressure	hPa	0	150
Ca	CO_2 concentration in the air	ppm	50	1000
Ta	Air temperature	°C	−10	50
Rin	Incoming shortwave radiation	W/m^2	0	1400
Rli	Incoming longwave radiation	W/m^2	0	400
Aerodynamic				
rwc	Within-canopy-layer resistance	s/m	0	20

See Verrelst et al. (2015a) for fixed values.

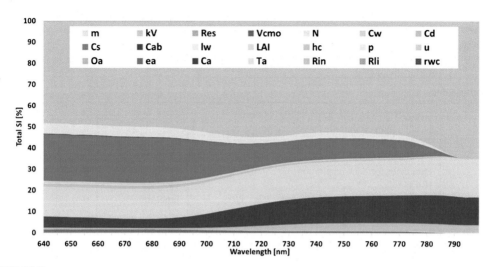

FIGURE 16.6

S_{Ti} results for SCOPE v.153 solar-induced fluorescence outputs.

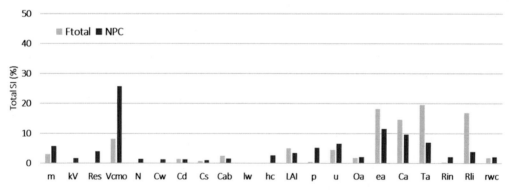

FIGURE 16.7

S_{Ti} results for SCOPE v.153 fluxes F_{total} (integral of the fluorescence broadband signal) and net photosynthesis of the canopy (NPC).

Consequently, to enable a proper interpretation of SIF maps, the GSA analysis reveals that spatially explicit information of variables that drive the canopy-leaving SIF emission is of key importance (see also Van Wittenberghe et al., 2015; Verrelst et al., 2015a). Although this requires a dedicated data assimilation approach, in fact most of the key variables can already be accurately retrieved, some of them even on a global and operational basis (e.g., Rin, Ta, LAI; e.g., Baret et al., 2013). For other surface variables (e.g., C_{ab}), operational retrieval schemes are only in their infancy (e.g., Verrelst et al., 2013).

Apart from photosynthesis, other fluxes generated by SCOPE such as FPAR, net radiation of the canopy, and average canopy temperature can be analyzed in the same way. With the use of the GSA toolbox, improved insights can be gained in identifying key variables that should be quantified to process satellite data toward these kinds of high-level products.

6. DISCUSSION

The here presented ARTMO GSA toolbox allows analyzing leaf, canopy, and combined RTMs through two GUIs. In the "configuration module" it requires selecting an RTM, input variables boundaries, number of samples, and the RTM output variables. Optionally, an optical sensor can be selected within the 400−2400 nm spectral range, which enables GSA calculation for a predefined Sensor band setting. The implemented GSA method is based on Saltelli et al. (2010). Once the configuration is completed, the GSA procedure starts running. First, the RTM simulations over the predefined variable space are generated, and subsequently, the Sobol's sensitivity indices are calculated. These results can then be accessed in the "GSA results" module, where output visualization options are provided.

The toolbox can facilitate remote sensing scientists with an improved understanding of radiation−vegetation interactions and the role of underlying mechanisms. The generated results enable the user to immediately visualize the driving input variables along the spectral range, which can be beneficial for a diversity of applications, for instance, (1) to find out the necessary variables to be sampled in the field, (2) to identify the most sensitive spectral wavelengths for a targeting biophysical variable, and (3) to identify invertible biophysical variables.

For instance, the driving variables across the spectral range of the most widely used leaf and canopy models have been identified, i.e. PROSPECT and SAIL, respectively. The results intuitively visualized in which spectral window driving variables are the most sensitive. At the canopy scale, PROSAIL results show that TOC reflectance is a function of both leaf optical (e.g., C_{ab}, water content) and canopy structural variables (e.g., LAI, leaf angle distribution). Accordingly, these results can lead to an improved understanding of remotely sensed reflectance data from vegetated surfaces. For instance, the user can apply such analysis given its own sensor settings, i.e., only for bands corresponding to an optical satellite.

Moreover, when applying GSA to more advanced RTMs such as the SVAT model SCOPE, not only the driving variables of spectral output such as reflectance and fluorescence along the spectral range but also the driving input variables of fluxes can be derived. Fluxes include a wide range of outputs that are calculated over the canopy as a whole. They include net radiation of soil and canopy, fraction of absorbed PAR (photosynthetically active radiation), average canopy temperature, total fluorescence, and photosynthesis of the canopy. In this chapter, the driving variables of total fluorescence (F_{total}) and NPC were identified. By decomposing these fluxes into its key variables it can be understood why they are related and through which spatially explicit biochemical, biophysical, and micrometeorological variables relationships are made possible. Accordingly, decomposing fluxes can help in understanding the vegetation system functioning and reveal underlying complex mechanisms between plant physiology and their environment.

On a more practical basis, the GSA identified that noninfluential variables can be safely kept fixed in subsequent simulations. Consequently, the as-such simplified version of a model reduces the computational load. This technique can critically facilitate the use of advanced, computationally intensive models. For instance, the SVAT model SCOPE is equipped with over 30 input variables and offers a wide range of output products (organized according to fluxes, radiation, reflectance, spectrum, surface temperature, fluorescence, vertical profiles). However, not all input variables play a role when one is interested in a specific output attribute. Varying all input variables would consequently lead to a large number of redundant simulations. Applying a GSA can therefore assist in identifying the input variables that matter, and those that are noninfluential. In this way, by safely fixing noninfluential variables, lighter models can be customized for a specific output variable. Simplified models possess several advantages over full models when aiming to retrieve biophysical parameters through inversion against a remote sensing image. For instance, the inversion algorithm has to resolve fewer possible solutions in the variable space, which mitigates the ill-posed problem. At the same time, since GSA enables identification of the input variables that exert most impact in a spectral region, it suggests that those variables are potentially the most promising to be successfully retrieved through inversion routines. Such GSA-based inversion approach has been implemented into a prototype satellite data processing strategy to improve the retrieval of biophysical variables (Mousivand et al., 2015). This is also the way forward for implementing SCOPE into a FLEX fluorescence retrieval inversion scheme (Verrelst et al., 2015a). Also, with respect to future optical missions, with ARTMO's capability of customizing sensor settings, GSA studies can become beneficial for assessing the sensitivity of newly designed optical sensors. Accordingly, assessments can be made about the predictive power of existing sensor band settings toward a diversity of vegetation properties. In turn, band positioning can be optimized in the spectral domain with respect to targeting specific surface variables.

When proceeding beyond the version presented here, progress is underway in making the toolbox generically applicable to any multi—input—output model. In a next version, it will become possible to import external LUTs, i.e., as generated by models outside the ARTMO environment. As such, the full parameter space of the imported LUT will be sampled through interpolation techniques in the same way as is currently done for the ARTMO RTMs. This new utility will open opportunities to analyze any kind of data, such as advanced, computationally expensive RTMs (e.g., ray tracing or atmospheric RTMs). Similarly, it will also become possible to analyze the driving variables of experimental data sets that have been collected in the field. With these developments in the pipeline, the GSA toolbox can become a benchmark in future GSA studies, not only applicable to complex models in EO science but also to other fields dealing with advanced deterministic models.

7. CONCLUSIONS

GSA enables to gain insight into the functioning of RTMs by identifying the driving input variables of RTM spectral outputs such as reflectance, fluorescence, or radiance. Noninfluential variables can also be identified and safely set to arbitrary values under a relatively wide ranges of conditions. This chapter introduced ARTMO's new "Global Sensitivity Analysis (GSA) toolbox." ARTMO is a modular scientific toolbox that consists of a suite of leaf, canopy, and combined RTMs. With the GSA toolbox, these RTMs can be analyzed intuitively on input—output relationships. Essentially, the GSA toolbox calculates the relative importance of RTM input variables through first-order and total-order Sobol' indices according to the method of Saltelli et al. (2010). Currently, the analysis can be employed along the spectral domain for any kind of optical sensor setting within the 400–2500 nm range. Accordingly, in combination with ARTMO's Sensor tool, the sensitivity of band settings of existing and forthcoming sensors toward vegetation properties can be assessed. The toolbox allows processing multiple RTM outputs within the same analysis, which is beneficial in case of advanced, multioutput RTMs such as SCOPE. By using SCOPE, driving input variables can be identified for spectral output as well for canopy fluxes such as photosynthesis of the canopy.

REFERENCES

Baker, N.R., 2008. Chlorophyll fluorescence: a probe of photosynthesis in vivo. Annual Review of Plant Biology 59, 89–113.

Baret, F., Weiss, M., Lacaze, R., Camacho, F., Makhmara, H., Pacholcyzk, P., Smets, B., 2013. GEOV1: LAI and FAPAR essential climate variables and FCOVER global time series capitalizing over existing products. Part1: Principles of development and production. Remote Sensing of Environment 137, 299–309.

Cukier, R.I., Fortuin, C.M., Shuler, K.E., Petschek, A.G., Schaibly, J.H., 1973. Study of the sensitivity of coupled reaction systems to uncertainties in rate coefficients. I: theory. The Journal of Chemical Physics 59, 3873–3878.

Feret, J.B., François, C., Asner, G.P., Gitelson, A.A., Martin, R.E., Bidel, L.P.R., Ustin, S.L., Guerric le Maire, G., Jacquemoud, S., 2008. PROSPECT-4 and 5: advances in the leaf optical properties model separating photosynthetic pigments. Remote Sensing of Environment 112, 3030–3043.

Frankenberg, C., Fisher, J.B., Worden, J., Badgley, G., Saatchi, S.S., Lee, J.E., Toon, G.C., Butz, A., Jung, M., Kuze, A., Yokota, T., 2011. New Global Observations of the Terrestrial Carbon Cycle from GOSAT: Patterns of Plant Fluorescence with Gross Primary Productivity. Geophysical Research Letters 38.

Guanter, L., Frankenberg, C., Dudhia, A., Lewis, P.E., Gómez-Dans, J., Kuze, A., Suto, H., Grainger, R.G., 2012. Retrieval and global assessment of terrestrial chlorophyll fluorescence from GOSAT space measurements. Remote Sensing of Environment 121, 236–251.

Ireland, G., Petropoulos, G.P., Carlson, T.N., Purdy, S., 2015. Addressing the ability of a land biosphere model to predict key biophysical vegetation characterisation parameters with Global Sensitivity Analysis Environmental Modelling & Software 65, 94–107. http://dx.doi.org/10.1016/j.envsoft.2014.11.010.

Jacquemoud, S., Baret, F., Andrieu, B., Danson, F.M., Jaggard, K., 1995. Extraction of vegetation biophysical parameters by inversion of the PROSPECT + SAIL models on sugar beet canopy reflectance data. Application to TM and AVIRIS sensors. Remote Sensing of Environment 52 (3), 163–172.

Jacquemoud, S., Verhoef, W., Baret, F., Bacour, C., Zarco-Tejada, P., Asner, G., Francois, C., Ustin, S., 2009. PROSPECT + SAIL models: a review of use for vegetation characterization. Remote Sensing of Environment 113 (Suppl. 1), S56–S66.

Mousivand, A., Menenti, M., Gorte, B., Verhoef, W., 2014. Global sensitivity analysis of the spectral radiance of a soil-vegetation system. Remote Sensing of Environment 145, 131–144.

Mousivand, A., Menenti, M., Gorte, B., Verhoef, W., 2015. Multi-temporal, multi-sensor retrieval of terrestrial vegetation properties from spectral-directional radiometric data. Remote Sensing of Environment 158, 311–330.

Nossent, J., Elsen, P., Bauwens, W., 2011. Sobol's sensitivity analysis of a complex environmental model. Environmental Modelling and Software 26, 1515–1525.

Papageorgiou, G.C., Govindjee, 2004. Chlorophyll a Fluorescence – A Signature of Photosynthesis. Springer, p. 818.

Petropoulos, G., Wooster, M.J., Kennedy, M., Carlson, T.N., Scholze, M., 2009c. A global sensitivity analysis study of the 1d SimSphere SVAT model using the GEM SA software. Ecological Modelling 220 (19), 2427–2440. http://dx.doi.org/10.1016/j.ecolmodel.2009.06.006.

Petropoulos, G.P., Ireland, G., Griffiths, H., Kennedy, M.C., Ioannou-Katidis, P., Kalivas, D.K.P., 2015. Extending the Global Sensitivity Analysis of the SimSphere Model in the Context of its Future Exploitation by the Scientific Community. Water MDPI 7, 2101–2141. http://dx.doi.org/10.3390/w7052101.

Rivera, J.P., Verrelst, J., Muñoz-Marí, J., Moreno, J., Camps-Valls, G., 2014a. Toward a semiautomatic machine learning retrieval of biophysical parameters. IEEE Journal of Selected Topics in Applied Earth Observation and Remote Sensing 7 (4), 1249–1259.

Rivera, J., Verrelst, J., Delegido, J., Veroustraete, F., Moreno, J., 2014b. On the semi-automatic retrieval of biophysical parameters based on spectral index optimization. Remote Sensing 6 (6), 4924–4951.

Rivera, J., Sabater, N., Tenjo, J., Vicent, N., Alonso, L., Moreno, 2014c. Synthetic Scene Simulator for Hyperspectral Spaceborne Passive Optical Sensors. Application to ESA's Flex/Sentinel-3 Tandem Mission (WHISPERS).

Rivera, J., Verrelst, J., Leonenko, G., Moreno, J., 2013. Multiple cost functions and regularization options for improved retrieval of leaf chlorophyll content and LAI through inversion of the PROSAIL model. Remote Sensing 5 (7), 3280–3304.

Saltelli, A., 2002. Making best use of model evaluations to compute sensitivity indices. Computer Physics Communications 145, 280–297.

Saltelli, A., Annoni, P., 2010. How to avoid a perfunctory sensitivity analysis. Environmental Modelling and Software 25, 1508–1517.

Saltelli, A., Annoni, P., Azzini, I., Campolongo, F., Ratto, M., Tarantola, S., 2010. Variance based sensitivity analysis of model output. Design and estimator for the total sensitivity index. Computer Physics Communications 181, 259–270.

Saltelli, A., Ratto, M., Andres, T., Campolongo, F., Cariboni, J., Gatelli, D., Saisana, M., Tarantola, S., 2008. Global Sensitivity Analysis: The Primer. John Wiley & Sons, Ltd.

Saltelli, A., Tarantola, S., Chan, K.P.S., 1999. A quantitative model-independent method for global sensitivity analysis of model output. Technometrics 41, 39–56.

Sobol', I.M., 1967. On the distribution of points in a cube and the approximate evaluation of integrals. USSR Computational Mathematics and Mathematical Physics 7, 86–112.

Sobol', I.M., 1990. On sensitivity estimation for nonlinear mathematical models. Matematicheskoe Modelirovanie 2, 112–118.

Song, X., Bryan, B.A., Paul, K.I., Zhao, G., 2012. Variance-based sensitivity analysis of a forest growth model. Ecological Modelling 247, 135–143.

Van der Tol, C., Berry, J.A., Campbell, P.K.E., Rascher, U., 2014. Models of fluorescence and photosynthesis for interpreting measurements of solar induced chlorophyll fluorescence. Journal of Geophysical Research: Biogeosciences 119, 2312–2327.

Van der Tol, C., Verhoef, W., Timmermans, J., Verhoef, A., Su, Z., 2009. An integrated model of soil-canopy spectral radiances, photosynthesis, fluorescence, temperature and energy balance. Biogeosciences 6, 3109–3129.

Verhoef, W., 1984. Light scattering by leaf layers with application to canopy reflectance modeling: the SAIL model. Remote Sensing of Environment 16, 125–141.

Verhoef, W., Jia, L., Xiao, Q., Su, Z., 2007. Unified optical-thermal four-stream radiative transfer theory for homogeneous vegetation canopies. IEEE Transactions on Geoscience and Remote Sensing 45, 1808–1822.

Verrelst, J., Alonso, L., Rivera Caicedo, J.P., Moreno, J., Camps-Valls, G., 2013. Gaussian process retrieval of chlorophyll content from imaging spectroscopy data. IEEE Journal of Selected Topics in Applied Earth Observations and Remote Sensing 867–874, 6365271.

Verrelst, J., Rivera, J.P., Van Der Tol, C., Magnani, F., Mohammed, G., Moreno, J., 2015a. Global sensitivity analysis of SVOPE model: what drives simulated canopy-leaving sun-induced fluorescence? Remote Sensing of Environment 166, 8–21.

Verrelst, J., Rivera, J.P., Veroustraete, F., Muñoz-Marí, J., Clevers, J.G.P.W., Camps-Valls, G., Moreno, J., 2015b. Experimental Sentinel-2 LAI estimation using parametric, non-parametric and physical retrieval methods — a comparison. ISPRS Journal of Photogrammetry and Remote Sensing 108, 260–272.

Verrelst, J., Romijn, E., Kooistra, L., 2012. Mapping vegetation density in a heterogeneous river floodplain ecosystem using pointable CHRIS/PROBA data. Remote Sensing 4 (9), 2866–2889.

Van Wittenberghe, S., Alonso, L., Verrelst, J., Moreno, J., Samson, R., 2015. Bidirectional sun-induced chlorophyll fluorescence emission is influenced by leaf structure and light scattering properties — a bottom-up approach. Remote Sensing of Environment 158, 169–179.

Wainwright, H.M., Finsterle, S., Jung, Y., Zhou, Q., Birkholzer, J.T., 2014. Making sense of global sensitivity analyses. Computers and Geosciences 65, 84–94.

Yang, J., 2011. Convergence and uncertainty analyses in Monte-Carlo based sensitivity analysis. Environmental Modelling and Software 26, 444–457.

GEM-SΛ: THE GAUSSIAN EMULATION MACHINE FOR SENSITIVITY ANALYSIS

17

M.C. Kennedy[1], G.P. Petropoulos[2]

Fera Science Ltd., York, United Kingdom[1]; University of Aberystwyth, Wales, United Kingdom[2]

CHAPTER OUTLINE

Sensitivity Analysis in Earth Observation Modelling. http://dx.doi.org/10.1016/B978-0-12-803011-0.00017-3

1. BAYESIAN ANALYSIS OF COMPUTER MODELS

Various methods have been developed for statistical analysis of data generated from computer experiments. Monte-Carlo simulation methods are very popular, not least because the basic methods are easily implemented. However, these rely on large numbers of code runs generated at randomly selected input points rather than more efficient designs. For very complex models it may not be practical to generate enough code runs for accurate estimates, especially when exploring a high-dimensional input space. Therefore, more efficient solutions are required. Theory based on the Bayesian analysis of functions provides a convenient solution for many computer models. The first step is to create a statistical *emulator* of the computer code of interest. This essentially makes use of some key properties that are satisfied in many computer codes:

1. The relationship between inputs and outputs is smooth. Small changes in input values lead to similar values of the output. Each code run therefore provides information not only about the output for its own input but also for the surrounding points. Sampling plans can be designed to be space filling, in some cases dramatically increasing the efficiency.
2. Even when a computer code is parameterized by many inputs, most of the output variation can be explained by a relatively small number.

The emulator is much cheaper to evaluate than the code itself, and also accounts for the uncertainty in the approximation. Furthermore, a convenient mathematical formulation allows various important outputs to be derived directly from the emulator:

- *Prediction* of the code output at any untried inputs
- *Main effect* of each individual input, showing the degree and nature of the influence of the input, averaged over the variations due to all other inputs
- *Joint effect* of each pair of inputs, showing the joint influence of inputs again after averaging
- *Total effect* of each input, which is a measure of influence that includes higher order interactions to which the input contributes.

These Bayesian methods assume that the output is deterministic. In other words, running the code repeatedly with the same input always gives the same output. In practice, however, the results can be useful even if the output has some numerical error, for example, caused by stochastic algorithms or numerical solvers. By including an explicit error term in the model (the "nugget" term), the emulator can be made to smooth out this random noise, rather than interpolate the training output points exactly. Mathematically, the computer code output is treated as an unknown function of its inputs and modeled using a Gaussian process. A brief overview is provided below to introduce the Gaussian Emulation for Sensitivity Analysis (GEM-SA) software. The theory is well documented elsewhere. A useful introduction is given in O'Hagan (2006), and more details can be found in Santner et al. (2013), Sacks et al. (1989), Kennedy and O'Hagan (2001), and Oakley and O'Hagan (2004). References and resources are also available from the website of MUCM (Managing Uncertainty in Computer Models, http://www.mucm.ac.uk).

2. GAUSSIAN PROCESS PRIOR DISTRIBUTION FOR A CODE OUTPUT

Each output is considered separately. We suppose that an output from the code is a function of p inputs, expressed as $y = f(x)$, where $x = (x_1,...,x_p)$. In GEM-SA the prior expectation of $f(x)$ can be constant, $E(f(x)) = 1$, or a linear function of each of the inputs

$$E(f(x)) = \beta_0 + \beta_1 x_1 + ... + \beta_p x_p \tag{17.1}$$

where β_0, β_1,..., β_p are unknown coefficients. Although this may be a crude estimate of the output, given enough runs of the model, the emulator can adapt to the specific output form, even if it is highly nonlinear. A covariance function is also specified to characterize the smoothness of the output. A product of smooth univariate functions is assumed:

$$Cov\left(f(x), f\left(x'\right)\right) = \sigma^2 \prod_{i=1}^{p} \exp\left\{- r_i \left(x_i - x_i'\right)^2\right\} \tag{17.2}$$

where r_i is a scaling parameter determining how rough the function is with respect to the ith input, i.e., how quickly the correlation between two code runs tends to 0 as the distance between their inputs increases. The joint probability distribution of any $f(x)$, $f(x')$ pair is assumed to be Gaussian. Note that by using Eq. (17.2), we assume the output is stationary, meaning that correlation between two outputs depend on the *distance* between the points and not their locations. The parameter σ^2 determines the overall level of variation around the expected response. The prior specification is completed by assigning noninformative priors for these parameters; specifically we use $p(\beta_0, ..., \beta_p, \sigma^2) \propto \sigma^{-2}$ and $p(r_i) \sim \text{Exp}(0.01)$ independently for $i = 1,...,p$.

3. POSTERIOR DISTRIBUTION AFTER OBSERVING CODE RUNS

Running the computer code at selected input points $(s_1,...,s_N)$ provides data in the form of observed outputs $d = (f(s_1),...,f(s_N))$. Fitting the emulator amounts to estimating the parameters $\theta = (\beta_0, ...,\beta_p, \sigma^2, r_1,...,r_p)$ of the mean and covariance functions. Conditional on the data d and the fitted emulator parameters, the posterior distribution of $f(\bullet)$ is found using Bayes' theorem to be Gaussian. The posterior distribution of $f(\bullet)$ conditional on the roughness parameters only is derived by integrating this distribution with respect to the posterior distribution of β_0, ...,β_p, σ^2. The posterior mean function is the emulator approximation of the code, which passes exactly through the observed runs unless numerical error is included. The posterior variance can be used to assess emulator accuracy. It is 0 at the observed runs, plus any numerical error variance, but increases for input values further away from the observed runs. Details can be found with the GEM-SA model documentation.

4. FUNCTIONALITY INCLUDED WITHIN GAUSSIAN EMULATION FOR SENSITIVITY ANALYSIS

GEM-SA was released in 2003 as a general tool to implement a number of Bayesian methods. It is a stand-alone Microsoft Windows application developed in C/C++ using the WxWindows toolbox (www.wxwidgets.org) for the graphical user interface. The application can be downloaded from www.tonyohagan.co.uk/academic/GEM and is free to use for noncommercial purposes. The website also includes presentations and documentation explaining the method and examples using the software. The main window of GEM-SA once it is started is shown in Fig. 17.1.

This software generates a statistical emulator of a computer code. The training data required to build the emulator include a set of inputs to the code together with the corresponding outputs from the code. GEM-SA also includes functions to generate sensible space-filling input designs using the LP-tau algorithm (Sobol, 1977) (Fig. 17.2) or maximin Latin hypercube algorithm (Morris and Mitchell, 1995). The LP-tau design generates a sequence in which each new point fills in gaps in the space. This design is therefore useful if a sequential strategy is required, where the analyst is not sure how many code runs are required in advance but is able to run the analysis with the first n available points and to repeat the process with larger n as further runs become available. The maximin Latin hypercube design

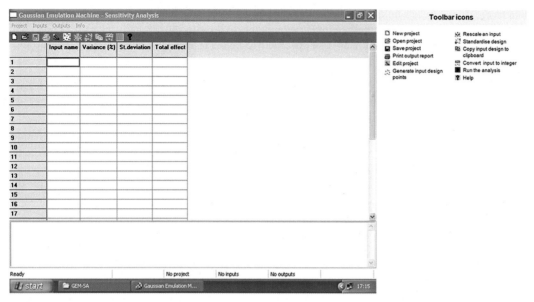

FIGURE 17.1

Central menu of Gaussian Emulation for Sensitivity Analysis (right) with the different options in the main menu explained (left).

iteratively generates a sequence of Latin hypercube designs and at each iteration selects the one with the largest minimum distance between pairs of points. It therefore generally gives good coverage of the input space and is uniformly distributed in each component. Individual components of the generated designs can be rounded to integers or rescaled in GEM-SA. Many more designs are available, which may give more accurate emulators. GEM-SA can be run for whichever input set the user chooses. It is the responsibility of the user to then generate corresponding model outputs and store the results in a text file.

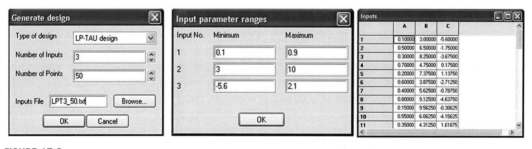

FIGURE 17.2

The sample generation options and workflow in Gaussian Emulation for Sensitivity Analysis, here as an example for the case of 3 inputs and 50 points using the LP-tau design.

4.1 STANDARDIZATION OF INPUTS

An automated process standardizes the inputs and outputs prior to fitting the emulator. It is convenient within GEM-SA to work with these standardized data as it allows for a wide range of different codes to be treated within a generic framework for implementation and helps to prevent numerical problems. The final results of GEM-SA are back transformed to the original scale, so the user only needs to consider this if interpreting or using the fitted emulator parameters. The Gaussian process model described previously is applied to transformed inputs x and output y, rather than the original scale, which we denote x' and y'. For each input component x_i we standardize to [0−1] by setting $x_i = (x'_i - x_i^{min})/(x_i^{max} - x_i^{min})$, where x_i^{min}, x_i^{max} define the minimum and maximum values in the training inputs for the ith component. The output is also scaled so that the mean and standard deviation of the resulting variable are approximately (0, 1). This is achieved by setting $y = (y' - \bar{y})/s$, where $\bar{y}$ and s are the sample mean and standard deviation of the training outputs.

4.2 SENSITIVITY AND UNCERTAINTY ANALYSES OF THE MODEL PREDICTION

If provided with a file of input prediction points, GEM-SA will give the emulator predictions and uncertainties for those predictions, based on the posterior distribution. Two options are available: prediction at each point independently or prediction of the joint distribution. The latter is useful to consider the correlation between output points, whereas the former simply uses the individual marginal distributions at each point separately.

Uncertainty analysis refers to the assessment of the distribution of the code output, caused by uncertainties in the input parameters. In GEM-SA, the default input distributions are uniform distributions between the minimum and maximum values in the training runs. These ranges can be modified, or normal distributions can be specified, for each input.

GEM-SA reports estimates of the mean code output and the variance, with respect to the distribution of the inputs. Uncertainty in these summaries is also reported, arising from:

- Emulator approximation
- Uncertainty in parameters ($\beta_0, \ldots, \beta_p, \sigma^2$) used to build the emulator
- Numerical error, if selected, in the code output

Uncertainty in the roughness parameters is not quantified by default, but an option is available to include this if required (see below).

The GEM-SA software can produce various measures of sensitivity of the model output to variations in single inputs or pairs of inputs. Details are provided in Oakley and O'Hagan (2004). Figs. 17.3 and 17.4 illustrate the main menu options provided by GEM-SA. Global sensitivity analysis is performed, in which all parameters other than the one (or the pair) under consideration is integrated out. The integration accounts for the distribution of the other inputs, so that the result is an expected value.

Main and joint effect simulations are output to files, allowing for further investigations. These contain multiple realizations simulated from the posterior distribution of the output at a regularly spaced grid of input points. Simulated main effect realizations are also displayed on screen with an individual panel for each selected output. As with the uncertainty analysis results, each of the sensitivity measures is subject to uncertainty. Multiple realizations are therefore generated to reflect the combined uncertainty, and this can clearly be seen in the spread of main effect lines.

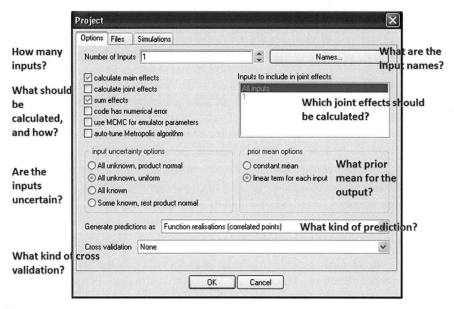

FIGURE 17.3

The main options menu for the implementation of sensitivity analysis within Gaussian Emulation for Sensitivity Analysis.

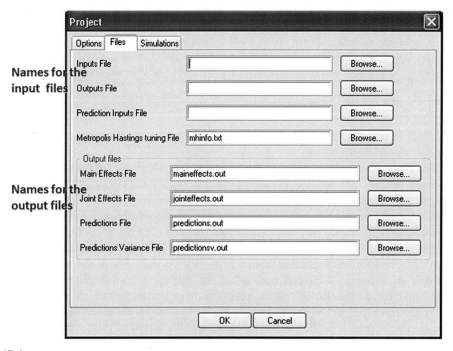

FIGURE 17.4

Further parameterization options in Gaussian Emulation for Sensitivity Analysis for the implementation of sensitivity analysis.

Sensitivity indices due to main and joint effects, as well as total effects, are listed in tabular form. When displayed as a percentage contribution to total output, these are relative to the total output uncertainty as calculated for uncertainty analysis. Because total effect indices are sums of all effects to which an input contributes, there is some duplication, and therefore the totals do not necessarily sum to 100%.

4.3 CROSS-VALIDATION

Diagnostics generated by GEM-SA, such as graphs showing the uncertainty bands and estimates of roughness parameters, provide a useful indication of emulator accuracy. For a more formal test, cross-validation can also be performed to measure how well the emulator fits the original code. GEM-SA includes two options: leave-one-out and leave final 20% out. With leave-one-out the emulator is used to estimate (and calculate emulator variance for) each of the training runs in turn. A slightly modified emulator is used to estimate each training point, built using all training points except the one being estimated. GEM-SA reports several diagnostic values, based on the results of the cross-validation. These can be used to assess the quality of the emulator approximation and can highlight potential violations of the assumptions underlying the emulator theory. The following are computed, where n is the number of runs, y_i is the true output for the ith training run, and $\widehat{y}_i, s_i$ are the corresponding emulator approximation and standard deviation calculated with the ith training point removed. Leave-one-out can sometimes give a misleading assessment of the fit because the roughness parameters and emulator fit are tuned specifically for the data being tested. The leave final 20% validation option is considered a more conservative estimate of the emulator accuracy, as part of the training data is used as an *independent* test and is not used to fit the emulator or estimate the roughness parameters. Whichever option is chosen, the $\widehat{y}_i$ and s_i^2 values are output to files cvpredmeans.txt and cvpredvars.txt, respectively, so that they can be analyzed further. The summaries generated are:

$$\text{CVRMSE} = \sqrt{\sum_{i=1}^{n} (y_i - \widehat{y}_i)^2 / n} \text{ is the cross-validation root mean squared error,}$$

$$\text{CVRMSRE} = \sqrt{\sum_{i=1}^{n} \left\{ (y_i - \widehat{y}_i / y_i) \right\}^2 / n} \text{ is the cross-validation root mean squared relative error, and}$$

$$\text{CVRMSSE} = \sqrt{\sum_{i=1}^{n} \left\{ (y_i - \widehat{y}_i / s_i) \right\}^2 / n} \text{ is the cross-validation root mean squared standardized error.}$$

Examination of the cross-validation results and quantification of the uncertainty of the emulator acts as a built-in consistency check. These diagnostics can in some cases highlight those inputs or input regions for which the emulator may be unreliable. Typical examples are where the output has discontinuities or extreme local variations within particular regions of the input space. In this case, further studies are recommended using additional code runs targeted in the regions where the roughness or threshold effects occur. If enough points can be located around the input values of a change point/discontinuity, then the accuracy can be improved.

5. UNCERTAINTY IN EMULATOR ROUGHNESS PARAMETERS

The roughness parameter r_i of a Gaussian process emulator for input x_i is a measure of how quickly the output can deviate from the expected level, such as a linear trend, as the input is varied. These were introduced in Eq. (17.2) as scaling parameters associated with individual inputs and are very important parameters in determining the spatial uncertainty in the emulator across the input space. As most simulators have multiple inputs, GEM-SA requires separate roughness parameters, and these are treated independently. In terms of model behavior, this means that the scale of variations with respect to an input is not affected by the variations due to another input. Given a set of training runs, the roughness can be estimated by maximizing the joint likelihood of all the r_i GEM-SA also has an option to include uncertainty in roughness parameters using a Markov Chain Monte Carlo algorithm. The GEM-SA documentation has more details, but for most applications it is recommended not to use this as the simpler default approach of plug-in roughness approximations performs equally well in practice.

6. USING THE GAUSSIAN EMULATION FOR SENSITIVITY ANALYSIS INTERFACE

All information related to a particular analysis is stored in a project file, which can be created through the interface. This specifies the location of input and output files, distributions for the input parameters, and options for the various models described later. Projects can be created, saved, and loaded from the interface. Figs. 17.5 and 17.6 illustrate this by showing an example setup for the SimSphere example. Here the user has selected uniform distributions, and so at the final stage, the user is asked to provide (*min*, *max*) ranges for each input (Fig. 17.6). Selecting "Defaults from input ranges" uses the minimum and maximum values from the extremes seen in the training input file.

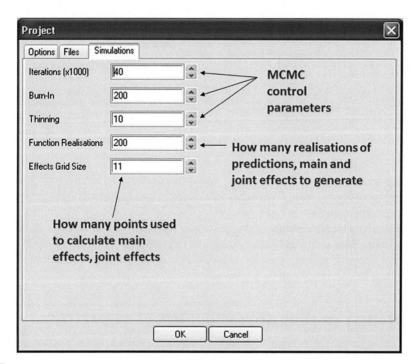

FIGURE 17.5

Gaussian Emulation for Sensitivity Analysis files input for the SimSphere example.

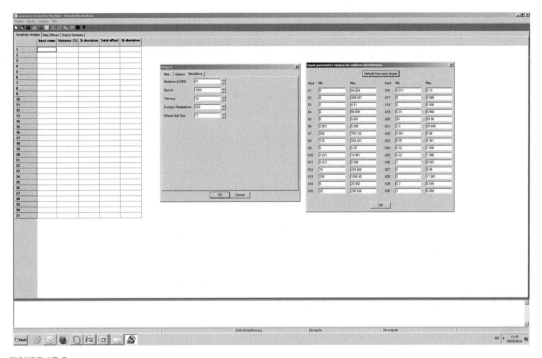

FIGURE 17.6

Gaussian Emulation for Sensitivity Analysis screenshot showing completion of the SimSphere example setup.

This ability to read, modify, and save project files is designed to make it easy to repeat or adapt the analysis as part of an iterative process to investigate a given computer code or to perform similar analyses in future. Some options determine the type of analyses to carry out or model parameters such as parameters for the code input distributions. Others indicate how many simulations to perform or how many points to include in the grid for simulated output. For example, when calculating main effect or joint effect plots, the default is to display output at 11 points equally spaced between the minimum and maximum values for each variable, but this may be altered.

In models that have many parameters, simulating joint effects for all possible pairs can be very time consuming. In practice most inputs usually have very little influence on these effects. An efficient strategy is therefore to perform an initial sensitivity analysis with main effects only ("calculate joint effects" option disabled). Then after examining the table of sensitivity analysis results, carry out a joint effects analysis but only include the inputs with total effects above some threshold value. Remember that main effects might be small even if interaction effects exist for an input, so it is the total effect that should be used for this selection.

7. SUMMARY OF INPUTS/OUTPUTS

All input and output files are in plain text format, so they can be generated from or read into external programs.

7.1 INPUTS

The project file includes the names of each of the remaining input files: code inputs, code outputs, and prediction inputs. Alternative filenames for these can be set when creating or editing a project. The main files are explained.

- Code inputs: space- or tab-delimited line for each run in the training data. Each line should therefore have entries for all unknown inputs.
- Code outputs must be prepared off-line by running the computer model at the code input points. The file should consist of a single entry on each line, which is the code output obtained using the corresponding line in the code inputs file.
- Prediction inputs must have the same number of columns as code inputs. If no prediction input file is specified within the project file, no predictions will be generated. The file should consist of a space-delimited line for each point at which you want to predict the output. The number of columns will match the number of columns in the code input file.

7.2 OUTPUTS

The main output files are: maineffects.out, jointeffects.out, predictions.out, cvpredmeans.txt, and cvpredvars.txt.

Each of the outputs is written as a single stacked vector. The length and sequence of the output depends on selections made for the project, as listed later.

7.2.1 Main Effect Realizations

The file maineffects.out has {number of samples × number of inputs × effects grid size} elements. Each block of [number of inputs × effects grid size] elements in the file corresponds to a realization from all the inputs, with inputs ordered as in the code inputs file, computed on the equally spaced points of the appropriate main effect grid. Note that main effects are only generated if the "calculate main effects" check box is selected in the project dialog.

7.2.2 Joint Effect Realizations

The file jointeffects.out has {number of samples × number of inputs × (number of inputs − 1)/ 2 × effects grid size × effects grid size}. Each block of [number of inputs × (number of inputs − 1)/ 2 × effects grid size × effects grid size] elements in the file corresponds to a realization from all the possible input pairs, ordered as (1, 2)...(1, number of inputs), (2, 3) ... (2, number of inputs),..., (number of inputs − 1, number of inputs). For every pair, the joint effect realization is computed on the product grid formed from the pair's two main effects grids. Note that joint effects are only generated if the "calculate joint effects" check box is selected in the project dialog.

7.2.3 Prediction Realizations

The file predictions.out has {number of samples × number of lines in prediction input file} elements. Each block of {number of lines in prediction input file} elements corresponds to a realization from the posterior distribution of the code output at the points specified in the prediction input file. Note that predictions are only generated if a valid prediction input file is specified in the project dialog.

7.2.4 Cross-Validation Results

These files represent results from cross-validation, if selected (CVRMSE, CVRMSRE, CVRMSSE as described earlier). Mean and variance of individual validation inputs are output as separate files. The values can be compared against the true values (training outputs) if required. Cross-validation summaries are also output in the log window of GEM-SA, based on these values.

8. CASE STUDY: SIMSPHERE

8.1 INTRODUCTION

Being able to understand Earth's system and also the interactions between its components including feedback processes has been identified as a topic of key importance in which future research should be directed (North et al., 2015). Some efforts have been concentrated in exploring the synergies of Earth Observation (EO) data and land biosphere models for this purpose, allowing delivering spatiotemporal estimates of key parameters characterizing land surface interactions. Indeed, a wide spectrum of techniques have been proposed, which essentially aim to provide improved estimates of such variables by combining the horizontal coverage and spectrally rich content of EO data with the vertical coverage and high temporal continuity of simulation process models (Petropoulos, 2013).

Petropoulos et al. (2009a) has pointed out in particular the promising potential of the group of such techniques that are based on the biophysical properties encapsulated in a satellite-derived scatter plot developed between the surface temperature (Ts) and vegetation index (VI) maps with a Soil Vegetation Atmosphere Transfer (SVAT) model, named SimSphere, for deriving spatiotemporal estimates of latent (LE) and sensible (H) heat fluxes as well as of surface soil moisture. Indeed, several variants of it are currently used or being developed for deriving operational global estimates of key parameters characterizing the Earth's water budget (Piles et al., 2016).

SimSphere was developed initially by Carlson and Boland (1978) and subsequently modified to its current state by Gillies (1993) and Petropoulos et al. (2013). SVAT models form a special category of deterministic simulation process models that attempt to describe the physical processes controlling energy and mass transfers in the soil/vegetation/atmosphere continuum and provide estimates of the time course of soil and vegetation state variables with a fine time step (Olioso et al., 1999). SimSphere provides predictions of key parameters characterizing land surface interactions simulating the physical processes, which occur as a function of time in a column that extends from the root zone below the soil surface up to a level higher than the surface canopy. It performs simulations over a 24-h cycle at a time step of 5 min or better, starting from a set of initial conditions given at 0600 h local time. An overview of SimSphere architecture and use so far can be found in Petropoulos et al. (2009b), whereas the software tool is at present distributed globally from the University of Aberystwyth (http://www.aber.ac.uk/simsphere).

8.2 SENSITIVITY ANALYSIS APPLIED TO SIMSPHERE

Herein, GSA was performed using GEM-SA with SimSphere to evaluate the sensitivity of the predicted Daily Average Air Temperature at 50 m $\left(\overline{Tair_{50m_{daily}}} \right)$. The sensitivity of this parameter was considered to demonstrate the use of GEM-SA due to its importance in characterizing land surface interaction processes and relevance to the Ts/VI group of methods when SimSphere is linked to EO data. A design space of 400 model simulations was prepared using the LP-tau sampling method. All input parameters were varied, except those of the geographical location and atmospheric profile, for which a priori values were taken from the Borgo Cioffi Italian CarboEurope site (40° 31′ 25.5″ N, 14° 57′ 26.8″ E) available for November 17, 2004. All 30 model inputs (Table 17.1) were assumed to be uniformly distributed, with their probability distribution functions defined using the mean and variance taken from the entire possible theoretical range they could take in SimSphere. Emulator performance was also evaluated herein using the "leave final 20% out" cross-validation method offered in GEM-SA.

8.3 RESULTS

8.3.1 Emulator Validation

The series of all the statistical measures (mentioned in an earlier section of this chapter) computed internally by GEM-SA were used to quantify the uncertainty of the SA due to the emulator approximation. These include the "cross-validation root mean square error," the "cross-validation root mean squared relative error," and the "cross-validation root mean squared standardized error." In addition, the emulator "roughness values" and the "sigma-squared" statistical parameter were also computed.

Tables 17.2 and 17.3 summarize the resulting statistics. As Table 17.2 shows, the roughness values for most of the input parameters are very low suggesting that the emulator is a very good approximation of the true model. Most roughness values obtained are less than 1.0, which suggests a smooth response of the emulator to the inputs variation. The rare roughness values greater than 1.0 are highlighted in bold in Table 17.2 (including roughness values for aspect, slope vegetation height, and fractional vegetation cover) and suggested the presence of a degree some nonlinearity in the relationships between inputs and outputs. Sigma squared values and cross-validation root mean squared-error values (shown in Table 17.2) are indicating moderate nonlinearity and a satisfactory build of emulator accuracy.

8.3.2 Results and Discussion

The relative contributions of the model input parameters with respect to the sensitivity of $\overline{Tair_{50m_{daily}}}$ to variations in those inputs are summarized in Table 17.4. Input parameters with a sensitivity of $>1\%$ main effect and/or $>1\%$ total effect are also highlighted in bold. These data are also displayed on main effects plots in Fig. 17.7, whereas the decomposition of variance (main effects only) of the model inputs is illustrated in Fig. 17.8. Evidently as it can be observed from the results, there is a large range in both main effect values and total effect values. The contribution of variance to the main effects is yet controlled by a fraction of the model inputs. For example, slope and aspect are two of the three largest contributors of variance to all the model outputs examined in terms of main effects and total effects. For the main effects, aspect contributes by 35.86% for $\overline{Tair_{50m_{daily}}}$. Aspect is, in fact, the largest contributor of percentage variance to main effects and total effects. Fractional vegetation cover contributed by 15.14% and 21.31% to main and total effects variance, respectively, for $\overline{Tair_{50m_{daily}}}$. Similarly, vegetation height is a significant parameter, contributing by 3.12% to the main effects of $\overline{Tair_{50m_{daily}}}$. Other important parameters in the sensitivity of $\overline{Tair_{50m_{daily}}}$ include the surface moisture availability and surface roughness.

The results reported herein are in agreement with those published in previous relevant studies (e.g., Petropoulos et al., 2009c; Ireland et al., 2015). Indeed, the present study has helped in identifying that a small number of the model inputs appear to affect the models' outputs. In particular, slope, aspect,

Table 17.1 SimSphere inputs considered in the GSA; units of each of the model inputs are also provided in the brackets

Model Input Short Name	Actual Name of the Model Input	Process in Which Each Parameter is Involved	Min Value	Max Value
X1	Slope (degrees)	Time & location	0	45
X2	Aspect (degrees)	Time & location	0	360
X3	Station height (m)	Time & location	0	4.92
X4	Fractional vegetation cover (%)	Vegetation	0	100
X5	LAI (leaf area index) (m^2/m^2)	Vegetation	0	10
X6	Foliage emissivity (unitless)	Vegetation	0.951	0.990
X7	[Ca] (external $[CO_2]$ in the leaf) (ppmv)	Vegetation	250	710
X8	[Ci] (internal $[CO_2]$ in the leaf) (ppmv)	Vegetation	110	400
X9	[O3] (ozone concentration in the air) (ppmv)	Vegetation	0.0	0.25
X10	Vegetation height (m)	Vegetation	0.021	20.0
X11	Leaf width (m)	Vegetation	0.012	1.0
X12	Minimum stomatal resistance (s/m)	Plant	10	500
X13	Cuticle resistance (s/m)	Plant	200	2000
X14	Critical leaf water potential (bar)	Plant	−30	−5
X15	Critical solar parameter (W/m^2)	Plant	25	300
X16	Stem resistance (s/m)	Plant	0.011	0.150
X17	Surface moisture availability (vol/vol)	Hydrological	0	1
X18	Root zone moisture availability (vol/vol)	Hydrological	0	1
X19	Substrate maximum volume water content (vol/vol)	Hydrological	0.01	1
X20	Substrate climatological mean temperature (°C)	Surface	20	30
X21	Thermal inertia $(W/m^2/K)$	Surface	3.5	30
X22	Ground emissivity (unitless)	Surface	0.951	0.980
X23	Atmospheric precipitable water (cm)	Meteorological	0.05	5
X24	Surface roughness (m)	Meteorological	0.02	2.0
X25	Obstacle height (m)	Meteorological	0.02	2.0
X26	Fractional cloud cover (%)	Meteorological	1	10
X27	RKS (saturated thermal conductivity) Cosby et al. (1984)	Soil	0	10
X28	CosbyB (see Cosby et al., 1984)	Soil	2.0	12.0
X29	THM (saturated volume water content) Cosby et al. (1984)	Soil	0.3	0.5
X30	PSI (saturated water potential) Cosby et al. (1984)	Soil	1	7

GSA, *global sensitivity analysis;* LAI, *leaf area index.*

Table 17.2 Summarized statistics concerning the emulator accuracy evaluation for the simSphere model outputs whose sensitivity was examined

	Model Input	$\overline{Tair_{50m_{daily}}}$
X1	Slope	**1.584**
X2	Aspect	**9.192**
X3	Station height	0.008
X4	Fractional vegetation cover	**1.122**
X5	LAI	0.297
X6	Foliage emissivity	0.015
X7	[Ca]	0.037
X8	[Ci]	0.015
X9	[O₃] in the air	0.000
X10	Vegetation height	**1.288**
X11	Leaf width	0.071
X12	Minimum stomatal resistance	0.039
X13	Cuticle resistance	0.057
X14	Critical leaf water potential	0.124
X15	Critical solar parameter	0.018
X16	Stem resistance	0.000
X17	Surface moisture availability	0.279
X18	Root zone moisture availability	0.000
X19	Substrate maximum volume water content	0.000
X20	Substrate climatological mean temperature	0.047
X21	Thermal inertia	0.000
X22	Ground emissivity	0.005
X23	Atmospheric precipitable water	0.078
X24	Surface roughness	0.976
X25	Obstacle height	0.000
X26	Fractional cloud cover	0.012
X27	RKS	0.000
X28	CosbyB	0.092
X29	THM	0.657
X30	PSI	0.000

*Bold text highlights the roughness values of the model inputs with values
>1.0. Rows X1 to X30 show roughness values for the model outputs examined. LAI, leaf area index.*

fractional vegetation cover, and to a lesser extent, surface soil moisture availability and vegetation height have been shown to be important. It is also reasonable that air temperature will be sensitive to the parameters to which incoming solar radiation is also sensitive (Fig. 17.8, Table 17.4). In addition, the influence of fractional vegetation cover is to be expected since the proportion of vegetation cover

Table 17.3 Additional statistics concerning the emulator performance for the SA tests conducted

	$\overline{Tair_{50m_{daily}}}$
Fitted Model Parameters	
Sigma-squared	0.998
Cross-Validation Results	
Cross-validation root mean squared error (W/m^2)	0.855
Cross-validation root mean squared relative error (%)	5.678
Cross-validation root mean squared standardized error	1.270
SA, *sensitivity analysis.*	

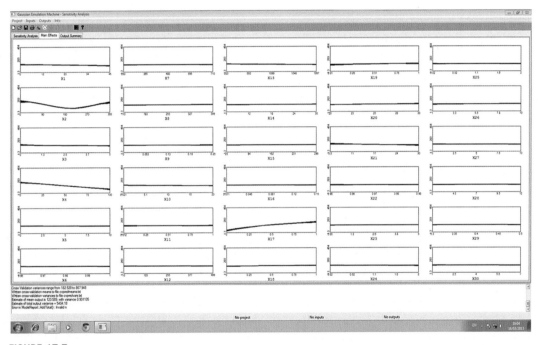

FIGURE 17.7

Outputs from Gaussian Emulation for Sensitivity Analysis showing the main effect relationships between the output and individual inputs.

may influence both LE and H heat flux and thus the air temperature, as well as the amount of radiation reflected and emitted. Other SA studies have also identified the significance of surface moisture availability on air temperature due in large part to its significant control on evapotranspiration (Carlson and Boland, 1978; Lockart et al., 2012) and thus its control on the partitioning of net radiation into LE

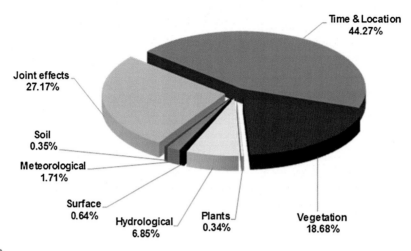

FIGURE 17.8

An illustration of the decomposition of variance (main effects only) of the SimSphere model input parameters for the different groups of model input parameters (as those defined in the SimSphere menu) for the case of examining the models' $\overline{Tair_{50m_{daily}}}$ sensitivity.

and H heat. This study has contributed to further extending the sensitivity analysis of SimSphere SVAT model by analyzing the sensitivity of a key parameter simulated by the model, namely, Tair. In the context of developing SimSphere itself as a stand-alone model as well as a tool that can be integrated with EO data, this study has further pointed out the fact that if accurate, EO measurements of a small number of key input parameters can be obtained, its potential for contributing to our understanding of regional-scale land—surface interactions is potentially important.

9. USING GAUSSIAN EMULATION FOR SENSITIVITY ANALYSIS EMULATORS WITH YOUR OWN SOFTWARE

In addition to the outputs generated by the GEM-SA interface, all mathematical constants required to generate further simulations or predictions from the emulator are stored in files. These provide much more flexibility in the use of the emulator than the analyses built into GEM-SA. A typical case would be a Monte Carlo—based uncertainty analysis using the emulator as a surrogate for a complex code, in which the input parameters are correlated and have much more interesting distributions than those allowed within GEM-SA. In 2012, emulators of pesticide exposure from crop spraying were prefitted and embedded within a larger simulation of bystander and resident exposure (Kennedy et al., 2012).

A set of plain text files produced by GEM-SA and required to generate emulator predictions (each with filenames beginning emulator_) are listed below.

emulator_ainv.txt
emulator_training_inputs.txt
emulator_mu_out.dat
emulator_precision_out.dat

Table 17.4 Summarized SA results from the implementation of the BACCO GEM-SA tool examining the sensitivity of $\overline{Tair_{50m_{daily}}}$

	Model Inputs	$\overline{Tair_{50m_{daily}}}$ Main Effects (%)	Total Effects (%)
X1	Slope	**8.351**	**24.897**
X2	Aspect	**35.859**	**56.144**
X3	Station height	0.061	0.142
X4	Fractional vegetation cover	**15.137**	**21.318**
X5	LAI	0.155	**1.769**
X6	Foliage emissivity	0.030	0.171
X7	[Ca]	0.027	0.371
X8	[Ci]	0.108	0.244
X9	[O3] in the air	0.028	0.029
X10	Vegetation height	**3.139**	**8.518**
X11	Leaf width	0.055	0.472
X12	Minimum stomatal resistance	0.024	0.327
X13	Cuticle resistance	0.135	0.668
X14	Critical leaf water potential	0.096	**1.059**
X15	Critical solar parameter	0.069	0.226
X16	Stem resistance	0.014	0.015
X17	Surface moisture availability	**6.760**	**8.722**
X18	Root zone moisture availability	0.055	0.056
X19	Substrate maximum volume water content	0.034	0.035
X20	Substrate climatological mean temperature	0.567	0.909
X21	Thermal inertia	0.029	0.030
X22	Ground emissivity	0.040	0.093
X23	Atmospheric precipitable water	0.042	0.632
X24	Surface roughness	**1.599**	**6.091**
X25	Obstacle height	0.035	0.036
X26	Fractional cloud cover	0.034	0.145
X27	RKS	0.018	0.019
X28	CosbyB	0.164	0.907
X29	THM	0.152	0.750
X30	PSI	0.018	0.019
Total % variance (main effects only)		72.385	
Total % variance contribution from first-order interactions		11.997	
Total % variance (main effects only)		72.385	

Computed main and total effect indices by the GEM tool (expressed as %) for each of the model parameters are shown. The last three lines summarize the percentages of the explained total output variance of the main effects alone and after including the interaction effects. Highlighted in bold are the model input parameters with greater than 1% variance decomposition and main and total effects greater than 1. GEM-SA, Gaussian emulation for sensitivity analysis; LAI, leaf area index.

emulator_ainvh.txt
emulator_rough_out.dat
emulator_g.txt
emulator_inv_hainvh.txt
emulator_minmax.txt
emulator_scale.txt

These contain all the internal parameters and matrix constants used to fit the emulator. The following is a script in R that will read in the emulator files and initialize the variables. Any subsequent calls to nextpoint(xpred), where xpred is a vector of emulator inputs, will return the emulator mean (prediction) and standard deviation. The standard deviation measures the uncertainty in the prediction due to the emulation approximation. Values of the prediction inputs xpred outside the range of the training data are permitted, but not recommended. Standard deviations at such points will be large. The algorithm can be adapted in various ways. Suppose, for example, that the emulator is believed to be sufficiently accurate that the calculation of standard deviation is not required. The following files/arrays can then be left out and the routine can be simplified:

```
emulator_inv_hainvh.txt, emulator_ainvh.txt, ainvh, tainvh, ainvt, inv_hainvh, hdiff
# Stand alone emulator
# using setup files produced by GEM-SA.exe
# this code is adapted from emulator.doc supplied with the GEM-SA help files
setwd("Insert here the path to a folder containing GEM-SA .txt output files")

# read GEM-SA emulator data from files
# open emulator setup files
fileInputs < - file("emulator_training_inputs.txt", "r") # training data inputs
header < - as.numeric(unlist(strsplit(readLines(fileInputs, 1), " "))) # reads header
line and splits by space
nmod < - header[1] # number of training runs
ninputs < - header[2] # number of inputs to the emulator
nreg < - header[3] # number of regression parameters
close(fileInputs)

ainv < - matrix(scan("emulator_ainv.txt"), byrow = T, ncol = nmod) # inverse correlation
matrix
minmax < - matrix(scan("emulator_minmax.txt"), byrow = T, ncol = ninputs) # mins and maxs
of each input
min < - minmax[1,]
max < - minmax[2,]
tmp < - scan("emulator_scale.txt") # mean and sd of output scaling
scalemean < - tmp[1]
scalesd < - tmp[2]
xmod < - matrix(scan("emulator_training_inputs.txt", skip = 1), byrow = T, ncol =
ninputs)
ainvh < - matrix(scan("emulator_ainvh.txt"), byrow = T, ncol = nreg) # ainv.H
prec < - scan("emulator_precision_out.dat") # estimated GP prec = 1/variance
```

```
g < - scan("emulator_g.txt") # ainv.(y-Hb)
betahat < - scan("emulator_mu_out.dat") # estimated regression parameters
roughness < - scan("emulator_rough_out.dat") # estimated function roughness (1
element per input)
inv_hainvh < - matrix(scan("emulator_inv_hainvh.txt"), byrow = T, ncol = nreg) #
(H'.ainv.H)^{-1}

tvec < - vector("numeric", nmod)
output < - vector("numeric", nVarSimulations)
xsd < - vector("numeric", 2)
outputsd < - vector("numeric", nVarSimulations)
h < - vector("numeric", nreg)
probExceed < - vector("numeric", nUncertainSimulations)
nextpoint = function(xpred){
 # evaluates the emulator at the vector point xpred
 # mean and standard deviation returned through xsd

 # rescale inputs exactly as in the emulator building code
 xpred < - (xpred - min)/(max-min)

 h[1] <- 1.0
 if (nreg >1){
  for (i in 1:ninputs){
   h[i+1] <- xpred[i] # assumes linear regression terms only
   }
  }

 # compute t vector of correlations between xpred and training data inputs
 for(i in 1:nmod){
  tvec[i] <- exp(-sum(roughness*(xpred-xmod[i,])*(xpred-xmod[i,])))
  }

 # compute mean
 xsd[1] <- scalemean + scalesd*(h %*% betahat + tvec %*% g)
 # compute standard deviation
 ainvt <- ainv %*% tvec
 tainvh <- tvec %*% ainvh
 hdiff <- h - tainvh
 aux <- (1.0 - tvec %*% ainvt + hdiff %*% (inv_hainvh %*% t(hdiff)))/prec
 if(aux <= 0.0){
  xsd[2] <- 0.0
  }else{
 xsd[2] <- scalesd*sqrt(aux)
  }
 xsd
}
```

10. CONCLUSIONS

In this study it was shown through practical application to the SimSphere model, how the GEM-SA software can be used to identify individual parameters and groups of parameters contributing significantly to the uncertainty or variation of a particular model output. By exploiting certain common features in model behavior such as smoothness in the output and additivity of the output it is possible to use a statistical model to dramatically improve efficiency compared with the traditional Monte Carlo simulation approaches. The required assumptions were described previously in the chapter. Violation of these assumptions is often highlighted within the diagnostic measure based on cross-validation or is visible in the roughness parameter estimates and main effect uncertainties. These therefore provide some level of self-checking when the emulator is built.

GEM-SA was designed to be as general and user-friendly as possible for general purpose application. Inevitably, this means that it cannot accommodate the unique features of all possible model types. It considers single outputs only, rather than multivariate outputs, and the prior mean and covariance specifications (1) and (2) are rather simple. The user interface is limited to independent Gaussian or uniform distributions for the inputs. As approximations, these generally lead to useful results because the Gaussian process model is able to adapt to the shape of the input/output form, and more flexibility can be added if the analyst uses the external emulator option described previously.

The practical usefulness of GEM-SA tool in a sensitivity analysis context that was demonstrated herein also evidenced the ability of the software to perform robustly and accurately a complete SA mapping of most sensitivity inputs of the model in predicting in this particular case Tair. The produced output is of key importance to SimSphere users and developers alike, in the context of the model use either as a stand-alone tool or synergistically with EO data. Results like the ones reported in this study on SimSphere can provide important assistance to improving the model architectural design and provide important assistance toward identifying directions of key priority in parameterizing the model.

ACKNOWLEDGMENTS

GEM-SA was developed while Marc Kennedy worked at the University of Sheffield as part of the NERC-funded Center for Terrestrial Carbon Dynamics (CTCD). Participation of G.Petropoulos to this work was supported by the TRANSFORM-EO Marie Curie EU-funded project.

REFERENCES

Carlson, T.N., Boland, F.E., 1978. Analysis of urban-rural canopy using a surface heat flux/temperature model. Journal of Applied Meteorology 17, 998–1014.

Cosby, B.J., Hornberger, G.M., Clapp, R.B., Ginn, T., 1984. A statistical exploration of the relationships of soil moisture characteristics to the physical properties of soils. Water Resources Research 20, 682–690.

Gillies, R.R., 1993. A Physically-Based Land Sue Classification Scheme Using Remote Solar and Thermal Infrared Measurements Suitable for Describing Urbanisation (Ph.D. thesis). University of Newcastle, UK, 121 pp.

Ireland, G., Petropoulos, G.P., Carlson, T.N., Purdy, S., 2015. Addressing the ability of a land biosphere model to predict key biophysical vegetation characterisation parameters with Global Sensitivity Analysis. Environmental Modelling & Software 65, 94–107. http://dx.doi.org/10.1016/j.envsoft.2014.11.010.

Kennedy, M.C., O'Hagan, A., 2001. Bayesian calibration of computer models. Journal of the Royal Statistical Society: Series B (Statistical Methodology) 63, 425−464 (With discussion).

Kennedy, M.C., Ellis, M.C.B., Miller, P.C.H., 2012. BREAM: a probabilistic Bystander and Resident Exposure Assessment Model of spray drift from an agricultural boom sprayer. Computers and Electronics in Agriculture 88, 63−71.

Lockart, N., Kavetski, D., Franks, S.W., 2012. On the role of soil moisture in daytime evolution of temperatures. Hydrological Processes. http://dx.doi.org/10.1002/hyp.9525.

Morris, M.D., Mitchell, T.J., 1995. Exploratory designs for computational experiments. Journal of Statistical Planning and Inference 43, 381−402.

North, M.R., Petropoulos, G.P., Rentall, D.V., Ireland, G.I., McCalmont, J.P., 2015. Quantifying the prediction accuracy of a 1-D SVAT model at a range of ecosystems in the USA and Australia: evidence towards its use as a tool to study Earth's system interactions. Earth Surface Dynamics Discussions 6, 217−265. http://dx.doi.org/10.5194/esdd-6-217-2015.

Oakley, J., O'Hagan, A., 2004. Probabilistic sensitivity analysis of complex models: a Bayesian approach. Journal of the Royal Statistical Society, Series B 66, 751−769.

O'Hagan, A., 2006. Bayesian analysis of computer code outputs: a tutorial. Reliability Engineering and System Safety 91, 1290−1300.

Olioso, A., Chauki, H., Courault, D., Wigneron, J.-P., 1999. Estimation of evapotranspiration and photosynthesis by assimilation of remote sensing data into SVAT models. Remote Sensing of Environment 68, 341−356.

Petropoulos, G.P., 2013. Remote sensing of surface Turbulent energy fluxes. Chapter 3. In: Petropoulos, G.P. (Ed.), Remote Sensing of Energy Fluxes and Soil Moisture Content. Taylor and Francis, ISBN 978-1-4665-0578-0, pp. 49−84.

Petropoulos, G., Carlson, T.N., Wooster, M.J., Islam, S., 2009a. A Review of T_s/VI remote sensing based methods for the retrieval of land surface fluxes and soil surface moisture content. Progress in Physical Geography 33 (2), 1−27.

Petropoulos, G., Carlson, T.N., Wooster, M.J., 2009b. An overview of the use of the SimSphere soil vegetation atmosphere transfer (SVAT) model for the study of land-atmosphere interactions. Sensors 9 (6), 4286−4308. http://dx.doi.org/10.3390/s90604286.

Petropoulos, G., Wooster, M.J., Kennedy, M., Carlson, T.N., Scholze, M., 2009c. A global sensitivity analysis study of the 1d SimSphere SVAT model using the GEM SA software. Ecological Modelling 220 (19), 2427−2440. http://dx.doi.org/10.1016/j.ecolmodel.2009.06.006.

Petropoulos, G.P., Konstas, I., Carlson, T.N., 2013. Automation of SimSphere land surface model use as a standalone application and integration with EO data for deriving key land surface parameters. In: European Geosciences Union, April 7−12th, 2013, Vienna, Austria.

Piles, M., Petropoulos, G.P., Sanchez, N., González-Zamora, A., Ireland, G., 2016. Towards improved spatio−temporal resolution soil moisture retrievals from the synergy of SMOS and MSG SEVIRI spaceborne observations. Remote Sensing of Environment 180, 403−417. http://dx.doi.org/10.1016/j.rse.2016.02.048.

Sacks, J., Welch, W.J., Mitchell, T.J., Wynn, H.P., 1989. Design and analysis of computer experiments. Statistical Science 4 (4), 409−435 (With comments and a rejoinder by the authors).

Santner, T.J., Wiliams, B.J., Notz, W.I., 2013. The Design and Analysis of Computer Experiments. Springer.

Sobol, I.M., 1977. Uniformly distributed sequences with an additional uniform property. USSR Computational Mathematics and Mathematical Physics 16, 236−242.

AN INTRODUCTION TO THE SAFE MATLAB TOOLBOX WITH PRACTICAL EXAMPLES AND GUIDELINES

18

F. Sarrazin, F. Pianosi, T. Wagener
University of Bristol, Bristol, United Kingdom

CHAPTER OUTLINE

1. INTRODUCTION

SAFE (Sensitivity Analysis For Everybody) is an open source sensitivity analysis software that implements key global sensitivity analysis (GSA) methods for the analysis of numerical models in any application area. SAFE can be freely obtained from the authors for academic and noncommercial purposes (http://bristol.ac.uk/cabot/resources/safe-toolbox/). The software can be used not only in Matlab but also in the Octave (www.gnu.org/software/octave/) environment, and also under any operating system (Windows, Linux, and Mac OS X). An R-version with limited functionality is also available.

SAFE was designed with a focus on supporting good practice in the implementation of GSA and to help users make appropriate choices when performing GSA (e.g. choice of sample size, model output, parameter ranges,). Such choices are indeed critical, since they can have a significant impact on the results (Pianosi et al., 2016). SAFE was also developed to enable access to a wide range of GSA

Sensitivity Analysis in Earth Observation Modelling. http://dx.doi.org/10.1016/B978-0-12-803011-0.00018-5

363

methods, and to facilitate a multimethod approach, as illustrated in Chapter 7. The toolbox is valuable to both nonspecialist and specialist users. On the one hand, commented workflow scripts aim to guide users through the different steps of a GSA. On the other hand, the different functions are extensively commented and provide in-depth information to more experienced users, who can then easily modify the code and tailor it to their specific needs. SAFE has been adopted for a large number of applications and in many different countries (Fig. 18.1).

We believe that tools such as SAFE are very valuable to support the design of efficient measurement systems (e.g., satellite observation systems) and to enhance the effective use of Earth observations (EO) for model calibration, testing, and validation. First, since the implementation and maintenance of measurement systems entail high financial costs, it is necessary to identify trade-offs between such costs

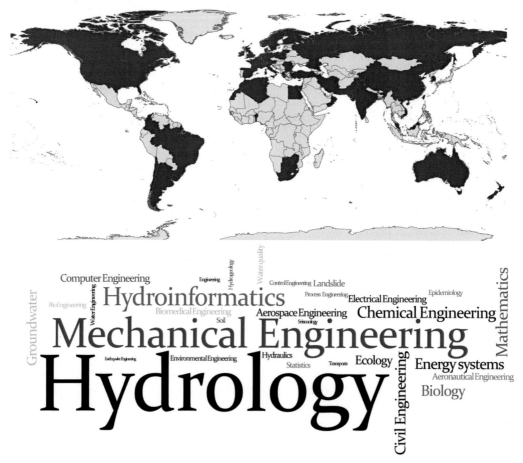

FIGURE 18.1

Analysis of download requests for SAFE (period: December 2014–May 2016). Top: origin countries of the researchers who requested SAFE. Bottom: main areas of expertise.

(Top) World Countries Map (country.gdb) from Esri, DeLorme Publishing Company, CIA World Factbook.

and the benefits arising from increased spatiotemporal resolution and accuracy of the measurements. To this end, GSA can facilitate the assessment of data requirement for a specific model application. For instance, Reed et al. (2015) used GSA to determine the necessary temporal resolution of the forcing data (satellite data rainfall measurements) that would provide acceptable performance of a flood prediction model. This information is then used to identify appropriate strategies for the replenishment of rainfall-observing satellite systems. Second, in the context of increasing availability of EO data sets, GSA provides a structured approach to assess the value of these data for model calibration, validation, and diagnostic evaluation. For example, Chapter 7 demonstrates the use of GSA to inform the calibration of a land surface model based on different heat fluxes and soil moisture observations. For distributed models, GSA can be used to investigate the spatiotemporal dynamics of the model processes and the parameter identifiability given spatially varying EO input data (e.g., Herman et al., 2013). Furthermore, GSA can help to assess the impact of measurement errors on the model performance and to identify "disinformative" data (e.g., Pianosi and Wagener, 2016).

The objective of this chapter is to complement a previous introductory article (Pianosi et al., 2015) about SAFE. That article mainly focused on software aspects (the rationale behind SAFE architecture, documentation, etc.), whereas here we aim to provide examples of SAFE applications and practical advice for its use. In Section 2 of this chapter, we provide an overview of the structure of the toolbox. In Section 3, we briefly present some of the GSA methods implemented in SAFE, focusing in particular on qualitative methods based on visualization tools, which are covered less in the literature although they might provide very interesting insights. In Section 4 we provide guidelines for performing GSA with SAFE by addressing some of the frequently asked questions from SAFE users. Finally, Section 5 outlines our views on the future evolution of the toolbox.

2. STRUCTURE OF THE TOOLBOX

This section describes the structure of the toolbox (Fig. 18.2). The toolbox implements the three basic steps of GSA, namely, (1) sampling of the input factor space, (2) model evaluation (the most computationally demanding step), and (3) post processing of the input−output samples by different qualitative and quantitative GSA methods. These three steps are performed separately in SAFE, using specific Matlab functions organized in different folders:

- "sampling" folder: contains functions to perform sampling of the input space and for model evaluation;
- "EET," "RSA," "VBSA," etc.: specific folders for each of the quantitative GSA methods implemented in SAFE. Each folder contains functions to compute the sensitivity indices, visualize the results, analyze the convergence, and possibly sample the parameter space when the method requires a tailored sampling strategy.
- "visualization" folder: includes functions for visual analysis of the input/output samples (qualitative GSA methods, see Section 3), for the visualization of sensitivity indices and for the validation of GSA results.
- "util" folder: contains functions that perform generic statistical operations that might be useful for other functions across folders.
- "examples" folder: includes test case studies, namely, two hydrological models, HyMod (Boyle, 2001; Wagener et al., 2001) and HBV (Bergström, 1995; Seibert, 1997), and some benchmark

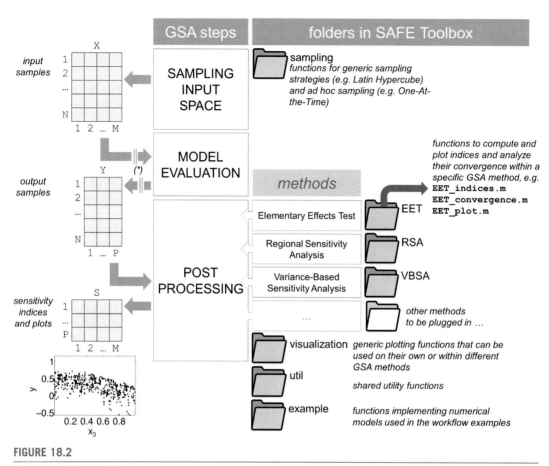

FIGURE 18.2

Structure of the SAFE toolbox.

Taken from Pianosi, F., Sarrazin, F., Wagener, T., 2015. A Matlab toolbox for global sensitivity analysis. Environmental Modelling and Software 70, 80–85. http://dx.doi.org/10.1016/j.envsoft.2015.04.009.

functions widely used in the GSA literature, such as the Ishigami–Homma function (Ishigami and Homma, 1990) and the Sobol' g-function (Saltelli et al., 2008).

This modular structure allows some flexibility in the implementation of GSA. Specifically, it is possible to:

- Perform one of the three steps, e.g., model evaluation, in a different computing environment (if the model executable is written in another language than Matlab).
- Add new sampling and postprocessing methods to the toolbox.
- Apply different GSA methods to the same input/output samples (an example is given in Chapter 7.)

Additionally, the toolbox includes a set of workflow scripts to help users to get started. An example of workflow script is reported in Fig. 18.3. For each GSA method, the workflow script shows how to

```
%% Step 1: set paths

%% Step 2: setup the model and define input ranges

% Load data:
load -ascii LeafCatch.txt
rain = LeafCatch(1:365,1)    ;
evap = LeafCatch(1:365,2)    ;
flow = LeafCatch(1:365,3)    ;

% Define input distribution and ranges:
M = 5 ; % number of uncertain parameters [ Sm beta alfa Rs Rf ]
DistrFun = 'unif'   ; % Parameter distribution
DistrPar = { [ 0 400 ]; [ 0 2 ]; [ 0 1 ]; [ 0 0.1] ; [ 0.1 1 ] } ;
% Parameter ranges

%% Step 3: Compute first-order and total-order variance-based indices

myfun = 'hymod_nse' ;

% Sample parameter space using the resampling strategy proposed by
% (Saltelli, 2008; for reference and more details, see help of
% functions vbsa_resampling and vbsa_indices)
SampStrategy = 'lhs' ;
N = 3000 ; % Base sample size.
% Comment: the base sample size N is not the actual number of input
% samples that will be evaluated. In fact, because of the resampling
% strategy, the total number of model evaluations to compute the two
% variance-based indices is equal to N*(M+2)
X = AAT_sampling(SampStrategy,M,DistrFun,DistrPar,2*N);
[ XA, XB, XC ] = vbsa_resampling(X) ;

% Run the model and compute selected model output at sampled parameter
% sets:
YA = model_evaluation(myfun,XA,rain,evap,flow) ; % size (N,1)
YB = model_evaluation(myfun,XB,rain,evap,flow) ; % size (N,1)
YC = model_evaluation(myfun,XC,rain,evap,flow) ; % size (N*M,1)

% Compute main (first-order) and total effects:
[ Si, STi ] = vbsa_indices(YA,YB,YC) ;

% Plot results:
X_Labels = {'Sm','beta','alfa','Rs','Rf'} ;
figure % plot main and total separately
subplot(121); boxplot1(Si,X_Labels,'main effects')
subplot(122); boxplot1(STi,X_Labels,'total effects')
figure % plot both in one plot:
boxplot2([Si; STi],X_Labels)
legend('main effects','total effects')% add legend
```

FIGURE 18.3

Example of workflow script for the application of variance-based sensitivity analysis (VBSA) to the 5 parameters of the hydrological HyMod model.

implement the method for a test case study, how to use the visualization tools, and how to test some of the choices that have to be made when applying GSA (e.g., choice of sample size or of the model output analyzed). Users can then easily modify the code (workflow) and adapt it to their case study.

3. GLOBAL SENSITIVITY ANALYSIS METHODS AND EXAMPLES OF APPLICATION

Two types of GSA methods are implemented in SAFE: (1) visual methods that support a qualitative analysis of the model sensitivities and (2) quantitative methods that provide a quantification of the model sensitivities through the computation of sensitivity indices.

3.1 VISUAL AND QUALITATIVE GLOBAL SENSITIVITY ANALYSIS METHODS

The visual and qualitative methods included in the toolbox allow preliminary assessment of input factor sensitivity, often using only a limited number of model evaluations. They are particularly useful in the early phases of model development or of getting to know a new model. Modelers can use them, for example, for code debugging to check that the model behavior is consistent with their expectations, before applying more rigorous and computationally demanding GSA methods. Fig. 18.4 reports an example of application of some of these methods to the three input factors x_i, $i = 1, 2, 3$ of the Ishigami—Homma test function (Ishigami and Homma, 1990):

$$y = \sin(x_1) + a \sin^2(x_2) + bx_3^4 \sin(x_1) \tag{18.1}$$

This function is widely used in the sensitivity analysis literature since the variance-based first-order (S_i) and total-order (S_{T_i}) sensitivity indices (Sobol', 1993; Saltelli, 2002) for its three input factors can be derived analytically. For example, using the equations given in Saltelli et al. (2008, pp. 179—182), and assuming that all input factors x_i are uniformly distributed over $[-\pi, \pi]$ and that a $= 2$ and b $= 1$, we obtain:

$$S_{T_1} = 0.9991; \quad S_{T_2} = 0.0009; \quad S_{T_3} = 0.6161$$
$$S_{T_1} - S_1 = S_{T_3} - S_3 = 0.6161; \quad S_{T_2} - S_2 = 0 \tag{18.2}$$

From Eq. (18.2), we can infer that x_2 is almost uninfluential since its total-order index S_{T_2} is very close to 0. Moreover, x_1 and x_3 are interacting since the difference $S_{T_i} - S_i$ between their total-order and first-order sensitivity indices is well above 0.

In the next paragraphs, we show how some of these insights might be obtained by a visual analysis of a data set of input/output samples (Fig. 18.4). The data set was created using a Latin hypercube sampling of 500 factor combinations.

3.1.1 Scatter Plots

The top panel of Fig. 18.4 presents three one-dimensional scatter plots produced by the *scatter_plots* function in SAFE. Influential input factors are identified when a structured distribution of points is observed while moving along the horizontal axis on the scatter plot. For example, input factor x_1 (Fig. 18.4A) is influential; in fact the sign of the output y changes from negative to positive when moving in the direction of increasing x_1. Input factor x_3 (Fig. 18.4C) is also influential, since the

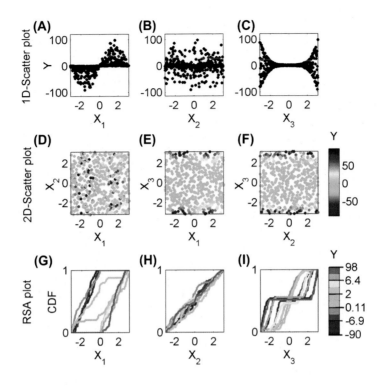

FIGURE 18.4

Example of visualization tools for qualitative GSA implemented in SAFE: one-dimensional scatter plot (A–C), two-dimensional scatter plots (D–F), and RSA with grouping (G–I), applied to the Ishigami–Homma test function.

variability of the output increases when moving toward the extreme ends of the variability range of x_3. Instead, input factor x_2 (Fig. 18.4B) appears not to be influential since output values are scattered uniformly when moving along the horizontal axis and no pattern can be identified.

The middle panel of Fig. 18.4 shows three two-dimensional scatter plots produced by the *scatter_plots_col* function. In these plots, the marker colors correspond to output value magnitudes. These plots can be used to look for interactions between pairs of factors while all factors are varied. For example, Fig. 18.4E shows that x_1 and x_3 are interacting since we observe that high values of the output occur with high values of x_1 and extreme values of x_3, and low values of the output occurs with low values of x_1 and extreme values of x_3. On the contrary, Fig. 18.4D and F reveal no interactions involving x_2. In fact, patterns only appear when moving along a horizontal line in Fig. 18.4D (i.e., when varying x_1) and when moving along a vertical line in Fig. 18.4F (i.e., when varying x_3).

The conclusions drawn by this visual analysis are therefore consistent with the analytical results of Eq. (18.2). For this specific case study, the one-dimensional scatter plot for x_3 (Fig. 18.4C) reveals the effect of this parameter even if it has an impact on the output through interactions only. However, in general, there is no guarantee that interactions can be detected in one-dimensional scatter plots or that high-order interactions can be detected in two-dimensional scatter plots.

3.1.2 Regional Sensitivity Analysis Based on Grouping

Another type of visual analysis implemented in SAFE is a variant of the regional sensitivity analysis (RSA) technique (Young et al., 1978; Spear and Hornberger, 1980). This variant was proposed by Wagener et al. (2001). Its advantage is that it does not require the selection of a single threshold to separate "good" and "bad" output values, as in the original version of RSA (more information regarding the choice of the threshold in RSA is provided in Table 18.2). This variant, that we called "RSA based on grouping," consists of splitting the input factor sample into a given number of groups (e.g., 10) according to the associated output value (e.g., 10 intervals of increasing output value, designed so to have an equal number of samples in each group). The Cumulative Distribution Function (CDF) of each input factor in each group is then computed (Fig. 18.4G–I). Influential input factors are identified when differences are observed between the 10 CDFs. When instead input factors are not influential, all distributions are very close to the distribution from which the input was sampled (e.g., a uniform distribution in the example of Fig. 18.4). For example, in Fig. 18.4G and I we see that the 10 CDFs are distinct for x_1 and x_3, whereas they are almost identical for x_2 (Fig. 18.4H). Therefore, x_1 and x_3 are considered influential, whereas x_2 is not, which is in accordance with the analytical results of Eq. (18.2). However, it should be noted that whereas a positive result (i.e., CDF separation) is sufficient to conclude that an input factor is influential, a negative result (CDFs overlapping) does not necessarily imply that the factor is noninfluential, since it may still have an effect on the output through interactions. One would have to investigate additional multidimensional plots (e.g., two-dimensional scatter plots for identifying interactions between pairs of inputs) to separate the two options (noninfluential or influential via interactions only).

Furthermore, RSA plots can allow identification of relationships between relevant output values and subranges of the input space and therefore provides additional information to other analyses (e.g., variance-based sensitivity analysis). For instance, we see from Fig. 18.4G that output values higher than 6.4 are systematically produced by positive values of x_1 and output values lower than 0.11 correspond to negative values of x_1.

3.1.3 Parallel Coordinate Plot

Another useful visualization tool implemented in SAFE is the Parallel Coordinate Plot (*parcoor* function) (e.g., Inselberg, 2009), which visualizes the input factor space and helps the modeler to identify subregions associated with particular output values. The plot can also provide insight into factor interactions. We do not provide more details here since an example of application for this tool is given in Chapter 7.

3.2 QUANTITATIVE GLOBAL SENSITIVITY ANALYSIS METHODS

SAFE includes the most widely used quantitative GSA methods, namely, the elementary effect test (EET, or method of Morris; Morris, 1991; Campolongo et al., 2011), RSA (Young et al., 1978; Spear and Hornberger, 1980), variance-based sensitivity analysis (VBSA; Sobol', 1993; Saltelli, 2002), Fourier amplitude sensitivity test (FAST, Cukier et al., 1978; Saltelli et al., 1999), and the new PAWN method (Pianosi and Wagener, 2015). Moreover, the toolbox also enables application of time-varying GSA using all previously mentioned methods. Time-varying GSA can be used to assess the output sensitivities at each time step of the simulation when the model output is a time series of a variable. The workflow *workflow_tvsa_hymod* demonstrates how it can be performed, using a generic GSA method. SAFE also includes a specific time-varying GSA called Dynamic Identifiability Analysis, which allows the tracking of factor sensitivity and optimal values simultaneously (Wagener et al.,

2003). GSA methods based on correlation or regression analysis (Saltelli et al., 2008; Pianosi et al., 2016) are not implemented in SAFE because they can be easily applied using functions already available in Matlab. Chapter 7 provides an example of correlation analysis used within the context of GSA. In that case, input/output samples were generated and postprocessed via RSA using functions implemented in SAFE, while the subsequent correlation analysis used the "*corr*" function from the Matlab Statistics and Machine Learning Toolbox.

4. GUIDELINES FOR THE IMPLEMENTATION OF GLOBAL SENSITIVITY ANALYSIS

Here we present guidelines for the implementation of GSA using the SAFE toolbox based on feedback from its users. A more in-depth discussion of the workflow for GSA application is provided in Pianosi et al. (2016).

An initial choice for any GSA study is the method a user wants to apply. Three criteria must be considered for selecting the appropriate GSA method for a specific case study:

1. the available computational resources;
2. the objective(s) of the analysis;
3. the fit to the method's underlying assumptions (e.g., linear model, symmetric output distribution)

To help with the identification of a method consistent with the available computational resources (criterion 1), we provide indications regarding the number of model evaluations that must be executed as a function of the number of input factors (denoted by M). Typically, as reported in Pianosi et al. (2016), a minimum of $10 \times M$ model evaluations has to be performed for EET, $100 \times M$ for RSA, and $1000 \times M$ for VBSA (this number can be considerably reduced when using FAST to approximate variance-based sensitivity indices) and PAWN. However, the number of model evaluations necessary to achieve reliable results might be significantly larger depending on the specific case study and it must be carefully checked (for more detailed guidance see Sarrazin et al., 2016). Ways to revise the adequacy of the chosen sample size and the reliability of GSA results are further discussed in Table 18.2. A trade-off has often to be made between computational cost and rigor of the analysis. For example, although VBSA results are often considered as benchmarks (e.g., Yang, 2011), the semiquantitative EET method is usually applied when computational resources are limited. In fact, EET can generally provide a good approximation of VBSA results using a significantly smaller number of model evaluations (e.g., Campolongo et al., 2007).

Regarding criterion 2, GSA can be used for different objectives such as to identify model dominant controls, to support model calibration, to prioritize efforts for uncertainty reduction, to perform a model diagnostic evaluation (verification of the model structure), and to analyze parameter identifiability. A discussion of how different GSA methods can be used for such different purposes can be found in Pianosi et al. (2016). A set of examples of applications of SAFE for these different objectives is given in Table 18.1. The table demonstrates the breadth of purposes and model complexities that can be handled by GSA.

Regarding criterion 3, methods implemented in SAFE can generally be applied to any case study independently of the model input—output response. However, the form of the output distribution must be checked when applying VBSA method, since it can give unreliable results when the distribution is skewed or multimodal (e.g., see Liu et al., 2006; Pianosi and Wagener, 2015).

Since the different methods rely on different assumptions and different rationales, we recommend multimethod analysis. For instance, a correlation analysis can be performed in addition to RSA to

Table 18.1 Examples of quantitative GSA applications using SAFE

Objective of GSA	GSA Method	Number of Input Factors	Type of Input Factors	Model	References
Identification of dominant controls	RSA	4	Parameters	Semidistributed hydrological model (VarKarst-R)	Hartmann et al. (2015)
Identification of dominant controls	PAWN	6	Parameters	Surrogate phase-field model	Hamdia et al. (2015)
Identification of dominant controls	VBSA	8	Parameters	Epidemiological model	López et al. (2016)
Identification of dominant controls	EET	54	Parameters	Distributed hydrological model (DHSVM)	Kelleher et al. (2015)
Design of data sampling strategy for model calibration	DYNIA	2	Parameters	Hydrochemical model (Birkenes)	Wang et al. (2016)
Support of model calibration	Multimethod (RSA, VBSA, PAWN)	12	Parameters & initial conditions	Land surface model (JULES)	Chapter 7
Effort prioritization for uncertainty reduction	PAWN (time varying)	6	Forcing inputs, parameters & output observations	Lumped hydrological model (HBV)	Pianosi and Wagener (2016)
Effort prioritization for uncertainty reduction and model diagnostic evaluation	VBSA	5	Spatial resolution, parameters, boundary condition & digital elevation model	Flood inundation model (LISFLOOD-FP)	Savage et al. (2016)
Analysis of parameter identifiability	VBSA	9–11	Parameters	3 Lumped hydrological models (NAM, SMARG, SMART)	Mockler et al. (2015)

study input interactions, or RSA can provide insight into relevant factor values to complement VBSA. Thanks to its modular structure, SAFE facilitates this. In fact, different GSA methods can be applied using the same model input—output sample, thus limiting the computational cost. In this regard, Chapter 7 proposes an approach for applying different GSA methods to the same input—output sample based on numerical approximations.

In Table 18.2, we briefly address other frequently asked questions from SAFE users and provide references where these issues are further discussed. Many of these questions concern critical choices that have to be made throughout the application of GSA. SAFE provides tools to test the adequacy of these choices and to validate the GSA results. In fact, it supports a multicriteria and multimethod analysis and implements the bootstrap approach to derive confidence intervals on the sensitivity

---

Table 18.2 Frequently asked questions from SAFE users regarding the implementation of the three main steps of GSA

Sampling

Q: How many input factors can be handled by SAFE?
A: *There is no upper limit to the number of input factors handled by SAFE. However, when the number is large, the computational cost of the analysis may be prohibitive (see Section 4 and* Pianosi et al., 2016*).*

Q: How can spatially or temporally distributed inputs be analyzed?
A: *Since GSA indices are defined for scalar input factors, a possible approach is to define an input factor for each spatial unit or time step of the nonscalar data. For example,* Berezowski et al. (2015) *analyze patterns of spatial sensitivity of a hydrological model to uncertainty in spatially distributed input data (remotely sensed snow cover fraction) and* Herman et al. (2013) *investigate the spatial sensitivity to distributed parameters. However, the computational cost of such analysis may be prohibitive when the model resolution is high.*
An alternative approach that preserves spatial or temporal relationships is to define a multiplicative or additive factor to be applied to the spatially or temporally varying input (e.g., Van Griensven et al., 2006*).*
Another strategy is to create a catalog of possible realizations of the temporally or spatially distributed input, and perform the sampling for GSA from such catalog (e.g., Baroni and Tarantola, 2014 *and references therein). For example,* Savage et al. (2016) *apply this technique to analyze the relative importance of spatial resolution and digital elevation model on the predictions of a flood inundation model, whereas* Pianosi and Wagener (2016) *use it to analyze the temporal sensitivity of lumped hydrological models to errors in precipitation, potential evapotranspiration, and streamflow time series.*
Finally, when analyzing potentially complex spatial patterns of sensitivity it is sometimes helpful to start by simpler virtual experiments using synthetic series so as to focus on the impact of specific spatial uncertainties. An example is given by Van Werkhoven et al. (2008b) *who analyze the change in parameter sensitivity when forcing the model with different virtual precipitation scenarios having distinct spatiotemporal patterns.*

Q: How can the input probability distributions (and input ranges) be defined?
A: *Input distributions quantify the uncertainty in the input factors and can be defined based on information available prior to performing GSA. When inputs are model parameters, uncertainty can be characterized based on their physical meaning, experimental data (when applicable), or previous studies. When inputs are forcing data, uncertainty can be quantified based on the characteristics of the data measurement and data processing techniques. Distributions and ranges can also be elicited from experts. When no specific information is available that suggests otherwise, then uniform distributions are typically chosen. A more detailed discussion of this issue can be found in (*Pianosi et al., 2016*). Sometimes, the assumed distribution can be tested and refined using site-specific data. For example, for model parameters, the comparison of model output simulations with observations*

Continued

Table 18.2 Frequently asked questions from SAFE users regarding the implementation of the three main steps of GSA—Continued

(when available) or with the modeler's expectation can lead to removing subranges that produce implausible model behavior (an example is given in Chapter 7). If the definition of input distributions is not univocal, then we recommend to test the impact of different choices on GSA results.

Q: Should very wide input ranges be chosen to be sure that all possible input values are included?

A: *No, input ranges should realistically characterize the uncertainty in the input factors. The existence of implausible values in the input ranges (and thus in the model response) could bias GSA results. If the input ranges are poorly known, then it might be necessary to refine them in a stepwise manner to avoid that unreasonable values control the study result (see for example Kelleher et al., 2013).*

Q: Which sampling strategy should be used to explore the input space?

A: *Some GSA methods (e.g., EET) require a tailored sampling strategy. Other methods (e.g., RSA) can be applied to input/output data sets generated by any sampling techniques like Latin hypercube sampling or quasirandom sampling. For more information on these sampling techniques we refer to Forrester et al. (2008) and Press et al. (1992). At present, SAFE implements random sampling and Latin hypercube sampling; however, its modular structure allows the user to plug-in other sampling functions (for instance, quasirandom sampling). When using random sampling, it may happen that several samples be clustered in one region of the sampling space while other regions might not be covered at all. Latin hypercube sampling addresses this issue and aims to provide a uniform coverage of the input space. In particular in SAFE, Latin hypercube samples are built so as to maximize the minimum interpoint distance, or in other words the spread between the points.*

Q: How many samples are needed?

A: *The appropriate value of the sample size depends on the specific case study, and one fit-for-all recommendation cannot be given (see guidance in Sarrazin et al., 2016). A practical approach to assess the robustness of GSA results to the chosen sample size is the estimation of confidence bounds for the sensitivity indices via bootstrapping (Archer et al., 1997). All quantitative GSA methods implemented in SAFE provide functionality to this end.*

Q: Can new samples be added to an already existing sample?

A: *Yes. When using random sampling, a new random sample can be simply added to the already existing sample. When using Latin hypercube sampling, SAFE provides specific functions (AAT_sampling_extend and OAT_sampling_extend) that increase an already existing Latin hypercube sample while trying to maximize the coverage of the input space.*

Model Evaluation

Q: Can the toolbox be connected to an external simulation model?

A: *Yes, the model can be executed in any computing environment outside Matlab/Octave. The connection between SAFE and the external model can be automatized by using the "system" command, which makes it possible to call external programs from Matlab/Octave. Alternatively, it can be established "manually" by exchanging input and output samples through text files. An example of the latter approach is given in a specific workflow script (workflow_external_model).*

Q: When models are evaluated using time series of input data, how long should the simulation time horizon be?

A: *The choice of the simulation time horizon depends on the information content of the input time series. Since different input factors may have an influence on the output over different types of event (e.g., dry or wet period), the input time series should include a sufficient variety of events. The adequacy of the length of input time series can be assessed by analyzing the variability of the GSA results over different data periods as proposed in Shin et al. (2013).*

Q: Are observations of the model output needed to perform a GSA?

A: *No, it is not required. The scalar model output used for GSA can be either a statistic of the simulated time series (e.g., Hartmann et al., 2015) or an objective function when output observations are available (e.g., van Werkhoven et al., 2008a).*

Table 18.2 Frequently asked questions from SAFE users regarding the implementation of the three main steps of GSA—Continued
Q: Which output should be analyzed?
A: *The choice of the output should be consistent with the purpose of the analysis. If needed, multiple outputs can be analyzed, either by examining different outputs individually (an example is given in Chapter 7 and in van Werkhoven et al., 2008a) or by directly integrating GSA with a multicriteria analysis (e.g., Rosolem et al., 2012).*
Postprocessing
Q: Should the sensitivity indices be computed over all sampled input sets or over selected sets only (e.g., sets for which the output value is considered to be physically plausible or for which the output value is close to observations)?
A: *This choice depends on the objective of the GSA. It is important to bear in mind that filtering out some input sets can significantly impact the GSA results (see, for instance, Chapter 7).*
Q: If bootstrapping is adopted to estimate the confidence bounds of sensitivity indices, how many bootstrap resamples should be used?
A: *In the literature, 1000 bootstrap resamples are typically used (e.g., Nossent et al., 2011). The adequacy of this choice can be tested as reported in* Archer et al. (1997).
Q: Why may it happen that the results have still not converged even for very large sample sizes?
A: *A lack of convergence of the results could be due to high nonlinearities in the model, for instance, when a small number of points in the input space generate extreme output values. In our experience, the sample size needed for convergence of GSA results can significantly differ from one application to another, and general indications about the appropriate number of samples are difficult to give (Sarrazin et al., 2016).*
Q: How can insensitive inputs be identified?
A: *It is common practice to set threshold values on the sensitivity indices, below which input factors are deemed to be insensitive (e.g., Tang et al., 2007). However, since there seems to be no fit-for-all threshold value, it is recommended to check the adequacy of this choice (Sarrazin et al., 2016). SAFE implements two methods to validate the choice of the threshold value: (1) the method proposed by Andres (1997) (function Andres_plot) and (2) the KS-test (Kolmogorov, 1933; Smirnov, 1939) proposed by Pianosi and Wagener (2015) (function pawn_ks_test), as discussed in Sarrazin et al., 2016.*
Q: [For RSA] How can the threshold value be chosen?
A: *The threshold is commonly chosen so to separate "behavioral" input sets (that produce simulations consistent with observations when available or with the modeler's expectation) from "nonbehavioral" ones. For example, in Chapter 7 the "behavioral" parameter sets are those that improve the model accuracy with respect to the default setup. When the output is binary, the choice of the threshold is straightforward. For instance, the threshold separates minor from excessive growth of an alga species in Spear and Hornberger (1980). When no clear threshold value can be easily identified, it can be constructed by trial and error. For instance, Hartmann et al. (2015) use the median of the simulated outputs. The grouped RSA approach also available in SAFE is an easy way to analyze the impact of choosing different thresholds and can be used as a preliminary step.*
Q: [For VBSA] What if negative sensitivity indices are found, since in theory they should always be positive?
A: *Variance-based sensitivity estimates are a numerical approximation of their theoretical value (which is by definition between 0 and 1). When approximation errors are large (and especially if the exact, unknown, value of the sensitivity index is small), the sum of the exact value and the approximation error can go below 0. Negative variance-based sensitivity estimates are thus evidence of large approximation errors, and hence of a too small sample size.*

indices values and validation functions for the identification of insensitive factors. Overall, when performing GSA, we recommend:

- Clear definition of the objective of the GSA, which is fundamental to address the critical choices.
- Implementation of a multicriteria and multimethod analysis to obtain robust results.

5. OUTLOOK

The main characteristics of the SAFE toolbox are the flexibility enabled by its modular structure, the extensive documentation embedded in the code (workflow scripts and extensive comments), the availability of tools to support the implementation of a robust sensitivity analysis (e.g., functions to analyze convergence, validate GSA results, test and refine input factors ranges), and the availability of visualization tools to perform qualitative sensitivity analysis as well as to visualize and communicate the results of quantitative GSA.

We plan to release new versions of SAFE on a regular basis, to include new methods for sampling, postprocessing, and visualization, and improve existing methods following feedback and suggestions from users. The new versions will be automatically provided to all registered users.

ACKNOWLEDGMENTS

A University of Bristol Alumni Postgraduate Scholarship to F.S supported this work. Partial support for F.P. and T.W. was provided by the Natural Environment Research Council [Consortium on Risk in the Environment: Diagnostics, Integration, Benchmarking, Learning and Elicitation (CREDIBLE); grant number NE/J017450/1].

REFERENCES

Andres, T.H., 1997. Sampling methods and sensitivity analysis for large parameter sets. Journal of Statistical Computation and Simulation 57 (1–4), 77–110. http://dx.doi.org/10.1080/00949659708811804.

Archer, G.E.B., Saltelli, A., Sobol, I.M., 1997. Sensitivity measures, ANOVA-like techniques and the use of boostrap. Journal of Statistical Computation and Simulation 58 (2), 99–120. http://dx.doi.org/10.1080/00949659708811825.

Baroni, G., Tarantola, S., 2014. A general probabilistic framework for uncertainty and global sensitivity analysis of deterministic models: a hydrological case study. Environmental Modelling and Software 51, 26–34. http://dx.doi.org/10.1016/j.envsoft.2013.09.022.

Berezowski, T., Nossent, J., Chormański, J., Batelaan, O., 2015. Spatial sensitivity analysis of snow cover data in a distributed rainfall-runoff model. Hydrology and Earth System Sciences 19, 1887–1904. http://dx.doi.org/10.5194/hess-19-1887-2015.

Bergström, S., 1995. The HBV model (chapter 13). In: Singh, V.P. (Ed.), Computer Models of Watershed Hydrology. Water Resources Publications, Highlands Ranch, Colorado, USA, pp. 443–476.

Boyle, D., 2001. Multicriteria Calibration of Hydrological Models (Ph.D. thesis). University of Arizona, Tucson.

Campolongo, F., Cariboni, J., Saltelli, A., 2007. An effective screening design for sensitivity analysis of large models. Environmental Modelling and Software 22 (10), 1509–1518. http://dx.doi.org/10.1016/j.envsoft.2006.10.004.

Campolongo, F., Saltelli, A., Cariboni, J., 2011. From screening to quantitative sensitivity analysis. A unified approach. Computer Physics Communications 182 (4), 978–988. http://dx.doi.org/10.1016/j.cpc.2010.12.039.

Cukier, R.I., Levine, H.B., Shuler, K.E., 1978. Nonlinear sensitivity analysis of multiparameter model systems. Journal of Computational Physics 26 (1), 1–42. http://dx.doi.org/10.1016/0021-9991(78)90097-9.

Forrester, A., Sobester, A., Keane, A., 2008. Engineering Design via Surrogate Modelling: A Practical Guide. John Wiley & Sons.

Hamdia, K.M., Msekh, M.A., Silani, M., Vu-Bac, N., Zhuang, X., Nguyen-Thoi, T., Rabczuk, T., 2015. Uncertainty quantification of the fracture properties of polymeric nanocomposites based on phase field modeling. Composite Structures 133, 1177–1190. http://dx.doi.org/10.1016/j.compstruct.2015.08.051.

Hartmann, A., Gleeson, T., Rosolem, R., Pianosi, F., Wada, Y., Wagener, T., 2015. A large-scale simulation model to assess karstic groundwater recharge over Europe and the Mediterranean. Geoscientific Model Development 8 (6), 1729–1746. http://dx.doi.org/10.5194/gmd-8-1729-2015.

Herman, J.D., Kollat, J.B., Reed, P.M., Wagener, T., 2013. From maps to movies: high-resolution time-varying sensitivity analysis for spatially distributed watershed models. Hydrology and Earth System Sciences 17, 5109–5125. http://dx.doi.org/10.5194/hess-17-5109-2013.

Inselberg, A., 2009. Parallel Coordinates: Visual Multidimensional Geometry and Its Applications. Springer, New York. http://dx.doi.org/10.1007/978-0-387-68628-8.

Ishigami, T., Homma, T., 1990. An importance quantification technique in uncertainty analysis for computer models. In: Proceedings of the ISUMA 90', First International Symposium on Uncertainty Modelling and Analysis. University of Maryland, pp. 398–403. http://dx.doi.org/10.1109/ISUMA.1990.151285.

Kelleher, C., Wagener, T., McGlynn, B., 2015. Model-based analysis of the influence of catchment properties on hydrologic partitioning across five mountain headwater subcatchments. Water Resources Research 51 (6), 4109–4136. http://dx.doi.org/10.1002/2014WR016147.

Kelleher, C., Wagener, T., McGlynn, B., Ward, A.S., Gooseff, M.N., Payn, R.A., 2013. Identifiability of transient storage model parameters along a mountain stream. Water Resources Research 49 (9), 5290–5306. http://dx.doi.org/10.1002/wrcr.20413.

Kolmogorov, A., 1933. Sulla Determinazione Empirica Di Una Legge Di Distribuzione. Giornale dell'Istituto Italiano Degli Attuari 4, 83–91.

Liu, H., Chen, W., Sudjianto, A., 2006. Relative entropy based method for probabilistic sensitivity analysis in engineering design. Journal of Mechanical Design 128 (2), 326–336. http://dx.doi.org/10.1115/1.2159025.

López, L., Izquierdo, A., Manzoli, D., Beldoménico, P., Giovanini, L., 2016. A myiasis model for *Philornis Torquans* (Diptera: Muscidae) and *Pitangus Sulphuratus* (Passeriformes: Tyrannidae). Ecological Modelling 328, 62–71. http://dx.doi.org/10.1016/j.ecolmodel.2016.02.001.

Mockler, E.M., O'Loughlin, F.E., Bruen, M., 2015. Understanding hydrological flow paths in conceptual catchment models using uncertainty and sensitivity analysis. Computers and Geosciences 90, 66–77. http://dx.doi.org/10.1016/j.cageo.2015.08.015.

Morris, M.D., 1991. Factorial sampling plans for preliminary computational experiments. Technometrics 33 (2), 161–174. http://dx.doi.org/10.2307/1269043.

Nossent, J., Elsen, P., Bauwens, W., 2011. Sobol' sensitivity analysis of a complex environmental model. Environmental Modelling & Software 26 (12), 1515–1525. http://dx.doi.org/10.1016/j.envsoft.2011.08.010.

Pianosi, F., Wagener, T., 2016. Understanding the time-varying importance of different uncertainty sources in hydrological modelling using global sensitivity analysis. Hydrological Processes. http://dx.doi.org/10.1002/hyp.10968.

Pianosi, F., Beven, K., Freer, J., Hall, J.W., Rougier, J., Stephenson, D.B., Wagener, T., 2016. Sensitivity analysis of environmental models: a systematic review with practical workflow. Environmental Modelling & Software 79, 214–232. http://dx.doi.org/10.1016/j.envsoft.2016.02.008.

Pianosi, F., Sarrazin, F., Wagener, T., 2015. A Matlab toolbox for global sensitivity analysis. Environmental Modelling and Software 70, 80–85. http://dx.doi.org/10.1016/j.envsoft.2015.04.009.

Pianosi, F., Wagener, T., 2015. A simple and efficient method for global sensitivity analysis based on cumulative distribution functions. Environmental Modelling & Software 67, 1–11. http://dx.doi.org/10.1016/j.envsoft.2015.01.004.

Press, W., Teukolsky, S., Vetterling, W., Flannery, B., 1992. Numerical Recipes in C. The Art of Scientific Computing, second ed. Cambridge University Press. http://dx.doi.org/10.2307/1269484.

Reed, P.M., Chaney, N.W., Herman, J.D., Ferringer, M.P., Wood, E.F., 2015. Internationally coordinated multimission planning is now critical to sustain the space-based rainfall observations needed for managing floods globally. Environmental Research Letters 10, 024010. http://dx.doi.org/10.1088/1748-9326/10/2/024010.

Rosolem, R., Gupta, H.V., Shuttleworth, W.J., Zeng, X., De Gonçalves, L.G.G., 2012. A fully multiple-criteria implementation of the Sobol' method for parameter sensitivity analysis. Journal of Geophysical Research 117 (7), 1–18. http://dx.doi.org/10.1029/2011JD016355.

Saltelli, A., 2002. Making best use of model evaluations to compute sensitivity indices. Computer Physics Communications 145 (2), 280–297. http://dx.doi.org/10.1016/S0010-4655(02)00280-1.

Saltelli, A., Ratto, M., Andres, T., Campolongo, F., Cariboni, J., Gatelli, D., Saisana, M., Tarantola, S., 2008. Global Sensitivity Analysis. The Primer. Wiley, Chichester, West Sussex, England.

Saltelli, A., Tarantola, S., Chan, K.P.-S., 1999. A quantitative model-independent method for global sensitivity analysis of model output. Technometrics 41 (1), 39−56.

Sarrazin, F., Pianosi, F., Wagener, T., 2016. Global sensitivity analysis of environmental models: convergence and validation. Environmental Modelling & Software 79, 135−152. http://dx.doi.org/10.1016/j.envsoft.2016.02.005.

Savage, J.T.S., Pianosi, F., Bates, P., Freer, J., Wagener, T., 2016. Quantifying the importance of spatial resolution and other factors through global sensitivity analysis of a flood inundation model. Water Resources Research. In Review.

Seibert, J., 1997. Estimation of parameter uncertainty in the HBV model. Nordic Hydrology 28 (4/5), 247−262. http://dx.doi.org/10.2166/nh.1997.015.

Shin, M.J., Guillaume, J.H.A., Croke, B.F.W., Jakeman, A.J., 2013. Addressing ten questions about conceptual rainfall-runoff models with global sensitivity analyses in R. Journal of Hydrology 503, 135−152. http://dx.doi.org/10.1016/j.jhydrol.2013.08.047.

Smirnov, N., 1939. On the estimation of the discrepancy between empirical curves of distribution for two independent samples. Bulletin Mathématique de l'Université de Moscou 2, 3−14.

Sobol', I.M., 1990. Sensitivity estimates for nonlinear mathematical models. Matematicheskoe Modelirovanie 2, 112−118 (in Russian), Translated in English (1993). In: Mathematical Modelling and Computational Experiments 1: 407−14.

Spear, R.C., Hornberger, G.M., 1980. Eutrophication in peel inlet − II. Identification of critical uncertainties via generalized sensitivity analysis. Water Research 14, 43−49. http://dx.doi.org/10.1016/0043-1354(80)90040-8.

Tang, Y., Reed, P., Wagener, T., Van Werkhoven, K., 2007. Comparing sensitivity analysis methods to advance lumped watershed model identification and evaluation. Hydrology and Earth System Sciences 11, 793−817. http://dx.doi.org/10.5194/hess-11-793-2007.

Van Griensven, A., Meixner, T., Grunwald, S., Bishop, T., Diluzio, M., Srinivasan, R., 2006. A global sensitivity analysis tool for the parameters of multi-variable catchment models. Journal of Hydrology 324 (1−4), 10−23. http://dx.doi.org/10.1016/j.jhydrol.2005.09.008.

Van Werkhoven, K., Wagener, T., Reed, P., Tang, Y., 2008a. Characterization of watershed model behavior across a hydroclimatic gradient. Water Resources Research 44 (1), 1−16. http://dx.doi.org/10.1029/2007WR006271.

Van Werkhoven, K., Wagener, T., Reed, P., Tang, Y., 2008b. Rainfall characteristics define the value of streamflow observations for distributed watershed model identification. Geophysical Research Letters 35, L11403. http://dx.doi.org/10.1029/2008GL034162.

Wagener, T., Boyle, D.P., Lees, M.J., Wheater, H.S., Gupta, H.V., Sorooshian, S., 2001. A framework for development and application of hydrological models. Hydrology and Earth System Sciences 5, 13−26. http://dx.doi.org/10.5194/hess-5-13-2001.

Wagener, T., McIntyre, N., Lees, M.J., Wheater, H.S., Gupta, H.V., 2003. Towards reduced uncertainty in conceptual rainfall-runoff modelling: dynamic identifiability analysis. Hydrological Processes 17 (2), 455−476. http://dx.doi.org/10.1002/hyp.1135.

Wang, L., van Meerveld, H.J., Seibert, J., 2016. Which stream water samples are most informative for event-based model calibration? Hydrology Research. In Review.

Yang, J., 2011. Convergence and uncertainty analyses in Monte-Carlo based sensitivity analysis. Environmental Modelling and Software 26 (4), 444−457. http://dx.doi.org/10.1016/j.envsoft.2010.10.007.

Young, P.C., Spear, R.C., Hornberger, G.M., 1978. Modelling badly defined systems: some further thoughts. In: Proceedings of SIMSIG Simulation Conference. Australian National University, Canberra, pp. 24−32.

CHALLENGES AND FUTURE OUTLOOK

SENSITIVITY IN ECOLOGICAL MODELING: FROM LOCAL TO REGIONAL SCALES

X. Song[1], B.A. Bryan[2], L. Gao[2], G. Zhao[3], M. Dong[4]

Zhejiang University, Hangzhou, China[1]; CSIRO Land and Water, Glen Osmond, SA, Australia[2]; University of Bonn, Bonn, Germany[3]; Beijing Institute of Surveying and Mapping, Haidian, Beijing, China[4]

CHAPTER OUTLINE

1. INTRODUCTION

Over the past few decades, ecological models have emerged as essential tools to help understanding, simulating, and predicting the outcomes of key processes in terrestrial ecosystems, including complex social-ecological systems (Stocker and Change, 2014). This endeavor could be attributed to more readily available computing resources, more thorough understanding of ecological mechanisms, and, most importantly, the complexity and challenges of the environmental problems we are facing nowadays. The desire to capture the complexity of environmental problems, both in depth and breadth, has greatly increased the modeling sophistication. This is reflected in the high temporal and spatial resolution and the interplay of dozens of model parameters. Ecological modeling practices in many fields, e.g., the land surface modeling (Henderson-Sellers et al., 1993; Bonan et al., 2002; Oleson et al., 2013), are becoming more integrated with an ambition to expand the system boundary by including more detailed parameterization schemes, processes, and submodels and to gradually approach the real state of the target terrestrial ecosystems.

With the increased complexity in social-ecological models, more parameters are generally needed to delineate the additional details in model processes. Given a complex social-ecological model, the main concerns are the correctness of model structure, i.e., whether the parameters work in a theoretically correct way including their interactions with other parameters, and the influence of uncertainty of the parameters or model forcing data on the model outcome. However, the closeness of model outcome to the observation does not naturally guarantee the correctness of model processes due to the canceling-out effect, and the uncertainties of parameters and data are as important as data values themselves in determining the model outcome (Raupach et al., 2005; Dong et al., 2015). So, what we expected from the global sensitivity analysis (GSA) is the capability to quantitatively disclose the interaction effect between parameters, and to measure the contribution of uncertainty in input parameters to the variance in model outputs (Saltelli et al., 2008).

As suggested by Williams et al. (2009), the typical life cycle of an ecological model encompasses a set of procedures, e.g., calibration, evaluation, testing, and structural improvement, following a data-centric approach (Fig. 19.1). The parameter or state estimation is not our focus here, but in both cases the model structure needs to be characterized through consistency checks and the sensitivity of model output to parameters should also be tested as a prerequisite. For a complex ecological model, the nonlinear and high-dimensional properties of the parameter space could easily make the analytical

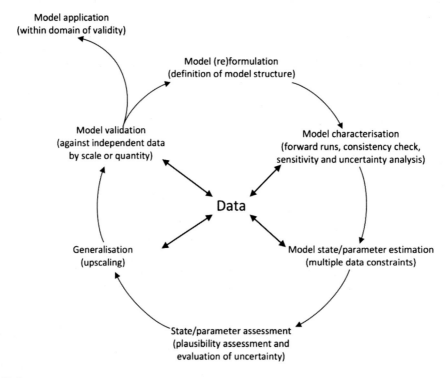

FIGURE 19.1

A conceptual diagram showing the multistage process of model data fusion.

Adapted from Williams, M., Richardson, A.D., Reichstein, M., Stoy, P.C., Peylin, P., Verbeeck, H., Carvalhais, N., Jung, M., Hollinger, D.Y., Kattge, J., Leuning, R., Luo, Y., Tomelleri, E., Trudinger, C.M., Wang, Y.P., 2009. Improving land surface models with FLUXNET data. Biogeosciences 6(7), 1341–1359.

methods fail to identify the interactions among parameters, and as a consequence, the model output will be under the joint influence of more parameters. As opposed to the traditional "one-factor-at-a-time" local sensitivity analysis, more and more ecological modelers have started using the GSA in model consistency checking and model calibration (Makler-Pick et al., 2011; Song et al., 2012; Lu et al., 2013; Zhao et al., 2014; Gao et al., 2015).

As we will show here, GSA has been successfully applied to process-based ecological models, e.g., forestry, agricultural, and land surface models (Song et al., 2012; Lu et al., 2013; Zhao et al., 2014), with parameters ranging from dozens to a few hundred. However, the aforementioned models are similar in nature from the perspective of GSA. That is, they are all about identifying the sensitive parameters for certain model output by simply varying the target parameters simultaneously with a specific sampling algorithm and then collecting the corresponding model output and decomposing the total variance to each parameter. This is the classical GSA routine up to now.

However, we realize in practice that, for a specific ecological model, it is often hard to draw a general conclusion about the sensitivity of model outputs to variation in input parameters, and the main causes could be attributed to (1) the different features of the forcing data in various model application scenarios and (2) the time-varying sensitivities of model outputs (Song et al., 2013). For terrestrial ecological models, the most common forcing data are climate data. Climate data have very strong spatial variability and great influence on soil, water, and plant physiology. Hence, it is common to find that some parameters' sensitivities are quite discrepant in different climatic conditions. Parameter sensitivities in ecological model may also change with time (Song et al., 2013). For example, it is a common practice to include an aging- or phenology-related factor in terrestrial ecological models to simulate the growth or life cycle of plants. The value of this parameter changes with model time to purposely revise the corresponding assimilation rates of key plant physiological processes. This kind of model setting will inevitably result in the parameter sensitivity having a strong correlation with time in model simulation. Naturally, the initial time or model state should also be taken into consideration when analyzing the result of GSA.

Highly integrated models of complex social-ecological systems, e.g., the Australian continental Land Use Trade-Offs (LUTO) model (Bryan et al., 2014b; Connor et al., 2015), are increasingly being used as essential tools to assess the evolution and consequences of land use change under future climate change, resource allocation strategies, and population dynamics, including considerations about market behavior relating food, energy, and carbon (Bryan et al., 2015, 2016b; Grundy et al., 2016). Compared to the traditional data-driven ecological models, complex social-ecological systems are usually subjected to even greater uncertainty in future conditions, i.e., scenarios about the specific natural, economic, and social trends (Moss et al., 2010; Gao et al., 2015). The probabilities of occurrence of the scenarios are unknown, and the consequent prediction uncertainty is characterized as deep uncertainty (Walker et al., 2010; Cox, 2012). The GSA of social-ecological models under the influence of deep uncertainty is similar to the case of forcing data, e.g., the climate data we mentioned earlier in the context of traditional ecological models. However, the difference is that the scenarios in social-ecological models are often enumerable, so that we can calculate the model sensitivity by performing sensitivity analysis under different scenarios separately. We argue that the parameter sensitivities under deep uncertainty could be framed as a problem of making a robust decision (Stephen et al., 2011), and a combination of scenarios and decision theory can be used to expand the possible state space and associated model outputs, and to make decisions (i.e., quantifying parameter sensitivities) given current understanding (Gao et al., 2015).

In social-ecological systems, land use maps are a fundamental data source for modeling land use change and ecosystem services at local, regional, national, and global scales (Raudsepp-Hearne et al., 2010; Bryan and Crossman, 2013; Crossman et al., 2013a; Verburg et al., 2013; Bryan et al., 2014a,b; Dong et al., 2015). Many of the land use maps used in social-ecological modeling systems are derived from remote sensing classification, but the classification accuracy is limited due to the spatial resolution and spectral mixture problems (Gong et al., 2013). So, it is important to quantitatively assess the influence of land use mapping error on the outputs of social-ecological models of land use and ecosystem services, i.e., the sensitivity to land use data uncertainty.

The objectives of this chapter are to describe our experiences and learning from undertaking GSA to better understand the uncertainty and sensitivity of a variety of social-ecological models. We present general guidelines about typical applications of GSA in ecological models, including methods, GSA result explanation, and computing considerations. We emphasize the importance of time-dependent sensitivities in ecological model simulation in GSA. We also describe a robust GSA method suitable for the social-ecological models featuring deep uncertainty. We provide an example of uncertainty in land use mapping and the corresponding model output sensitivities in a specific social-ecological model. Last, we describe computing challenges in undertaking GSA for complex models and discuss strategies for overcoming them, for example, by combining high-performance computing and an improved screening method suitable for the ecological models with deep uncertainty.

2. SENSITIVITY IN PROCESS-BASED ECOLOGICAL MODELS

The sensitivity of model outputs to parameters can be seen as, from another perspective, the contribution of certain parameter to the variances or uncertainties of the model outputs. As a common practice in GSA, the sampling strategy generally assumes that parameters follow certain probability distribution functions, e.g., the uniform distribution if the parameter's prior distribution probability is unknown (Song et al., 2012). However, we argue that, if the observed data are available, the probability distribution functions of parameters should be explicitly quantified as it will determine the concentration of parameter values, and consequently, the sensitivities of model outputs.

Process-based modeling approaches constitute the mainstream of ecological modeling by virtue of their flexibility in incorporating dynamic responses to altered environmental conditions, compared with the traditional statistical or rule-based models (Cuddington et al., 2013). Despite complex processes and numerous parameters, the GSA practices in typical process-based models are generally similar in nature, but the success of GSA largely depends on the analyst's familiarity with the theoretical basis of the specific model. In the following discussion we explore this in more detail in the context of crop modeling.

Crop models are frequently used for understanding the biophysical processes in agricultural systems and optimizing management practices to improve crop performance under specific field conditions (Keating et al., 2003; Holzworth et al., 2010, 2011). They simulate and integrate the complex interactive effects of climate—soil factors on plant growth as well as management interventions such as irrigation, fertilization, tillage, variety selection, and herbicide/pesticide application. Since the interactions of environmental and management effects on plant physiology and crop growth are of high complexity, hundreds to thousands of parameters are introduced to represent and constrain the various biophysical processes.

When models are applied to new varieties, the relevant parameters need to be calibrated. Calibration is the process of fitting parameter estimates to data. The parameters that are derived from model calibration are critical to the robustness of the simulation results. The accuracy of the calibration is mainly determined by the calibration method as well as the quality and quantity of reference data from field trials. Brute force and Bayesian calibration (BC) are the two frequently used calibration approaches (Wilkinson, 2010). Due to the complex structure as well as the large number of parameters and outputs in crop models, calibration is often prohibitive with the brute force approach as it increases the running time of the models exponentially with an extremely large number of parameters (Oakley and O'Hagan, 2004). BC exclusively uses probability for each parameter and uses reference data to update the probability density function (PDF) of the parameters. This also makes BC computationally expensive.

Fortunately, the development of the sampling-based methods [e.g., Markov Chain Monte Carlo (MCMC) and Gibbs] lessened the computing demand to an acceptable range. If data quality is high, i.e., high frequency and low error, the posterior PDF should be narrowed (Gallagher and Doherty, 2007). BC has been applied to a wide range of ecological models and is the most widely used method to estimate the posterior PDF for the model parameters. For example, Lehuger et al. (2009) implemented MCMC to calibrate the CERES-EGC model and reduced the root-mean-square error by 73% when estimating N_2O emissions. Van Oijen et al. (2005) reduced the uncertainty of 39 parameters by BC of a forest model against observed biomass and leaf area index data. Van Oijen et al. (2013) also demonstrated that BC can be used as a quantitative method in benchmarking process-based forest models in simulating the greenhouse gas emissions in Norway spruce forest. Minunno et al. (2013) found that the ability of BC can be improved by only calibrating the most influential parameters for the key output variables, which can be selected by GSA (Saltelli et al., 2008; Zhao et al., 2014; Specka et al., 2015).

Zhao et al. (2014) used the extended Fourier Amplitude Sensitivity Test (FAST) (eFAST) variance-based sensitivity analysis method to investigate the sensitivity of four Agricultural Production Systems sIMulator (APSIM) outputs (e.g., grain yield, aboveground biomass, flowering day, and maturity day) to variations in 10 cultivar parameters. They studied the effects of management and growth conditions of climate and soil on sensitivity indices across five sites in the Australian croplands. The uncertainty derived from the variation of the cultivar parameters for the four outputs was quantified. The results indicated that the grain yield was mainly determined by the yield component parameters (grains per gram stem, max grain size, and potential grain filling rate) and the length of the key reproductive stages. All cultivar parameters influenced the accumulation of aboveground biomass with the vernalization sensitivity of the variety and thermal time requirement for floral initiation being most influential. The phenological outputs of both flowering and maturity days were sensitive to vernalization and thermal time requirement for floral initiation. The effects of fertilization were stronger than climate—soil condition on the order of the parameters' impacts on yield and biomass. Fertilization significantly increased the variation in yield and biomass, but had negligible effects on flowering and maturity days. The phenology outputs were found insensitive to photoperiod. This may be sourced from the parameter sampling strategies of the study, i.e., the varying range of the parameter might be too narrow to cover the proper parameter value distributions.

Zhao et al. (2014) concluded that the phenology-related parameter should be calibrated before the other parameters. The GSA can reduce the number of parameters to calibrate a crop model and target reducing the uncertainty of a specific output variable, thus the calibration efficiency can be improved.

The calibration time demand will also be reduced. Generally, the GSA should be conducted earlier than the calibration.

3. TIME-DEPENDENT SENSITIVITY AND ITS IMPLICATIONS

The life cycles of different ecological components in a specific ecosystem may have a great influence on parameter sensitivities from the temporal perspective (Song et al., 2013). We argue that the typical GSA practice as commonly reported, i.e., running the ecological model for a certain time period (e.g., one growing season for crops or several decades for the common terrestrial models) and then collecting the model outcomes for GSA, might provide a biased or unrepresentative picture about model sensitivities for a certain ecological model, and we termed this as *snapshot* GSA.

One purpose of GSA is to validate the correctness of model structure. However, considering that the typical ecological models are dynamic and many of the state variables evolve as the simulation steps through, what we could expect from the snapshot GSA is just a slice from the whole picture of model sensitivities. For model development and verification, this kind of information about parameter sensitivity is of limited practical value and might be misguided. Another common purpose of GSA is to rank parameters according to their sensitivity index, and then determine the high-priority parameters for model calibration and uncertainty reduction (Esprey et al., 2004; Yang, 2011). However, the parameter sensitivities (corresponding to a specific model output), as we will show later, might exhibit great variations, e.g., from insensitive to highly sensitive, depending on the time period of model simulation. Another purpose of GSA is model parameter calibration. GSAs are often used to screen out the less influential parameters to alleviate the cost of model calibration and data collection. For this kind of application, we stress that, omitting the time dependence of parameter sensitivities and loosely setting the parameters to default values, which might be sensitive in a certain period of model simulation, could cause significant bias in model outcomes.

Time-dependent and nonlinear processes are common in the process-based ecological models (Landsberg and Waring, 1997; Thornton et al., 2002). In the forest growth model Physiological Principles Predicting Growth (3-PG2) (Landsberg and Waring, 1997), for example, the stomatal conductance of leaves nonlinearly decreases with tree aging, and this will in turn affect the other key physiological processes. This parameter might be influential at the young age of the forest stand as the trees are undergoing rapid growth and the demand for carbohydrate is very strong, but might be less influential for older forest stands. To illustrate the importance of time-dependent GSA in model verification, Song et al. (2013) assessed the sensitivities of the 3-PG2 model using data of a typical forest stand and analyzed the dynamics of parameter sensitivities over time. The site-specific parameters were set as default, and only the end age of the simulation was changed from 5 to 50 years, with the model run in annual steps. The authors found that the sensitivities of most of the parameters changed significantly during the simulation period, and the patterns of parameter sensitivities over time were increasing, decreasing, and peaked. Only a few parameters showed no obvious trends with time, and the authors suggested that only these kinds of parameters could be safely set at default values. For the other parameters, care must be paid during model calibration and verification, i.e., the time series parameter sensitivity during the simulation period should be cross-checked to validate the correctness of model structure and theoretical assumptions underpinning the model. Song et al. (2013) further analyzed the influence of forcing data, e.g., climate data, and site-specific data such as the soil properties (soil texture and fertility rate), on parameter sensitivities over time. The influence of climate data change was made possible by changing the simulation start time. The authors found that the climate data change had less influence on the time series parameter sensitivities compared with soil properties, especially fertility. And it is not surprising to find that the sensitivities of the state variable, root biomass, to key parameters were significantly affected by changing the soil texture.

Time-dependent parameter sensitivity may be common in ecological models, and we suggest that the temporal factor must be considered in GSA. Also, the influence of forcing or site-specific data should not be omitted during this process. So, it is reasonable to conclude that the GSA in ecological models is case-specific to some extent and extrapolation of the results, especially to regions with distinct environmental conditions, should be done with care.

4. GLOBAL SENSITIVITY ANALYSIS IN SOCIAL-ECOLOGICAL SYSTEMS

Highly uncertain future drivers, such as climatic, technological, socioeconomic, and political change, are more than ever incorporated into social-ecological models for projecting multiple plausible future trajectories and assisting in designing policies that are robust to the future projections (Steve et al., 2015; Bryan et al., 2016a,b). Knowledge of the state of these future drivers is very limited, and probabilities of the occurrence of these drivers are largely unknown and often cannot even be ranked (Walker et al., 2013; Maier et al., 2016). This kind of uncertainty cannot be dealt with by traditional uncertainty analysis approaches that are based on frequency-based probabilities or subjective (Bayesian) probabilities, and is characterized as deep uncertainty (Knight, 1921; Lempert and Collins, 2007; Walker et al., 2013).

As a model diagnostic technique, GSA is challenged by the presence of deep uncertainty. The impact of deep uncertainty on the sensitivity of model parameters needs quantifying. When faced with deep uncertainty, traditional predict-then-act modeling approaches are no longer appropriate for social-ecological assessments. Scenario planning, a popular and effective way of coping with deep uncertainty, identifies a range of plausible future scenarios to help decision makers recognize and respond more effectively to changes (Peterson et al., 2003; Wilkinson and Kupers, 2013; Kirby et al., 2014). Scenarios can capture the key ingredients of deep uncertainty about future drivers. However, each scenario is essentially internally consistent, and varying the scenario inputs within a specified range of uncertainty independently (the way of GSA treating model input parameters) may invalidate the internal consistency, resulting in invalid parameter combinations. Gao and Bryan (2016) found that the influence of scenarios on output uncertainty and parameter sensitivity was significant. Different scenarios provided different drivers of change to social-ecological systems and led to significant variation in influential and non-influential parameters, and in parameter influence ranking and magnitude. These findings are important contributions to model diagnostics under deep uncertainty. The existing exercises of scenario planning (Moss et al., 2010; Bateman et al., 2013; Howells et al., 2013; Kirby et al., 2015) rarely explored the influence of different scenarios on model outcomes resulting from variation in model inputs.

Since scenarios have significant influence on model diagnostics, modelers need a set of quantified sensitivity indices that are robust to deep uncertainty and can provide reliable information on data collection and analysis for refining model parameters. Gao and Bryan (2016) employed four criteria from decision sciences (maximax, weighted average, minimax regret, and limited degree of confidence) to the calculation of synthesized sensitivity indicators that are robust to deep uncertainty. These robust sensitivity indicators incorporated different attitudes of decision makers to risk and worked acceptably well under all scenarios in the diagnosis of a complex social-ecological system model. The choice of a decision criterion is based on the objective and the risk preference of the decision maker. If the decision maker wants to include any possible influential parameter, the "maximax" criterion fits this purpose. The "weighted average" criterion is a way of converting "deep uncertainty" to "probabilistic uncertainty" and suitable for the decision maker who has confidence in occurrence probabilities of the futures described by different scenarios. The "minimax regret" criterion is recommended for the decision maker who wants to minimize the cost of choosing the worst case. The "limited degree of confidence" criterion provides a balanced measure of criteria "weighted average" and "minimax regret." The resulting robust sensitivity indicators can be used as a reliable and robust guide for collecting data, modeling, and refining parameter estimates.

Gao and Bryan (2016) further investigated the potential of the elementary effects (EE) method (Morris, 1991) as a sensitivity analysis approach that is robust to deep uncertainty. The benefit of the EE method is to provide a measure of the total effects with greatly reduced computational demand. Gao and Bryan (2016) modified the EE method, creating a new variant they called *robust elementary effects* (rEE), which performs robust sensitivity analysis under different scenarios. Internally consistent scenarios were incorporated into p-level parameter space (in the sampling process of the EE method, each dimension of the parameter space is divided into p levels). The robust GSA performance of rEE was assessed by statistically comparing the sensitivity effects from rEE with the robust sensitivity indicators obtained by applying four decision criteria from Gao and Bryan (2016) described earlier. Gao and Bryan (2016) found that when social-ecological system models have strong nonlinear interactions, rEE did not provide accurate sensitivity quantification under deep uncertainty. But the sensitivity indicators from rEE could be used to screen out noninfluential input parameters under deeply uncertain conditions. This method is particularly suitable for computationally expensive models. The two robust sensitivity analysis methods are both valuable additions to GSA and can help social-ecological system modelers deal with deep uncertainty and better prepare for a range of possible futures.

Scenarios that represent plausible futures have been increasingly incorporated into social-ecological systems models. However, concern that a small number of scenarios are insufficient to describe the full range of future uncertainty was raised (Bryant and Lempert, 2010). The generation of many (e.g., thousands of) scenarios has been advocated to explore robust strategies under deep uncertainty (Bryant and Lempert, 2010; Kasprzyk et al., 2013; Gao et al., 2014; Herman et al., 2015). Applying many scenarios to the diagnostics of social-ecological systems models will lead to a significant computational challenge—GSAs of social-ecological models are usually computationally expensive, let alone applying the analyses under many scenarios. Further research efforts are required to overcome this barrier.

5. SENSITIVITY OF SOCIAL-ECOLOGICAL MODELS TO LAND USE MAPPING ERROR

The outputs of ecological model are sensitive to parameters, forcing data, and site-specific data, e.g., the soil texture for terrestrial ecological models, as illustrated earlier (Song et al., 2012, 2013). For social-ecological models, e.g., the LUTO model, land use mapping data provide an essential foundation on which model projections are based and can be seen as a special kind of "site-specific data."

Land use data are not fixed and can be transformed into other land use types according to certain model rules (De Groot et al., 2010; Shuang et al., 2010; Crossman et al., 2013a,b). However, land use or land cover maps have significant thematic and spatial uncertainties (Foody, 2002; Giri et al., 2005; Fritz et al., 2010) and these errors may propagate to model outputs (Fang et al., 2006; Schulp and Alkemade, 2011; Verburg et al., 2011). Due to the increasing influence of models of land use and ecosystem services in policy and management, it is important to assess the impact of land use mapping error propagation through these models including the sensitivity of estimates of land use change and impacts on ecosystem services to land use mapping error.

Global- and continental-scale land use maps are commonly generated by the classification of remotely sensed imagery (Hansen et al., 2000; Schneider et al., 2010; Gong et al., 2013). The Landsat Thematic Mapper series of sensors are multispectral satellite-based remote sensing platforms that have been purposely designed for terrestrial remote sensing and have been widely used for this purpose. Landsat and other multispectral satellite platforms use wavelengths from the visible and infrared parts of the light spectrum to readily distinguish broad land cover categories such as forest, grassland, cropland, water bodies, urban development, snow/ice, and bare soil. However, for use in social-ecological models, more detail and resolution is required in the land use classification. For example, cropland needs to be broken into the component crops such as wheat, rice, sugarcane, vegetables, citrus, and grapevines. To do this, the Land Use of Australia mapping program (ABARES, 2010) combined fortnightly time series satellite imagery from the Advanced Very High Resolution Radiometer (AVHRR), field survey data, and agricultural census data to map individual agricultural commodities. Normalized Difference Vegetation Index data from AVHRR indicate the greenness of vegetative land cover, and different crops have different greenness signatures over the year. By matching satellite-derived greenness of each grid cell to a library of field-based greenness signatures, and constraining this to census-reported areas of agricultural commodities using the Spatial Reallocation of Aggregated Data (SPREADII) method—a Bayesian MCMC technique (Bryan et al., 2009), the likelihood (i.e., posterior probability) of each cell supporting each commodity can be mapped. Actual land use mapping then occurred using a maximum likelihood allocation routine. Hence, although we end up with a map in which each grid cell is assigned to one specific land use, this is uncertain, with greater uncertainty occurring in areas with more diverse land use (Dong et al., 2015).

Assessing the uncertainty in land use mapping on social-ecological modeling has largely involved comparing estimates of model outputs (e.g., ecosystem services impacts) derived from different land use maps (Benítez et al., 2007; Schulp and Alkemade, 2011). However, beyond identifying a significant effect, estimates of ecosystem services from different land use maps tell us little about how mapping error propagates through social-ecological models. To overcome this shortcoming, Foody (2015) and Livne and Svoray (2011) simulated error in land use maps based on a confusion matrix from remote sensing classification, and then compared the value of ecosystem services using the mapped area and the misclassification error-adjusted area. However, the simulation based on confusion matrices only provides one level of mapping error and does not assess the relationship between the degree of land use error and the corresponding variance in model outputs. Dong et al. (2015) developed a probability-based Monte Carlo method to simulate baseline land use mapping error at four error levels, i.e., 10%, 20%, 30%, and 40%. Based on the posterior probability layers it indicates the confidence of land use classification. The baseline land use map and error-simulated land use maps were then used as input to a complex social-ecological model, i.e., LUTO model (Bryan et al., 2014a,b; Connor et al., 2015). The estimated land use transition and ecosystem services supply were then compared with the estimates generated by the baseline map. The probability-based Monte Carlo method enables the simulation of error land use maps and uncertainty analysis thereafter.

The uncertainty of land use mapping error on social-ecological model could be analyzed spatially and systematically. The spatial uncertainties usually represent the likelihood of occurrence or discrepancy of land use conversions. For example, Verburg et al. (2013) aggregated the model projected land use maps to probability maps by occurrences. Dong et al. (2015) calculated the frequency of discrepancies in land use projections by the LUTO model when using simulated land use maps compared with when using the baseline map cell-by-cell; then they defined high likelihood, low likelihood, and uncertain ranges to represent the uncertainties spatially. Systematic uncertainties represent the similarity or discrepancy in ecosystem services, wherein the grid square or administrative regions are usually chosen as analysis unit. Schulp and Alkemade (2011) analyzed pair-wise ecosystem services using fuzzy numerical similarity for each grid cell and calculated the average similarity over all cells. Schulp et al. (2014) calculated the relative difference for each NUTS2 region and then calculated the overall mean value. Dong et al. (2015) used four spatial scales, i.e., national, regional, subregional, and local, as analysis units to represent different spatial aggregations and analyzed the differences in ecosystem services modeled based on the error-simulated land use maps under the four spatial scales and to those modeled on the baseline map.

Dong et al. (2015) found that increasing land use mapping error increased the area of uncertainty in land use transition and altered its spatial arrangement, and differentially affected the supply of ecosystem services. This influence was highly localized, with concentrations in mixed cropping areas rather than in extensive grazing areas. The error effects also varied with spatial scales. Although barely discernible when aggregated at the national level, error effects increased markedly with granularity of assessment. Sensitivity to land use mapping error also differed between ecosystem services. For example, under the 40% error level and at the local scale, the percentage difference in biofuel production and agricultural production were about 36% and 27%, respectively; whereas biodiversity services, carbon emission abatement, and water resources were about 12—18%. Compared to LUTO estimates with the baseline land use map, agricultural production estimates based on the 40% error map differed by 3% when assessed at the national scale, 12% at the regional scale, 18% at subregional scale, and 27% at the local scale. The finding that land use mapping error introduces strong localized, scale-dependent uncertainty into land use and ecosystem services has important implications for policy and decision making. This suggests that whereas national- and global-scale ecosystem service assessments are insensitive to this error, one needs to be aware of the potential impact of catchment- and regional-scale assessments on the uncertainty of outputs. These results emphasize the need for locally tailored and targeted policy and management strategies.

6. COMPUTING STRATEGY

As mentioned earlier, the social-ecological models are usually equipped with dozens or hundreds parameters, and we call this the dimensionality of parameter space. If we follow a strict equal space sampling strategy of model parameters in GSA, the number of sampling points will grow exponentially with the dimensionality of parameter space, i.e., the so-called curse of dimensionality (Bishop, 2006). Even if we follow some optimized sampling strategy, e.g., the eFAST, the computing burden is still daunting. In this section, we describe several methods for dealing with these computation demands.

Over the past few years, the rate of increase in computer processing speeds has slowed due to physical limitations such as heat dissipation. Instead, the number of cores on each central processing unit is now increasing. In addition, high-performance computing resources have become more accessible, characterized by multiple cores on multiple nodes, connected by high-speed networks in cluster and grid architectures. However, to make use of this high-performance computing power, processing jobs must be structured such that they can be executed concurrently across multiple cores and nodes. This can be done as either data parallel where each processor takes part of the data and executes the full program or task parallel where each processor takes the full data set and executes part of the program. As GSA requires multiple runs using different parameterizations on the full data set, they can be set up naturally as task parallel, and because they do not require communication between runs, they can be termed *embarrassingly parallel* (Bryan, 2013).

The other key consideration in managing the computational demands of undertaking GSA of complex ecological models is the efficiency of the GSA technique itself. Different GSA techniques (see Introduction section of this chapter) have different requirements for the number of model runs required. The development of GSA techniques over time has resulted in increasingly efficient methods for exploring the parameter space. However, there can be a trade-off between search efficiency and quality of solution. Early methods such as the EE method of Morris (1991) require far fewer model runs. Although they can only calculate the main effects, they are useful for screening out noninfluential parameters, thereby reducing the problem size for other methods such as the variance-based FAST (Cukier et al., 1973). However, for complex ecological models, FAST is too computationally intensive and smart sampling methods such as the Random Balance Design (Tarantola et al., 2006) have been applied to reduce the number of runs required. More advanced methods such as the eFAST (Saltelli et al., 1999) have been able to increase the efficiency of parameter space search without unduly affecting the quality of the solution.

In practice, managing the high computational demand of GSA of complex ecological models requires attention to both the method and the computing environment. Next, we describe some of our experiences. In conducting GSA of the tree growth model 3-PG2, Song et al. (2012) screened out noninfluential parameters using Morris' EEs method, and then they applied Saltelli's method to calculate both the main and total sensitivity effects. Over 360,000 3-PG2 model runs were then undertaken in the GSA for each site, processed in batch mode on a Linux-based computing cluster. Song et al. (2013) also screened using EEs, but then they used the FAST method to calculate parameter sensitivities, which required only around 2000 model runs to achieve convergence. Zhao et al. (2014) evaluated four methods including Monte Carlo brute force, Saltelli's method, FAST, and eFAST in calculating the sensitivity of the Agricultural Production Systems Simulator (Keating et al., 2003). They ran the simulations on grid-computing infrastructure, and found the eFAST to perform best. Due to its good convergence ability, it requires only small sample sizes (around 1000 for each parameter), which reduces the computing demands. Gao and colleagues (Connor et al., 2015; Gao and Bryan, 2016) undertook GSA of a national land use model for Australia—the LUTO model. LUTO is a complex environmental-ecological model of land use and ecosystem services and runs at an annual time step and 1.1-km grid cell resolution, and each run takes about 40 h. They used eFAST, which required 7250 simulations covering 50 parameters. Too big a job even for a high-performance compute cluster, Gao and Bryan (2016) employed column aggregation methods (Nazari et al., 2015) to reduce the spatial resolution to 3.3 km.

The presence of deep uncertainty further challenges present-day computational limits. To capture a full range of a highly uncertain future resulting from changes in climate, technology, and socioeconomics, many scenarios (each represents a plausible future) should be applied to the social-ecological model to explore robust/adaptive strategies. This inevitably puts great pressure on the already high computational demands in the sensitivity analyses of complex social-ecological models. Further advances are needed to address this challenge. Only by combining model speed-ups, for example, via efficiencies gained through smart problem size reduction (Nazari et al., 2015), with smart and efficient GSA methods can this new frontier be approached.

7. CONCLUDING REMARKS

Global climate change and underlying feedbacks of the terrestrial ecosystems constitute one of the most challenging environmental issues of the present age. The endeavor to understand the complex interactions between climate system and land surface, including the impacts of human activities, has stimulated active researches in terrestrial ecological modeling. Compared with the fast-growing model complexity, the observed data are always scarce. In this kind of situation, the role of GSA is more prominent in model correctness validation and parameter calibration.

In this chapter, we discuss some typical applications of GSA in ecological modeling, e.g., parameter sensitivity analysis, including its temporal feature, spatial application of social-ecological modeling, with scenario settings, and specific computing strategies. We also discuss the sensitivity of social-ecological modeling to land map errors, a very common error source, although it does not belong to GSA. We demonstrate that model sensitivity has close relationship with model uncertainty as the variances of model outputs are largely determined by the corresponding highly sensitive parameters. The model sensitivity is highly local to some extent due to the reliance on forcing or external data, e.g., the climate data, and in situ data such as soil properties for terrestrial ecological models. Model sensitivities may also be dynamic and dependent on time, with time series parameter sensitivity being especially useful for model validation during model development. For the social-ecological models, however, the GSA combined with future scenarios to resolve the deep uncertainty is still challenged by the representativeness of a limited number of scenarios and daunting computing resource requirement.

ACKNOWLEDGMENTS

The authors are grateful for the support of CSIRO and the Chinese Scholarship Council, which made our collaborations possible. X.D. Song was supported by National Science Foundation of China (No. 31270588) and the Fundamental Research Funds for the Central Universities, China. B.A. Brett and L. Gao thank the support from CSIRO Agriculture, and M. Dong was supported by the Beijing Key Laboratory of Urban Spatial Information Engineering (No. 2016103).

REFERENCES

ABARES, 2010. Land Use of Australia, Version 4, 2005−06 Dataset. Australian Bureau of Agricultural and Resource Economics, Canberra, Australia.

Bateman, I.J., Harwood, A.R., Mace, G.M., Watson, R.T., Abson, D.J., Barnaby, A., Amy, B., Andrew, C., Day, B.H., Steve, D., 2013. Bringing ecosystem services into economic decision-making: land use in the United Kingdom. Science 341 (6141), 45−50.

Benítez, P.C., Mccallum, I., Obersteiner, M., Yamagata, Y., 2007. Global potential for carbon sequestration: geographical distribution, country risk and policy implications. Ecological Economics 60 (3), 572–583.

Bishop, C.M., 2006. Pattern Recognition and Machine Learning. Springer, New York.

Bonan, G.B., Oleson, K.W., Vertenstein, M., Levis, S., Zeng, X., Dai, Y., Dickinson, R.E., Yang, Z.L., 2002. The land surface climatology of the community land model coupled to the NCAR community climate model. Journal of Climate 15 (22), 3123–3149.

Bryan, B.A., 2013. High-performance computing tools for the integrated assessment and modelling of social–ecological systems. Environmental Modelling & Software 39 (1), 295–303.

Bryan, B.A., Barry, S., Marvanek, S., 2009. Agricultural commodity mapping for land use change assessment and environmental management: an application in the Murray–Darling Basin, Australia. Journal of Land Use Science 4 (3), 131–155.

Bryan, B.A., Crossman, N.D., 2013. Impact of multiple interacting financial incentives on land use change and the supply of ecosystem services. Ecosystem Services 4, 60–72.

Bryan, B.A., Crossman, N.D., Nolan, M., Li, J., Navarro, J., Connor, J.D., 2015. Land use efficiency: anticipating future demand for land-sector greenhouse gas emissions abatement and managing trade-offs with agriculture, water, and biodiversity. Global Change Biology 21 (11), 4098–4114.

Bryan, B.A., King, D., Zhao, G., 2014a. Influence of management and environment on Australian wheat: information for sustainable intensification and closing yield gaps. Environmental Research Letters 9 (4).

Bryan, B.A., Nolan, M., Harwood, T.D., Connor, J.D., Navarro-Garcia, J., King, D., Summers, D.M., Newth, D., Cai, Y., Grigg, N., 2014b. Supply of carbon sequestration and biodiversity services from Australia's agricultural land under global change. Global Environmental Change 28 (1), 13–19.

Bryan, B.A., Nolan, M., McKellar, L., Connor, J.D., Newth, D., Harwood, T., King, D., Navarro, J., Cai, Y., Gao, L., 2016a. Land-use and sustainability under intersecting global change and domestic policy scenarios: trajectories for Australia to 2050. Global Environmental Change 38, 130–152.

Bryan, B.A., Runting, R., Capon, T., Cunningham, S., Perring, M.P., Kragt, M., Nolan, M., Law, E.A., Renwick, A., Eber, S., Christian, R., Wilson, K.A., 2016b. Designer policy for carbon and biodiversity co-benefits under global change. Nature Climate Change 6, 301–305.

Bryant, B.P., Lempert, R.J., 2010. Thinking inside the box: a participatory, computer-assisted approach to scenario discovery. Technological Forecasting & Social Change 77 (1), 34–49.

Connor, J.D., Bryan, B.A., Nolan, M., Stock, F., Gao, L., Dunstall, S., Graham, P., Ernst, A., Newth, D., Grundy, M., 2015. Modelling Australian land use competition and ecosystem services with food price feedbacks at high spatial resolution. Environmental Modelling & Software 57 (C), 141–154.

Cox, L.A., 2012. Confronting deep uncertainties in risk analysis. Risk Analysis 32 (10), 1607–1629.

Crossman, N.D., Bryan, B.A., Groot, R.S.D., Lin, Y.P., Minang, P.A., 2013a. Land science contributions to ecosystem services. Current Opinion in Environmental Sustainability 5 (5), 509–514.

Crossman, N.D., Burkhard, B., Nedkov, S., Willemen, L., Petz, K., Palomo, I., Drakou, E.G., Martín-Lopez, B., Mcphearson, T., Boyanova, K., 2013b. A blueprint for mapping and modelling ecosystem services. Ecosystem Services 4, 4–14.

Cuddington, K., Fortin, M.-J., Gerber, L., Hastings, A., Liebhold, A., O'Connor, M., Ray, C., 2013. Process-based models are required to manage ecological systems in a changing world. Ecosphere 4 (2), 1–12.

Cukier, R.I., Fortuin, C.M., Shuler, K.E., Petschek, A.G., Schaibly, J.H., 1973. Study of the sensitivity of coupled reaction systems to uncertainties in rate coefficients. I Theory. The Journal of Chemical Physics 59 (8), 3873–3878.

De Groot, R., Alkemade, R., Braat, L., Hein, L., Willeman, L., 2010. Challenges in integrating the concept of ecosystem services and values in landscape planning, management and decision making. Ecological Complexity 7 (3), 260–272.

Dong, M., Bryan, B.A., Connor, J.D., Nolan, M., Gao, L., 2015. Land use mapping error introduces strongly-localised, scale-dependent uncertainty into land use and ecosystem services modelling. Ecosystem Services 15, 63–74.

Esprey, L.J., Sands, P.J., Smith, C.W., 2004. Understanding 3-PG using a sensitivity analysis. Forest Ecology and Management 193 (1–2), 235–250.

Fang, S., Gertner, G., Wang, G., Anderson, A., 2006. The impact of misclassification in land use maps in the prediction of landscape dynamics. Landscape Ecology 21 (2), 233–242.

Foody, G.M., 2002. Status of land cover classification accuracy assessment. Remote Sensing of Environment 80 (1), 185–201.

Foody, G.M., 2015. Valuing map validation: the need for rigorous land cover map accuracy assessment in economic valuations of ecosystem services. Ecological Economics 111, 23–28.

Fritz, S., See, L., Rembold, F., 2010. Comparison of global and regional land cover maps with statistical information for the agricultural domain in Africa. International Journal of Remote Sensing 31 (9), 2237–2256.

Gallagher, M., Doherty, J., 2007. Parameter estimation and uncertainty analysis for a watershed model. Environmental Modelling & Software 22 (7), 1000–1020.

Gao, L., Bryan, B.A., 2016. Incorporating deep uncertainty into the elementary effects method for robust global sensitivity analysis. Ecological Modelling 321, 1–9.

Gao, L., Bryan, B.A., Nolan, M., Connor, J.D., Song, X., Zhao, G., 2015. Robust global sensitivity analysis under deep uncertainty via scenario analysis. Environmental Modelling & Software 76, 154–166.

Gao, L., Connor, J.D., Dillon, P., 2014. The economics of groundwater replenishment for reliable urban water supply. Water 6 (6), 1662–1670.

Giri, C., Zhu, Z., Reed, B., 2005. A comparative analysis of the Global Land Cover 2000 and MODIS land cover data sets. Remote Sensing of Environment 94 (1), 123–132.

Gong, P., Wang, J., Yu, L., Zhao, Y., Zhao, Y., Liang, L., Niu, Z., Huang, X., Fu, H., Liu, S., 2013. Finer resolution observation and monitoring of global land cover: first mapping results with Landsat TM and ETM+ data. International Journal of Remote Sensing 34 (7), 2607–2654.

Grundy, M.J., Bryan, B.A., Nolan, M., Battaglia, M., Hatfield-Dodds, S., Connor, J.D., Keating, B.A., 2016. Scenarios for Australian agricultural production and land use to 2050. Agricultural Systems 142, 70–83.

Hansen, M.C., Defries, R.S., Townshend, J.R.G., Sohlberg, R., 2000. Global land cover classification at 1 km spatial resolution using a classification tree approach. International Journal of Remote Sensing 21 (6), 1331–1364.

Henderson-Sellers, A., Yang, Z.L., Dickinson, R.E., 1993. The Project for intercomparison of land-surface parameterization schemes. Bulletin of the American Meteorological Society 74 (7), 1335–1349.

Herman, J.D., Reed, P.M., Zeff, H.B., Characklis, G.W., 2015. How should robustness be defined for water systems planning under change? Journal of Water Resources Planning & Management 141 (10).

Holzworth, D.P., Huth, N.I., Devoil, P.G., 2011. Simple software processes and tests improve the reliability and usefulness of a model. Environmental Modelling & Software 26 (4), 510–516.

Holzworth, D.P., Huth, N.I., Voil, P.G.D., 2010. Simplifying environmental model reuse. Environmental Modelling & Software 25 (2), 269–275.

Howells, M., Hermann, S., Welsch, M., Bazilian, M., Alfstad, T., Gielen, D., Rogner, H., Fischer, G., 2013. Integrated analysis of climate change, land-use, energy and water strategies. Nature Climate Change 3 (7), 621–626.

Kasprzyk, J.R., Nataraj, S., Reed, P.M., Lempert, R.J., 2013. Many objective robust decision making for complex environmental systems undergoing change. Environmental Modelling & Software 42 (2), 55–71.

Keating, B.A., Carberry, P.S., Hammer, G.L., Probert, M.E., Robertson, M.J., Holzworth, D., Huth, N.I., Hargreaves, J.N., Meinke, H., Hochman, Z., 2003. An overview of APSIM, a model designed for farming systems simulation. European Journal of Agronomy 18 (3), 267–288.

Kirby, J.M., Connor, J., Ahmad, M.D., Gao, L., Mainuddin, M., 2014. Climate change and environmental water reallocation in the Murray–Darling Basin: impacts on flows, diversions and economic returns to irrigation. Journal of Hydrology 518 (2), 120–129.

Kirby, M., Connor, J., Ahmad, M.D., Gao, L., Mainuddin, M., 2015. Irrigator and environmental water management adaptation to climate change and water reallocation in the Murray–Darling basin. Water Economics & Policy 1 (3), 1550009.

Knight, F.H., 1921. Risk, Uncertainty and Profit. Houghton Mifflin, Boston, MA.

Landsberg, J.J., Waring, R.H., 1997. A generalised model of forest productivity using simplified concepts of radiation-use efficiency, carbon balance and partitioning. Forest Ecology and Management 95 (3), 209–228.

Lehuger, S., Gabrielle, B., Van Oijen, M., Makowski, D., Germon, J.-C., Morvan, T., Hénault, C., 2009. Bayesian calibration of the nitrous oxide emission module of an agro-ecosystem model. Agriculture, Ecosystems & Environment 133 (3), 208–222.

Lempert, R.J., Collins, M.T., 2007. Managing the risk of uncertain threshold responses: comparison of robust, optimum, and precautionary approaches. Risk Analysis 27 (4), 1009–1026.

Livne, E., Svoray, T., 2011. Components of uncertainty in primary production model: the study of DEM, classification and location error. International Journal of Geographical Information Science 25 (3), 473–488.

Lu, X., Wang, Y.P., Ziehn, T., Dai, Y., 2013. An efficient method for global parameter sensitivity analysis and its applications to the Australian community land surface model (CABLE). Agricultural & Forest Meteorology 182–183 (22), 292–303.

Maier, H.R., Guillaume, J.H.A., van Delden, H., Riddell, G.A., Haasnoot, M., Kwakkel, J.H., 2016. An uncertain future, deep uncertainty, scenarios, robustness and adaptation: how do they fit together? Environmental Modelling & Software 81, 154–164.

Makler-Pick, V., Gal, G., Gorfine, M., Hipsey, M.R., Carmel, Y., 2011. Sensitivity analysis for complex ecological models — a new approach. Environmental Modelling & Software 26 (2), 124–134.

Minunno, F., van Oijen, M., Cameron, D., Cerasoli, S., Pereira, J., Tomé, M., 2013. Using a Bayesian framework and global sensitivity analysis to identify strengths and weaknesses of two process-based models differing in representation of autotrophic respiration. Environmental Modelling & Software 42, 99–115.

Morris, M.D., 1991. Factorial sampling plans for preliminary computational experiments. Technometrics 33 (2), 161–174.

Moss, R.H., Edmonds, J.A., Hibbard, K.A., Manning, M.R., Rose, S.K., Vuuren, D.P., Van Carter, T.R., Seita, E., Mikiko, K., Tom, K., 2010. The next generation of scenarios for climate change research and assessment. Nature 463 (7282), 747–756.

Nazari, A., Ernst, A., Dunstall, S., Bryan, B., Connor, J., Nolan, M., Stock, F., 2015. Combined aggregation and column generation for land-use trade-off optimisation. Environmental software systems. In: Denzer, R., Argent, R.M., Schimak, G., Hřebíček, J. (Eds.), Infrastructures, Services and Applications: 11th IFIP WG 5.11 International Symposium, ISESS 2015, Melbourne, VIC, Australia, March 25-27, 2015. Proceedings. Springer International Publishing, Cham, pp. 455–466.

Oakley, J.E., O'Hagan, A., 2004. Probabilistic sensitivity analysis of complex models: a Bayesian approach. Journal of the Royal Statistical Society: Series B (Statistical Methodology) 66 (3), 751–769.

Oleson, K.W., Lawrence, D.M., Bonan, G.B., Drewniak, B., Huang, M., Koven, C.D., Levis, S., Li, F., Riley, W.J., Subin, Z.M., 2013. Technical description of version 4.5 of the community land model (CLM). Geophysical Research Letters 37 (7), 256–265.

Peterson, G.D., Cumming, G.S., Carpenter, S.R., 2003. Scenario planning: a tool for conservation in an uncertain world. Conservation Biology 17 (2), 358–366.

Raudsepp-Hearne, C., Peterson, G.D., Bennett, E.M., 2010. Ecosystem service bundles for analyzing tradeoffs in diverse landscapes. Proceedings of the National Academy of Sciences of the United States of America 107 (11), 5242–5247.

Raupach, M.R., Rayner, P.J., Barrett, D.J., DeFries, R.S., Heimann, M., Ojima, D.S., Quegan, S., Schmullius, C.C., 2005. Model–data synthesis in terrestrial carbon observation: methods, data requirements and data uncertainty specifications. Global Change Biology 11 (3), 378–397.

Saltelli, A., Ratto, M., Andres, T., Campolongo, F., Cariboni, J., Gatelli, D., Saisana, M., Tarantola, S., 2008. Global Sensitivity Analysis: The Primer. John Wiley & Sons, Ltd.

Saltelli, A., Tarantola, S., Chan, K.P.S., 1999. A quantitative model-independent method for global sensitivity analysis of model output. Technometrics 41 (1), 39–56.

Schneider, A., Friedl, M.A., Potere, D., 2010. Mapping global urban areas using MODIS 500-m data: new methods and datasets based on 'urban ecoregions'. Remote Sensing of Environment 114 (8), 1733–1746.

Schulp, C.J., Burkhard, B., Maes, J., Van, V.J., Verburg, P.H., 2014. Uncertainties in ecosystem service maps: a comparison on the European scale. PLoS One 9 (10), e109643.

Schulp, C.J.E., Alkemade, R., 2011. Consequences of uncertainty in global-scale land cover maps for mapping ecosystem functions: an analysis of pollination efficiency. Remote Sensing 3 (9), 2057–2075.

Shuang, L., Costanza, R., Troy, A., D'Aagostino, J., Mates, W., 2010. Valuing New Jersey's ecosystem services and natural capital: a spatially explicit benefit transfer approach. Environmental Management 45 (6), 1271–1285.

Song, X., Bryan, B.A., Almeida, A.C., Paul, K.I., Zhao, G., Ren, Y., 2013. Time-dependent sensitivity of a process-based ecological model. Ecological Modelling 265, 114–123.

Song, X., Bryan, B.A., Paul, K.I., Zhao, G., 2012. Variance-based sensitivity analysis of a forest growth model. Ecological Modelling 247, 135–143.

Specka, X., Nendel, C., Wieland, R., 2015. Analysing the parameter sensitivity of the agro-ecosystem model MONICA for different crops. European Journal of Agronomy 71, 73–87.

Stephen, P., Carpenter, S.R., Carl, F., Bonnie, K., 2011. Decision-making under great uncertainty: environmental management in an era of global change. Trends in Ecology & Evolution 26 (8), 398–404.

Steve, H.D., Heinz, S., Adams, P.D., Baynes, T.M., Brinsmead, T.S., Bryan, B.A., Chiew, F.H.S., Graham, P.W., Mike, G., Tom, H., 2015. Australia is 'free to choose' economic growth and falling environmental pressures. Nature 527 (7576), 49–53.

Stocker, T., Change, I., 2014. Climate Change 2013 : The Physical Science Basis. Working Group I contribution to the Fifth Assessment Report of the Intergovernmental Panel on Climate Change. Cambridge University Press, Cambridge.

Tarantola, S., Gatelli, D., Mara, T.A., 2006. Random balance designs for the estimation of first order global sensitivity indices. Reliability Engineering & System Safety 91 (6), 717–727.

Thornton, P.E., Law, B.E., Gholz, H.L., Clark, K.L., Falge, E., Ellsworth, D.S., Goldstein, A.H., Monson, R.K., Hollinger, D., Falk, M., Chen, J., Sparks, J.P., 2002. Modeling and measuring the effects of disturbance history and climate on carbon and water budgets in evergreen needleleaf forests. Agricultural and Forest Meteorology 113 (1–4), 185–222.

Van Oijen, M., Reyer, C., Bohn, F., Cameron, D., Deckmyn, G., Flechsig, M., Härkönen, S., Hartig, F., Huth, A., Kiviste, A., 2013. Bayesian calibration, comparison and averaging of six forest models, using data from Scots pine stands across Europe. Forest Ecology and Management 289, 255–268.

Van Oijen, M., Rougier, J., Smith, R., 2005. Bayesian calibration of process-based forest models: bridging the gap between models and data. Tree Physiology 25 (7), 915–927.

Verburg, P.H., Asselen, S.V., Zanden, E.H.V.D., Stehfest, E., 2013. The representation of landscapes in global scale assessments of environmental change. Landscape Ecology 28 (6), 1067–1080.

Verburg, P.H., Neumann, K., Nol, L., 2011. Challenges in using land use and land cover data for global change studies. Global Change Biology 17 (2), 974–989 (916).

Walker, W.E., Lempert, R.J., Kwakkel, J.H., 2013. Deep uncertainty. Encyclopedia of Operations Research & Management Science 395–402.

Walker, W.E., Marchau, V.A.W.J., Swanson, D., 2010. Addressing deep uncertainty using adaptive policies: introduction to section 2. Technological Forecasting & Social Change 77 (6), 917–923.

Wilkinson, A., Kupers, R., 2013. Living in the futures. Harvard Business Review 91 (86), 118–127.

Wilkinson, R.D., 2010. Bayesian calibration of expensive multivariate computer experiments. Large-Scale Inverse Problems and Quantification of Uncertainty 195–215.

Williams, M., Richardson, A.D., Reichstein, M., Stoy, P.C., Peylin, P., Verbeeck, H., Carvalhais, N., Jung, M., Hollinger, D.Y., Kattge, J., Leuning, R., Luo, Y., Tomelleri, E., Trudinger, C.M., Wang, Y.P., 2009. Improving land surface models with FLUXNET data. Biogeosciences 6 (7), 1341–1359.

Yang, J., 2011. Convergence and uncertainty analyses in Monte-Carlo based sensitivity analysis. Environmental Modelling & Software 26 (4), 444–457.

Zhao, G., Bryan, B.A., Song, X., 2014. Sensitivity and uncertainty analysis of the APSIM-wheat model: interactions between cultivar, environmental, and management parameters. Ecological Modelling 279, 1–11.

CHALLENGES AND FUTURE OUTLOOK OF SENSITIVITY ANALYSIS

20

H. Gupta[1], S. Razavi[2]

The University of Arizona, Tucson, AZ, United States[1]; University of Saskatchewan, Saskatoon, SK, Canada[2]

CHAPTER OUTLINE

1. INTRODUCTION

Sensitivity analysis (SA) is an important tool in the development and application of dynamic models of Earth and Environmental System (EES) observations and is becoming even more important as such models become progressively more realistic. Although the objectives of an SA can be varied (see Razavi and Gupta, 2015), by far the most common purpose is to identify (and prioritize and screen) which of the parameters in the model are the most important/influential, in that perturbations to these parameters most strongly affect the magnitude, variability, and dynamics of model response. This is commonly referred to as a "parameter sensitivity analysis" and (although not quite grammatically correct) a parameter that causes a "stronger" change in the model response when perturbed is typically

Sensitivity Analysis in Earth Observation Modelling. http://dx.doi.org/10.1016/B978-0-12-803011-0.00020-3

said to be a "more sensitive parameter," whereas a parameter that causes a "weaker" change in the model response when perturbed is typically said to be a "less sensitive parameter" (note that sensitivity is therefore a relative concept).

The importance of a parameter SA is that it enables us to assess how much care must be taken in the specification of the values of various model parameters when using the model to help make decisions in a specific situation. If parameter A is relatively "more sensitive" than parameter B (meaning that the model-based decision D is more sensitive to unit changes in parameter A than to unit changes in parameter B), then greater care will need to be taken in specifying the value of parameter A.

As a corollary to this, if substantial perturbations to some parameter C (such as varying it across its entire feasible range of values) result in little or no effect on the model-based decision D, then such a parameter is typically referred to as "insensitive." Once detected, it is common practice to fix such a parameter at some reasonable nominal value and not give it further attention; this knowledge can sometimes also be used to guide structural simplifications of the model.

The ability to understand the importance and role of the various parameters in a dynamic model of the observations related to an EES cannot be understated. Such models can cover a wide range of spatiotemporal scales (from ground- to satellite-based), involve many different sources of data, and deal with many different types of processes. Parameter SA therefore provides an essential tool that can help the Earth observation modeler navigate the myriad questions/problems that can arise while seeking answers that will inform a better model-based decision. This ability becomes even more relevant when dealing with climate- and human-induced nonstationarities in the structure and functioning of EESs (Razavi et al., 2015).

2. BRIEF REVIEW OF SOME COMMONLY USED SENSITIVITY ANALYSIS METHODS

Consider that the model of interest has a set of parameters $\mathbf{\Theta} = \{\theta_1, \ldots, \theta_p\}$, where p represents the dimension of the parameter space (the number of parameters in the model). The value of each parameter θ_j can vary across the range $\left[\theta_{j, \text{ min}}, \theta_{j, \text{ max}}\right]$, such that $\mathbf{\Psi}$ defines the feasible space of values that the model parameters can jointly take on; in the case that the parameters can vary independently over their full range, $\mathbf{\Psi}$ defines a hypercube in the parameter space. Hence, $\mathbf{\Theta}^i = \{\theta_1^i, \ldots, \theta_p^i\}$ represents a single point in $\mathbf{\Psi}$ where each parameter θ_j^i takes on a specific value.

By running the model using the parameter values indicated by $\mathbf{\Theta}^i$, we generate a model output $Y(\mathbf{\Theta}^i)$. Note that Y can be as simple as a single value, or it can be considerably more complex, such as a time series of values computed for some specific location in space, a spatially distributed field of values computed for some specific time, or (most generally) a spatiotemporal varying field of values.

The behavior of this model output is then typically summarized in terms of a small set of scalar quantities $Z(\mathbf{\Theta}^i)$ that are assumed to represent the "response" of the model. For example, Z is often taken to be a model performance metric that measures the (scalar) distance between the model output $Y(\mathbf{\Theta}^i)$ and some data representing observations made on the real system. Alternatively, Z can represent a model-based decision (or set of decisions) D. For purposes of further discussion we will assume that Z is a single scalar value; vector values will require a similar but more complex analysis.

To summarize, the model parameters $\mathbf{\Theta}$ can vary over a feasible space $\mathbf{\Psi}$, and at each point in this feasible space we can compute a value (or set of values) Z that represents some model-based quantity (or quantities) of interest. If perturbation of one of the parameters θ_j causes a relatively large change in **Z**, then that parameter will be considered to be (relatively) sensitive, and if perturbation of another parameter θ_k causes a relatively small change in **Z**, then that parameter will be considered to be (relatively) insensitive.

2.1 LOCAL ASSESSMENT OF PARAMETER SENSITIVITY

At any specific point $\mathbf{\Theta}^i$, the idea of parameter sensitivity is mathematically well defined in terms of the vector of local partial derivatives $d\mathbf{Z}/d\mathbf{\Theta}|_{\mathbf{\Theta}^i} = \{d\mathbf{Z}/d\theta_1, \ldots, d\mathbf{Z}/d\theta_p\}|_{\mathbf{\Theta}^i}$, which indicates how sensitive the model response **Z** is to infinitesimally small individual changes in each of the parameters $\{\theta_1, \ldots, \theta_p\}$ at the point $\mathbf{\Theta}^i$ (Fig. 20.1). In practice, the value of $d\mathbf{Z}/d\mathbf{\Theta}|_{\mathbf{\Theta}^i}$ is often computed via a finite difference approximation using small parameter perturbations around the nominal value.

Now, in the unlikely case of a perfectly linear relationship between **Z** and $\mathbf{\Theta}$ (this does not generally happen in EES models), the magnitude of each term in $d\mathbf{Z}/d\mathbf{\Theta}|_{\mathbf{\Theta}^i}$ remains constant throughout the feasible space $\mathbf{\Psi}$, and hence the vector $d\mathbf{Z}/d\mathbf{\Theta}$ (evaluated at any point in $\mathbf{\Psi}$) can be considered to be a complete characterization of the sensitivity of **Z** to $\mathbf{\Theta}$. However, in EES models, the relationship is typically nonlinear (and can be quite complex). In general, therefore, $d\mathbf{Z}/d\mathbf{\Theta}$ changes from point to point and hence a proper (i.e., informative) characterization of sensitivity requires a more "global" investigation, as discussed in the next section.

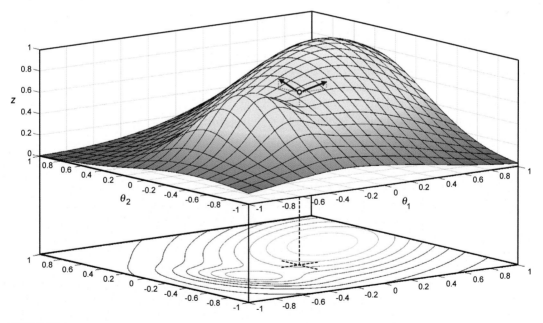

FIGURE 20.1

A two-parameter surface "**R1**" showing how the response **Z** varies with values of the parameters θ_1 and θ_2. The directional gradients (partial derivatives) $d\mathbf{Z}/d\mathbf{\Theta} = \{d\mathbf{Z}/d\theta_1, d\mathbf{Z}/d\theta_2\}$ are shown for a selected nominal point $\mathbf{\theta}^i$.

The exception to needing a global characterization would be when we have already precisely determined the "best" values Θ^* for Θ and are only interested in the characterization of the relative sensitivity of Z to Θ at that specific location. But, in the vast majority of cases, these "best" values are not known before conducting the parameter calibration process, and we would like to know how sensitive (in a relative sense) the overall response is to each of the parameters so that we can decide which parameters to optimize and which to keep fixed. Alternatively, there may simply exist more than one point of interest in the feasible space Ψ.

2.2 GLOBAL ASSESSMENT OF PARAMETER SENSITIVITY

So, for all but the trivial case of a perfectly linear relationship between Z and Θ, our assessment of parameter sensitivity must take into consideration the fact that the nature of the $\Theta \rightarrow Z$ relationship varies throughout the feasible space Ψ. Over the past several decades, several attempts have been made to represent the "global" nature of the sensitivity of model response. Some of the early ones were based on concepts such as the statistical design of experiments (factorial design) and regression and correlation analyses (see discussion in Razavi and Gupta, 2015); these approaches were typically based on strong assumptions regarding the mathematical form (e.g., linear or polynomial) of the $\Theta \rightarrow Z$ relationship. For EES models, such assumptions about the form of response surfaces are generally unjustifiable, and we will therefore not discuss those methods here.

Instead, we consider the following five different (but related) strategies that have been proposed in the literature and that are generally applicable to EES models. Each of these strategies is based on a somewhat different philosophical approach to the definition of sensitivity.

1. The *"One-Dimensional Cross Section"* Strategy
2. The *"Distribution of Derivatives"* Strategy
3. The *"Analysis of Variance"* Strategy
4. The *"Analysis of Cumulative Distributions"* Strategy
5. The *"Variogram Analysis of Response Surface"* Strategy

We briefly discuss each of these strategies below, with a view to comparing and contrasting their similarities and differences.

2.2.1 The "One-Dimensional Cross Section" Strategy

The simplest and computationally cheapest way to conduct a "quick and dirty" assessment of relative parameter sensitivity across the feasible space is to examine the nature of the different one-dimensional cross-sections of the response Z as each parameter θ_j is varied between its minimum and maximum values $\left[\theta_{j,\ min}, \ \theta_{j,\ max}\right]$, whereas the other parameters are fixed at nominal values (Fig. 20.2). If the cross-section for parameter θ_j tends to have steeper slopes across the parameter range than the cross-section for parameter θ_k, and/or the cross-section for parameter θ_j covers a larger range of the response Z than the cross-section for parameter θ_k, then we can conclude that the parameter sensitivity is higher for the former than for the latter. Such plots are, therefore, intuitively easy to understand.

However, although relatively simple and computationally inexpensive to perform, this strategy has some serious weaknesses. First, the analysis is really only valid for a particular point Θ^i in the parameter space, as this point must be specified before the cross-sections can be conducted; in this sense it is really just a slight extension of the local parameter sensitivity approach. Second, the

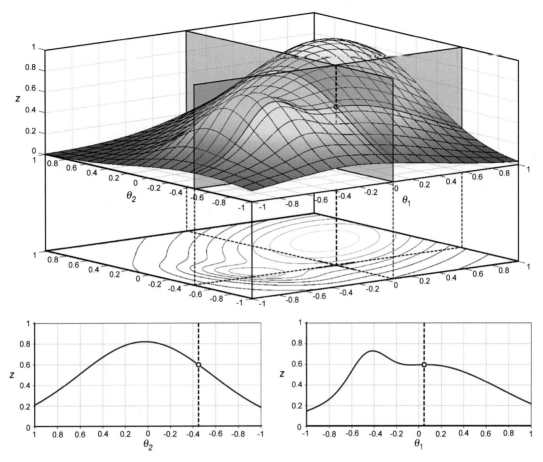

FIGURE 20.2

The main plot illustrates one-dimensional cross-sections in each of the parameter directions θ_1 and θ_2 for the two-parameter response surface "**R1**." The sections have been constructed to pass through the nominal point θ^i (indicated by the vertical dashed line). The two subplots show the two θ versus **Z** cross-sectional plots so obtained.

assessment relies upon a subjective visual examination of the nature of the cross-sections. To establish an objective analysis, one would have to define a mathematical metric to characterize the nature of each cross-section in terms of a single numerical "sensitivity" value. In doing so, one must necessarily accept the fact that, whatever the strategy used for such characterization, many different cross-sectional shapes will inevitably have the same numerical "sensitivity" value (e.g., see cross-sections **C1** and **C2** in Fig. 20.3).

Third, to be truly globally representative, the analysis would need to be conducted at a representative sample of locations Θ^i in the parameter space, in which case the assessment is neither simple nor computationally inexpensive. Such a "sampled cross section" strategy is actually embedded in the

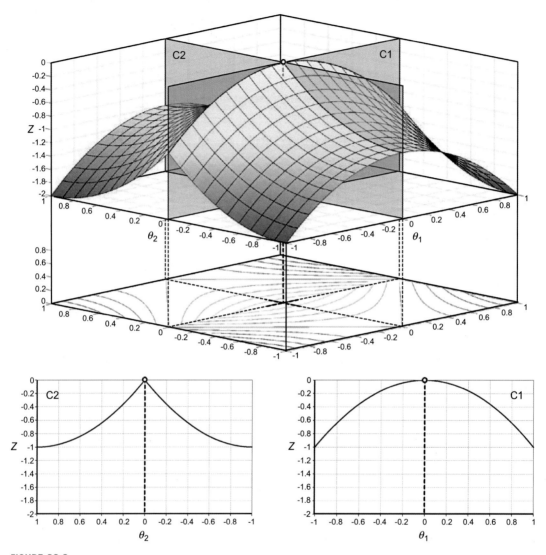

FIGURE 20.3

A cartoon illustration of a two-parameter response surface "**R2**," for which the one-dimensional θ versus **Z** cross-sections **C1** and **C2**, corresponding to parameters θ_1 and θ_2, respectively, have been constructed. These cross-sections have different shapes but identical *average* slope (and also average absolute slope and average squared slope) over the parameter ranges. If a metric based on averaging some function of the slope is used to characterize parameter sensitivity associated with each cross-section, then the response will be deemed to be equally sensitive to both parameters.

STAR−VARS implementation (Razavi and Gupta, 2016b) of the variogram analysis of response surfaces (VARS) framework introduced in Section 2.2.5 below, but we leave the details for the reader to explore.

2.2.2 The "Distribution of Derivatives" Strategy

Given that local parameter sensitivity at a point $\mathbf{\Theta}^i$ can be mathematically characterized by the vector of partial derivatives $d\mathbf{Z}/d\mathbf{\Theta}|_{\mathbf{\Theta}^i}$, a logical extension is to define "global" parameter sensitivity in terms of the distribution of different values of $d\mathbf{Z}/d\mathbf{\Theta}$ obtained as $\mathbf{\Theta}^i$ is varied throughout the feasible space $\mathbf{\Psi}$ (Fig. 20.4). Since each directional partial derivative $d\mathbf{Z}/d\theta_j$ is now characterized by a probability distribution $p(d\mathbf{Z}/d\theta_j)$, we can compactly characterize global parameter sensitivity in terms of the statistics of this distribution.

Morris (1991) proposed simply using the mean $\mu_j = E\{d\mathbf{Z}/d\theta_j\}$ and standard deviation $\sigma_j = Std\{d\mathbf{Z}/d\theta_j\}$ of the distribution of derivatives to characterize global sensitivity, with higher parameter sensitivity corresponding to larger values of both μ_j and σ_j. Subsequently, Campolongo et al. (2007) pointed out that since the derivatives $d\mathbf{Z}/d\theta_j$ can take on both positive and negative values, it would be better to use instead the mean of absolute derivatives $E\{|d\mathbf{Z}/d\theta_j|\}$, and Sobol and Kucherenko (2009) suggested the mean of the squared derivatives $E\{(d\mathbf{Z}/d\theta_j)^2\}$. Note, however, that these approaches are based on selecting summary statistics that are *globally aggregated measures of local sensitivities*; in 2014, Rakovec et al. reiterated the importance of considering the entire distribution of derivatives, instead of only the first two moments, to better characterize the available information.

In practice, the Distribution of Derivatives strategy is applied by obtaining a sample of $d\mathbf{Z}/d\mathbf{\Theta}$ values computed at a representative set of locations distributed uniformly across the feasible parameter space, and an experimental design strategy can be employed to minimize the computational cost associated with having to perform a large number of model runs.

2.2.3 The "Analysis of Variance" Strategy

An alternative global SA strategy, which has the advantage of not requiring partial derivatives to be computed, was proposed by Sobol' (1990) based on the concept of an analysis of variance. The basis for this approach is the recognition (Fig. 20.5) that as the model parameters $\mathbf{\Theta}$ vary throughout the feasible space $\mathbf{\Psi}$, the model response $\mathbf{Z}$ takes on a distribution of values $p(\mathbf{Z})$ that can be characterized in terms of its "total" unconditional variance $V(\mathbf{Z})$.

Now, if we fix one of the parameters θ_j at some nominal value $\widehat{\theta}_j$ while allowing all the other parameters to vary over their feasible ranges, we will obtain a conditional distribution $p(\mathbf{Z}|\theta_j = \widehat{\theta}_j)$ that has a different variance $V_j = V(\mathbf{Z}|\theta_j = \widehat{\theta}_j)$; in general, the variance of this conditional distribution may be smaller or larger than the unconditional variance $V(\mathbf{Z})$.

Repeating this computation for different nominal values $\widehat{\theta}_j$ of the "fixed" parameter θ_j sampled across its feasible range will result in a probability distribution of values $p(V_j)$ for the conditional variance. Sobol' proposed that the *expected value* (average) of the conditional variance will always be smaller than (or equal to) the total variance, and that their difference—the average reduction in variance $\Delta V_j = V - E(V_j)$ due to fixing parameter θ_j—can serve as a meaningful characterization of the sensitivity of the model response $\mathbf{Z}$ to that parameter. In other words, if fixing a parameter does not result in a sizable expected reduction ΔV_j in total variance, then clearly variations of that parameter (over its feasible range) did not contribute substantially to the total variance $V(\mathbf{Z})$.

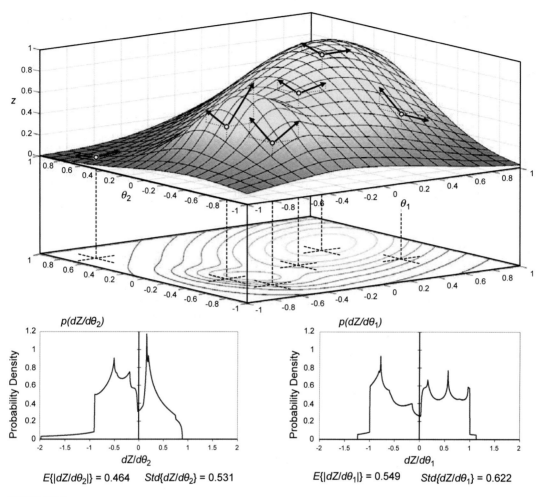

FIGURE 20.4

The main plot shows the two-parameter surface **"R1"** on which directional gradients (partial derivatives) $d\boldsymbol{Z}/d\boldsymbol{\Theta}$ are indicated at a sample of different nominal points in the feasible parameter space $\boldsymbol{\Psi}$. The two subplots show the probability distributions $p(d\boldsymbol{Z}/d\theta_1)$ and $p(d\boldsymbol{Z}/d\theta_2)$ of the directional gradients, evaluated over all the parameter locations in $\boldsymbol{\Psi}$. Note that the distribution for parameter θ_1 has a larger mean ($E\{d\boldsymbol{Z}/d\theta_1\} > E\{d\boldsymbol{Z}/d\theta_2\}$), indicating a larger average local sensitivity, and a larger standard deviation ($Std\{d\boldsymbol{Z}/d\theta_1\} > Std\{d\boldsymbol{Z}/d\theta_2\}$), indicating a larger variability of the local sensitivity.

This "expected" reduction in variance ΔV_j is called the "main effect" associated with parameter θ_j. Sobol' further points out that this main effect may, however, only be a part of the contribution by that parameter θ_j to the total unconditional variance V (therefore called the individual or independent contribution), and that further contributions can come through its interdependence (called "interactions") with other parameters. For example, if we fix two of the parameters at nominal values so

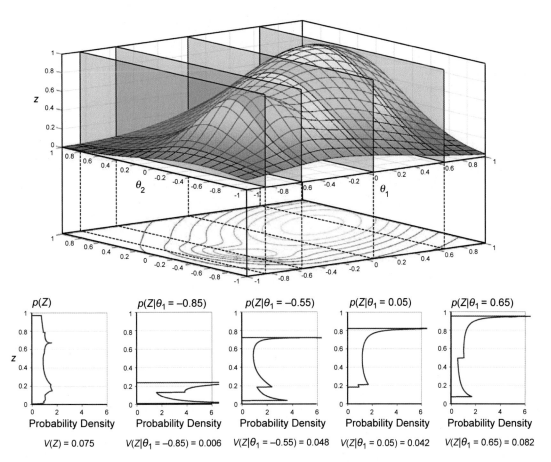

FIGURE 20.5

The main plot shows the two-parameter surface "**R1**" and illustrates one-dimensional cross-sections where parameter θ_2 is varied across its range, whereas parameter θ_1 is fixed at four different nominal values. The first subplot (lower left) shows the probability distribution $p(Z)$ of the response Z over the entire parameter domain Ψ, for which the variance $V(Z) = 0.075$. The subsequent four subplots show the conditional probability distributions $p(Z|\theta_1)$ evaluated for each of the cross-sectional response surfaces indicated in the main plot; note that the high peaks correspond to regions where the response surfaces are relatively flat. The conditional variances $V(Z|\theta_1)$ for these four cross-sections are {0.006, 0.048, 0.042, 0.082}, for which the (sampled) average change in variance due to conditioning on θ_1 at only these four different nominal values is 0.0305.

that $\theta_j = \widehat{\theta}_j$ and $\theta_k = \widehat{\theta}_k$ while allowing the remaining parameters to vary over their feasible ranges, we will obtain a conditional distribution $p\left(Z|\theta_j = \widehat{\theta}_j, \ \theta_k = \widehat{\theta}_j\right)$ with variance $V_{jk} = V\left(Z|\theta_j = \widehat{\theta}_j, \ \theta_k = \widehat{\theta}_k\right)$. Repeating this in a representative sample of nominal points $\left\{\widehat{\theta}_j, \ \widehat{\theta}_k\right\}$ will result in a probability distribution of values $p(V_{jk})$ from which the expected reduction in variance $\Delta V_{jk} = V - E(V_{jk})$ due to simultaneously fixing the two parameters $\{\theta_j, \ \theta_k\}$ can be computed. Now the second-order interaction effect, $INT_{jk} = \Delta V_{jk} - \Delta V_j - \Delta V_k$, between parameters θ_j and θ_k becomes clear.

More generally, one can simultaneously fix three or more parameters to study higher order interaction effects, and Sobol' (1990) has shown that the overall variance V can be decomposed into its constituent elements as:

$$V = \sum \Delta V_j + \sum \sum INT_{jk} + \sum \sum \sum INT_{jkl} + higher\ order\ terms$$

where the ΔV_j terms represent the main effects associated with individual parameters, the INT_{jk} terms represent second-order interaction effects, the INT_{jkl} terms represent third-order interaction effects, and so on.

To summarize, the sensitivity of response Z to parameter θ_j is characterized by the so-called *total-order* effect, consisting of the summation of all contributions due to parameter θ_j, including the main effect ΔV_j and all the second-order and higher order interaction effects (INT_{jk}, INT_{jkl}, ...) associated with parameter θ_j. Computationally, a simple way to estimate the total-order effect is by fixing all the parameters except θ_j (several times at a sufficiently large representative sample of locations) and computing $\Delta V_{\sim j} = V - E(V_{\sim j})$; here the notation $V_{\sim j}$ means that parameter θ_j is *not* fixed.

As with the Morris distribution of derivatives strategy, sampling strategies have been proposed that enable efficient computation of the various Sobol sensitivity metrics with minimal computational cost (Saltelli et al., 2008).

2.2.4 The "Analysis of Cumulative Distributions" Strategy

A major critique that can be leveled against the *analysis of variance* strategy is that the "variance" can often be a poor summary statistic of variability (or more accurately the "change" in variability) associated with a distribution of response values, particularly when the distribution is either highly skewed or multimodal (i.e., when it deviates significantly from a symmetrical and unimodal distribution form). For such cases, it has been proposed that alternative measures of the variability expressed by the distribution be applied, such as its entropy (e.g., Park and Ahn, 1994). However, accurate numerical estimates of entropy are difficult to obtain when working with small sample sizes, and a promising alternative strategy has been proposed based on the use of *cumulative* probability distribution functions (CDFs) that are much easier to construct and evaluate numerically than actual probability distribution functions (PDFs).

As with the analysis of variance approach, the *"Analysis of Cumulative Distributions"* strategy is based on the probability distribution $p(Z)$ obtained by representative sampling of the model parameters Θ in the feasible space Ψ, and on the corresponding conditional distributions $p(Z|\theta_j = \hat{\theta}_j)$ obtained by fixing each of the parameters θ_j at nominal values $\hat{\theta}_j$ while allowing all the other parameters to vary over their feasible ranges. However, rather than characterizing the difference between the unconditional to conditional distributions by the reduction in variance, the "Analysis of Cumulative Distributions" strategy uses some measure of the "distance" between the associated *cumulative* distributions (0 distance indicates no sensitivity, whereas a larger distance indicates a strong sensitivity). Since the strategy is similar to the Sobol' approach with two important differences (working with cumulative distributions and using some alternative statistic than the variance), all the different order and interaction terms can, in principle, be estimated by fixing one or more combinations of parameters.

A specific implementation of this strategy is the PAWN approach proposed by Pianosi and Wagener (2015), in which the *Kolmogorov–Smirnov* statistic is used to measure the distance between the two (unconditional and conditional) one-dimensional cumulative probability distributions. The authors point out that the advantages of working with CDFs (rather than PDFs) include the ability to test for statistical significance, the ease with which bootstrapping can be performed to evaluate robustness, and the ability to focus the evaluation on segments of the distribution such as the extremes. Note, however, that Pianosi and Wagener (2015) limit their analysis to only first-order effects (second-order and higher order interaction effects are not addressed).

2.2.5 The "Variogram Analysis of Response Surface" Strategy

A major weakness of the *"Distribution of Derivatives"* (e.g., Morris), *"Analysis of Variance"* (e.g., Sobol'), and *"Analysis of Cumulative Distributions"* strategies (e.g., PAWN) is that none of them consider or accounts for the spatially ordered structure of the response Z in the parameter space (Razavi and Gupta, 2015). In other words, they ignore the fact that the response values are not randomly distributed throughout the feasible parameter space, but instead there is a *spatially continuous* correlation structure to the values of Z, and hence also to the values of $dZ/d\Theta$. As a result, such approaches can actually assign *identical* sensitivity estimates to response surfaces having totally *different* spatial correlation structures (shapes), even though such an assignment of relative sensitivity may run counter to intuition (Fig. 20.3; see examples and detailed discussion in Razavi and Gupta, 2015).

To address this problem, Razavi and Gupta (2016a,b) proposed an SA approach based on the properties of the directional variograms of the response Z in the parameter space (Figs. 20.6 and 20.7); the approach is called VARS. Note that a variogram $\gamma(h)$ is constructed by computing the variance of the differences ΔZ between values of the response computed at (a large number of) pairs of points at different locations spaced a distance $h = \Delta\theta$ apart throughout the parameter space. So, a directional variogram $\gamma(h_j)$ represents the variance $Var\{\Delta Z(h_j)\}$ computed at points spaced $h_j = \Delta\theta_j$ apart in the direction of parameter θ_j. Accordingly, a higher value of $\gamma(h_j)$ for distance h_j corresponds to a higher

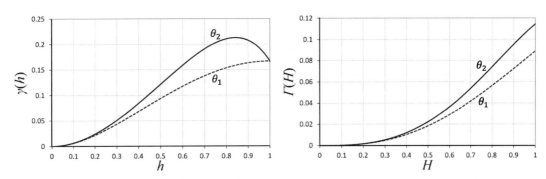

FIGURE 20.6

The left plot shows the directional variograms $\gamma_1(h)$ and $\gamma_2(h)$ corresponding to parameters θ_1 and θ_2 for the two-parameter response surface "**R2**" (see Fig. 20.3). Intuitively we consider the response sensitivity to parameters θ_1 to be lower than that for parameter θ_2. This is reflected in the variogram $\gamma_1(h)$ for parameter θ_1, which tends to be below the variogram $\gamma_2(h)$ for parameter θ_2. The right plot shows the corresponding integrated directional variograms $\Gamma_1(h)$ and $\Gamma_2(h)$; for this case of response surface "**R2**," the overall sensitivity to parameter θ_1 is higher than to parameter θ_2 regardless of the scale distance h. For a more complex case see Fig. 20.7.

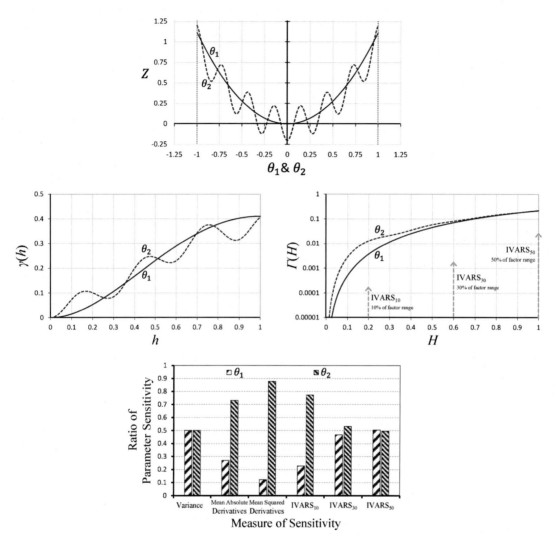

FIGURE 20.7

The top plot shows individual parameter cross-sections for the response surface $Z = 1.11\theta_1^2 + \theta_2^2 - 0.2\cos(7\pi\theta_2)$ evaluated through the nominal point [0, 0]; the two cross-sections have identical response variance (and therefore identical Sobol' sensitivity) but different structural organization. The corresponding directional variograms $\gamma_1(h)$ and $\gamma_2(h)$ (second row left) clearly indicate that relative parameter sensitivity varies with scale distance h; sensitivity is higher to parameter θ_2 at shorter scale distance h. However, the corresponding directional integrated variograms $\Gamma_1(h)$ and $\Gamma_2(h)$ (second row right) show that integrated parameter sensitivity is larger for parameter θ_2 at all but the very large scales (at $h = 1$ the results agree with Sobol'). The final plot shows relative parameter sensitivities evaluated using different methods (Variance → the Sobol' approach; Mean Absolute and Mean Squared Derivatives → the Morris' approach; IVARS$_{10}$, IVARS$_{30}$, and IVARS$_{50}$ → the VARS approach at 10%, 30%, and 50% scale lengths). Note that IVARS stands for integrated variogram across a range of scales.

Based on Figs. 20.2 and 20.3 in Razavi, S., Gupta, H.V., 2016a. A new framework for comprehensive, robust, and efficient global sensitivity analysis: Part I — theory. Water Resources Research 52, 423–439. http://dx.doi.org/10.1002/2015WR017558.

"rate of variability" (and therefore "sensitivity") of Z at that distance in the direction of parameter θ_j. Notably, this rate of variability at a particular distance in the problem domain represents the "scale-dependent sensitivity" of the response surface to the corresponding parameter. To account for this variation in relative sensitivity across scales (distances), the VARS approach uses the integrated variogram to represent the average sensitivity over all distances up to a given scale length.

The variogram-based SA approach has several interesting properties:

1. *Multiscale dependency*: Because variograms are defined over the full range of distances/scales, they provide a spectrum of information about the sensitivity of the model response to the factor of interest, i.e., they highlight the fact that the relative sensitivity of parameters in EES models is a "scale-dependent" concept.

2. *Mathematical foundation:* If we take the square root of $\gamma(h_j)$, which gives us $Std\{\Delta Z(h_j)\}$, and divide by the distance h_j, we get a quantity that corresponds to the average finite difference approximation of $\Delta Z(\Delta\theta_j)/\Delta\theta_j$, which approaches $dZ/d\theta_j$ as $\Delta\theta_j \rightarrow 0$, and is therefore recognizable as being related to the mathematical definition of local sensitivity.

3. *Theoretical basis:* It can be formally shown (see Razavi and Gupta, 2016a) that the sensitivity ranking provided by VARS converges to that provided by the Morris (distribution of derivatives) strategy at small h_j distances in the parameter space, and to that provided by Sobol' (analysis of variance) strategy at large h_j distances in the parameter space. In other words, the Morris and Sobol' strategies are special cases of the more general variogram-based approach. Furthermore, the STAR-VARS implementation (Razavi and Gupta, 2016b) of VARS is based on the use of cross-sections sampled at numerous locations in the parameter space, and so is theoretically also related to the "*One-Dimensional Cross Section*" strategy.

4. *Efficiency and robustness:* Because construction of the variogram is based on *pairs of points* (rather than individual points) sampled across the parameter space, and because the number of pairs grows rapidly as $\sim n^2$ (where n is the rate of increase of number of points sampled in the feasible space), the VARS approach is computationally efficient and statistically robust, even for high-dimensional response surfaces. Accordingly, it is able to provide relatively stable estimates of parameter sensitivity with much fewer samples than are required by the Morris and/or Sobol' strategies (Razavi and Gupta, 2016a,b found it to be as much as two orders of magnitude more efficient on two application cases).

5. *Confidence intervals:* The practical implementation of VARS presented by Razavi and Gupta (2016b) includes a bootstrap procedure that provides estimates of confidence intervals and reliabilities of the VARS-based estimates of relative parameter sensitivity.

3. CHALLENGES AND FUTURE OUTLOOK

Over the past two decades, EES models have rapidly become more complex and computationally intensive, growing in parameter dimensionality as they reflect our growing understanding about the nature and functioning of the world. Accordingly, the need for robust, informative, and computationally efficient SA techniques and tools has become even more pressing. There are, however, several challenges that need critical attention; we identify and discuss some of these challenges below.

3.1 COMPUTATIONAL EFFICIENCY

Perhaps the most obvious challenge is that the growing computational expense associated with increasingly more complex models tends to inhibit the widespread applicability of rigorous approaches to SA, and so practitioners are often forced to rely upon simpler ad hoc strategies (such as *One-Dimensional Cross Sections*) that are relatively inexpensive to implement. This is understandable, since SA methods such as Morris and Sobol' require very large numbers of model runs to provide relatively stable parameter sensitivity rankings, and practitioners may instead prefer to reserve their computational budget for the parameter optimization stage of their analysis. In general, we can expect that the computational efficiency associated with EES models can be improved by:

1. More efficient/intelligent extraction of information from EES models runs
2. Use of more powerful and efficient computational resources

In regard to the former, there have already been significant advances. For example, the VARS approach can be used to obtain more efficient estimates of both Morris (short scale) and Sobol' (large-scale) parameter sensitivity rankings, while also providing information about the structure of the response surface at a range of intermediate scales. There are also other approaches such as the fourier amplitude sensitivity test (FAST) (Cukier et al., 1973, 1975; Schaibly and Shuler, 1973; Cukier et al., 1975) and "extended FAST" (Saltelli et al., 1999) methodologies based on use of Fourier series expansion that are designed to be more computationally efficient than the Sobol'-type approaches based on Monte Carlo sampling. Finally, surrogate modeling strategies can be used to develop cheaper-to-run surrogates of computationally intensive models (Razavi et al., 2012a,b).

In regard to the latter, high-performance computing (HPC) resources are rapidly becoming widespread, making it increasingly feasible to run large numbers of EES model configurations (e.g., different parameter values) simultaneously in a parallel manner. Because SA techniques are essentially based on sampling (with no feedback from the model response surface required), they are perfectly parallelizable to take advantage of the full capacity of HPC resources. As HPC resources become more readily available, their use for performing rigorous SA of large-scale computationally expensive EES models will grow accordingly.

3.2 RELIABILITY (ACCURACY AND ROBUSTNESS)

Closely related to the issue of computational efficiency is that of accuracy and robustness. Practical implementations of the various strategies to compute parameter sensitivity require a "sufficient" number of representative samples of model response behavior distributed throughout the feasible parameter space Ψ. As model complexity (and hence parameter dimensionality) increases, the number of samples required for statistically accurate and robust sensitivity estimates grows geometrically. Moreover, although SA is generally most useful when the dimensionality of the problem is large (e.g., for factor screening or prioritization), it is precisely in such problems that the ability to generate that required number of samples becomes restrictive.

So, with limited computational budgets, the confidence that can be placed in the accuracy of estimated parameter sensitivity rankings can decline very rapidly, to the point of being virtually useless. For example, Fig. 20.8 illustrates the reliability of factor rankings obtained when different SA approaches were applied to a 45-parameter EES model using two different computational budgets

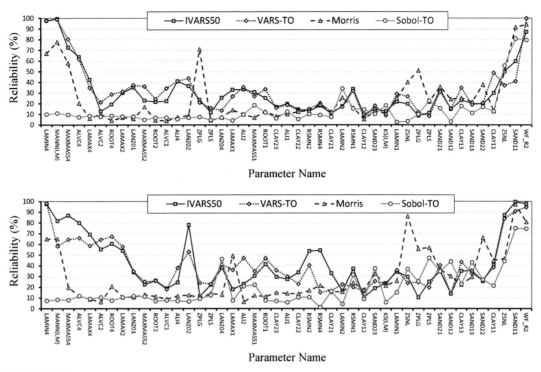

FIGURE 20.8

Both plots show percent reliability of parameter sensitivity ranking obtained by different methods for a 45-parameter computationally intensive physically based land surface hydrology model; the results in the upper plot were obtained using a sample size of ~20,000 function evaluations (samples) and the results in the lower plot were obtained using a sample size of ~100,000 function evaluations (a fivefold increase). Although reliability increases with sample size, it can be still be quite low (around 10–50%). The plots also show that the VARS-based sensitivity rankings (IVARS$_{50}$ and VARS-TO) tend to be more reliable than those provided by Morris and Sobol'.

Based on Figs. 12 and 13 in Razavi, S. Gupta, H.V., 2016b. A new framework for comprehensive, robust, and efficient global sensitivity analysis: Part II — applications. Water Resources Research 52, 440–455. http://dx.doi.org/10.1002/2015WR017559.

(sample sizes). Although the reliability of parameter rankings tends to increase with sample size, the results obtained using any of these methods can still be quite low (below 50%). In general, however, due to its efficient use of information from pairs of points (rather than individual points), the VARS-based approach tends to provide more reliable results than those provided by Morris and/or Sobol'.

The challenge for future SA methods is, therefore, to enable the reliable analysis of parameter sensitivity for high-dimensional problems. Complex models with hundreds, or even thousands, of uncertain parameters are likely to become more widespread, and as such, the role of SA will become more important than ever. It is important, therefore, that the development and application of practical

SA tools be designed to *maximize the use of information* provided by the $\Theta \rightarrow Z$ response surface (while being based in robust statistical approaches). For example, the VARS approach is able to exploit the information about differential response provided by pairs of points at various distances in the feasible space. Strategies (such as VARS) that can exploit information about response surface structure need to be explored in further depth.

In this regard, note that the statistical approaches mentioned in this chapter are based primarily on the first two moments (mean and variance/covariance) of the distribution of model response. Because these two statistics can be insufficient to statistically characterize the information contained in the underlying PDF (i.e., skewness and other higher order moments are ignored), it seems desirable to explore how the information contained in the entire distribution of model response can be better exploited (e.g., via more powerful implementations of the "*Analysis of Cumulative Distributions*" strategy).

3.3 AMBIGUITY IN THE DEFINITION OF "SENSITIVITY"

As explained earlier, the Morris (*Distribution of Derivatives*) strategy is based on extending the concept of local sensitivity to the entire feasible space, whereas the Sobol' (analysis of variance) strategy is based on the concept of attributing portions of the variance in the overall model response Z to individual parameters. Although each strategy seems (on face value) to be intuitively meaningful, they can result in quite different interpretations of relative parameter sensitivity (and therefore different conclusions about which parameters are to be deemed "sensitive" and "insensitive").

Razavi and Gupta (2016a) provide a theoretical analysis demonstrating that the reason for the different interpretations of sensitivity is primarily the "scale" at which sensitivity is evaluated; i.e., Morris corresponds to a "small-scale" viewpoint, whereas Sobol' corresponds to a "large-scale" viewpoint. Importantly, neither approach takes into consideration the fact that the continuous structural form of response surface must be taken into account for an evaluation of parameter sensitivity to be meaningful.

Although the VARS approach helps to address this problem, it opens up the new challenge that sensitivity ranking must be treated as a *scale-dependent* concept, which makes the task of interpreting sensitivity results more challenging; i.e., it becomes necessary to acknowledge that a parameter might show high sensitivity at small scales but low sensitivity at longer scales (and vice versa). The application of this new information to the model development process becomes model/problem dependent and (given the relative newness of this finding) the manner in which to interpret this additional information remains to be properly understood.

The need for a comprehensive characterization of sensitivity becomes even more pressing given the utility of SA for applications that go beyond the need for parameter screening/prioritization (Razavi and Gupta, 2015) as a prerequisite to optimization for model calibration. For example, SA is likely to see extensive application as a diagnostic tool for the evaluation of similarities between the functioning of the model and the underlying system, so as to assess fidelity of the model structure and conceptualization.

Related to the this, the development of improved visualization techniques and tools that more effectively communicate the sensitivity information to the modeler/user is much needed. Accordingly, SA techniques that can comprehensively characterize the complex nature of the high-dimensional EES model response surfaces in terms of a lower dimensional number of metrics that can be readily visualized will be required. For example, the anisotropic variogram and covariogram functions of high-dimensional response surfaces presented by Razavi and Gupta (2016a,b) provide a way to represent the full spectrum of sensitivity information via directional (unidimensional) variograms.

3.4 SPECIFICATION OF THE CRITERION REPRESENTING MODEL "RESPONSE"

In general, all approaches to assessing parameter sensitivity described previously rely upon the use of a single preselected model performance criterion as "the meaningful and informative" representative of model response. However, any aggregate average measure of model performance (such as the Mean Squared Error criterion) unavoidably acts as a filter to emphasize some aspects of model behavior while deemphasizing others.

For example, use of the Mean Squared Error criterion in watershed modeling emphasizes the reproduction of flood peaks, and so the aquifer transmissivity parameters will typically be "found" to be relatively insensitive. In contrast, use of the Mean Squared Error criterion applied to log-transformed flows emphasizes the reproduction of recession periods, and so the aquifer transmissivity parameters will appear to be more sensitive than those that control the formation of flood peaks.

As EES models grow in complexity and capability, there are actually a great many aspects of model response that characterize its behavior (e.g., water balance, latent and sensible heat fluxes, carbon and nitrogen fluxes, spatial patterns of the aforementioned). In such models, a single aggregate measure of model performance will fail to provide useful information about parameter sensitivity. To address this challenge, Rosolem et al. (2012) present a multiple criteria approach that helps to preserve and highlight the important distinctions regarding sensitivity of different aspects of model response to various parameters.

It is the authors' belief that further work in the direction of multivariate SA should be encouraged and, in particular, the use of "signature properties" (Gupta et al., 2008) as robust and informative measures of model response should be pursued, given their ability to provide diagnostic information about the structure and functioning of the model (see, e.g., Yilmaz et al., 2008; de Vos et al., 2010; Martinez and Gupta, 2010, 2011; Pokhrel et al., 2012; He et al., 2015; Guse et al., 2016).

4. CONCLUSIONS

SA is a critical tool in the development and application of EES models. However, its application can be inhibited by computational expense, and its usefulness is limited by the inability to extract useful diagnostic information from the model response. It is of paramount importance that ongoing research seeks to develop strategies for SA that are both *effective and efficient*. As illustrated by the various chapters in this book, a variety of different SA approaches (including Morris' and Sobol') have been proposed in the literature. As discussed in this chapter, each approach has its strengths and weaknesses.

The *"Variogram Analysis of Response Surface"* strategy proposed by Razavi and Gupta (2016a,b), *"Analysis of Cumulative Distributions"* strategy proposed by Pianosi and Wagener (2015), and the *"Multiple Criteria"* approach presented by Rosolem et al. (2012) represent different aspects of progress toward resolving some of those weaknesses, but more work needs to be done. Perhaps most important is the need to establish a clear definition of *how to "compactly" characterize the sensitivity of EES model responses to perturbations in their causal factors, and to do so in a manner that maximizes the "diagnostic" information provided by the analysis.*

ACKNOWLEDGMENTS

The first author received partial support from the Australian Research Council through the Centre of Excellence for Climate System Science (grant CE110001028), and from the EU-funded project "Sustainable Water Action (SWAN): Building Research Links Between EU and US" (INCO-20011-7.6 grant 294947). The second author is thankful to the University of Saskatchewan's Global Institute for Water Security and Howard Wheater, the Canada Excellence Research Chair in Water Security, for encouragement and support.

REFERENCES

Campolongo, F., Cariboni, J., Saltelli, A., 2007. An effective screening design for sensitivity analysis of large models. Environmental Modelling and Software 22 (10), 1509−1518.

Cukier, R.I., Fortuin, C.M., Shuler, K.E., Petschek, A.G., Schaibly, J.H., 1973. Study of the sensitivity of coupled reaction systems to uncertainties in rate coefficients. I Theory. The Journal of Chemical Physics 59 (8), 3873−3878.

Cukier, R.I., Schaibly, J.H., Shuler, K.E., 1975. Study of the sensitivity of coupled reaction systems to uncertainties in rate coefficients. III Analysis of the approximations. The Journal of Chemical Physics 63 (3), 1140−1149.

de Vos, N.J., Rientjes, T.H.M., Gupta, H.V., 2010. Diagnostic evaluation of conceptual rainfall−runoff models using temporal clustering. Hydrological Processes. http://dx.doi.org/10.1002/hyp.7698. Published online in Wiley InterScience. www.interscience.wiley.com.

Gupta, H.V., Wagener, T., Liu, Y.Q., 2008. Reconciling theory with observations: towards a diagnostic approach to model evaluation. Hydrological Processes 22 (18), 3802−3813. http://dx.doi.org/10.1002/hyp.6989.

Guse, B., Pfannerstill, M., Strauch, M., Reusser, D., Lüdtke, S., Volk, M., Gupta, H., Fohrer, N., 2016. On characterizing the temporal dominance patterns of model parameters and processes. Hydrological Processes. http://dx.doi.org/10.1002/hyp.10764.

He, Z., Tian, F., Gupta, H.V., Hu, H.C., Hu, H.P., 2015. Diagnostic calibration of a hydrological model in an alpine area by hydrograph partitioning. Hydrology and Earth Systems Science 19, 1807−1826. http://dx.doi.org/10.5194/hess-19-1807-2015. www.hydrol-earth-syst-sci.net/19/1807/2015/.

Martinez, G.F., Gupta, H.V., 2010. Toward improved identification of hydrological models: a diagnostic evaluation of the "abcd" monthly water balance model for the conterminous United States. Water Resources Research 46, W08507. http://dx.doi.org/10.1029/2009WR008294.

Martinez, G.F., Gupta, H.V., 2011. Hydrologic consistency as a basis for assessing complexity of water balance models for the continental United States. Water Resources Research. http://dx.doi.org/10.1029/2011WR011229.

Morris, M.D., 1991. Factorial sampling plans for preliminary computational experiments. Technometrics 33 (2), 161−174.

Park, C., Ahn, K., 1994. A new approach for measuring uncertainty importance and distributional sensitivity in probabilistic safety assessment. Reliability Engineering and System Safety 46, 253−261.

Pianosi, F., Wagener, T., 2015. A simple and efficient method for global sensitivity analysis based on cumulative distribution functions. Environmental Modelling & Software 67, 1−11. http://dx.doi.org/10.1016/j.envsoft.2015.01.004.

Pokhrel, P., Yilmaz, K., Gupta, H.V., 2012. Multiple-criteria calibration of a distributed watershed model using spatial regularization and response signatures. Journal of Hydrology 418−419, 49−60. http://dx.doi.org/10.1016/j.jhydrol.2008.12.004. Special Issue on DMIP-2.

Rakovec, O., Hill, M.C., Clark, M.P., Weerts, A.H., Teuling, A.J., Uijlenhoet, R., 2014. Distributed evaluation of local sensitivity analysis (DELSA), with application to hydrologic models. Water Resources Research 50, 409−426. http://dx.doi.org/10.1002/2013WR014063.

Razavi, S., Elshorbagy, A., Wheater, H., Sauchyn, D., 2015. Toward understanding nonstationarity in climate and hydrology through tree ring proxy records. Water Resources Research 51, 1813–1830. http://dx.doi.org/10.1002/2014WR015696.

Razavi, S., Gupta, H.V., 2015. What do we mean by sensitivity analysis? the need for a comprehensive characterization of 'global' sensitivity in earth and environmental systems models. Water Resources Research. http://dx.doi.org/10.1002/2014WR016527.

Razavi, S., Gupta, H.V., 2016a. A new framework for comprehensive, robust, and efficient global sensitivity analysis: Part I – theory. Water Resources Research 52, 423–439. http://dx.doi.org/10.1002/2015WR017558.

Razavi, S., Gupta, H.V., 2016b. A new framework for comprehensive, robust, and efficient global sensitivity analysis: Part II – applications. Water Resources Research 52, 440–455. http://dx.doi.org/10.1002/2015WR017559.

Razavi, S., Tolson, B.A., Burn, D.H., 2012a. Numerical assessment of metamodelling strategies in computationally intensive optimization. Environmental Modelling and Software 34, 67–86.

Razavi, S., Tolson, B.A., Burn, D.H., 2012b. Review of surrogate modelling in water resources. Water Resources Research 48, W07401. http://dx.doi.org/10.1029/2011WR011527.

Rosolem, R., Gupta, H.V., Shuttleworth, W.J., Zeng, X., de Goncalves, L.G.G., 2012. A fully multiple-criteria implementation of the Sobol method for parameter sensitivity analysis. Journal of Geophysical Research 117, D07103. http://dx.doi.org/10.1029/2011JD016355.

Saltelli, A., Tarantola, S., Chan, K., 1999. A quantitative model-independent method for global sensitivity analysis of model output. Technometrics 41 (1), 39–56.

Saltelli, A., Ratto, M., Andres, T., Campolongo, F., Cariboni, J., Gatelli, D., Saisana, M., Tarantola, S., 2008. Global Sensitivity Analysis: The Primer. John Wiley, Hoboken, NJ.

Schaibly, J.H., Shuler, K., 1973. Study of the sensitivity of coupled reaction systems to uncertainties in rate coefficients. II Applications. The Journal of Chemical Physics 59 (8), 3879–3888.

Sobol', I.M., 1990. On sensitivity estimation for nonlinear mathematical models. Matematicheskoe Modelirovanie 2 (1), 112–118 (in Russian).

Sobol', I.M., Kucherenko, S., 2009. Derivative based global sensitivity measures and their link with global sensitivity indices. Mathematics and Computers in Simulation 79 (10), 3009–3017.

Yilmaz, K.K., Gupta, H.V., Wagener, T., 2008. A process-based diagnostic approach to model evaluation: application to the NWS distributed hydrologic model. Water Resources Research 44, W09417. http://dx.doi.org/10.1029/2007WR006716.

Index